Information Photonics

Fundamentals, Technologies, and Applications

Information Photonics

Fundamentals, Technologies, and Applications

Asit Kumar Datta
Soumika Munshi

CRC Press
Taylor & Francis Group
Boca Raton London New York

CRC Press is an imprint of the
Taylor & Francis Group, an **informa** business

CRC Press
Taylor & Francis Group
6000 Broken Sound Parkway NW, Suite 300
Boca Raton, FL 33487-2742

First issued in paperback 2020

ISBN 13: 978-0-367-57418-5 (pbk)
ISBN 13: 978-1-4822-3641-5 (hbk)

Library of Congress Cataloging-in-Publication Data

Names: Datta, Asit Kumar, author. | Munshi, Soumika, author.
Title: Information photonics : fundamentals, technologies, and applications / Asit Kumar Datta and Soumika Munshi.
Description: Boca Raton : CRC Press, 2017. | Includes bibliographical references and index.
Identifiers: LCCN 2016009792 | ISBN 9781482236415 (alk. paper)
Subjects: LCSH: Optical communications. | Optical data processing. | Photonics.
Classification: LCC TK5103.59 .D38 2017 | DDC 621.36/5--dc23
LC record available at https://lccn.loc.gov/2016009792

Visit the Taylor & Francis Web site at
http://www.taylorandfrancis.com

and the CRC Press Web site at
http://www.crcpress.com

To our Teachers, Peers, Friends, and Families

Contents

Preface   xvii

List of Figures   xxi

List of Tables   xxix

1   Information Communication   1

   1.1   Information   1
   1.2   Probability   3
      1.2.1   Random variables   4
         1.2.1.1   Expectation, variance, and covariance   5
      1.2.2   Conditional probability and independence   5
      1.2.3   Cumulative distribution function   6
      1.2.4   Probability density function   6
   1.3   Entropy and information   7
   1.4   Information communication   9
   1.5   Source encoding: noiseless coding   11
   1.6   Shannon's theorems   12
      1.6.1   Shannon's first fundamental theorem   13
      1.6.2   Shannon's second fundamental theorem   15
   1.7   Communication channels   16
      1.7.1   Continuous channel   17
         1.7.1.1   Entropy of continuous channel   17
         1.7.1.2   Mutual information of a continuous noisy channel   18
         1.7.1.3   Discrete channel with discrete noise   19
      1.7.2   Band-limited channel   20
      1.7.3   Gaussian channel   20
         1.7.3.1   Capacity of band-limited channel with average white Gaussian noise   21
   1.8   Detection of information   22
      1.8.1   Bayes criteria   24
      1.8.2   Minimax criteria   25
      1.8.3   Neyman-Pearson criteria   25
   1.9   Estimation of information   26

1.9.1   Bayes estimate   27
1.9.2   Minimax estimate   27
1.9.3   Maximum likelihood estimate   28

Bibliography   31

2 Introduction to Photonics   33

2.1   Information and physics   33
      2.1.1   Photonics   34
2.2   Photons and the nature of light   35
      2.2.1   Properties of photons   38
              2.2.1.1   Photon energy and momentum   38
              2.2.1.2   Photon mass   39
      2.2.2   Speed of light   40
              2.2.2.1   Speed of light and signal propagation .   41
      2.2.3   Coherence of light waves in space and time . . .   42
              2.2.3.1   Temporal coherence   44
              2.2.3.2   Spatial coherence   44
              2.2.3.3   Partial coherence 45
              2.2.3.4   Coherent and incoherent light sources . 45
2.3   Electromagnetic wave propagation   46
      2.3.1   Maxwell's equations 46
              2.3.1.1   Constitutive relations 47
              2.3.1.2   Negative refractive index material . . . 49
      2.3.2   Electromagnetic wave equation 49
              2.3.2.1   Plane wave solution 52
2.4   Electromagnetic spectrum 54
      2.4.1   Infrared region 56
      2.4.2   Visible region 57
      2.4.3   Ultraviolet region 57
      2.4.4   Terahertz region 57
2.5   Optical imaging systems 58
      2.5.1   Ray optic theory of image formation 59
              2.5.1.1   Fermat's principle 61
              2.5.1.2   Law of refraction and reflection 62
              2.5.1.3   Matrix formulation of an optical system . . . 62
      2.5.2   Wave optics: diffraction and interference 63
              2.5.2.1   Scalar wave equation 64
              2.5.2.2   Huygens-Fresnel principle 65
              2.5.2.3   Diffraction 66
              2.5.2.4   Interference 69
              2.5.2.5   Polarisation 71
2.6   Interaction of light and matter 75
      2.6.1   Absorption 75

  2.6.2 Scattering 76
 2.7 Photometric image formation 77
  2.7.1 Reflectance and shading 78

Bibliography **81**

3 Vision, Visual Perception, and Computer Vision **85**

 3.1 Introduction 85
  3.1.1 Human visual system 85
   3.1.1.1 Structure of the human eye 85
   3.1.1.2 Visual signal processing in retina and brain . 91
   3.1.1.3 Photopic, mesopic, and scotopic vision 94
   3.1.1.4 The CIE V(λ) function 96
 3.2 Visual perception 97
  3.2.1 Perception laws 98
   3.2.1.1 Weber's psychophysical law 98
   3.2.1.2 Fechner's law 98
   3.2.1.3 Steven's power law 99
   3.2.1.4 Grouping laws and gestalt principles 99
   3.2.1.5 Helmholtz principle 101
  3.2.2 Seeing in three dimensions 101
  3.2.3 Bayesian interpretation of visual perception 102
 3.3 Computer vision 103
  3.3.1 Image formats 104
  3.3.2 Image acquisition 106
 3.4 Geometric primitives and transformations 106
  3.4.1 Geometrical primitives 107
  3.4.2 Geometrical transformation 108
   3.4.2.1 Geometric model of camera 110

Bibliography **115**

4 Photonic Sources and Detectors for Information Processing 117

 4.1 Introduction 117
 4.2 Elements of semiconductor physics 117
  4.2.1 Current density 121
  4.2.2 Semiconductor materials for photonic devices 122
  4.2.3 Semiconductor junction 123
   4.2.3.1 Homojunction under bias voltage 124
   4.2.3.2 Current flow in homojunction under bias . . 127
   4.2.3.3 Semiconductor heterojunction 128
  4.2.4 Quantum well structure 129
 4.3 Photonic sources 131
  4.3.1 Laser amplifiers 133

          4.3.1.1   Population inversion   135
          4.3.1.2   Rate equation   135
          4.3.1.3   Characteristics of laser amplifier   138
    4.3.2   Laser oscillators   139
    4.3.3   Light-emitting diodes   143
          4.3.3.1   Light-current characteristics   147
          4.3.3.2   Spectral width and modulation bandwidth .   148
    4.3.4   Semiconductor lasers: Injection laser diodes   149
    4.3.5   Organic light-emitting diode (OLED)   154
    4.3.6   Typical lasers used in photonics   155
          4.3.6.1   Gas lasers   155
          4.3.6.2   Chemical lasers   156
          4.3.6.3   Liquid dye lasers   157
          4.3.6.4   Solid-state lasers   157
  4.4 Photodetectors   158
    4.4.1   Photoconductors   160
    4.4.2   Junction photodiodes   162
          4.4.2.1   p-i-n photodiode   163
          4.4.2.2   Avalanche photodiode   164
    4.4.3   Charge-coupled devices   166
    4.4.4   Silicon photonic devices   169
          4.4.4.1   Silicon photonic technologies for light sources   170
          4.4.4.2   Silicon photonic technologies for light modu-
                    lators   171
          4.4.4.3   Silicon photonic technologies for light detec-
                    tion   172

Bibliography   **175**

5 Photonic Devices for Modulation, Storage, and Display   **179**

  5.1 Photonic devices for light-beam modulation   179
    5.1.1   Electro-optic effect   179
          5.1.1.1   Electro-optic modulators   180
    5.1.2   Liquid crystal spatial light modulators   182
          5.1.2.1   Electrically addressed liquid crystal-based
                    spatial light modulator   186
          5.1.2.2   Optically addressed spatial light modulator
                    (OASLM)   187
    5.1.3   Magneto-optic spatial light modulator   188
    5.1.4   Acousto-optic modulator   190
    5.1.5   Deformable mirror devices for light modulation   191
  5.2 Photonic switching devices   192
    5.2.1   Etalon switching devices   195
    5.2.2   Self-electro-optic effect device   196

5.3 Photonic storage in three dimensions (3D) 198
    5.3.0.1 Storage based on photo-refractive material . 198
    5.3.0.2 Storage based on photochromic materials . . 199
    5.3.0.3 Switching using two-photon absorption material 199
    5.3.0.4 Storage based on bacteriorhodopsin 200
  5.3.1 Holographic storage devices 200
5.4 Flat panel displays 203
  5.4.1 Liquid crystal flat panel displays 206
  5.4.2 LED and OLED display panels 209
  5.4.3 Plasma display panels 211
  5.4.4 Flexible flat panel display 212
  5.4.5 Electrophoresis and e-book displays 213

Bibliography **215**

6 Photonics in Transform Domain Information Processing **217**

6.1 Introduction 217
6.2 Fourier transform 217
  6.2.1 Fourier transform properties 219
  6.2.2 Discrete Fourier transform 220
    6.2.2.1 Fast Fourier transform 221
  6.2.3 Three dimensional Fourier transform 222
  6.2.4 Fourier transform in log-polar coordinate system . . . 222
  6.2.5 Fourier transform properties of lens 223
  6.2.6 Fractional Fourier transform 225
    6.2.6.1 Properties of fractional Fourier transform . . 226
    6.2.6.2 Discrete fractional Fourier transform 226
6.3 Hartley transform 227
6.4 Wavelet analysis and wavelet transform 227
  6.4.1 Wavelet transform 229
    6.4.1.1 Continuous wavelet transform (CWT) 230
    6.4.1.2 Wavelet transform in frequency domain . . . 231
    6.4.1.3 Time-frequency analysis 231
    6.4.1.4 Discrete wavelet transform (DWT) 232
    6.4.1.5 Haar wavelet 234
    6.4.1.6 Daubechies wavelet 235
    6.4.1.7 Multiresolution analysis 236
  6.4.2 2D wavelet transform and image processing 237
6.5 Radon transform 239
  6.5.1 Central slice theorem 240
6.6 Hough transform 241
  6.6.1 Line detection 242
  6.6.2 Circle detection 244

   6.7   Photonic implementation of linear transforms 245
       6.7.1   Implementation of wavelet transform 246
       6.7.2   Photonic implementation of Hough transform 247

Bibliography   **249**

7 Low-Level Photonic Information Processing   **251**

   7.1   Introduction 251
   7.2   Low-level image processing 252
       7.2.1   Point operators 252
       7.2.2   Group operations 254
       7.2.3   Neighbourhood operations 255
   7.3   Morphological operation 255
       7.3.1   Binary morphological image processing 256
       7.3.2   Gray-level morphological processing 257
       7.3.3   Morphological processing by photonic technique . . . 258
   7.4   Photonic techniques of transformation of images into graphics 260
       7.4.1   Range image acquisition 261
       7.4.2   Image registration and integration 261
            7.4.2.1   Laser-based range acquisition by time-of-flight 262
       7.4.3   Photonic profilometric techniques 264
            7.4.3.1   Laser triangulation technique 265
            7.4.3.2   Fringe projection techniques 266
            7.4.3.3   Phase-measuring profilometry 268
            7.4.3.4   Fourier transform profilometry 270
       7.4.4   Flow measurement by photonic technique 270
            7.4.4.1   Laser Doppler anemometry 270

Bibliography   **275**

8 Photonics in Networking and Communication   **279**

   8.1   Light propagation in optical fibre 279
       8.1.1   Optical fibre 280
            8.1.1.1   Types of optical fibre 284
            8.1.1.2   Signal distortion in optical fibre 286
       8.1.2   Point-to-point fibre link 288
            8.1.2.1   Loss budget 289
            8.1.2.2   Rise-time budget 290
            8.1.2.3   Modulation format 290
            8.1.2.4   System performance 291
       8.1.3   Long-haul photonic system 292
   8.2   Photonic free-space communication 292
       8.2.1   Near-ground free-space photonic link 293
            8.2.1.1   Diffraction 293

		8.2.1.2	Absorption	294
		8.2.1.3	Scattering	294
		8.2.1.4	Scintillation	294
8.3	Photonic networks			295
	8.3.1	Wavelength division multiplexing (WDM)		298
		8.3.1.1	Routing topologies	299
		8.3.1.2	Photonic cross-connects	301
		8.3.1.3	Add-drop multiplexer	303
		8.3.1.4	Add-drop filter	303
		8.3.1.5	Optical clock and bus	304
		8.3.1.6	Access network	304
		8.3.1.7	Transport network	305
	8.3.2	Erbium-doped fibre amplifier (EDFA)		306
8.4	Photonic secure communication			309
	8.4.1	Phase and polarisation encryption		309
	8.4.2	Decryption techniques and error control		310
	8.4.3	Photonic image encryption techniques		310

Bibliography   **315**

9 Photonic Computing   **319**

9.1	Introduction			319
9.2	Requirements of high-speed computing and parallel processing			319
	9.2.1	Single-instruction multiple data architecture		321
	9.2.2	Multiple-instruction multiple-data architecture		322
	9.2.3	Pipelined and parallel architecture		322
9.3	Limitations of electronic computation			323
	9.3.1	Physical limitations		324
	9.3.2	Topological limitations		324
	9.3.3	Architectural bottleneck		325
	9.3.4	Fundamental limits		325
9.4	Photonics computing			326
	9.4.1	Throughput issues		329
9.5	Digital optics			330
	9.5.1	Array logic, multivalued logic, and programmable logic		330
		9.5.1.1	Optical programmable array logic (OPAL)	331
9.6	Photonic multistage interconnection			333
9.7	Number systems for parallel photonic computing			335
	9.7.1	Residue number system		337
	9.7.2	Binary number system		338
	9.7.3	Signed-digit (SD) numbers		338
		9.7.3.1	Modified signed-digit (MSD) representation	339
		9.7.3.2	Mixed modified signed-digit (MSD) representation	340

            9.7.3.3   Trinary signed-digit (TSD) representation . .   340
            9.7.3.4   Quaternary signed-digit (QSD) representation   341
      9.7.4 Negabinary number system and its variants   341
            9.7.4.1   Unsigned negabinary number system   342
            9.7.4.2   Mixed negabinary number system   343
            9.7.4.3   Signed negabinary number system   343
            9.7.4.4   Floating point negabinary notation   344
9.8   Photonic computing architectures 344
      9.8.1 Optical shadow-casting architecture   345
      9.8.2 Optical shadow-casting-based unified arithmetic and
             logic-processing architecture 347
            9.8.2.1   Negabinary carry-less arithmetic processor .   349
            9.8.2.2   Negabinary-based logic processor 351
            9.8.2.3   Sequential photonic flip-flop processor 353
      9.8.3 Convolution-based multiplier architecture 355
            9.8.3.1   Matrix-vector multiplication architecture . . 357
      9.8.4 Hybrid photonic multiprocessor 360
      9.8.5 All-optical digital multiprocessor 360
      9.8.6 S-SEED-based all-photonic multiprocessor 362

Bibliography   **365**

10 Photonic Pattern Recognition and Intelligent Processing   **371**

10.1 Introduction 371
10.2 Correlation filters for pattern recognition 371
      10.2.1 Vander Lugt correlation 372
10.3 Frequency domain correlation filters 373
      10.3.1 Performance evaluation metrics of correlation 375
            10.3.1.1 Peak-to-correlation energy ratio 375
            10.3.1.2 Signal-to-noise ratio (SNR) 375
            10.3.1.3 Discrimination ratio 375
            10.3.1.4 Peak-to-sidelobe ratio 376
      10.3.2 Types of frequency domain correlation filters 376
            10.3.2.1 SDF filter design 380
            10.3.2.2 MACE filter design 381
            10.3.2.3 MVSDF design 382
            10.3.2.4 Optimal trade-off (OTF) filter design 382
            10.3.2.5 MACH filter design 382
            10.3.2.6 UMACE filter design 385
            10.3.2.7 OTMACH filter design 385
      10.3.3 Hybrid digital-photonic correlation 385
      10.3.4 Joint transform correlation 386
            10.3.4.1 Photonic joint transform correlator 389
10.4 Intelligent photonic processing 392

10.5 Artificial neural network (ANN) 393
    10.5.1 Training of the ANN 396
    10.5.2 Linear pattern classifier and multilayer perceptron . . 398
        10.5.2.1 Hopfield model 400
        10.5.2.2 Boltzmann machine 403
    10.5.3 Unsupervised ANN 404
        10.5.3.1 Kohonen's self-organising feature map 404
        10.5.3.2 Carpenter-Grossberg model 405
    10.5.4 Photonic implementation of artificial neural network . 406
10.6 Associative memory model for 2D pattern recognition 408
    10.6.1 Auto-associative memory model 409
        10.6.1.1 Holographic storage of interconnection weight
                matrix 412
    10.6.2 Hetero-associative memory 416
    10.6.3 Photonic hetero-associative memory 422
    10.6.4 Winner-take-all model 423

Bibliography   **427**

11 Quantum Information Processing   **431**

11.1 Introduction 431
11.2 Notations and mathematical preliminaries 432
11.3 Von Neumann entropy 434
11.4 Quantum bits or qubits 435
    11.4.1 Multiple qubits 438
    11.4.2 Conditional probabilities of qubits 439
    11.4.3 Operations on qubits 439
        11.4.3.1 Unitary transformation 439
        11.4.3.2 Measurement 440
    11.4.4 Realisation of qubits 440
        11.4.4.1 Atom and ion qubits 441
        11.4.4.2 Electronic qubits 442
        11.4.4.3 Josephson junction 442
        11.4.4.4 Nuclear magnetic resonance (NMR) devices . 444
        11.4.4.5 Photonic qubits 444
11.5 Entangled states and photon entanglement 445
11.6 Quantum computing 446
    11.6.1 Quantum logic gates and circuits 447
        11.6.1.1 Classical logic circuits 448
        11.6.1.2 Quantum gates 448
        11.6.1.3 Quantum gate circuits 450
    11.6.2 Quantum computing architecture 452
11.7 Quantum communication 454
    11.7.1 Quantum cryptography 455

            11.7.1.1 Quantum key distribution(QKD)   457
            11.7.1.2 Generic features of a protocol   458
       11.7.2 Quantum teleportation   459
   11.8 Concluding remarks   461

Bibliography   **463**

12 Nanophotonic Information System   **467**

   12.1 Introduction   467
   12.2 Nanophotonic devices   467
       12.2.1 Photonic crystal   469
            12.2.1.1 Photonic bandgap   470
            12.2.1.2 Planar photonic crystal   476
            12.2.1.3 Defects in photonic crystal structure   478
            12.2.1.4 Photonic crystal fibre   479
       12.2.2 Plasmonic devices   485
            12.2.2.1 Surface plasmon polariton   486
            12.2.2.2 Surface plasmon polaritons at two-interface
                      system   490
            12.2.2.3 Coupling between free electromagnetic waves
                      and surface plasmon polaritons   491
   12.3 Photonic integration and nanophotonic system   494

Bibliography   **501**

Index   **505**

Preface

Information photonics is an emerging area of interest as it interfaces with the scientific and technological developments in photonics, and the ever-expanding all pervasive applications of information technology. Photonics deals with the generation, transmission, and detection of light in the broader sense, and mostly encompassing the spectra from ultraviolet and visible to the far-infrared wavelengths. On the other hand, information technology deals with the applications of computers and telecommunication equipment to store, retrieve, transmit, and manipulate data. Since the application areas of both photonics and information technology have merged along with the advancement of technology, the area of information photonics is identified as an approach to developing photonic information processing systems, innovatively created by the fusion between photonics technology and information technology.

The word photonics appeared in the late 1960s to describe a research field whose technological goal in those days, was to use light to perform functions that traditionally fell within the typical domain of electronics, such as telecommunication, computation, and other information processing systems. Moreover, photonics was identified as a subset of physics, particularly semiconductor physics, when interest grew in the development of devices after the evolution of laser and optical fibre. As the photonics technologies are entering and expanding into every conceivable domain of public use, the individual boundaries of subject areas are becoming merged. This scenario is unfolding the advantages of today's information age, marked by the high speed of information communication with large bandwidth, which cannot be made without the applications of photonics technology. Information photonics is therefore identified as an approach to developing photonic information processing systems, which are innovatively created as an enabling technology of our information age.

At the outset it is necessary to establish a generic terminology that might be advantageously used in information photonics. During the last 50 years or so, every then-conceivable development in the area of today's photonics was designated with the name of applied optics, optical technology, or optical engineering. When the domain of photonics began to evolve from optics and electronics, various terminologies such as optical-electronic, optronics, and optoelectronics were interchangeably used. Optoelectronics is now considered to be the study and application of electronic devices that can source, detect, and control light, and so optoelectronics is now a sub-field of photonics.

In this book we shall only consider the devices and systems which are governed by the laws of optics to be optical devices and system. On the other hand, when the role of photons and the laws governing them are in play we shall consider those devices and systems to be photonic devices and photonic systems. Therefore, optical fibres are optical devices, but when a system is studied with the addition of sources and detectors, and is used for communication, those systems will be termed photonic communication systems instead of optical communication systems. In the same vein, optical computers will be termed photonic computers, the sources, detectors, and modulators shall be termed photonic detectors, photonic sources, and photonic modulators. A system evolved with the combination of optical and photonic devices shall be called a photonic system.

The main aim of this book is to introduce, to post-graduate students, researchers, engineers, and scientists, the concept of photonic information processing technologies as applied in various fields. It will help readers to get insight into the concepts of photonic information processing as a system, and photonic devices as system components required and applied in the areas of communication, computation, and intelligent pattern recognition. As an extension of these concepts, the areas of quantum communication and computation, where photons have a major role to play, shall also be introduced. The book converges in the last chapter by introducing the emerging areas of nanophotonics, where over and above the systems, photonic crystal and plasmonic devices will have an important role to play in the near future.

The first chapter deals with the basic philosophy of information communication in linear channels in the light of probability and Shannon's theorems. The second chapter introduces the properties of photons, concepts of electromagnetic theory, and an introduction to optical systems and laws. The third chapter deals with the human vision system and the ideas of visual perception, which is an integral part of human interface with systems. Photonic devices play a crucial role in establishing photonic system architectures and therefore the devices required for the generation of photonic signals (sources), those required for the detection of photons, are introduced in the fourth chapter. This chapter, expectedly, will describe the building blocks of the systems. The fifth chapter continues to introduce the devices required for the manipulation of photonic signals (the modulators and switches) and ultimately the photonic storage and display devices.

An important aspect of information processing is the analysis of two dimensional images in space and frequency domains. Chapter 6 addresses this aspect by discussing space and time transform domain information processing techniques by Fourier, fractional Fourier and wavelet transforms. Some other transforms which are helpful in detection problems are also discussed. Chapter 7 deals with low-level information processing of images with the help of operators and morphological processing. This chapter also elaborates photonic instrumentations required for range acquisition and profiling of an object.

Having introduced the photonic devices and basic processing techniques in space and time domain, Chapter 8 of the book deals with the major application of photonic processing in communication and networking. Propagation of light waves in optical fibres, communication in free space, and techniques of secure communication are discussed. Another major area of photonic information processing is photonic computation. Chapter 9 introduces the ideas and architectures of photonic computations, bringing out the arithmetic and logic required for carry-less two-dimensional operations. Since interconnects and switching are integral parts of photonic computing, a portion of this chapter is devoted to the interconnection and switching methods. Chapter 10 deals with the application areas involving photonic information processing in intelligent pattern recognition, mostly using neural network models and image correlation filters. Chapter 11 deals with quantum communication and computation as an aspect of photonic information processing, which is treated as a fusion between photonics and quantum mechanics. The mathematical preliminaries of qubits and devices required are also discussed. The last chapter (Chapter 12) introduces the latest developments in nanophotonic devices and systems. The theories and applications of photonic crystals and plasmonic devices form an integral part of this chapter. The book concludes with a note on the evolution of nanophotonic information systems.

Nowhere in the book are attempts made to derive the equations required to establish a theory behind photonic devices and information processing techniques. Only major equations are mentioned. A large number of books are referred to as additional and supporting reading at the end of each chapter, where the derivations of the equations can be found. Important references are also included at the end of each chapter.

Thanks are due to our colleagues who have assisted us during many phases of preparing the manuscript of the book. Our special gratitude goes to Gagandeep Singh, CRC Press editorial manager and senior editor (engineering/environmental sciences) without whose support and constant encouragement this book would not have been possible.

Asit Kumar Datta, Soumika Munshi

List of Figures

1.1   Basic probabilistic model of a channel   10
1.2   Basic model of a channel   13
1.3   A typical communication channel   16
1.4   Basic model of a channel with average white Gaussian noise   21
1.5   Condition probabilities and decision regions   24

2.1   Coordinate system for plane wave propagation   52
2.2   Ray propagation in constant $S(r)$ for plane wave and for
       spherical wave   61
2.3   Ray propagation in an optical system   63
2.4   Representative diffraction regime   66
2.5   Diffraction of a plane wave incident normally on an aperture   67
2.6   (a) Fresnel diffraction of a wire and (b) Fraunhofer diffraction
       pattern of a rectangular aperture   69
2.7   Young's experiment for the formation of interference pattern   70

3.1   Basic structure of an eye   86
3.2   Basic structure of retina   88
3.3   Distribution of rods and cones in retina   90
3.4   Visual pathways from eye to brain   91
3.5   Signal flow in visual system   92
3.6   Comparative performance of photopic and scotopic vision .   94
3.7   CIE $V-\lambda$ curve   96
3.8   Some examples of illusion   100
3.9   An example of visual interpretation and inference   102
3.10   (a) 2D line representation and (b) 3D plane representation .   108
3.11   Geometric 2D transformations   109
3.12   Perspective projection model of a pinhole camera   110
3.13   Coordinate system   112

4.1   Energy band diagram of an abrupt p-n junction   124
4.2   Energy band and potentials in (a) forward biased and (b)
       reverse biased abrupt p-n homojunction   125
4.3   V-I characteristics of a pn junction   128

4.4   Relative differences in bandgaps in 3D and quantum well structure are shown in (a) and (b), quantisations of absorption bands are shown in (c) and (d) 130

4.5   Transition processes in a two-level atomic system: (a) absorption, (b) spontaneous emission, and (c) stimulated emission   133

4.6   Two-level pump involving band 137

4.7   Three-level and four-level transitions 138

4.8   Schematic diagram of a laser cavity 140

4.9   Output characteristics of a laser oscillator 143

4.10   Basic operation of LED 144

4.11   Basic homostructure and heterostructure of a representative LED 146

4.12   Optical power-current relationship of LED 147

4.13   Optical power-current relationship of ILD 150

4.14   Schematic diagram of a double heterostructure structure used in an ILD 151

4.15   Schematic diagram of structure of: (a) distributed feedback (DFB) laser, (b) conventional Fabry-Perot injection diode laser, and (c) distributed Bragg reflector (DBF) of multi-section laser 152

4.16   Schematic structure of a quantum well laser 153

4.17   Representative OLED structure 155

4.18   Current-voltage characteristics of a photodiode at two different power levels of optical illumination 163

4.19   Structure of a diffused p-i-n photodetector 164

4.20   Structure of an APD and its electric field distribution 165

4.21   Depletion-type MOSFET CCD 167

4.22   Shift of potential wells with sequential clocking of gates . . 168

5.1   Phase modulation using a Pockels cell 181

5.2   Intensity modulation using a Mach-Zehender interferometer   182

5.3   Phases of liquid crystals: (a) nematic phase, (b) smectic A, (c) smectic B, and (d) cholesteric phase 183

5.4   Molecular orientation of an FLC molecule 184

5.5   (a) A generic electrically addressed liquid crystal cell, and (b) its addressing mode 185

5.6   An electrically addressed liquid crystal cell 186

5.7   Schematic structure of a reflective OASLM 187

5.8   OASLM in incoherent to coherent image conversion 188

5.9   Structure of MOSLM 189

5.10   Acousto-optic diffraction 191

5.11   Deformable mirror device 192

5.12   (a) Torsion type deformable mirror device (side view) and (b) deflection in OFF pixels 193

5.13   Characteristics of intensity bistability 194

5.14   A Fabry-Perot etalon   196
5.15   (a) Schematic structure of a SEED and (b) input/output char-
       acteristics of a SEED   196
5.16   Structure of a symmetric SEED (s-SEED)   197
5.17   Hologram (a) recording and (b) read-out   201
5.18   Active matrix addressing   206
5.19   Layers of LC flat panel display   207
5.20   Active matrix addressing of OLED panel (one individual cell
       is shown)   210
5.21   Flat panel display using OLED   211
5.22   Plasma flat panel display   212

6.1   Log-polar transformation of images   223
6.2   Lens producing an FT at the back focal plane   224
6.3   (a) Translation of wavelet and (b) scaling of wavelet   228
6.4   (a) Morlet wavelet and (b) Mexican hat wavelet   228
6.5   Time-frequency analysis with STFT and continuous wavelet   229
6.6   Heisenberg boxes of (a) wavelet $\psi_{s,\tau}$ and (b) Heisenberg boxes
       of STFT   232
6.7   Localisation of the discrete wavelets in the time-scale space
       on a dyadic grid   233
6.8   (a) Time-frequency tiling of FT, (b) time-frequency tiling of
       STFT, and (c) time-frequency tiling of DWT   234
6.9   Wavelet processing of image   238
6.10   2D wavelet decomposition of image   239
6.11   Two-dimensional projection of image at an angle θ to the
       vertical axis   240
6.12   Flowchart of reconstruction of image   241
6.13   (a) Projection from 2D image $f(x, Y)$ and (b) image recon-
       struction from projected data   241
6.14   (a) A straight line in (x_1, y_1) space and (b) a straight line
       passing through (x_1, y_1) in (m, c) space   242
6.15   (a) A straight line represented by r, θ in (x, y) space and (b)
       representing lines passing through point (x, y) in the (r, θ)
       space   243
6.16   Circle detection through the creation of a cone   244
6.17   4f setup   245
6.18   Schematic of photonic setup for the implementation of WT   247
6.19   Arrangement for photonic implementation of Hough trans-
       form   247

7.1   Mapping of one pixel with a certain gray value to a different
       value in a point operation   252
7.2   Mapping of one pixel with a certain gray value to a different
       value in a point operation   253

7.3   Scaling operation   254
7.4   Fourier plane morphological operation   258
7.5   Morphological operation on test fabric for defect detection .   259
7.6   Photonic architecture for morphological operation   259
7.7   Schematic of morphological erosion in shadow-casted technique   260
7.8   Schematic diagram of laser triangulation technique   265
7.9   Schematic diagram of a fringe projection technique   266
7.10   Some of the fringes used in profilometry   267
7.11   Geometry of the phase-measuring system   269

8.1   Fibre optic communication system   279
8.2   Launching of light and its propagation inside the fibre . . .   281
8.3   Different TE modes in an optical fibre   283
8.4   Types of fibre with their profiles and ray-paths within the
       fibre: (a) multimode step-index, (b) monomode step-index,
       and (c) graded index   284
8.5   Pulse broadening due to dispersion in a fibre   286
8.6   Attenuation in optical fibre at different wavelengths   288
8.7   Some of the modulation codes used in optical communication   290
8.8   Various network topologies: (a) point-to-point, (b) bus, (c)
       star, and (d) ring type   296
8.9   (a) A two-fibre ring pair in fibre distributed data interface
       (FDDI) network and (b) use of primary and secondary fibres
       in the case of a fibre break in an FDDI ring   297
8.10   WDM in an optical fibre communication system   298
8.11   Wavelength routing network   300
8.12   Routing schemes: (a) unicast, (b) broadcast, (c) anycast, (d)
       multicast, and (e) geocast   300
8.13   Modeling a photonic cross-connect   302
8.14   Block diagram of an add/drop multiplexer   303
8.15   Add/drop multiplexer using optical switch or grating   303
8.16   Add/drop filter using fibre grating, couplers, and isolators .   304
8.17   An access network   305
8.18   Synchronous digital hierarchy (SDH) ring-network   306
8.19   Generic optical amplifier   307
8.20   Energy levels in erbium   307
8.21   Orthogonal states or basis of BB84 protocol   310
8.22   Encoding by double random phase code in a 4f setup   311
8.23   Encoding by double random phase code in a JTC architecture   312

9.1   Schematic of SIMD architecture   321
9.2   Schematic of MIMD architecture   322
9.3   Schematic of a photonic computing architecture   327
9.4   Schematic of an OPAL architecture   332

9.5   Schematic diagram of (a) space variant and (b) space invariant
       networks   333
9.6   Schematic diagram of (a) space variant and (b) space invariant
       networks   334
9.7   Schematic diagram of the perfect shuffle network with $N = 8$ 334
9.8   Schematic of (a) banyan network and (b) crossover network   335
9.9   (a) Single cell operation of optical shadow-casting architecture
       and (b) coding of input A and B   346
9.10   Multicell operation of optical shadow casting technique . . .   347
9.11   Schematic diagram of (a) architecture for implementing arith-
       metic and logical operations and (b) partition of SLM into
       three areas R, S, and T distances   348
9.12   (a) Cell structure generated effectively modulates the source
       array to perform signed Ex-OR operation on the operands.
       (b) (4×4) cell decoding mask for each bit of the projected
       superimposed input.   350
9.13   Bit-wise superimposition of the (3×3) patterns of the two
       input operands   350
9.14   Resultant superimposed input combinations of each bit of A
       and B when (a) operands A and B are converted to signed and
       unsigned negabinary forms, respectively, and (b) operands A
       and B are converted to unsigned and signed negabinary forms,
       respectively   351
9.15   Resultant superimposed input combinations of each bit of A
       and B: (a) superimposed patterns of the addend and augend in
       two channels, (b) projected pattern of superimposed pattern
       for addition in the two channels (thick-bordered cells represent
       the cells to be scanned), and (c) scanned coefficients in the
       thick-bordered cells   352
9.16   Design of cell-structure in front of the lenslet array for some
       logic operations   353
9.17   Coding of the three inputs to flip-flop   354
9.18 (3×3) source modulated by SLM 354
9.19   Source pattern and decoding masks for different flip-flops: (a)
       type of flip-flop, (b) source pattern, and (c) decoding mask . 355
9.20   Schematic diagram of the matrix-vector multiplication pro-
       cess. S: laser source; LA: lenslet array; P1: source control SLM;
       CL: cylindrical lens; L: condensing lens; P2: SLM; D: detector
       array   360
9.21   Schematic diagram of a hybrid photonic multiprocessor . . .   361
9.22   Schematic diagram of an all-optical digital multiprocessor .   361
9.23   Block diagram of AT&T digital optical multiprocessor . . .   362

10.1   Schematic diagram of a VLC architecture   372
10.2   Optical arrangement of the VLC architecture   373

10.3   Basic frequency domain correlation technique for face recognition 374

10.4   Pictorial representation of PSR metric evaluation from correlation plane output 376

10.5   Optical arrangement of the hybrid digital-optical correlation architecture 386

10.6   Schematic diagram of the JTC architecture 387

10.7   Flowchart for obtaining JTC 390

10.8   Schematic diagram of photonic JTC 391

10.9   Schematic diagram of single SLM photonic JTC 391

10.10 Optical arrangement of the JTC architecture 392

10.11 A simple processing element (PE) 394

10.12 Transfer functions: (a) step function, (b) linear threshold function, (c) sigmoid function 395

10.13 Perceptron model 397

10.14 Ex-OR classification problem 398

10.15 Multilayer perceptron model 399

10.16 Hopfield model of ANN 401

10.17 Hopfield auto-associative model 402

10.18 Boltzmann machine 403

10.19 Kohonen's self-organising map 404

10.20 Carpenter-Grossberg model of adaptive resonance theory . . 405

10.21 Holographic architecture for implementing perceptron network 407

10.22 Hybrid photonic architecture for implementing perceptron network 407

10.23 (a) Set of training input patterns where the associations are (A,S) and (P,D), and (b) output of associated pattern S even if the given input pattern A is distorted 412

10.24 (a) Frame T^i showing the outer-product matrix for the vector $(A^1), C^i$, (b) frame T^j showing the outer-product matrix for the vector $(A^j), C^j$, and (c) sum of the outer product matrix $(T^i + T^j)$ 413

10.25 Scheme of (a) recording and (b) readout of outer product matrix 414

10.26 Scheme of (a) recording and (b) readout of outer product matrix 415

10.27 Stored pattern pairs 420

10.28 Hetero-associative recall of patterns 421

10.29 Operational flowchart of the hetero-associative architecture 422

10.30 Three-layer associative memory model 423

10.31 Photonic architecture for WTA model 424

10.32 Photonic architecture for three-layer hetero-associative memory with WTA 425

11.1 Visualisation of (a) classical bits and (b) qubits 436
11.2 Representation of qubit states in a spherical surface 437
11.3 Notation used for representing qubit gates 451
11.4 Notation used for representing some quantum gate circuits . 452
11.5 Register operation using (a) classical bits and (b) qubits . . 452
11.6 (a) Classical and (b) quantum cryptography system 455
11.7 Teleportation model by a 3-bit system 461

12.1 Representative 1D, 2D, and 3D photonic crystal structure (redrawn from [11]) 470
12.2 (a) 1D structure with unidirectional periodicity of permittivity, (b) photonic band structure for propagation perpendicular
to the structure for the same permittivity, and (c) formation
of photonic bandgap when permittivity is periodic, as shown
in (a) 474
12.3 (a) Square lattice of photonic crystal and (b) Brillouin zone
of square lattice (irreducible zone is shaded) 474
12.4 Dispersion diagram for light propagation along x-axis for various hole sizes 474
12.5 Band diagram calculated along the high-symmetry direction
of the 2D photonic crystal 475
12.6 Band diagram calculated along the high-symmetry direction
of 2D photonic crystal 476
12.7 Schematic diagram of planar photonic crystal 477
12.8 Drawings of various structures: (a) birefringent PCF; (b)
ultra-small core PCF; (c) hollow core PCF; and (d) hollow
core Bragg fibre. The white regions represent silica, the black
regions are hollow, and the gray regions are other materials
(glasses or polymers) 479
12.9 Microstructured optical fibres with air holes 479
12.10 Behaviour of glass core/air hole fibre 481
12.11 Behaviour of glass core/air hole fibre (d is the hole width and
λ is the pitch) 481
12.12 Sketch of a typical density-of-photonic state (DOPS) diagram 483
12.13 Comparison of a PCF based supercontinuum source with a
fibre coupled white light source 484
12.14 (a) Schematic of the SPP electromagnetic field at an interface
defined by $z = 0$, and (b) the SPP field component $|E_z|$ decays
exponentially into both conductor ($z < 0$) and dielectric ($z >
0$) while propagating along the y-direction 487
12.15 Schematic of the geometry used for deriving the SPP electromagnetic field 488
12.16 Dispersion relation of SPP 489
12.17 Schematic of the geometry used for deriving the SPP in a
two-interface system 490

12.18 General schematic of prism coupling of free electromagnetic waves to SPPs 492
12.19 Dispersion relation of SPP for prism coupling 492
12.20 The shifted dispersion relations (dashed lines) are depicted for three values of n 493

List of Tables

2.1   Ranges of electromagnetic spectrum   56
2.2   Differences between electromagnetic and optical waves . . .   64

10.1   Different values of diagonal preprocessor matrices and corresponding filters   377
10.2   Unconstrained linear filter designs   379
10.3   Some advanced correlation filters   380

11.1   Technologies for qubit realisation   440

Chapter 1

Information Communication

1.1 Information

Information is one of the prime drivers of human society. It is all-pervasive in technology, commerce, and industry. In this context, it is imperative to discuss the issue in more detail, highlighting the interplay between the terms data, information, and knowledge.

The word, information, was most probably derived from the Latin verb *informare* (to inform), which in turn may have derived from either a French word *informer* or from the Latin verb *informare*, meaning a quantity to give form to, or to form an idea from. At the end of the last century, the term 'information' was defined as *knowledge*, which can be transmitted without loss of integrity, thus suggesting that information is one form of knowledge. In the same sense, the knowledge was identified as a state of preparedness built up partly by personal commitment, interests, and experiences and partly by the legacy of the tradition of the society. Knowledge is therefore fundamentally tacit [1], [2].

At the outset, it is necessary to distinguish between the three words data, information, and knowledge. In everyday discourse, the distinction between data and information, on the one hand, and between information and knowledge, on the other, remains typically vague. The terms data and information are used interchangeably and the notion of information is conflated with knowledge. Fundamentally, data are sensory stimuli that are perceived through our senses. Information is that subset of data that has been processed into a meaningful form for a user. Knowledge is what has been understood and evaluated by the user. The following three points summarise the essential differences between the three words and concepts [3]:

1. *Data* are syntactic entities with no meaning; they are input to an interpretation process, i.e., to the initial step of decision-making.

2. *Information* is interpreted data. It is the output from data interpretation as well as the input to, and output from, the knowledge-based process of decision-making.

3. *Knowledge* is learned information incorporated in the user's reasoning

resources, and made ready for active use within a decision process; it is the output of a learning process.

Therefore, the data is information when used in decision-making, but goes beyond this definition since it links the use of data to the underlying interpretation process that enables its use. Knowledge, then, is what is needed in order to perform the interpretation and what gets learned from new information. The role of knowledge, in general, is therefore to play an active part in the processes of transforming data into information, deriving other information, and acquiring new knowledge [4]. The following process gives the route to transform data into knowledge.

1. Transform data into information (referred to as data interpretation).

2. Derive new information from existing information (referred to as elaboration).

3. Acquire new knowledge (referred to as learning).

Information constitutes the significant regularities residing in data that the user attempts to extract. Therefore, information is an extraction from data by modifying the relevant probability distributions and has the capacity to perform useful work on the user's knowledge base. A system cannot possess information without having knowledge, i.e., without being what is referred to as a knowledge-based system. The term 'information system', however, is usually used for systems that do not necessarily have knowledge and reasoning capabilities. They are the systems intended to store and process structures that are to be interpreted as information for the user. An important distinction between a knowledge-based system and an information system can be set [5]. The frame of reference of information in an information system is the system user, while the frame of reference of knowledge in a knowledge-based system is the system itself. Therefore, information systems and information technology, as their names imply, are meant to deal with information in the domain of computer science and engineering [6], [7]. Over and above the articles cited, archival publications are assembled in [8], [9].

Major theoretical development in the area of information communication process started with Shannon's classical theorems on code capacity and channel capacity. Shannons ideas, first presented in his seminal papers [10], [11], [12] have been crucial in enabling the advances in information and communication technology. In fact, his works were referred to as the Magna Carta of the information age [13]. However, Shannon's theorems did not provide the methods of achieving ideal channel capacity for a given statistics of the channel, with which all practical communication systems can be evaluated with respect of their efficiency. Shannon's theory showed how to design more efficient information communication and storage systems by demonstrating the enormous gains achievable by coding, and by providing the intuition for the correct design of coding systems. While developing the theorems and subsequent theories related to various aspects of information, the question of uncertainty of information either generated or received have become an important issue which can be studied in the context of probability theory. Probability theory is mainly related to the study of uncertainty. The uncertainty of an event is measured by its probability of occurrence and is inversely proportional to that. The more uncertain an event, the more information is required to resolve uncertainty of that event. Information is also a kind of event that affects the state of a dynamic system that can interpret the information.

1.2 Probability

Probability theory is the study of uncertainty. Two different points of view are expressed about what probability actually means: (a) *relative frequency*: this is exercised by sampling a random variable a great many times and tallying up the fraction of times that each of its different possible values occurs, to arrive at the probability of each. (b) *degree of belief*: probability is expressed as the plausibility of a proposition or the likelihood that a particular state (or value of a random variable) might occur, even if its outcome can only be decided once. In this and subsequent section the basic equations are only used in the context of information technology. The first view, is the one that predominates in information theory.

The reason for using probability is the deficiency of information to study with certainly an event or a process. Probabilistic approaches are also necessary to learn with some certainty, since the process of learning is the process of inference. The processing of partial information which yields answers with uncertainty are probabilities. Additionally, if more information is available for reprocessing, a new probability is arrived at. However, probabilities by themselves may not always be practically useful. They are mostly useful when probabilities are compared. If one of the alternative decisions has a higher probability of benefits, then that decision has to be considered for further action. One may assign equal probabilities to the other decisions, if no other information is present. If anything about various possible benefits of a decision is known, then each benefit will have the same probability. This was called the *principle of insufficient reason* (PIR).

Some basic axioms of probability are necessary to establish its relation with entropy and information processing. In probability theory, a sample space Ω is defined as a set of all outcomes of a random experiment. Each outcome can be thought of as a complete description of the state of the real world at the end of the experiment. Next, an event space Γ is defined as a set whose elements $A \in \Gamma$ are called events, which are subsets of Ω, that is, $A \subseteq \Omega$. Events are collections of possible outcomes of an experiment. A function P that satisfies the following axioms is called a probability measure or simply the probability. The three axioms for event A are:

1. $0 \leq P(A) \leq 1$, for all $A \in \Gamma$

2. $P(\Omega) = 1$

3. If, $A_1, A_2....$ are disjoint events ($A_i \cap A_j = \phi$, when $i \neq j$, then $P(\cup_i A_i) = \sum_i P(A_i)$

While considering two events A and B, the union of A and B is $A \cup B$ denotes the points of sample space Ω that are in A or B or both. $A \cap B$ is the intersection of A and B and is the set of points that are in both A and B. ϕ denotes the empty set. $A \subset B$ means that A is contained in B and A^c is the complement of A, that is, the points in Ω that are not in A. The following properties of a probability measure hold good:

1. If $A \subseteq B \Rightarrow P(A) \leq P(B)$

2. $P(A \cap B) \leq min(P(A), P(B))$

3. $P(A \cup B) \leq P(A) + P(B)$ (called union bound).

4. $P(\Omega|A) = 1 - P(A) = P(A^c)$

5. If $A_1, ...A_k$ are a set of disjoint events, such that $\cup_{i=1}^{k} A_i = \Omega$, then $\sum_{i=1}^{k} P(A_k) = 1$ (called the law of total probability).

In general, probability theory rests upon two rules:

1. Product rule: This rule indicates the joint probability of both A and B, and is given by $P(A, B) = P(A|B)P(B)$ or, $P(B|A)P(A)$. In case A and B are independent events, $P(A|B) = P(A)$ and $P(B|A) = P(B)$. In this case joint probability is simply $P(A, B) = P(A)P(B)$.

2. Sum rule: If event A is conditionalised on a number of other events B, then the total probability of A is the sum of its joint probabilities with all B,as indicated by $P(A) = \sum_B P(A, B) = \sum_B P(A|B)P(B)$. From the product rule, it may be noted that $P(A|B)P(B) = P(B|A)P(A)$, since $P(A, B) = P(B, A)$.

1.2.1   Random variables

Typically, a random variable X in $\Re$ space is a function of $X : \Omega \rightarrow \Re^2$. Generally, a random variable is denoted by upper case X and the value that a random variable may take on is denoted by lower case x. Random variables can be continuous or discrete. Two random variables X and Y are independent, if the knowledge of value of one variable has no effect on the conditional probability distribution of the other variable. That is, if X and Y

are independent, then for any subsets $A, B \subseteq \Re$, the probability equation can be written as $P(X \in A, Y \in B) = P(X \in A)P(Y \in B)$. This explains that if X is independent of Y, then any function of X is independent of any function of Y.

1.2.1.1 Expectation, variance, and covariance

The expectation $E[X]$ of a real valued random variable X is defined as

$$E[X] = \sum_x P(x)x \tag{1.1}$$

It is an average of the values x taken by X weighted by their probabilities $P(X = x)$. The above sum is finite by the assumption of discreteness.

The variance $var[X]$ of a random variable X is a measure of how concentrated the distribution of the random variable is around its mean and is defined as

$$var[X] = E[(X - E(X))^2] = E[X^2] - E[X]^2 \tag{1.2}$$

As the variance is the expected quadratic deviation of X from its mean $E[X]$, it is a measure of spread of the distribution of X around $E[X]$.

The covariance of two random variables X and Y is defined as

$$cov[X, Y] = E[(X - E[X])(Y - E[Y])] = E[XY] - E[X]E[Y] \tag{1.3}$$

When $cov[X, Y] = 0$, X and Y are uncorrelated.

1.2.2 Conditional probability and independence

The conditional probability of an event is a probability obtained with the additional information that some other event has already occurred. If B is an event with non-zero probability, then the conditional probability of any event A, given B is defined as

$$P(A|B) \triangleq \frac{P(A \cap B)}{P(B)} \tag{1.4}$$

Therefore, $P(A|B)$ is the probability measure of the event A after observing the occurrence of event B and this is read as, the probability of A, given B. If $P(A|B)$ is known, then $P(B|A)$ can be determined as

$$P(B|A) = \frac{P(A|B)P(B)}{P(A|B)P(B) + P(A|B^c)P(B^c)} \tag{1.5}$$

Two events are treated as independent, if and only if, $P(A \cap B) = P(A)P(B)$, or $P(A|B) = P(A)$. Therefore, independence is equivalent to stating that measurement of B does not have any effect on the probability of A. Further, if A and B are independent, then A and B^c are also independent.

The formula given above, which is often used while trying to derive expression for the conditional probability of one variable, given another, is called Bayes theorem or Bayes' rule [14].

In the case of discrete random variables X and Y, conditional probability in the case of $P_Y(Y = y) \neq 0$, is given by

$$P(X = x|Y = y) = \frac{P_{XY}(XY = xy)}{P_Y(Y = y)} \qquad (1.6)$$

Evidently, if $P_Y(Y = y) = 0$, then conditional probability is undefined. Similar expression can be obtained for continuous random variables.

The Bayes' theorem allows us to reverse the conditionalising of events and provides a simple mechanism for repeatedly updating our assessment of the hypothesis as more data continues to arrive. Therefore, we can compute $P(Y|X)$ from knowledge of $P(Y|X)$, $P(X)$, and $P(Y)$. Often these are expressed as prior and posterior probabilities. By definition, a prior probability is an initial probability value originally obtained before any additional information is obtained. Similarly, a posterior probability is a probability value that has been revised by using additional information that is obtained at a later stage.

1.2.3   Cumulative distribution function

Cumulative distribution function is a probability measure used while dealing with random variables. It is often convenient to specify functions from which the probability measure governing an experiment immediately follows. A cumulative distribution function (CDF) is such a function, $F_X : \Re \rightarrow [0, 1]$, which specifies a probability measure as $F_X(x) \triangleq P(X \leq x)$. By using this function the probability of any event can be calculated.

1.2.4   Probability density function

For some continuous random variables, the cumulative distribution function $F_X(x)$ is differentiable everywhere. In such situations, the probability density function or PDF is defined as the derivative of CDF given by

$$f_X(x) \triangleq \frac{dF_x(x)}{dx} \qquad (1.7)$$

For very small Δx,

$$P(x \leq X \leq x + \Delta x) \approx f_X(x)\Delta x \qquad (1.8)$$

When they exist, both CDFs and PDFs can be used for calculating the probabilities of different events. However, the value of PDF at any given point x is not the probability of that event, i.e., $f_X(x) \neq P(X)$. For example, $f_X(x)$ can take on values larger than 1, but the integral of $f_X(x)$ over any subset will be at most 1.

1.3   Entropy and information

Entropy and information can be considered as measures of uncertainty of probability distribution. To gain information is to lose uncertainty by the same amount. No information is gained (no uncertainty is lost) by the appearance of an event or the receipt of a message that was completely certain. However, the functional relationship between entropy and the associated probability distribution has long been a question of debate in statistical and information science [15], [16]. There are many relationships established on the basis of the properties of entropy. Among all these definitions and interpretations of entropies, the much-discussed one is Shannon's informational entropy.

The quantitative measure of information is based on the intuitive notion of information, which can be expressed in terms of probability. When an event occurs, the posteriori probability is 1. The relation between probability of occurrence $P(x)$ of an event x and information $I(x)$ is given by

$$I(x) = \log \frac{1}{P(x)} = -\log P(x) \tag{1.9}$$

$I(x)$ is a continuous function of $P(x)$ and the unit of $I(x)$ depends on the base of $\log[.]$. In the case of a digital system

$$I(x) = -\log_2 P(x) \text{ bits} \tag{1.10}$$

The logarithmic measure is justified, as information is additive. When independent packets of information arrive, the total information received is the sum of all individual pieces. But the probabilities of independent events multiply to give their combined probabilities. The logarithm has to be taken in order to derive the joint probability of independent events. Therefore, the equation can be generalized as

$$I(x) = -\log_D P(x) \text{ D-ary units} \tag{1.11}$$

The information measure can be a monotonically decreasing function of P and the information content of a message increases as $P(x)$ decreases and vice versa. The total information of two or more independent messages or events is the sum of the individual information. Therefore, if the two independent events x_1 and x_2 have *a priori* probabilities of P_1 and P_2, then their joint probability is $P = P_1 P_2$ and the information associated with the event is

$$I(x_1, x_2) = -\log P = -\log P_1 - \log P_2 = -\sum_i \log P_i \tag{1.12}$$

A situation may be considered where an information source S is delivering messages from its alphabet $X = \{x_1, x_2, ...x_m\}$ with probabilities

$P\{x\} = \{P_1, P_2, ...P_m\}$. If N messages are delivered, then a symbol x_i occurs NP_i times and each occurrence conveys information $-\log P_i$. Hence, the total information due to NP_i messages is $-NP_i \log P_i$ and total information due to all N messages is given by $-N \sum_1^m P_i \log P_i$, where $\sum P_i = 1$.

This equation allows us to speak of the information content or the entropy of a random variable, from knowledge of the probability distribution that it obeys. Evidently, entropy does not depend upon the actual values taken by the random variable, it depends only upon their relative probabilities.

The Shannon's entropy $H(X)$ of the source is now defined as the average amount of information per source symbol and is given by

$$H(X) \triangleq -\sum_1^m P_i \log P_i. \tag{1.13}$$

where, $\sum P_i = 1$. Evidently, Shannon's entropy in information theory is the measure of the amount of information that is missing before reception. Shannon's entropy is associated with a source, whereas information is associated with a message.

There is a link between Shannon's entropy and thermodynamic entropy. The expressions for the two entropies are similar. The information entropy H for equal probabilities $P_i = P = \frac{1}{n}$ is given by

$$H = k \log \left(\frac{1}{P}\right) \tag{1.14}$$

where k is a constant that determines the units of entropy.

The entropy as defined has some important properties, such as:

1. $H(X) = 0$, if and only if, all $P_is = 0$ except 1. This lower bound on entropy corresponds to no uncertainty.

2. $H(X) = -\sum_1^m \frac{1}{m} \log \frac{1}{m} = \log m$.

3. If $P_1 = P_2 = ...P_m$ then $H(X) = \frac{1}{m}$.

4. $\log m$ is maximum for any possible set of P_is.

For a binary source with alphabets $\{0, 1\}$, and probabilities $(P, 1 - P)$,

$$H(X) = P \log \frac{1}{P} + (1 - P) \log \frac{1}{1 - P} \triangleq H(P) \tag{1.15}$$

It can be seen that $H(P)$ is maximum when $H(p)$ is 1 bit and occurs only when $P = (1 - P) = 0.5$. For m symbol source, the maximum entropy $H(X)_{max}$ increases as $\log m$ and for any other $P(i) \neq P(j)$, the entropy of the source is less.

Entropy of several alphabets can also be calculated. The case of two sources

S_1 and S_2 delivering symbols x_is and y_js with some correlation between them is considered. In conformity with the definition of joint probability $P(x_i, y_j)$, the entropy of the joint event (x_i, y_j), known as joint entropy of the sources S_1 and S_2 is defined as

$$H(X,Y) = -\sum_{i=1}^{m}\sum_{j=1}^{n} P(x_i, y_j) \log P(x_i, y_j) \qquad (1.16)$$

where $P(x_i, y_j) = P(x_i)P(y_j|x_i)$.

If $\{X\}$ and $\{Y\}$ are independent of each other, then $P(x_i, y_j) = P(x_i)P(y_j)$, and $H(X,Y) = H(X) + H(Y)$. The mutual information between two random variables measures the amount of information that one conveys about the other. Equivalently, it measures the average reduction in uncertainty about X that results from learning about Y. Similarly, the conditional entropy of Y given that X has occurred is given by

$$H(Y|X) = -\sum_{i=1}^{m}\sum_{j=1}^{n} P(x_i, y_j) \log P(y_j|x_i) \qquad (1.17)$$

If $\{X\}$ and $\{Y\}$ are independent, then $H(Y|X)$ can be shown as equal to $H(Y)$. The results for two variables can be extended to three variables, giving the relations as $H(X,Y,Z) = H(X,Y) + H(Z|X,Y)$ and $H(X,Y) \geq H(X,Y|Z)$. The results can be extended for q variables as

$$H(S_1, S_2....., S_q) = H(S_1, S_2....., S_{(q-1)}) + H(S_q|S_1, S_2, ..., S_{(q-1)}) \qquad (1.18)$$

where $H(S_1, S_2....., S_q)$ is the joint entropy of all sources S_1 to S_q. It is also possible to maximise entropy and conditional entropy [17].

1.4 Information communication

The useful interpretation of entropy given so far is used to form a model of an information transmission system in terms of input (source) and output (sink) and the channel. An information source is a mathematical model for a physical entity that produces a succession of symbols called outputs from the source in a random manner. According to the types of message the system can be discrete or continuous. In discrete system, the message and the signal both consist of a series of discrete (discontinuous) symbols. In a continuous system, the message and the signal are both continuously varying.

The source output is a sequence of finite random variables over a finite set known as source alphabets. The source is known if all the finite dimensional distributions of the random outputs are known. A source is therefore

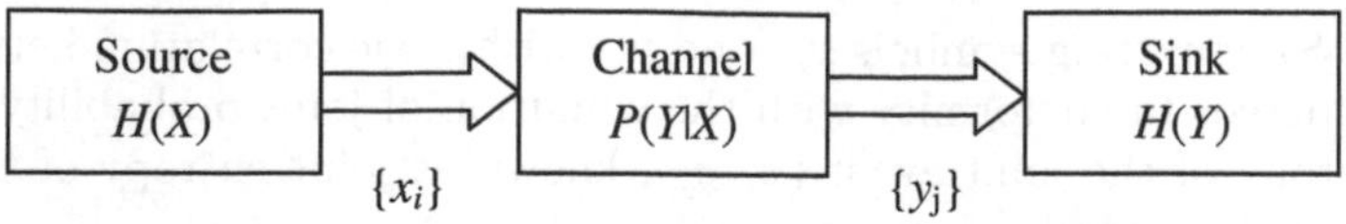

FIGURE 1.1: Basic probabilistic model of a channel

essentially an assignment of a probability measure to events consisting of sets of sequences of symbols from the alphabet. It is useful, however, to explicitly treat the notion of time as a transformation of sequences produced by the source.

The source output sequence needs to be converted into a sequence over another finite set called code alphabets. The method of converting is called encoding and it is necessary to reconstruct the original source output from the encoded version. A channel model is shown in Figure 1.1. The entropies of the source and receiver are denoted by $H(X)$ and $H(Y)$, respectively. The channel may consist of a coder which normally changes the statistics of the source symbol $\{x_i\}$ and is dependent on the transmission probability matrix $P(Y|X)$. Similarly, in an information transformation channel the transmitted symbols are changed due to the noise present and the symbols received by the sink are denoted by $\{y_j\}$. The received symbols can be interpreted in terms of inverse probability matrix $P(X|Y)$.

The presence of noise may change the source or transmitter statistics $P\{x_i\}$, though certain probabilistic information of the source statistics is available, once the probability $P\{x_i|y_j\}$ is calculated at the receiver or sink. The initial uncertainty in $\{x_i\}$ and the final uncertainty of $\{x_i\}$ after reception of $\{y_j\}$ are given by, $-\log P\{x_i\}$ and $-\log P\{x_i|y_j\}$, respectively. Thus, the information gain of the channel, known as trans-information of the channel, is given by

$$
\begin{aligned}
I(n_i, y_j) &= \log \frac{P\{x_i|y_j\}}{P\{x_j\}} \\
&= \log \frac{P\{x_i, y_j\}}{P\{x_i\}.P\{y_j\}} \\
&= \log \frac{P\{y_j|x_i\}}{P\{y_j\}} \\
&= I\{y_j, x_i\}
\end{aligned}
\tag{1.19}
$$

The mutual information between two random variables measures the amount of information that one conveys about the other. Equivalently, it measures the average reduction in uncertainty about X that results from learning about Y. Average information gained by the sink is obtained in terms of expectation

E, as

$$
\begin{aligned}
I(X,Y) &= E[I(x_i, y_j)] \\
&= \sum_{j=1}^{n} \sum_{i=1}^{n} P(x_i, y_j) I(x_i, y_j) \text{ bits/symbol} \\
&= \sum_{j} \sum_{i} P(x_i, y_j) \log \frac{P(x_i|y_j)}{P(x_i)} \qquad (1.20) \\
&= H(X) - H(X|Y) \qquad (1.21)
\end{aligned}
$$

Note that in case X and Y are independent random variables, then the numerator inside the logarithm equals the denominator. Then the log term vanishes, and the mutual information equals zero. Moreover mutual information is always ≥ 0.

By using the equation of $H(X,Y)$, $I(x,y)$ can be written as

$$
\begin{aligned}
I(X,Y) &= H(Y) - H(Y|X) \\
&= H(X) + H(Y) - H(X,Y) \text{ bits/symbol} \qquad (1.22)
\end{aligned}
$$

where $H(X|Y)$ is the loss of information due to channel noise called equivocation.

For a noise-free channel, $H(Y|X) = H(X|Y) = 0$ and

$$
I(X,Y) = H(X) = H(Y) = H(X,Y) \qquad (1.23)
$$

For a channel where $\{x_i\}$ and $\{y_j\}$ have no correlation and are independent, then $H(X|Y) = H(X)$, and $H(Y|X) = H(Y)$, hence,

$$
I(X,Y) = H(X) - H(X|Y) = 0 \qquad (1.24)
$$

Therefore, no information is transmitted through the channel.

1.5 Source encoding: noiseless coding

The main objective of source coding is to convert the information contained in a given realisation of the source in the most compact way. It allows in fact, to formalise the intuitions of information and uncertainty which are associated with the definition of entropy.

A probabilistic information source generates a sequence of symbols taking values in a finite alphabet set and any symbols are taken to be random variables. Let an m-ary source alphabets $\{x_1, \ldots x_m\}$ be transformed in a convenient form, say binary $\{0, 1\}$ digits. Since the rate of transmission of information in a noisy channel is maximum, if the source probability $P(x_i)$

is all equal, that is $H(x) = \log m$. It is desirable that the transformed code alphabets $\{C_1, C_2, ..., C_D\}$ have equal probability of occurrence. The method is thus known as entropy coding.

Often the sequence of symbols is a part of a longer stream. The compression of this stream is realised in three steps. First, the stream is broken into blocks of length. Then each block is encoded separately and finally, the code words are glued to form a new compact stream. To avoid the ambiguity or deciphering the codes and to minimise the time of transmission of the message, the codes should have the most important property, that is the codes should be uniquely decodable. Other properties are states as:

1. The average length of the code word is given by,

$$\bar{L} = \sum P_i L_i \qquad (1.25)$$

   where L_i is the length of ith code word and P_i is the probability of occurrence.

2. $\bar{L}$ should approach the value $H(X)/\log D$, for D-ary digits, with the condition that $\bar{L} \geq H(X)/\log D$.

3. Since $\sum P_i L_i$ needs to be minimum, the code word with larger L_i should have smaller P_i. This condition leads to,

$$L_i \propto \log(1/P_i) \qquad (1.26)$$

4. The code efficiency η is defined as the ratio of the average information per symbol of the encoded language to the maximum possible information per code symbol and therefore,

$$\eta = \frac{H(X)}{L \log D} \qquad (1.27)$$

   where the optimum value of η is 1, for compact codes of D-ary digits.

5. The redundancy of codes is defined as, $R = (1 - \eta)$

1.6   Shannon's theorems

Information theory was established by Claude E. Shannon for the study of certain aspects of information, mainly to analyse the impact of coding on the information transmission. His key observation was that the colloquial term 'information' needs no exact definition. Semantic aspects are irrelevant, rather it needs operational characterization through numerically measurable quantities. Further, information sequences are random processes, thereby probability theory needs to be applied when analysing problems concerned with information systems. Theoretical results, as a consequence of the ideas, are usually referred to as Shannon's theory. He introduced the notion of entropy as a measure of uncertainty, as entropy is a measure of the amount of information allowed for transmission. In general, the entropy equals the minimum number of binary digits needed on average to encode a given message in a uniquely decodable way.

Another important theorem with which Shannon's name is associated is known as the Nyquist-Shannon sampling theorem. Sampling theorem is a fundamental bridge between continuous-time signals and discrete-time signals (often called digital signals). It establishes a sufficient condition for a sample rate that permits a discrete sequence of samples to capture all the information from a continuous-time signal of finite bandwidth. It states simply that if a continuous time signal $y(t)$ contains no frequencies higher than B Hertz, it is completely determined by giving its ordinates at a series of points spaced $\frac{1}{2B}$ seconds apart. Sufficient sample-rate is therefore $2B$ samples/second, or anything larger. Equivalently, for a given sample rate f_s, perfect reconstruction is possible for a bandlimit $B \leq \frac{f_s}{2}$. Readers may refer to the historical issues along with generalization of the theorem in [18], [19], and [20].

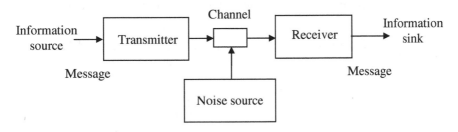

FIGURE 1.2: Basic model of a channel

1.6.1 Shannon's first fundamental theorem

Shannon's first fundamental theorem, known as Shannon's source coding theorem, is formalised in a system consisting of a zero memory source S with entropy $H(X)$ per symbol. Further, the channel is a noiseless channel as shown in Figure 1.2, with a capacity C bits per message. According to source coding theorem, it is possible to encode the output of S, with an encoder delivering messages to the channel, so that all information generated in S is transmitted through the channel without loss, if and only if $C \geq H(X)$.

By noiseless encoding, the average length $\bar{L}$ of the code words at the coder output approaches $H(X)$ per symbols of the source, particularly with the nth extension of the source, where the average information per source symbol is $H(X)$ bits. A message of length n source symbols gives $nH(X)$ bits of information and equivalently the source generates $M_s = 2^{nH(X)}$ messages. If the duration of each symbol is t_0 seconds, then in time T seconds, the number of source messages generated is $M_s = 2^{H(X)T/t_0}$, as $n = T/t_0$.

For D-ary digits, at the coder output, the average information per code word is $\bar{L} \log D$ bits and in T seconds, the number of messages generated is $M_c = 2^{n\bar{L}\log D}$. It can be intuitively stated that $M_c \geq M_s$. Thus, the bound can be written as

$$M_c = 2^{n\bar{L}\log D} \geq M_s = 2^{nH(X)} \tag{1.28}$$

and

$$\bar{L} \geq \frac{H(X)}{\log D} \tag{1.29}$$

Evidently, with $n \to \infty$, $\bar{L} \to H(X)/\log D$. With D-ary encoding, $D > 2$, the bound can be found as

$$\log 1/P(x_i) \leq L_i \log D \leq \log 1/P(x_i) + 1 \tag{1.30}$$

A necessary and sufficient condition for the existence of an instantaneous code with word length L_i in D-ary digits is $\sum_i^m D^{-L_i} \leq 1$. Multiplying by $P(x_i)$ and summing over all $i's$, we get for a zero memory source,

$$\sum_{i=1}^m P(x_i) \log 1/P(x_i) \leq \sum_{i=1}^m P_i L_i \, logD \leq \sum_{i=1}^m P(x_i) \log 1/P(x_i) + 1$$

or,

$$\frac{H(X)}{\log D} \leq \bar{L} \leq \frac{H(X)}{\log D} + 1 \tag{1.31}$$

In zero memory sources there are no inter symbol influences and the symbols occur independently in a long message, according to their $P(x_i)$. Using nth extension of S, better efficiency can be obtained with $\bar{L}_n$ as the new code length, having $H(S^n) = nH(x)$. The above equation then can be modified as,

$$\frac{H(X)}{\log D} \leq \frac{\bar{L}_n}{n} \leq \frac{nH(X)}{\log D} + \frac{1}{n} \tag{1.32}$$

where $\bar{L}_n/n$ is the average number of code symbols used per single symbol of S. When $n \to \infty$, the average code length for an efficient code $\bar{L}_n$ is obtained as

$$\lim_{n \to \infty} \bar{L}_n = \frac{H(X)}{\log D} \tag{1.33}$$

If the input to the coder is n symbol messages for the extended source S^n, $\bar{L}_n \neq \bar{L}$, where $\bar{L}$ is the average code length for the source S. The capacity of the channel is given by

$$(\bar{L}_n/n) \log D = C \ \text{bits/message} \tag{1.34}$$

and for successful transmission of messages through the channel, the following equation can be written,

$$H(X) \leq (\bar{L}_n/n) \log D = C \ \text{bits/message} \qquad (1.35)$$

The above equation is also valid for a source with finite memory called Markov source. In the case of Markov source, inter symbol influence occurs, that is the occurrence of x_i in the zero-th position s_0 of the message depends on the previous q symbols $\{s_1, s_2, ...s_q\}$. A qth order Markov source S_q is specified by the set of conditional probabilities. In a first order Markov source S_1 and its adjoint $\bar{S}_1$ have identical first order symbol probabilities $P(x_1), P(x_2), ...P(x_m)$ and $\bar{S}_1$ is a zero memory source, with $H(S_1) < \bar{S}_1$); the average code length for both the sources are identical. Thus,

$$\bar{L} \log D \geq H(\bar{S}_1) > H(S_1) \qquad (1.36)$$

Using, the n th extension of S_1, the related equation can now be modified as

$$H(\bar{S}_1^n) \leq \bar{L}_n \log D \leq H(\bar{S}_1^n) + 1 \qquad (1.37)$$

Dividing by n, the equation can now be written for large n as

$$\lim_{n \longrightarrow \infty} \frac{\bar{L}_n}{n} = \frac{H(S_1)}{\log D} \qquad (1.38)$$

The result can be extended to the qth order Markov source by replacing S_1 in the above equation by S_q.

1.6.2 Shannon's second fundamental theorem

This theorem is also known as the theorem on coding for the memoryless noisy channel. In such a channel, it is possible in principle, to devise a means whereby a communication system will transmit information with an arbitrary small probability of error, provided that the information rate R is less than or equal to channel capacity C. The technique used to achieve this objective is called coding. The theorem can be stated as: Given a source of M equally likely messages, with $M >> 1$, which is generating information at a rate R bits /symbol, and the capacity of the discrete noisy channel without memory is C bits/symbol, then it is possible to transmit the message through the channel with small probability of error P_e if $R < C$, and the coded message length $n \to \infty$. Thus, for $n \to \infty$, $R = (C - \epsilon), \epsilon > 0$, $P_e \to 0$, where ϵ is a small arbitrary error. Conversely, if $R > C$, then $P_e \to 1$, then reliable transmission is not possible.

This theorem indicates that $R \leq C$ transmission may be accomplished without error even in the presence of noise. Moreover, if the information rate R exceeds a specified value C, the error probability will increase towards unity as M increases. Also, increase in the complexity of the coding results in an increase in the probability of error.

Shannon has defined the channel capacity C of a communication channel as the maximum value of trans-information, $I(X,Y)$ as,

$$C = I(X,Y)_{max} = max[H(X) - H(Y|X)] \tag{1.39}$$

The maximisation is carried out with respect to all possible sets of probabilities that could be assigned to the input symbols.

An alternative approach to state the theorem is as follows. If a discrete memoryless source with an alphabet S has an entropy $H(S)$ and produces symbols every T_s seconds, and a discrete memoryless channel has a capacity $I(X,Y)_{max}$ and is used once every T_c seconds, then there exists a coding scheme for which the source output can be transmitted over the channel and be reconstructed with an arbitrarily small probability of error if,

$$\frac{H(S)}{T_s} \leq \frac{I(X,Y)_{max}}{T_c} \tag{1.40}$$

The parameter C/T_c is called the critical rate. When this condition is satisfied with the equality sign, the system is said to be signalling at the critical rate. Conversely, it is not possible to transmit information over the channel and reconstruct it with an arbitrarily small probability of error.

1.7 Communication channels

The typical block-diagram of a communication system is shown in Figure 1.3. It applies to all situations of communication between any source and sink.

It is assumed that the message m is an M bit sequence. This message is first encoded into a longer one, say an N bit message denoted by x with $N > M$, where the added bits provide the redundancy used to correct the transmission errors. The encoder is thus a map from $\{0,1\}^M \rightarrow \{0,1\}^N$ and the encoded message is sent through the communication channel. The output of the channel is a message y. In a noiseless channel, one would simply have $y = x$. In a realistic channel, y is, in general, a string of symbols different from

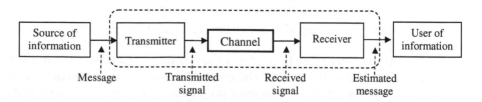

FIGURE 1.3: A typical communication channel

x, however, y is not necessarily a string of bits. The channel can be described by the transition probability $Q(y|x)$. This is the probability that the received signal is y, conditional to the transmitted signal being x. Different physical channels are described by different $Q(y|x)$ functions. The decoder takes the message y and deduces from it an estimate m_e of the sent message m.

In a memoryless channel, for any input $x = (x_1, ..., x_N)$, the output message is a string of N letters and $y = (y_1, ..., y_N)$, from an alphabet which may not necessarily be binary. In such a channel, noise acts independently on each bit of the input. This means that the conditional probability $Q(y|x)$ factorises as,

$$Q(y|x) = \prod_{i=1}^{N} Q(y_i|x_i) \tag{1.41}$$

where, the transitional probability $Q(y_i|x_i)$ is not dependent on i.

The probabilistic behaviour of information sources was formulated by Shannon and Wiener [21], where they both designated probabilistic behaviour by entropy due to the likeness with similar expressions in statistical mechanics. Shannon, furthermore, uses this concept in his general definition of channel capacity C as,

$$C = max[H(x) - H(y)] \tag{1.42}$$

This expression can be interpreted as the maximum of the difference of the uncertainty about the message before and after reception. The result also gives an upper bound of how much information can be transmitted without error through a channel.

1.7.1 Continuous channel

Until now we have considered discrete sources and discrete channels that pertain to an information communication system. However, a signal can be transmitted in analog form, where amplitude and frequency modulations are used. Here the modulating signal $X(t)$, is the set of messages to be transmitted. This message can be treated as equivalent to a continuous sample space whose sample points form a continuum, in contrast to the discrete case. Thus, a continuous channel is defined as one whose input is a sample point from a continuous sample space and the output is a sample point belonging to either the same sample space or to a different sample space. Further, a zero memory continuous channel is defined as the one in which the channel output statistically depends on the corresponding channels without memory.

1.7.1.1 Entropy of continuous channel

In the case of discrete channel $\{x_1, ...x_m\}$, the entropy is defined as,

$$H(X) = -\sum_{m} P_i \log P_i, \tag{1.43}$$

The entropy is maximum when all P_is are equal. Similarly, the entropy of a continuous channel is defined as,

$$H(X) = -\int_{-\infty}^{\infty} P(x) \log P(x) dx \tag{1.44}$$

In the case of continuous signal, entropy maximisation is related to the rectangular distribution of amplitude A leading to the entropy,

$$H(X) = -\frac{1}{A} \log \frac{1}{A} \int_{0}^{A} dx = \log A \tag{1.45}$$

where, $P(x) = 1/A, 0 \le x \le A$

Since the probabilities are bounded between $0 < P_i < 1$, in the discrete case, all entropies have positive values. In a practical system, either average power or peak power sources are used, and therefore the entropy has to be maximised under such restrictions.

1.7.1.2 Mutual information of a continuous noisy channel

The entropy definition for continuous signal is a relative one and using the logarithmic property, it follows that all entropy relations that have been studied for discrete signals do hold for continuous signals as well. Thus for continuous signal,

$$H(X,Y) = -\int_{-\infty}^{\infty} \int_{-\infty}^{\infty} P(x,y) \log P(x,y) dx dy \tag{1.46}$$

$$H(X|Y) = -\int_{-\infty}^{\infty} \int_{-\infty}^{\infty} P(x,y) \log P(x|y) dx dy \tag{1.47}$$

As in the case of discrete signal, the mutual information of continuous noisy signal is also given by,

$$I(X,Y) = H(X) + H(Y) - H(X,Y) \tag{1.48}$$

In contrast to the discrete signal case, in certain cases of continuous distribution the entropy may take on negative value, though mutual information can never be negative and remains invariant under all linear transformation. In the continuous case, the condition that needs to be satisfied is given as,

$$\int_{-\infty}^{\infty} P(x) dx = 1$$

The channel capacity is also given by, $C = I(X,Y)_{max}$. The amount of mutual information can be calculated for continuous channel, by assuming that the channel noise is additive, and statistically independent of the transmitted signal. Then, $P(Y|X)$ depends on $(Y - X)$, and not on $\{X\}$ or $\{Y\}$.

Since $Y = X + n$, where, n is the channel noise, then $y = x + n$ and $P(y|x) = P(n)$, subject to the normalisation of received signal $\{Y\}$ with respect to transmitted signal $\{X\}$. It follows that when $\{X\}$ has a given value, the distribution of $\{Y\}$ is identical to that of n, except for a translation of X. The conditional entropy can be written as,

$$
\begin{aligned}
H(Y|X) &= -\int_{-\infty}^{\infty}\int_{-\infty}^{\infty} P(X)p(Y|X)\log P(Y|X)dxdy \\
&= \int_{-\infty}^{\infty}\int_{-\infty}^{\infty} P(x)dx\, p(n)\log p(n)dn] \\
&= H(n) \text{ bits/symbol} \quad\quad (1.49)
\end{aligned}
$$

since, $\int_{-\infty}^{\infty} P(x)dx = 1$.

Therefore, the rate of transmission per sample is $H(Y) - H(n)$ and the equivocation is $H(X) - H(Y) + H(n)$. The channel capacity is denoted as R_{max} and is given by,

$$
C = R_{max} = [H(Y) - H(n)]_{max} \quad\quad (1.50)
$$

Under a given constraint, it is necessary to maximise $H(Y)$ when $H(n)$ is specified by its entropy power.

1.7.1.3 Discrete channel with discrete noise

Binary or multilevel (D-ary) signals are normally transmitted in the digital communication system and the channel noise produce error at the receiving end. The channel performance is generally defined by the bit error rate (BER) or equivalently by transition/ noise matrix of the channel. Mutual information $I(X,Y)$ as already defined is now given as,

$$
\begin{aligned}
I(X,Y) &= H(X) - H(X|Y) \\
&= H(Y) - H(Y|X) \\
&= H(X) + H(Y) - H(X,Y) \text{ bits/symbol} \quad\quad (1.51)
\end{aligned}
$$

Evidently, $I(X,Y)$ is the rate of transmission in bits/symbol and the rate of transmission in bits per second is given by,

$$
R(X,Y) = I(X,Y)/t_0 \quad\quad (1.52)
$$

where, t_0 is the duration of each symbol. It is assumed that all symbols have equal duration and there is no inter symbol probability constraint. However, the Shannon's theorem is still valid when these restrictions are removed.

The channel capacity C of a channel is defined as the maximum of the mutual information that can be transmitted through the channel, and is given by,

$$
\begin{aligned}
C &= I(X,Y)_{max} \text{ bits/symbol} \\
&= R(X,Y)_{max} \text{ bits/second} \quad\quad (1.53)
\end{aligned}
$$

Given a noise matrix, $P(Y/X)$, C is obtained by maximising $H(Y)$ and the corresponding $H(X)$ gives the value of $P\{X\}$. The transmission efficiency of the channel η can then be defined as,

$$\eta = \frac{I(X,Y)}{C} \tag{1.54}$$

and the redundancy of the channel is given by,

$$R = 1 - \eta = \frac{C - I(X,Y)}{C} \tag{1.55}$$

1.7.2   Band-limited channel

The information is actually carried by a time-waveform in a practical channel, instead of a random variable. A band-limited channel with white noise can be described by convolution operation $*$ as, $y(t) = [x(t) + n(t)] * h(t)$, where $x(t)$ is the signal waveform, $n(t)$ is the waveform of the white Gaussian noise and $h(t)$ is the impulse response of an ideal bandpass filter, which cuts all frequencies greater than B.

A key issue in connection with the transmission of time varying signal in a band limited channel is the sampling theorem which states that, if $f(t)$ is band-limited to B Hz, then the time signal is completely determined by samples of the signal taken every $\frac{1}{2B}$ seconds apart. This is the classical Nyquist sampling theorem. From this theorem, it is concluded that a band-limited signal has only $2B$ degrees of freedom (known as the dimensionality theorem). For a signal which has most of the energy in bandwidth B and most of the energy in a time T, then there are about $2BT$ degrees of freedom, and therefore the time- and band-limited signal can be represented using $2BT$ orthogonal basis functions.

As stated, the amount of information that may be transmitted over a system is proportional to the bandwidth of that system. Moreover, that information content is proportional to the product of time T and bandwidth B, and that one quantity can be traded for the other. This is explicitly stated by Hartley's law [22], where the amount of information is equal to a $KBT \log m$, where m is the number of levels or current values and K is a constant. However, a fundamental deficiency of the law established by Hartley, is that the formula does not include noise, which sets a fundamental limit to the number of levels that can be reliably distinguished by a receiver.

1.7.3   Gaussian channel

A Gaussian channel is a channel with output Y_i, input X_i and noise N_i, such that,

$$Y_i = X_i + N_i \tag{1.56}$$

where the noise N_i is zero-mean-Gaussian with variance N and is independent of X_i.

The Gaussian channel is the commonly used continuous alphabet channel used for modelling a wide range of communication systems. If the noise variance is zero or the input is unconstrained, the capacity of the channel is infinite. The channel however, may have constraint on the input power. That is, for an input codeword $(x_1, x_2, \ldots, x_n)$, the average power $\bar{P}$ is constrained so that,

$$\frac{1}{n}\sum_{i=1}^{n} x_{i^2} \leq \bar{P} \tag{1.57}$$

The probability of error P_e for binary transmission can be obtained by assuming that either $\sqrt{\bar{P}}$ or $-\sqrt{\bar{P}}$ can be sent over the channel. The receiver looks at the received signal amplitude Y and determines the signal transmitted with equal probability of high and low levels using a threshold test. Then probability of error P_e is given by,

$$\begin{aligned} P_e &= \frac{1}{2}P(Y < 0|X = +\sqrt{\bar{P}}) + \frac{1}{2}P(Y > 0|X = -\sqrt{\bar{P}}) \\ &= P(N > \sqrt{\bar{P}}) \end{aligned} \tag{1.58}$$

where P is the probability of occurrence. It may be noted that by using the scheme the Gaussian channel is converted into a binary symmetric two level channel. The main advantage of such a conversion is the processing of output signal at the cost of loss of information due to quantization.

1.7.3.1 Capacity of band-limited channel with average white Gaussian noise

Considering a situation of channel with average white Gaussian noise as shown in Figure 1.4, where the received signal is composed of the transmitted signal X corrupted with noise n.

The joint entropy at the transmitter end, assuming signal and noise are independent, is

$$H(X, n) = H(X) + H(n|X) = H(X) + H(n) \tag{1.59}$$

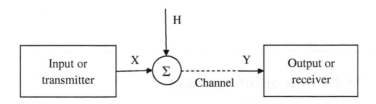

FIGURE 1.4: Basic model of a channel with average white Gaussian noise

The joint entropy at the receiver end is given by,

$$H(X, Y) = H(Y) + H(X|Y) \tag{1.60}$$

Since the received signal is $Y = X + n$, and the joint entropy over the channel is invariant, it follows that,

$$H(X, n) = H(X, Y) \tag{1.61}$$

and

$$I(X, Y) = H(Y) - H(n) \tag{1.62}$$

If the additive noise N is white and Gaussian, the statistic of received signal is also Gaussian. If the average signal power is limited to S and the average total power is $(S + N)$, the entropy $H(Y)$ is given by,

$$H(Y) = \frac{1}{2} \log 2\pi e(S + N) \text{ bits/sample} \tag{1.63}$$

and

$$H(n) = \frac{1}{2} \log 2\pi eN \text{ bits/sample} \tag{1.64}$$

Therefore, channel capacity C is given by,

$$C/\text{sample} = \frac{1}{2} \log \left(1 + \frac{S}{N}\right) \text{ bits} \tag{1.65}$$

and C per second in a bandwidth B is given by,

$$C/\text{second} = B \log_2 \left(1 + \frac{S}{N}\right) \text{ bits} \tag{1.66}$$

This result is known as Shannon-Hartley law or simply Shannon's formula of channel capacity. The primary significance of the formula is that it is possible to transmit over a channel of bandwidth B Hz perturbed by white Gaussian noise, at a rate of C bits/sec with an arbitrarily small probability of error if the signal is encoded in such a manner that the samples are all Gaussian signals. The Shannon-Hartley law predicts that a noiseless Gaussian channel with $(S/N = \infty)$ has an infinite capacity. However, the channel capacity does not become infinite even when the bandwidth is made infinite.

1.8   Detection of information

A signal detection problem arises in an information communication system consisting of a source and sink of information, and the received signal is corrupted due to non ideal characteristics of the channel and random noise. The receiver should have the capability of deciding on the correct reception of the signal irrespective of the above mentioned perturbations. Moreover, the receiving of signal is a random process and therefore the properties of statistical decision theories can be applied [23], [24].

Considering that a signal $s(t)$ is intermittently transmitted from a source and the receiver is receiving a signal $y(t) = s(t) + n(t)$, where $n(t)$ is the additive noise signal. The receiver is also required to decide whether the signal $s(t)$ is present or not. Two situations may then arise as (a) no signal is present at the receiver and (b) signal is present. These two situations may be statistically represented by two hypotheses, the null hypothesis H_0 and the alternative hypothesis H_1. Therefore,

$$H_0 : y(t) = n(t) \quad \text{no signal present}$$

and

$$H_1 : y(t) = s(t) + n(t) \quad \text{signal present} \tag{1.67}$$

The decision of $H_0|H_1$ is based on probabilistic criteria such as: select that hypothesis which is most likely to occur based on multiple observations. Thus the decision rule known as maximum *a posteriori* probability (MAP) criterion, is given by,

$$\text{If, } P(H_1|y) > P(H_0|y) \text{ decide on } H_1$$

and if otherwise decide on H_0. Here $P(H_1|y)$ and $P(H_0|y)$ are the *posteriori* probabilities.

The decision rule can also be stated in terms of *a priori* probabilities. If $P_1 = P(1) = P(H_1), P_0 = P(0) = P(H_0), (P_1 + P_0) = 1$ and conditional probabilities $P(y|1) = P(Y|H_1), P(y|0) = P(Y|H_0)$, then the decision rules can be stated as: If, $\dfrac{P_1.P(y|1)}{P_0.P(y|0)}$ is greater than 1, then the hypothesis H_1 is true and if the quantity is less than 1, then hypothesis H_0 is true. A test based on this premise is known as likelihood rational (LR) test. The ratio α_1 and α_0, known as likelihood ratio, is given by,

$$\alpha_1 = \frac{P(y|1)}{P(y|0)} > \frac{P_0}{P_1} \quad \text{for hypothesis } H_1$$

and

$$\alpha_0 = \frac{P(y|0)}{P(y|1)} > \frac{P_1}{P_0} \quad \text{for hypothesis } H_0 \tag{1.68}$$

The method is equivalent to MAP detection criterion and the receiver performing the test based on the likelihood rational (LR) test and is called an ideal observer. The sample y is sometimes referred to as test statistics for receiver working optimally.

A decision rule based on the selection of a threshold y_0 at the receiver to minimise the overall error probability P_e can also be used. y_0 divides the conditional probability curves into two regions R_0 and R_1 as shown in Figure

1.5. The error probabilities called false alarm probability P_f and miss alarm probability P_m are then given as,

$$P_f = P_0 \int_{y_0}^{\infty} P(y|0)dy$$

and

$$P_m = P_1 \int_{-\infty}^{y_0} P(y|1)dy \tag{1.69}$$

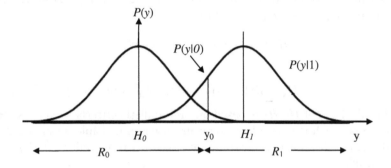

FIGURE 1.5: Condition probabilities and decision regions

The total probability of error $P_\epsilon = (P_f + P_m)$ is minimised by proper selection of y_0. Minimisation of P_ϵ is obtained by setting $\partial P_e/\partial y_0 = 0$ which yields to,

$$\frac{P(y_0|1)}{P(y_0|0)} = \frac{P_0}{P_1} = \alpha_{th} \tag{1.70}$$

Then, the decision rule is

$$\alpha_0 \geq \frac{1}{\alpha_{th}} \quad \text{for} \ \ H_0$$

and

$$\alpha_1 \geq \alpha_{th} \quad \text{for} \ \ H_1 \tag{1.71}$$

The same rules are obtained for minimum error criteria for ideal observer leading to same likelihood test. The probabilities P_0, P_1, P_f and P_m for H_1 and H_0 are, however, unweighed.

1.8.1   Bayes criteria

In actual practice, the probabilities are mostly weighted according to costs involved, which means that the decisions have to be biased with different risk values, known as Bayes risk factor [25]. Therefore, a cost matrix and the hypothesis H_1/H_0 decide on the decision rules which is now a decision matrix. The cost matrix $[C]$ can be written as,

$$[C] = \begin{bmatrix} C_{11} & C_m \\ C_f & C_{00} \end{bmatrix} \tag{1.72}$$

where, the elements c_{ij} of matrix $[C]$ are the cost of choice of H_0, when actually H_1 is true and vice versa. In most of the cases the costs of correct decisions of H_1 and H_0 is taken as zero and therefore $[C]$ is reduced to

$$[C] = \begin{bmatrix} 0 & C_m \\ C_f & 0 \end{bmatrix} \tag{1.73}$$

for error probabilities $[P]$, given by,

$$[P] = \begin{bmatrix} 0 & P_m \\ P_f & 0 \end{bmatrix} \tag{1.74}$$

Using the equations of P_m and P_f, the average risk per decision $\bar{C}(y_0)$ is deduced as,

$$\bar{C}(y_0) = P_0 C_f \int_{y_0}^{\infty} P(y|0)dy + P_1 C_m \int_{-\infty}^{y_0} P(y|1)dy \tag{1.75}$$

To obtain optimum threshold value of y_0, the average risk per decision $\bar{C}(y_0)$ is minimised to yield the threshold value as,

$$\alpha_{th} = \frac{P_0 C_f}{P_1 C_m} \tag{1.76}$$

The same decision rule can be applied with the weighted value of α_{th} and the value of $\bar{C}(\alpha_{th})$ is now minimum.

1.8.2   Minimax criteria

Minimax criteria is used when *a priori* probabilities P_0 and P_1 are not known and the Bayes risk factors cannot be calculated. In such cases Bayes minimum risks $\bar{C}_{min}(y_0)$ for various values of P_0 or P_1 are calculated and a curve for maximum of the $\bar{C}_{min}(P_0)$ vs P_0 is plotted. From the plot the maximum value $\bar{C}_{min}$ is considered as minimum risk. Average risks for any value of P_0 or P_1 can never exceed this value and is thus the value gives the worst risk condition.

1.8.3   Neyman-Pearson criteria

Neyman-Pearson criteria are a special case of Bayes criteria and are applied when the *a priori* probabilities and also the cost matrix are not known. In such case, one assigns pre-assigned value for the false alarm probability P_f which is likely to occur and seeks a decision strategy to yield a minimum value of probability for miss P_m. This corresponds to the maximisation of the detection probability $P_d = (1 - P_m)$ at receiver for a given false-alarm probability P_f. The maximisation problem can be solved by using the Lagrangian multiplier μ in the equation $Q = P_m + \mu P_f$, where Q needs to be minimised for a given P_f. Substituting $P_1 C_m = 1$ and $P_0 C_f = \mu$, the average cost $\bar{C}$ is given by,

$$\bar{C} = \int_{y_0}^{\infty} P(y|1)dy + \mu \int_{-\infty}^{y_0} P(y|0)dy = P_m + \mu P_f \qquad (1.77)$$

The conditions for Q_{min} are obtained as,

$$\alpha_1 = \frac{P(y|1)}{P(y|0)} \geq \mu \qquad (1.78)$$

Thus, the decision rule for a given P_f, which minimises P_m and equivalently maximises P_d is given by,

$$\alpha_1(y) > \alpha_{th} \quad \text{for} \quad H_1$$

and

$$\alpha_1(y) \leq \alpha_{th} \quad \text{for} \quad H_0$$

where α_{th} and y_0 are calculated form the given P_f.

Instead of binary type of decisions discussed so far, there are many information communication systems where, the source transmits M signals corresponding to M hypothesis and the receiver has to decide on one of the M hypotheses based on certain criteria. In general, only Bayes criteria is used for such problems since Minmax and Neymer-Pearson criteria deal with binary hypothesis only. The detailed calculations are beyond the scope of this book.

1.9   Estimation of information

The estimation techniques deal with building a function from given observations, which allows achieving a value for the unknown magnitude as accurately as possible [26]. A class of estimators known as parametric estimators are built from the knowledge or the assumption of probability density function of received signal and the transmitter parameters. The theory of these estimators uses techniques to find dependence or the adjustment of goodness of the data with the density functions. Another class of estimators known as non-parametric estimators, where no assumption on the behaviour of the data is carried out and hence these techniques are more robust.

1.9.1 Bayes estimate

In a binary decision problem, the number of true states are only two and the decision is taken for hypothesis H_1 or H_0. However, in an estimation problem, the number of true states and the number of observed states are infinitely large. Bayes estimates, where *a priori* probability density $P(\theta)$ is known gives minimum average cost, while with $p(\theta)$ unknown, maximum likelihood estimates are used.

While finding the best estimate $\hat{\theta}$ of θ of the transmitted signal, the average risk cost $\bar{C}[\hat{\theta}|Y]$ is minimised with respect to cost function $C(\hat{\theta}, \theta)$ and conditional probability density function $P(Y, \theta) = P(y_1, y_2.....y_n|\theta)$, where $(y_1, y_2.....y_n)$ are the measured quantity corresponding to θ. The cost function may be linear or squared as given by,

$$C(\hat{\theta}, \theta) = (\hat{\theta} - \theta) \tag{1.79}$$

or,

$$C(\hat{\theta}, \theta) = f(\theta)(\hat{\theta} - \theta)^2 \tag{1.80}$$

The squared cost function, similar to mean square error, is commonly used. Bayes estimate is obtained by minimising the average risk in terms of squared error cost function given by,

$$\bar{C} = \int \left[\int \{\hat{\theta}(Y) - \theta\}^2 P(\theta|Y) d\theta \right] P(Y) dY \tag{1.81}$$

The mean square variation of $\hat{\theta}$ is minimum if $\hat{\theta} = \bar{\theta}$ and the average risk $\bar{C}$ is minimised by obtaining $\partial \bar{C} / \partial \hat{\theta} = 0$, yielding,

$$\hat{\theta}(Y) = \int \theta P(\theta|Y) d\theta = E[\theta|Y] \tag{1.82}$$

where, $E[.]$ is the conditional expected value of θ.

For simultaneous estimation of many parameters $(\theta_1, \theta_2.....\theta_n)$ from the measurements of $Y = \{y_1, y_2, ...y_n\}$, the average risk is also modified.

1.9.2 Minimax estimate

If the prior probability distribution function of $P(\theta)$ is unknown, then the least favourable distribution is calculated. The Bayes estimate found by using least favourable distribution $\bar{P}(\theta)$ is called minimax estimates. This gives the maximum value of the Bayes risk $\bar{C}_{min}$. If the mean of $\hat{\theta}_j(Y) = \theta_j$, the estimate is said to be unbiased, if not the mean difference of $(\hat{\theta}_j - \theta_j)$ is called the bias of the estimate. The bias of the estimate should be small and also the spread of $\hat{\theta}_j$ about their mean should be small. Since the estimates $\hat{\theta}$ vary because of the random nature of variations in Y, the best estimate of $\hat{\theta}$ should be as close as possible to θ and $P(\hat{\theta}|\theta)$ should be peaked in the neighbourhood of θ.

1.9.3   Maximum likelihood estimate

A method of selecting an estimate, when the cost function is not known, is to maximise the *a posteriori* probability density function $P(\theta|Y)$. This is also equivalent to maximising the likelihood function $P(Y|\theta)$, when the dispersion of $P(\theta)$ is large. The probability density function of $P(\theta|Y)$ has a peak in the neighbourhood of the value of θ, that maximises it and the corresponding peaks of $P(Y|\theta)$ and $P(\theta|Y)$ become nearer to each other. As $P(\theta)$ becomes more uniform the initial values of the parameters become more uncertain. The value of θ at which maximisation of joint PDF $P(Y|\theta)$ and the observation $\{Y\}$ take place is adopted as the estimate of θ. The maximising of $P(Y|\theta)$ is equivalent to maximising the likelihood ratio $\alpha(Y|\theta)$. Sometimes, it is useful to consider average likelihood ratio α_{av} instead of α to obtain maximum likelihood (ML) estimates.

In many parameter cases, the ML estimates is found by solving the set equations, given by,

$$\frac{\partial}{\partial\theta_j}P(Y|\theta) = 0, \quad 1 \le j \le n \qquad (1.83)$$

The solution of this equation is difficult as there are many roots and one has to select the solution which gives the highest peaks in $P(Y|\theta)$.

Books for further reading

1. *Principles and Practice of Information Theory*: R. E. Blahut, Addison-Wesley Longman, Boston, 1987.

2. *Information and Information Systems*: M. Buckland, Greenwood Press, New York, 1991.

3. *Transmission of Information*: R. M. Fano, John Wiley, New York, 1961.

4. *Elements of Information Theory*: T. M. Cover and J. A. Thomas, John Wiley, New York, 2006.

5. *Information Technology in Theory*: P. Aksoy and L. DeNardis, Course technology, Thomson Learning Inc., Canada, 2008.

6. *Fundamentals of Information Technology*: S. Handa, Lexis Nexis Butterworths, Ohio, 2004.

7. *Information Technology: Principles and Applications*: A. K. Ray and T. Acharya, Prentice Hall of India, New Delhi, 2004.

8. *Information Theory, Inference, and Learning Algorithms*: D. J. C. MacKay, Cambridge University Press, Cambridge, UK, 2003.

9. *Philosophical Theories of Probability*: D. Gillies, Routledge, London, 2000.

10. *Probability Theory*: A. Renyi, Dover Publications, New York, 2007.

11. *Introduction to Probability Models*: S. M. Ross, Academic Press, Boston, 1989.

12. *Bayesian Inference in Statistical Analysis*: G. E. P. Box and G. C. Tiao, Addison Wesley, New York, 1992.

13. *On Measures of Entropy and Information*: A. Renyi, University of California Press, California, 1961.

14. *Mathematical Theory of Entropy*: N. G. F. Martin and J. W. England, Addison-Wesley, Reading, MA, 1981.

15. *Complexity, Entropy and the Physics of Information*: W. H. Zurek, Perseus, Reading, MA, 1990.

16. *An Introduction to Signal Detection and Estimation*: H. V. Poor, Springer-Verlag, New York, 1988.

17. *Detection and Estimation Theory and its applications*: T. A. Schonhoff and A. A. Giordano, Pearson Education, New York, 2006.

18. *Statistical Theory of Signal Detection*: C. W. Helstrom, Pergamon Press, Oxford, 1968.

19. *Bayesian Data Analysis*: A. Gelman. et al., CRC Press, London, 2004.

20. *Probability via Expectation*: P. Whittle, Springer-Verlag, New York, 2000.

21. *Fundamentals of Statistical Signal Processing: Estimation Theory*: S. M. Kay, Pearson Education, New Delhi, 2010.

22. *Entropy and Information Optics*: F. T. S. Yu, Marcel Dekker, New York, 2000.

Bibliography

[1] A. Aamodt and M. Nygard. Different roles and mutual dependencies of data, information, and knowledge. *Data and Knowledge Eng.*, 16:191–222, 1995.

[2] C. Zins. Conceptual approaches for defining data, information, and knowledge. *J. Am. Soc. for Information Sc. and Tech.*, 58(4):479–493, 2007.

[3] R. Capurro and B. Hjorland. The concept of information. *Annual Rev. Information Sc. and Techn.*, 37(8):343–411, 2003.

[4] C. Zins. Redefining information science: From information science to knowledge science. *J. Documentation*, 62(4):447–461, 2006.

[5] C. Zins. Knowledge map of information science. *J. Am. Soc. Information Sc. and Tech.*, 58(4):526–535, 2007.

[6] R. Landaue. The physical nature of information. *Physics Let.*, A:188–217, 1996.

[7] N. Gershenfeld. *The Physics of Information Technology.* Cambridge University Press, Cambridge, UK, 2000.

[8] D. Slepian (ed.). *Key papers in the development of information theory.* IEEE Press, New York, 1974.

[9] B. McMillan. The basic theorems of information theory. *Annals of Math. and Stat.*, 24:196–219, 1953.

[10] C. E. Shannon. A mathematical theory of communication. *Bell Syst. Tech. J.*, 27:379–423, 623–656, 1948.

[11] C. E. Shannon. Communication in the presence of noise. *Proc. IRE*, 37:10–21, 1949.

[12] C. E. Shannon. Coding theorems for a discrete source with a fidelity criterion. *IRE. National Conv. Rec.*, 1:142–163, 1959.

[13] N. J. A. Sloane and A. D. Wyner. (Ed.). *Collected Papers of Claude Shannon.* IEEE Press, New York, 1993.

[14] G. E. P. Box and G. C. Tiao. *Bayesian Inference in Statistical Analysis.* Addison Wesley, 1973.

[15] D. P. Ruelle. Extending the definition of entropy to nonequilibrium steady states. *Proc. of Nat. Aca. Sc.*, 100:3054–3058, 2003.

[16] I. Samengo. Estimating probabilities from experimental frequencies. *Physical Rev.*, E, 65:046–124, 2002.

[17] J. M. Van Campenhout and T. M. Cover. Maximum entropy and conditional entropy. *IEEE Trans. Inf. Th.*, 27:483–489, 1981.

[18] A. J. Jerri. Shannon sampling theorem-its various extensions and applications: a tutorial review. *Proc. IEEE*, 65:1565–1596, 1977.

[19] M. Unser. Sampling-50 years after Shannon. *Proc. IEEE*, 88:569–587, 2000.

[20] P. P. Vaidyanathan. Generalizations of the sampling theorem: Seven decades after Nyquist. *IEEE. Trans. Circuits and Syst.(1)*, 48(9), 2001.

[21] N. Wiener. *Cybernetics: or Control and Communication in the Animal and the Machine.* MIT Press, Cambridge, MA, USA, 1948.

[22] R. V. L. Hartley. Transmission of information. *Bell Sys. Tech. J.*, 7:535–563, 1928.

[23] H. V. Poor. *An Introduction to Signal Detection and Estimation.* Springer-Verlag, New York, USA, 1988.

[24] E. T. Jaynes and J. H. Justice (ed.). *Bayesian Methods: General Background, Maximum Entropy and Bayesian Methods in Applied Statistics.* Cambridge University Press, Cambridge, UK, 1985.

[25] A. Gelman et. al. *Bayesian Data Analysis.* CRC Press, London, UK, 2004.

[26] H. L. Van Trees, K. L. Bell, and Z. Tian. *Detection, Estimation, and Modulation Theory, Part I.* John Wiley, New York, USA, 2013.

Chapter 2

Introduction to Photonics

2.1 Information and physics

Information has a well-defined meaning in physics. The most fundamental analysis of the nature of information carried out so far originates from and in physics. In physics, the sub-discipline that comes closest to the derived meaning is communication theory. Originating traditionally as it does, communication theory concerns itself primarily with the challenge of information transmission rather than with the problems of information content or meaning. It is abstract in its approach to information, though concerned with the technical characteristics of communication channels independently of the nature of message sources, senders, receivers, or message destinations and coding strategies.

It is claimed that a growing trend in physics is to define the physical world as being made up of information itself. Data and knowledge (in terms of information) can therefore be treated as originating in discernible differences in the physical world, and in terms of space, time, and energy.

In physics, the most fundamental analysis of physical processes takes place at the quantum level. It is within the new field of quantum information theory [1] that we are confronted with the deepest level of analysis of information. Quantum information theory operates at the most abstract level, quite removed from any classical conception of information. An important breakthrough for the development of quantum information theory is the discovery that quantum states can be treated as if they are information. Thus, if information is physical, what is physical is also information; however, in classical terms the branch of physics which comes closest to the theory and practice of information communication and utilization is now termed photonics. Photonics as it is treated today encompasses the areas which go by terminologies like electro-optics, optoelectronics, and optical electronics. In fact, a large portion of modern optics and engineering optics are considered to be a part of photonics.

2.1.1 Photonics

Photonics is the science and technology of harnessing light, encompassing generation, transmission, and detection of light. Photonics is an enabling technology where management of light is executed for the benefit of society through guidance, modulation, switching, and amplification of light [2], [3]. Photonics bears the same relationship to light and photons as electronics does to electricity and electrons. In this century, photonics promises to be a vital part of the information revolution. Photonics covers all applications of lightwave technology, from the ultraviolet part of the electromagnetic spectrum, through the visible, to the near-, mid-, and far-infrared.

The word *photonics* is derived from the Greek word *photos* meaning light. The word appeared in the late 1960s to describe a research field whose goal was to use light to perform such functions, which traditionally fell within the typical domain of electronics and optics. The momentum in research started with the invention of the laser and was followed by the commercial production and deployment of optical fibres as a medium for transmitting information in the 1970s. In the following years, these developments in photonics led to the revolution in information communication technologies that provided the infrastructure of the Internet [4]. The information revolution that followed made a major contribution to the development of global communication system, and made possible the development of the global knowledge society. A knowledge society is now characterised by communication power and infrastructure that consumes less energy and is environmentally friendly.

The progress in information communication is based on the properties of light as well as on the understanding of the interactions between light and matter. In years to come, the key to success will be related to the photonics solutions that generate or conserve energy, cut greenhouse gas emissions, reduce pollution, yield environmentally sustainable outputs. The future focus might be on an area of so-called *green photonics* [5].

Apart from the all-pervasive applications of photonics in information technology, today photonics is related to wide ranges of scientific and technological applications, including laser-based manufacturing, biological and chemical sensing, medical diagnostics and therapy, lighting and display technology, as well as optical and quantum computing and signal processing. Development of the high-intensity laser allowed researchers to study extreme optical responses the so-called nonlinear effects. Exploring and harnessing this behaviour for futuristic technologies continues to be a major goal in photonics research. Photonics technology also potentially covers a broad range of applications for photovoltaic energy generation, highly efficient solid-state lighting, advanced sensing and instrumentation for environmental monitoring, and the like [6].

In recent years, new ideas on light and matter interactions resulted in the development of photonic crystal - a regular array of wavelength scale of light. Photonic crystals can block or guide specific bands of wavelengths via interference to create behaviour analogous to that of electrons in semiconductors.

Moreover, the photonic band-gap crystals offered the prospect of manipulating light at micrometre and nanometre scales. The idea is replicated in microstructured photonic crystal fibres. These fibres act as a nonlinear medium, producing intense white light, and in future may be utilised as high-power light sources.

Evolution of metamaterials is another recent achievement of photonics research. Metamaterials are composite materials designed to show electric and magnetic responses to light analogous to those of atoms, to produce bulk optical effects such as refraction. The metamaterials exhibit negative refraction - a remarkable property of controlling light [7]. Achieving negative refraction opens the way to totally novel applications such as the invisibility cloak, where light beams curve right around an object to create a kind of hole in space, in which the now-invisible object resides.

A contributing phenomenon in manipulating light via structured materials is the coupling of light with the oscillations of electrons at metal surfaces to form plasmons. Plasmonics encompasses behaviour that allows light to be influenced by structures that are smaller than its wavelength, and therefore offers the prospect of bridging the scale-gap between photonics and electronics. Physicists have also been investigating ingenious ways of slowing or even halting the passage of light through a material, an enabling technology for controlling optical data transmission and storage.

All these areas in photonics are moving rapidly through active research in a highly synergistic way. Photonics is truly coming of age.

2.2 Photons and the nature of light

The term 'photon' used for describing the particles of light was coined by Gilbert N. Lewis [8] in 1926 for an article, *The Conservation of Photons*. He wrote: *I therefore take the liberty of proposing for this hypothetical new atom, which is not light but plays an essential part in every process of radiation, the name photon.* Interestingly, his notion of a photon was different from the notion of photons we use today. He considered photons to be *atoms* of light, analogous to the atoms in electronics.

The scientific discourse on the nature of light can be traced back to ancient Greeks and Arabs. The belief revolved around the idea of the connection between light and vision and that vision is initiated by eyes reaching out to touch or feel the object at a distance. At a later date, mainly due to the work of the eleventh century Arab scientist Abu Al Hasan Ibn Al-Haitham, it was believed that vision resulted from an illuminated object emitting energy that was sensed by the eyes. He was also responsible for investigations on the refraction and dispersion of light. Later, Renaissance thinkers in Europe envisioned light as a stream of particles, which passes through ether (an invisible medium) and through all transparent materials.

In the seventeenth century, the phenomenon of refraction was explained by Pierre de Fermat in his theory of least line, which states that a ray of light takes the path that minimises the optical path length between two points. This indirectly points out that a light ray consists of particles that travel along a path as geometrical rays. The notion that the light is a conglomeration of particles was later adopted by Isaac Newton, who termed these particles as corpuscles.

On the other hand, Cristiaan Huygens advocated the wave nature of light to explain the phenomenon of interference, where a point source is thought to generate a wavefront and the interference is due to the interactions of secondary wavelets. Almost one and a half centuries later, Augustin-Jean Fresnel established the theory of wave optics to show that light does not show the characteristics of particles but bends around an obstacle or an aperture. The effects of diffraction of light were first carefully observed and characterized by Francesco Maria Grimaldi, who also coined the term diffraction, from the Latin *diffringere* (to break into pieces), referring to light breaking up into different directions. Consequently, Thomas Young demonstrated two-slit interference phenomenon to establish the wave nature of propagation of light. Major fallout of this experiment was the explanation of the phenomenon of destructive interference where the addition of two light beams under certain conditions can produce less illumination.

The wave theory of light [9] was established on firm theoretical evidence and analysis as a consequence of four equations of the theory of radiation established by James Clerk Maxwell in the late nineteenth century [10]. Maxwell theory unified electricity and magnetism and proposed that the propagation of light is nothing but the propagation of electromagnetic waves. Maxwell also answered an additional question about light. That is, if light is nothing but oscillation, then in which direction does this occur? The fields in electromagnetic waves oscillate in directions perpendicular to their motion. Therefore, even for identical frequency and phase, waves can still differ: they can have different polarisation directions. When the polarisation of light was discovered in 1808 by the French physicist Etienne-Louis Malus, this firmly established the wave nature of light where all possible polarisations of light form a continuous set. However, a general plane wave can be seen as the superposition of two orthogonal and linearly polarised waves with different amplitudes and different phases. Interestingly, a generally polarised plane wave can also be seen as the superposition of right and left circularly polarised waves. At the end of the twentieth century, the final confirmation of the wave nature of light became possible [11] through sophisticated experiments. Researchers measured the oscillation frequency of light directly as lying between 375 and 750 THz.

Interestingly, the particle theory of light was revived after the experiment of Albert Einstein on photoelectric effect. Einstein postulated that light consists of quanta of energy, $E = \hbar\nu$, where E is the energy, ν is the circular frequency of radiation, and $\hbar$ is Plank's constant divided by 2π. This relation again reintroduced the notion of the particulate nature of light, but the particles are not localised in space and thus are equivalent to the discreteness of energy. Interestingly, the discretization of energy was conceived by Neils Bohr in his atomic model.

Photoelectric effect is mathematically stated as $\hbar\nu = E + \phi$, where ϕ is the work function and E is the energy. From the experiment of photoelectric effect, three observations were noted. First, when light shines on a photo-emissive surface, electrons are ejected with kinetic energy E_0 equal to $\hbar$ times the frequency ν of the incident light less the work function ϕ. Secondly, the rate of electron emission is the square of the incident electric field. Thirdly, there is not necessarily a time delay between the instant the electric field is turned on and the time when the photoelectrons are ejected. However, the third corollary of the experiment is not that subtle and directly contradicts the classical expectations [12].

The interaction of radiation and matter is key to understanding the nature of light and the concept of photons [13], [14], [15]. Interesting collected works on photons are available in [16]. In fact, Plank's hypothesis and Einstein's photoelectric effect stem from the consideration of how energy is exchanged between radiation and matter. In the classical view, the matter is quantised. Einstein's idea of the photon does not require the quantization of the radiation field. Maxwell's electromagnetic equations, which stand on wave theory of radiation combined with Schrodinger's equation depicting wave theory of matter, gave remarkably accurate descriptions of photoelectric effect, stimulated emission, and absorption and pulse propagation phenomena. Moreover, many properties of laser light, such as frequency selectivity, phase coherence, and directionality can be explained within this framework.

Up until now, what has been described is that a photon can be considered as a true particle and a carrier of discrete energy, $E = h\nu$, a concept guided by the considerations of interactions between radiation and matter. The idea can be expanded further to establish the characteristics of the spatial discreteness of a photon, when it interacts with a finite-sized atom. The idea is a basic offshoot of the quantum theory of photo-detection, which tried to explain Young's double slit type of experiments, where a photon has no independent identity in going through one hole or the other. If a photon is *assumed* to be nothing more or less than a single quantum of excitation of appropriate normal mode, then the experiment can be explained as superposition of normal modes, which is nothing but the interference pattern of the classical explanation [17]. If the normal modes are quantised, then a photodetector capturing the peak of the interference pattern responds to single quantum excitation of a set of normal modes which are localized at the peaks of the interference pattern. The response is zero when the photodetector is placed at nodes. This leads to a question, whether a so-called photon wave function is equivalent to an electron wave function or electron probability density does exist!

Conservation of probability cannot be worked out for photons as the continuity equation for electron relating current density and probability density function cannot be worked out in the case of the photon stream. The wave function of an electron is the position state corresponding to the exact localisation of the electron at a given point in space. An equivalent statement for photons cannot be obtained as there is no particle creation operator that creates a photon at an exact point in space [18], [19]. However, the probability $P(r)dA$ of observing a photon at a point r, in an elementary area dA at a particular time is proportional to the optical intensity. The photon is more likely to be detected at the locations where the intensity is high.

2.2.1   Properties of photons

To summarize the discussion and without entering into current debates regarding the properties and characteristics of photons, some general statements can be made. Light consists of photons which are quanta of energies showing particle-like behaviour. The photons travel at the speed of light in vacuum c_0 (henceforth it will be represented as c unless specifically mentioned otherwise). The speed of light is retarded in matter. A photon has zero rest mass but carries electromagnetic energy and momentum. It has an intrinsic angular momentum (or spin) that governs its polarisation properties. Photons also show a wavelike character that determines their localisation properties in space and therefore they interfere and diffract.

2.2.1.1   Photon energy and momentum

Like each type of electromagnetic field, and like every kind of wave, light carries energy. The energy E of a photon is given as $E = h\nu = \hbar\omega$, where $h = 6.63 \times 10^{-34}$ J-s is the Planck's constant and $\hbar = \frac{h}{2\pi}$ and ν is the frequency. The energy can be added or taken only in the units of $h\nu$. However, a resonator system having zero photons carries an energy $E_0 = \frac{1}{2}h\nu$.

Photon energy can be calculated easily. For example a photon at wavelength $\lambda = 1\mu m$ (frequency $= 3 \times 10^{14}$ Hz) has energy $h\nu = 1.99 \times 10^{-19} J = 1.24 eV$. This is the same as the kinetic energy of an electron that has been accelerated through a potential difference of 1.24 V. Thus, photon energy can be converted into wavelength using a relation given by $\lambda = \frac{1.24}{E}$, where E is in electron volts.

The speed-energy dependence is also considered for photons [20]. In general, the energy flow T per unit of surface and per unit time is given by

$$\mathbf{T} = \frac{1}{\mu_0}\mathbf{E} \times \mathbf{B} \qquad (2.1)$$

where, μ_0 ($= 4\pi \times 10^{-7} N.s^2/C^2$) is the electric permeability of free space, and $\mathbf{E}$ and $\mathbf{B}$ are the electric and magnetic field vectors.

Therefore the average energy flow is given by

$$\langle T \rangle = \frac{1}{2\mu_0} E_{max} B_{max} \tag{2.2}$$

The photon travels in the direction of wave vector and the magnitude of momentum of photon P is related to energy E by the equation

$$P = \frac{E}{c} \tag{2.3}$$

where c is the speed of light.

In fact, it does not take much to deduce that if light has linear momentum, circularly polarised light also has angular momentum. For such a wave, the angular momentum P_a is given by

$$P_a = \frac{E}{\omega} = \frac{h}{\lambda} \tag{2.4}$$

Equivalently, the angular momentum of a wave is $\lambda/2\pi$ times its linear momentum. Therefore, a light beam can put certain materials into rotation [21] and can be used to deflect a beam of atoms travelling perpendicular to the photons.

In terms of energy flow T, the pressure P_r exerted by light on a body is given by

$$P_r = \frac{T}{c}(1 + r) \tag{2.5}$$

where r is the reflectivity of the surface. Incidentally, for black bodies $r = 0$ and for ideal mirrors, $r = 1$.

2.2.1.2 Photon mass

A photon is said to have no mass and no charge. It is a carrier of electromagnetic energy and interacts with other discrete particles (e.g., electrons, atoms, and molecules) [22]. A basic implication of Maxwell's electromagnetism is the constant speed of all electromagnetic radiation in a vacuum. The fact is experimentally confirmed to a high degree of accuracy that all electromagnetic radiation travels at the speed of light, c, over a wide range of frequencies. In turn, this implies that photons appear to be massless. The theories associated with quantum electrodynamics (QED) have led to acceptance of a concept of massless photons. However, despite this acceptance, a substantial experimental effort has been made to determine, either directly or indirectly, whether the photon mass is zero or nonzero. Moreover, a finite photon mass is perfectly compatible with the general principles of elementary particle physics, and an answer to the question of its size has to be obtained only through experiments and/or observations, though it seems to be impossible at the present time [23]. An upper limit of photon rest mass m_0 can be determined by setting the upper limit to its size with the help of the uncertainty principle, and is given as $m_0 \approx \frac{\hbar}{(\Delta t)}c^2$, or $\approx 10^{-66}$g. Implication of such an infinitesimal mass is far-reaching and includes the notion of wavelength dependence of the speed of light in free space. The particle data group currently accepts upper limit on the rest mass of a photon as $\leq 1 \times 10^{-49}$g [24].

The origin of and basis for the properties of elementary particles, and their existence, are some of the most challenging questions in physics and will have profound implications in deciding the properties of photons.

Since a photon has energy $(h\nu/c$ or $hc/\lambda)$, it may be *assumed* to have an inertial mass given by

$$m = \frac{h\nu}{c^2} \tag{2.6}$$

When a light beam passes near a heavy star, its trajectory gets deflected. Also, when a photon leaves a star, its energy decreases due to gravitational field. This manifests itself in the decrease of frequency which is referred to as the *gravitational red shift*. The red shift can be approximately calculated by noting the potential energy V on the surface of the star as

$$V \simeq -\frac{GM}{R} \cdot \frac{h\nu}{c^2} \tag{2.7}$$

where M is the mass of the star, R is its radius, and G is the gravitational constant.

When a light beam reaches earth, its frequency becomes

$$h\nu' = h\nu + V \tag{2.8}$$

or neglecting Earth's gravitational field,

$$\frac{\Delta\nu}{\nu} = \frac{GM}{Rc^2} \tag{2.9}$$

Using general theory of relativity, the limiting radius of a star, known as its Schwarzschild radius, can be found as

$$R_s = \frac{2GM}{c^2} \tag{2.10}$$

If the mass of a star is contained inside the sphere of R_s, a light beam may never leave the star. The star then can be called a *black hole*. Indeed, black holes of radius around 10 km have been detected for a mass of 10^{34} grams.

2.2.2   Speed of light

Light is an electromagnetic wave, and any electromagnetic wave must propagate with a speed c given by

$$c = \frac{1}{\sqrt{\varepsilon_0\mu_0}} \tag{2.11}$$

where, ε_0 ($= 8.854 \times 10^{-12}$ $C^2/N.m^2$) is the electric permittivity of free space and μ_0 ($= 4\pi \times 10^{-7}$ $N.s^2/C^2$) is the electric permeability of free space. Free space values of the permittivity ε_0 and permeability μ_0 can be related to respective values of the material medium ϵ and μ when multiplied by dielectric constant or relative permittivity and relative permeability. The right-hand side of Equation 2.11 contains electric and magnetic quantities, and the left-hand side is an optical quantity. The expression thus unifies electromagnetism and optics.

The modern value for the speed of light, usually called c from the Latin *celeritas*, is $c = 299792458$ m/s. The speed c is invariant and is the limiting speed in nature [25]. Incredibly, nobody explored the consequences of this invariance until Lorentz and a few others started doing so in the 1890s. The invariance of the speed of light is the essential point that distinguishes special relativity from Galilean physics.

2.2.2.1 Speed of light and signal propagation

A (physical) signal is the transport of information using the transport of energy. There are no signals without a motion of energy. Indeed, there is no way to store information without storing energy. To any signal, a propagation speed can be ascribed. The highest possible signal speed is also the maximal velocity of the general influences, or the maximal velocity with which effects spread causes. If the signal is carried by matter, such as by the written text in a letter, the signal velocity is then the velocity of the material carrier. For a wave carrier such as light or radio waves, the situation is less evident.

The phase velocity of a wave is given by the ratio between the frequency and the wavelength of a monochromatic wave, as

$$v_{ph} = \frac{\omega}{k} \tag{2.12}$$

Light in a vacuum has the same phase velocity $v_{ph} = c$ for all frequencies. On the other hand, there are cases where the phase velocity is greater than c, most notably when light travels through an absorbing substance, and when at the same time the frequency is near to an absorption maximum. In these cases, the phase velocity is not the signal velocity. For such situations, a better approximation to the signal speed is the group velocity, which is given by

$$v_{gr} = \frac{d\omega}{dk}|_{k_0} \tag{2.13}$$

where k_0 is the central wavelength of the wave packet.

It may be noted that $\omega = 2\pi v_{ph}\lambda$ implies the relation

$$v_{gr} = v_{ph} - \lambda \frac{dv_{ph}}{d\lambda} \tag{2.14}$$

This means that the sign of the last term determines whether the group velocity is larger or smaller than the phase velocity. For a travelling group, this means that new maxima appear either at the end or at the front of the group. This case happens for light passing through matter. For light travelling in a vacuum, the group velocity has the same value $v_{gr} = c$, for all values of the wave vector k.

The statement that the group velocity in a material is never greater than c, the speed of light in a vacuum, may not be correct. Actually, the group velocity in a material can be zero, infinite, or even negative. These conditions happen when the light pulse is very narrow, i.e., when it includes a wide range of frequencies, or again when the frequency is near an absorption transition. In many (but not all) cases the group is found to widen substantially, or even to split, making it difficult to define precisely the group maximum and thus its velocity. For example, the group velocity in certain materials has been measured to be ten times that of light. The refractive index is then smaller than 1. However, in all these cases the group velocity is not the same as the signal speed.

Sometimes it is conceptually easier to describe signal propagation with the help of the energy velocity. As previously mentioned, every signal when propagating, transports energy. The energy velocity v_{en} is defined as the ratio between the energy flow density and the energy density W, both taken in the direction of propagation. This ratio is given by

$$v_{en} = \frac{\langle T \rangle}{\langle W \rangle} \tag{2.15}$$

However, the underlying averaging procedure has to be specified, as denoted by $\langle . \rangle$, i.e., whether the energy is transported by the main pulse or by the pulse-front of it.

The progress in light detector technology allows one to detect even the tiniest energies. This has forced scientists to take the fastest of all these energy velocities as signal velocity. Using detectors of the highest possible sensitivity, the first point of the wave train whose amplitude is different from zero, i.e., the first tiny amount of energy arriving, should be detected as signal. This velocity, is commonly called the front velocity or the forerunner velocity and is simply given by

$$v_{fr} = \lim_{n \longrightarrow \infty} \frac{\omega}{k} \tag{2.16}$$

The forerunner velocity is never greater than the speed of light in a vacuum. In fact, it is precisely equal to c, since for extremely high frequencies, the ratio ω/k is independent of the material, and properties of vacuum take over. The forerunner velocity is the true signal velocity or the true velocity of light.

2.2.3   Coherence of light waves in space and time

Coherence was originally conceived in connection with Thomas Young's double-slit experiment in optics [26], and is a manifestation of the statistical nature of radiation produced by all optical sources [27]. At one extreme, thermal sources, such as incandescent lamps, produce chaotic and unordered emission of light; and at the other extreme, continuous-wave (CW) gas lasers produce comparatively ordered emission of light. Lasers come close to containing a single frequency, and travelling in a single direction. Any real laser emits light with statistical properties, in particular random fluctuations of both the amplitude and phase of the radiation. Moreover, the radiation travels through an intervening medium which has statistical properties, unless it is a perfect vacuum. Finally, the light reaches a detection apparatus (including our eyes), where the parameters and measurement techniques have some statistical and probabilistic issues. In general, the concepts of time and space coherence are related to the statistical properties of light waves produced by sources [28].

While describing the coherence of light waves, two types of coherence can be distinguished. They are temporal coherence and spatial coherence. Temporal coherence describes the correlation or predictable relationship between waves observed at different moments in time. Such waves have the ability to produce interference fringes (amplitude splitting). Spatial coherence describes the correlation between waves at different points in space; they are spatially shifted, but not delayed (wave-front shifting). The ideas can be generalized with the help of the concept of mutual coherent function. Over and above this, there can be spectral coherence of a source.

An arbitrary optical wave is described by a wave-function $u(\mathbf{r}, t) = Re[U(\mathbf{r}, t)]$, where $U(\mathbf{r}, t)$ is the complex wave-function and takes the form $U(\mathbf{r})e^{j2\pi\nu t}$ for monochromatic light, or it may be a sum of many similar functions of different ν for polychromatic light. In this case, the dependence of the wave-function on time and position is perfectly periodic and predictable. For random light, both functions, $u(\mathbf{r}, t)$, and $U(r, t)$, are random, and therefore the dependence of the wave-function on time and position is not totally predictable and cannot generally be described without resorting to statistical methods. It may be noted that the optical intensity $I(\mathbf{r}, t)$ of coherent light is $I(\mathbf{r}, t) = |U(\mathbf{r}, t)|^2$ and, for random light, $U(\mathbf{r}, t)$ is a random function of time and position. The intensity is therefore also random. The average intensity is then averaged over many realisations of the random intensity function. When the light is stationary, the statistical averaging operation can be determined by time averaging over a long time duration, instead of averaging over many realisations of the wave. As the coherence deals with the definitions of the statistical averages, light sources are classified as coherent, incoherent, and partially coherent.

Coherence is also a measure of the correlation that exists between the phases of the wave measured at different points. The coherence of two waves is quantified by the cross-correlation function, which describes the ability to predict the value of the second wave by knowing the value of the first. As a corollary, if two waves are perfectly correlated for all times, then at any time, if the first wave changes, the second will change in the same way. When combined, they can exhibit complete constructive interference or superposition at all times, and it follows that the second wave need not be a separate entity. In this case, the measure of correlation is the autocorrelation function (sometimes called self-coherence). Degree of correlation involves correlation functions [29], [30].

2.2.3.1 Temporal coherence

The stationary random wave function $U(\mathbf{r}, t)$ can be considered to have constant intensity $I(\mathbf{r})$, and it is possible to drop the r dependence (since r is fixed), so that $U(\mathbf{r}, t) = U(t)$, and $I(r) = I$. The autocorrelation function $G(\tau)$ of $U(t)$ is the average of the product of $U^*(t)$ and $U(t + \tau)$ and is given as

$$G(\tau) = \langle U^*(t)U(t + \tau)\rangle \qquad (2.17)$$

The phase of the product $U^*(t)U(t + \tau)$ is the angle between the phasors. When $\langle U(t)\rangle = 0$, the phase of the phasor $U(t)$ is equally likely to have any value between 0 and 2π. If $U(t)$ and $U(t + \tau)$ are uncorrelated then the angle between them varies randomly, and while taking average, the autocorrelation function vanishes. On the other hand, if $U(t)$ and $U(t+\tau)$ are correlated, their phasors will maintain some relationship and the average will not vanish.

The autocorrelation function $G(\tau)$ is called temporal coherence function. At $\tau = 0$, the intensity $I = G(0)$. Degree of temporal coherence $g(\tau)$ is expressed as

$$g(\tau) = \frac{G(\tau)}{G(0)} \qquad (2.18)$$

The value of $g(\tau)$ is a measure of degree of coherence and the absolute value of $g(\tau)$ lies between 0 and 1.

When $U(t) = Ae^{j2\pi\nu t}$, that is, the light is deterministic and monochromatic, then $g(\tau) = e^{j2\pi\nu\tau}$ and $|g(\tau)| = 1$. The waves $U(t)$ and $U(t + \tau)$ are then completely correlated. Further, if $|g(\tau)|$ monotonically decreases with time delay, the value τ_c at which the decrease is $\frac{1}{e}$ is called the coherence time τ_c. When $\tau < \tau_c$, the fluctuations are strongly correlated. The coherence time of monochromatic light source is infinite since $|g(\tau)| = 1$ everywhere. In general, τ_c is the width of the function $|g(\tau)|$. The light is effectively coherent if the distance $l_c = c\tau_c$ is much greater than all path difference encountered in any optical system. The distance l_c is known as the coherence length.

2.2.3.2 Spatial coherence

Spatial coherence is a measure of the correlation between the phases of a light wave at different points transverse in the direction of propagation. Spatial coherence describes the ability of two points in space to interfere when averaged over time. More precisely, the spatial coherence is the cross-correlation between two points in a wave for all times.

The cross-correlation function of $U(\mathbf{r_1}, t)$ and $U(\mathbf{r_2}, t)$ at positions $\mathbf{r_1}$ and

$\mathbf{r}_2$ is given by

$$G(\mathbf{r}_1, \mathbf{r}_1, \tau) = \langle U^*(\mathbf{r}_1, t) U(\mathbf{r}_2, t + \tau) \rangle \qquad (2.19)$$

where the time delay τ is known as mutual coherence function, and its normalised form with respect to intensities at $\mathbf{r}_1$ and $\mathbf{r}_1$ is called complex degree of coherence.

In a case where $G(\mathbf{r}_1, \mathbf{r}_1, \tau)$ vanishes, light fluctuations at two points are uncorrelated. The absolute value of complex degree of coherence is bounded by 0 and 1. It is therefore considered a measure of the degree of correlation between the fluctuations at $\mathbf{r}_1$ and those at $\mathbf{r}_2$ at a time τ later. When the two phasors fluctuate independently, their phases are also random.

2.2.3.3 Partial coherence

The partially coherent plane wave is spatially coherent within each transverse plane, but partially coherent in the axial direction [31]. The axial (longitudinal) spatial coherence of the wave has a one-to-one correspondence with the temporal coherence [32]. The ratio of the coherence length $l_c = c\tau_c$ to the maximum optical path difference l_{max} in the system governs the role played by coherence. If $l_c >> l_{max}$ then the wave is effectively completely coherent.

2.2.3.4 Coherent and incoherent light sources

An incandescent light bulb is an example of incoherent source. Coherent light from an incoherent source can be obtained at the cost of light intensity. This can be done by first spatially filtering the light from the incoherent source to increase the spatial coherence, and then spectrally filtering the light to increase the temporal coherence. Different forms of light sources with respect to their coherent properties are listed below:

1. Temporally incoherent and spatially coherent source: White light lamp spatially limited by a pinhole or a source located far away (e.g., stars, sun).

2. Temporally and spatially incoherent: White light source at a nearby distance without spatial limitations.

3. Temporally and spatially coherent: Monochromatic laser sources such as doubled Nd:YAG, He-Ne, Ar$^+$ lasers.

4. Temporally coherent and spatially incoherent: Monochromatic laser sources (He-Ne, doubled Nd:YAG) with a diffuser such as ground glass in the beam path.

An optical system behaves differently if illuminated by temporally or spatially coherent or incoherent light. Temporally incoherent illumination is typically associated with white light (or, generally, broadband) operation and limited by chromatic aberration. The degree of spatial coherence alters the description of an optical system as being a linear system.

2.3   Electromagnetic wave propagation

Electromagnetic theory forms a theoretical basis of optical propagation phenomena, capable of describing most of the known observations on the propagation of light. There are, however, experiments that cannot be explained by classical wave theory especially those conducted at short wavelengths or at very low light levels. Maxwell's equations predict the propagation of electromagnetic energy away from time-varying sources (current and charge) in the form of waves. The media through which an electromagnetic wave may travel can be distinguished as (a) homogeneous media and non-homogeneous media and (b) isotropic and anisotropic media. Homogeneous medium shows the same properties at all points. Isotropic medium shows the same properties along any direction.

2.3.1   Maxwell's equations

Maxwell's equations provide the mathematical basis of all (classical) electromagnetic phenomena [33], [34] and are given (in SI units) as

$$
\begin{aligned}
\nabla \times \mathbf{E} &= -\frac{\partial \mathbf{B}}{\partial t} \\
\nabla \times \mathbf{H} &= \frac{\partial \mathbf{D}}{\partial t} + \mathbf{J} \\
\nabla \cdot \mathbf{D} &= \rho \\
\nabla \cdot \mathbf{B} &= 0
\end{aligned}
\tag{2.20}
$$

The quantities $\mathbf{E}$ and $\mathbf{H}$ are the electric and magnetic field intensities and are measured in units of *volt/m* and *ampere/m*, respectively. The quantities $\mathbf{D}$ and $\mathbf{B}$ are the electric and magnetic flux densities and are in units of *coulomb/m^2* and *weber/m^2*, or Tesla. $\mathbf{D}$ is also called the electric displacement and $\mathbf{B}$, the magnetic induction. The quantities ρ and $\mathbf{J}$ are the volume charge density and electric current density (charge flux) of any external charges (that is, not including any induced polarisation charges and currents.) They are measured in units of *coulomb/m^3* and *ampere/m^2*.

The first equation is Faraday's law of induction, the second is Ampere's law as amended by Maxwell to include the displacement current $\frac{\partial \mathbf{D}}{\partial t}$, the third and the fourth are Gauss laws for the electric and magnetic fields. The displacement current term is essential in predicting the existence of propagating electromagnetic waves. The right-hand side of the fourth equation is zero because there are no magnetic monopole charges. The charge and current densities ρ, $\mathbf{j}$ may be thought of as the sources of the electromagnetic fields. For wave propagation problems, these densities are localized in space. The generated electric and magnetic fields are radiated away from the source and can propagate to large distances. Away from the sources, or in source-free regions of space, Maxwell's equations take the simpler form by putting in $\mathbf{J} = 0$ and $\rho = 0$.

2.3.1.1 Constitutive relations

When the sources are known, Maxwell's equations represent a system of six independent scalar equations (the two vector curl equations) in twelve unknowns (four field vectors with three scalar components each). This is true because third and fourth equations are not independent of first and second equations; they can be derived from these equations using the law of conservation of charge and magnetic charges. Therefore, six more scalar equations are needed to completely determine the fields. These are obtained by expressing two of the field vectors $\mathbf{D}$ and $\mathbf{B}$ as functions of $\mathbf{E}$ and $\mathbf{H}$.

The electric and magnetic flux densities $\mathbf{D}$, $\mathbf{B}$ are related to the field intensities $\mathbf{E}$ and $\mathbf{H}$ via the so-called constitutive relations, whose precise form depends on the material in which the fields exist. In a vacuum, they have the simplest form, given by

$$
\begin{aligned}
\mathbf{D} &= \varepsilon_0 \mathbf{E} \\
\mathbf{B} &= \mu_0 \mathbf{H}
\end{aligned}
\tag{2.21}
$$

where ε_0 and μ_0 are the permittivity and permeability of a vacuum, respectively.

The constitutive relationships and are completely determined by the material medium. The functional form can be deduced from the microscopic physics of the material. The material is then classified according to this functional dependence.

From the two quantities ε_0 and μ_0, two other physical constants, namely, the speed of light c_0 and the characteristic impedance of vacuum η_0, can be defined as

$$
c_0 = \frac{1}{\sqrt{\mu_0 \varepsilon_0}}
\tag{2.22}
$$

and

$$
\eta_0 = \sqrt{\frac{\mu_0}{\varepsilon_0}}
\tag{2.23}
$$

Taking the approximate value of $c_0 = 3 \times 10^8$ m/sec, the value of $\eta_0 = 377$ ohm.

In practice, isotropic material properties are specified relative to the free-space values using $\varepsilon \equiv \varepsilon_r \varepsilon_0$ and $\mu \equiv \mu_r \mu_0$, where ε_r is called the relative permittivity or dielectric constant, and μ_r is called the relative permeability. Materials for which the material parameters vary with frequency (or time) are called temporally dispersive. If the material parameters depend on position (as would be the case in a layered or stratified medium like the ionosphere), the medium is termed inhomogeneous, or spatially dispersive. There are few truly isotropic media in nature. Polycrystalline, ceramic, or amorphous materials, those for which the material constituents are only partially or randomly ordered throughout the medium, are approximately isotropic. Hence many dielectrics and substrates used in commercial antenna work are often made from ceramics or other disordered matter, or mixtures containing such materials. Other materials, and especially crystalline materials, interact with fields in ways that depend to some extent on the orientation of the fields; this is mostly evident when the atomic structure of the material is considered. Such materials are called anisotropic.

In Maxwell's equation, the densities ρ, $\mathbf{J}$ represent the external or free charges and currents in a material medium. The induced polarisation $\mathbf{P}$ and magnetization $\mathbf{M}$ may be made explicit in Maxwell's equations by using the constitutive relations given by

$$\begin{aligned} \mathbf{D} &= \varepsilon_0 \mathbf{E} + \mathbf{P} \\ \mathbf{B} &= \mu_0 \mathbf{E} + \mathbf{M} \end{aligned} \tag{2.24}$$

The Maxwell's equations are then modified in terms of the fields $\mathbf{E}$ and $\mathbf{B}$ as

$$\begin{aligned} \nabla \times \mathbf{E} &= -\frac{\partial \mathbf{B}}{\partial t} \\ \nabla \times \mathbf{B} &= \mu_0 \varepsilon_0 \frac{\partial \mathbf{B}}{\partial t} + \mu_0 [\mathbf{J} + \frac{\partial P}{\partial t} + \nabla \times \mathbf{M}] \\ \nabla \cdot \mathbf{D} &= \frac{1}{\varepsilon_0 (\rho - \nabla \times M)} \\ \nabla \cdot \mathbf{B} &= 0 \end{aligned} \tag{2.25}$$
$$\tag{2.26}$$

The current and charge densities due to the polarisation of the material is given as

$$\begin{aligned} \mathbf{J}_{pol} &= \frac{\partial \mathbf{P}}{\partial t} \\ \rho_{pol} &= -\nabla \cdot \mathbf{P} \end{aligned} \tag{2.27}$$

In nonlinear materials, ε may depend on the magnitude $\mathbf{E}$ of the applied electric field. Non-linear effects are desirable in some applications, such as various types of electro-optic effects used in light phase modulators and phase retarders for altering polarisation. In other applications, however, they are undesirable. A typical consequence of non-linearity is to cause the generation of higher harmonics.

Materials with a frequency-dependent dielectric constant $\varepsilon(\omega)$ are referred to as dispersive. The frequency dependence property comes into effect when the polarisation response of the material is not instantaneous at the application of time-varying electric field. All materials are, in fact, dispersive. However, $\varepsilon(\omega)$ typically exhibits strong dependence on ω only for certain frequencies. For example, water at optical frequencies has refractive index $n = \sqrt{\varepsilon}$ whose value is 1.33, but at radio frequency the value of n is 9.

One major consequence of material dispersion is pulse spreading, that is, the progressive widening of a pulse as it propagates through such a material. This effect limits the data rate at which pulses can be transmitted. There are other types of dispersion, such as intermodal dispersion in which several modes may propagate simultaneously, or waveguide dispersion, introduced by the confining walls of a waveguide. There exist materials that are both non-linear and dispersive that support certain types of non-linear waves called solitons, in which the spreading effect of dispersion is exactly cancelled by the nonlinearity. Therefore, soliton pulses maintain their shape as they propagate in such media.

2.3.1.2 Negative refractive index material

Maxwell's equations do not preclude the possibility that one or both of the quantities ε, μ be negative. For example, plasma below their plasma frequency, and metals up to optical frequencies, have $\varepsilon < 0$ and $\mu > 0$, with interesting applications in surface plasmons. Isotropic media with $\mu < 0$ and $\varepsilon > 0$ are more difficult to produce, although many of such media have been fabricated.

Negative-index media, also known as left-handed media, have ε, μ that are simultaneously negative, $\varepsilon < 0$ and $\mu < 0$. They have unusual electromagnetic properties, such as having a negative index of refraction and the reversal of Snell's law. Such media, termed metamaterials, have been constructed using periodic arrays of wires and split-ring resonators, and by transmission line elements. When $\varepsilon_{re} < 0$ and $\mu_{re} < 0$, the refractive index, $n^2 = \varepsilon_{re}\mu_{re}$, is defined by the negative square root of $n = -\sqrt{\varepsilon_{re}\mu_{re}}$. This is because then $n < 0$ and $\mu_{re} < 0$ implies that the characteristic impedance of the medium $\eta = \eta_0 \mu_{re}/n$ is positive, which also implies that the energy flux of a wave is in the same direction as the direction of propagation.

2.3.2 Electromagnetic wave equation

The simplest electromagnetic waves are uniform plane waves propagating along some fixed direction, say the z-direction, in a lossless medium. The assumption of uniformity means that the fields have no dependence on the transverse coordinates x, y and are functions only of z, t. Thus, the solutions of Maxwell's equations of the form $E(x, y, z, t) = E(z, t)$ and $H(x, y, z, t) = H(z, t)$. Because there is no dependence on x, y, the partial derivatives $\dfrac{\partial}{\partial x} = 0$ and $\dfrac{\partial}{\partial y} = 0$. Then, the gradient, divergence, and curl operations take the simplified forms, given by

$$\nabla = z\frac{\partial}{\partial z} \quad \nabla \cdot E = \frac{\partial E_z}{\partial z} \quad \nabla \times E = z \times \frac{\partial E}{\partial z} = -x\frac{\partial E_y}{\partial z} + y\frac{\partial E_x}{\partial z}$$

Assuming that $\mathbf{D} = \varepsilon\mathbf{E}$ and $\mathbf{B} = \mu\mathbf{H}$, the source-free Maxwell's equations become

$$\nabla \times \mathbf{E} = -\mu\frac{\partial \mathbf{H}}{\partial t}$$

$$\nabla \times \mathbf{H} = \sigma\mathbf{E} + \varepsilon\frac{\partial \mathbf{E}}{\partial t}$$

$$\nabla \cdot \mathbf{E} = 0$$

$$\nabla \cdot \mathbf{H} = 0 \qquad (2.28)$$

The conduction current in the source-free region is accounted for in the $\sigma\mathbf{E}$ term. An immediate consequence of uniformity is that $\mathbf{E}$ and $\mathbf{H}$ do not have components along the z-direction, that is, $\mathbf{E}_z = \mathbf{H}_z = 0$.

Taking *curl* of the first equation in Equation 2.28, the following relation is established as

$$\nabla \times \nabla \times \mathbf{E} = -\mu\frac{\partial}{\partial t}(\nabla \times \mathbf{H}) \qquad (2.29)$$

Inserting in the second equation of Equation 2.28 gives

$$\nabla \times \nabla \times \mathbf{E} = -\mu\frac{\partial}{\partial t}\left(\sigma\mathbf{E} + \varepsilon\frac{\partial \mathbf{E}}{\partial t}\right) = -\mu\sigma\frac{\partial \mathbf{E}}{\partial t} - \mu\varepsilon\frac{\partial^2 \mathbf{E}}{\partial t^2} \qquad (2.30)$$

A similar equation for $\nabla \times \nabla \times \mathbf{H}$ can also be obtained.

Using vector identity, $\nabla \times \nabla \times \mathbf{E} = \nabla(\nabla \cdot \mathbf{E}) - \nabla^2\mathbf{E}$, the following equations are obtained:

$$\nabla^2\mathbf{E} = \mu\sigma\frac{\partial \mathbf{E}}{\partial t} + \mu\varepsilon\frac{\partial^2 \mathbf{E}}{\partial t^2}$$

$$\nabla^2\mathbf{H} = \mu\sigma\frac{\partial \mathbf{H}}{\partial t} + \mu\varepsilon\frac{\partial^2 \mathbf{H}}{\partial t^2} \qquad (2.31)$$

For time-harmonic fields, the instantaneous (time-domain) vector is related to the phasor (frequency-domain) vector and therefore the instantaneous vector wave equations are transformed into the phasor vector wave equations, and are given by

$$\nabla^2\mathbf{E}_s = \mu\sigma(j\omega)\mathbf{E}_s + \mu\varepsilon(j\omega)^2\mathbf{E}_s = j\omega\mu(\sigma + j\omega\varepsilon)\mathbf{E}_s$$
$$\nabla^2\mathbf{H}_s = \mu\sigma(j\omega)\mathbf{H}_s + \mu\varepsilon(j\omega)^2\mathbf{H}_s = j\omega\mu(\sigma + j\omega\varepsilon)\mathbf{H}_s \qquad (2.32)$$

Defining a complex constant γ as the propagation constant given by

$$\gamma = \sqrt{j\omega\mu(\sigma + j\omega\varepsilon)} = \alpha + j\beta \qquad (2.33)$$

the phasor equation is reduced to

$$\nabla^2\mathbf{E}_s - \gamma^2\mathbf{E}_s = 0$$

$$\nabla^2\mathbf{H}_s - \gamma^2\mathbf{H}_s = 0 \qquad (2.34)$$

The operator in the above equations ∇^2 is the vector Laplacian operator.

The real part of the propagation constant α is defined as the attenuation constant while the imaginary part β is defined as the phase constant. The attenuation constant defines the rate at which the fields of the wave are attenuated as the wave propagates. An electromagnetic wave propagates in an ideal (lossless) media without attenuation $\alpha = 0$. The phase constant defines the rate at which the phase changes as the wave propagates.

Given the properties of the medium μ, ε, σ, the equations for the attenuation and phase constants can be determined as

$$\alpha = \omega \sqrt{\frac{\mu\varepsilon}{2} \left[\sqrt{1 + \left(\frac{\sigma}{\omega\varepsilon}\right)^2} - 1 \right]} \tag{2.35}$$

$$\beta = \omega \sqrt{\frac{\mu\varepsilon}{2} \left[\sqrt{1 + \left(\frac{\sigma}{\omega\varepsilon}\right)^2} + 1 \right]} \tag{2.36}$$

The properties of an electromagnetic wave, such as the direction of propagation, velocity of propagation, wavelength, frequency, and attenuation, can be determined by examining the solutions to the wave equations that define the electric and magnetic fields of the wave. In a source-free region, the phasor vector wave equations can be transformed in rectangular coordinates, by relating the vector Laplacian operator ∇^2 related to the scalar Laplacian operator as shown:

$$\nabla^2[.] = \frac{\partial^2[.]}{\partial x^2} + \frac{\partial^2[.]}{\partial y^2} + \frac{\partial^2[.]}{\partial z^2} \tag{2.37}$$

The phasor wave equations can then be written as

$$(\nabla^2 E_{xs})\mathbf{a}_x + (\nabla^2 E_{ys})\mathbf{a}_y + (\nabla^2 E_{zs})\mathbf{a}_z = \gamma^2 (E_{xs}\mathbf{a}_x + E_{ys}\mathbf{a}_y + E_{zs}\mathbf{a}_z) \tag{2.38}$$

$$(\nabla^2 H_{xs})\mathbf{a}_x + (\nabla^2 H_{ys})\mathbf{a}_y + (\nabla^2 H_{zs})\mathbf{a}_z = \gamma^2 (H_{xs}\mathbf{a}_x + H_{ys}\mathbf{a}_y + H_{zs}\mathbf{a}_z) \tag{2.39}$$

Individual wave equations for the phasor field components, E_{xs}, E_{ys}, E_{zs} and H_{xs}, H_{ys}, H_{zs} can be obtained by equating the vector components on both sides of each phasor wave equation.

$$\frac{\partial^2 E_{xs}}{\partial x^2} + \frac{\partial^2 E_{xs}}{\partial y^2} + \frac{\partial^2 E_{xs}}{\partial z^2} = \gamma^2 E_{xs}$$

$$\frac{\partial^2 E_{ys}}{\partial x^2} + \frac{\partial^2 E_{ys}}{\partial y^2} + \frac{\partial^2 E_{ys}}{\partial z^2} = \gamma^2 E_{ys}$$

$$\frac{\partial^2 E_{zs}}{\partial x^2} + \frac{\partial^2 E_{zs}}{\partial y^2} + \frac{\partial^2 E_{zs}}{\partial z^2} = \gamma^2 E_{zs}$$

$$\frac{\partial^2 H_{xs}}{\partial x^2} + \frac{\partial^2 H_{xs}}{\partial y^2} + \frac{\partial^2 H_{xs}}{\partial z^2} = \gamma^2 H_{xs}$$

$$\frac{\partial^2 H_{ys}}{\partial x^2} + \frac{\partial^2 H_{ys}}{\partial y^2} + \frac{\partial^2 H_{ys}}{\partial z^2} = \gamma^2 H_{ys}$$

$$\frac{\partial^2 H_{zs}}{\partial x^2} + \frac{\partial^2 H_{zs}}{\partial y^2} + \frac{\partial^2 H_{zs}}{\partial z^2} = \gamma^2 H_{zs} \qquad (2.40)$$

The component fields of any time-harmonic electromagnetic wave in rectangular coordinates must individually satisfy these six partial differential equations. In many cases, the electromagnetic wave will not contain all six components. In plane wave, **E** and **H** lie in a plane ⊥ to the direction of propagation and also ⊥ to each other. If the plane wave is uniform, **E** and **H** in the plane are ⊥ to the direction of propagation and vary only in the direction of propagation. The coordinate system for directions of **E**, **H** and the propagation are shown in Figure 2.1

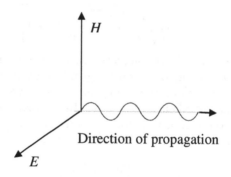

FIGURE 2.1: Coordinate system for plane wave propagation

2.3.2.1   Plane wave solution

The uniform plane wave has only a z-component of electric field and an x-component of magnetic field, which are both functions of only y. An electromagnetic wave which has no electric or magnetic field components in the direction of propagation (all components of **E** and **H** are perpendicular to the direction of propagation) is called a transverse electromagnetic (TEM) wave. All plane waves are TEM waves. The polarisation of a plane wave is defined as the direction of the electric field. For uniform plane wave, the wave equations for the only two field components E_{zs}, H_{xs} can be simplified significantly, given the field dependence on y only. The remaining single partial derivative in each wave equation becomes a pure derivative, since $\mathbf{E}_{zs}$ and $\mathbf{H}_{xs}$ are functions of y only and are linear, homogeneous second order differential equations, as given by

$$\frac{d^2 \mathbf{E}_{zs}}{dy^2} - \gamma^2 \mathbf{E}_{zs} = 0 \qquad (2.41)$$

$$\frac{d^2 \mathbf{H}_{xs}}{dy^2} - \gamma^2 \mathbf{H}_{xs} = 0 \qquad (2.42)$$

The general solutions to the reduced waves equations are given by

$$\mathbf{E}_{zs}(y) = \mathbf{E}_1 e^{\gamma y} + \mathbf{E}_2 e^{-\gamma y} \qquad (2.43)$$

$$\mathbf{H}_{zs}(y) = \mathbf{H}_1 e^{\gamma y} + \mathbf{H}_2 e^{-\gamma y} \qquad (2.44)$$

where $\mathbf{E}_1, \mathbf{E}_2$ are electric field amplitudes and $\mathbf{H}_1, \mathbf{H}_2$ are magnetic field amplitudes. It may be noted that $\mathbf{E}_{zs}$ and $\mathbf{H}_{xs}$ satisfy the same differential equation. Thus, other than the field amplitudes, the wave characteristics of the fields are identical.

The characteristics of the waves defined by the general field solutions can be determined by investigating the corresponding instantaneous fields as given by:

$$
\begin{aligned}
\mathbf{E}_z(y,t) &= Re[\mathbf{E}_{zs}(y)e^{j\omega t}] \\
&= Re[(\mathbf{E}_1 e^{\alpha y} e^{j\beta y} + \mathbf{E}_2 e^{-\alpha y} e^{-j\beta y})]e^{j\omega t} \\
&= \mathbf{E}_1 e^{\alpha y}\cos(\omega t + \beta y) + \mathbf{E}_2 e^{-\alpha y}\cos(\omega t - \beta y) \qquad (2.45)
\end{aligned}
$$

The focus is either the electric field or the magnetic field since they both have the same wave characteristics and, again they both satisfy the same differential equation.

The velocity at which a point of constant phase moves is the velocity of propagation for the wave. Solving for the position variable y in the equations defining the point of constant phase gives

$$y = \pm\frac{1}{\beta}(\omega t - constant) \qquad (2.46)$$

Given the y-coordinate of the constant phase point as a function of time, the vector velocity $u = \dfrac{\omega}{\beta}$ at which the constant phase point moves is found by differentiating the position function with respect to time. Moreover, given a wave travelling at a velocity u, the wavelength $\lambda = uT = \frac{2\pi}{\beta}$ during one period T.

For a uniform plane wave propagating in a given medium, the ratio of electric field to magnetic field is a constant. The units on this ratio has units of ohms and is defined as the intrinsic wave impedance for the medium. The travelling uniform plane wave defined by an electric field given by

$$\mathbf{E}_s = E_{zs}\mathbf{a}_z = E_0 e^{-\gamma y}\mathbf{a}_z \qquad (2.47)$$

The corresponding magnetic field can be obtained from the source free Maxwell's equations as

$$\nabla \times \mathbf{E}_s = -j\omega\mu\mathbf{H}_s \qquad (2.48)$$

and

$$
\begin{aligned}
\mathbf{H}_s &= -\frac{1}{j\omega\mu}\left[\frac{\partial}{\partial y}(E_0 e^{-\gamma y})\mathbf{a}_x\right] \\
&= \frac{\gamma}{j\omega\mu}E_0 e^{-\gamma y}\mathbf{a}_x \\
&= H_{xs}\mathbf{a}_x \qquad (2.49)
\end{aligned}
$$

The direction of propagation for this wave is in the same direction $E \times H (\mathbf{a}_z \times \mathbf{a}_x) = \mathbf{a}_y$. This characteristic is true for all plane waves.

The intrinsic impedance η of the wave is defined as the ratio of the electric field and magnetic field phasors (complex amplitudes) and is given by,

$$\eta = \frac{E_{zs}}{H_{xs}} = |\eta| e^{j\theta_\eta} = \sqrt{\frac{j\omega\mu}{\sigma + j\omega\varepsilon}} \qquad (2.50)$$

In general, the intrinsic wave impedance is complex. The lossless medium propagation constant is purely imaginary ($\gamma = j\beta$) while the intrinsic wave impedance is purely real. The electric field and magnetic field in a lossless medium are in phase. For a lossy medium, the only difference would be an exponential decay in both **E** and **H** in the direction of wave propagation. The propagation constant and the intrinsic wave impedance of a lossy medium are complex ($\gamma = \alpha + j\beta$) and also $\eta = |\eta| e^{j\theta_\eta}$ which yields the following electric field and magnetic fields equations

$$\mathbf{E}_s = E_0 e^{-\alpha y} e^{-j\beta y} \mathbf{a}_z \qquad (2.51)$$

$$\mathbf{H}_s = \frac{E_0}{\eta} e^{-\alpha y} e^{-j\beta y} \mathbf{a}_x \qquad (2.52)$$

and

$$\mathbf{E} = E_0 e^{-\alpha y} \cos(\omega t - \beta y) \mathbf{a}_z \qquad (2.53)$$

$$\mathbf{H} = \frac{E_0}{|\eta|} e^{-\alpha y} \cos(\omega t - \beta y - \theta_\eta) \mathbf{a}_z \qquad (2.54)$$

The electric and magnetic fields in a lossy medium are out of phase by an amount equal to the phase angle of the intrinsic impedance.

Wave propagation in air is guided by very low loss (negligible attenuation) with little polarisation or magnetisation. Thus, approximately, for free space (vacuum) $\sigma = 0, \varepsilon = \varepsilon_0 = 1$ and $\mu = \mu_0 = 1$. In a good conductor, displacement current is negligible in comparison to conduction current. Most good conductors have very high conductivities and negligible polarisation. The attenuation in a good conductor increases with frequency and the rate of attenuation in a good conductor can be characterized by a distance defined as the skin depth.

2.4   Electromagnetic spectrum

The electromagnetic spectrum is the distribution of electromagnetic radiation according to energy, frequency, or wavelength. The electromagnetic spectrum extends from below the low frequencies used   for modern radio communication to gamma radiation at the short-wavelength (high-frequency) end, thereby covering wavelengths from thousands of kilometres down to a fraction of the size of an atom. The limit for long wavelengths is the size of the universe itself, while it is thought that the short wavelength limit is in the vicinity of the Planck length. Maxwell's equations predicted an infinite number of frequencies of electromagnetic waves, all travelling at the speed of light. This was the first theoretical indication of the existence of the entire electromagnetic spectrum. Maxwell's predicted waves included waves at very low frequencies which in theory might be created by oscillating charges in an ordinary electrical circuit of a certain type.

Electromagnetic waves are typically described by any of the three physical properties: the frequency, wavelength, or photon energy. Frequencies observed in astronomy range from 2.4×1023 Hz (1 GeV gamma rays) down to the plasma frequency of the ionized interstellar medium ($\sim$1 kHz). Wavelength is inversely proportional to the wave frequency, so gamma rays have very short wavelengths that are fractions of the size of atoms, whereas wavelengths on the opposite end of the spectrum can be as long as the universe. As the energy is directly proportional to the wave frequency, gamma ray has the highest energy (around a billion electron volts), while radio wave has very low energy (around a femto-electron-volt). The behavior of EM radiation depends on its wavelength. When EM radiation interacts with single atoms and molecules, its behaviour also depends on the amount of energy per quantum (photon) it carries. Electromagnetic radiation interacts with matter in different ways across the spectrum. These types of interactions are so different that historically different names have been applied to different parts of the spectrum. Although these electromagnetic radiations form a quantitatively continuous spectrum of frequencies and wavelengths, the spectrum remains divided for practical reasons related to these qualitative interaction differences.

Generally, electromagnetic radiation is mainly classified by wavelength according to its use in different applications, such as the radio wave, microwave, millimetre wave, or terahertz (or sub-millimetre wave) regions. The spectrum, which are of major interest in photonics instrumentation and communications, are visible, infrared, and a portion of ultraviolet. Recently, the terahertz region, which is in between microwave and infrared, is gaining importance in photonics research.

The electromagnetic spectrum is classified based on the source, detector, and materials technologies employed in each of the spectrum regions. Because spectrum is an international commodity it is controlled by an international agreement. The ITU (International Telecommunications Union) is the United Nations organization which coordinates the allocation of frequencies to the activities which use them, such as radio navigation, satellite communications, radar systems, etc. ITU organizes world radio communication conferences (WRC) every two to three years to modify the frequency allotments as required (or requested). The property right to the use of the spectrum is defined in terms of a specified spectrum frequency band, a specified geographic area, a specified permitted maximum strength of the signal beyond the boundaries of the geographic area, and a specified time period. The property right (in perpetuity) is the right to transmit over the specified spectrum band, so long as the signals do not exceed a specified strength (expressed in volts/meter) beyond the specified geographic boundaries during the specified time period.

The whole electromagnetic spectrum is shown in Table 2.1.

TABLE 2.1: Ranges of electromagnetic spectrum

Types of radiation	Frequency range	Wavelength range
Gamma rays	$<3 \times 10^{20}$	< 1 fm
X rays	3×10^{17}- 3×10^{20}	1 fm- 1 nm
Ultraviolet	7.5×10^{14} - 3×10^{17}	1 nm - 400 nm
Visible	4×10^{14} - 7.5×10^{14}	$0.4 \ \mu$m
Near-infrared	10^{14}-7.5×10^{14}	$0.75 \ \mu$-$3.0 \ \mu$m
Midwave infrared	5×10^{13}- 10^{14}	$3 \ \mu$m-$6 \ \mu$m
Longwave infrared	2×10^{13}- 5×10^{13}	$6 \ \mu$m-$15 \ \mu$m
Extreme infrared	3×10^{11}- 2×10^{13}	$15 \ \mu$m-1 mm
Micro and radio waves	$<3 \times 10^{11}$	> 1 mm

2.4.1 Infrared region

Infrared (IR) is invisible radiant energy, electromagnetic radiation with longer wavelengths than those of visible light, extending from the nominal red edge of the visible spectrum at 700 nanometers (frequency 430 THz) to 1 mm (300 GHz). People can see infrared up to around 1050 nm as confirmed through experiments. Most of the thermal radiation emitted by objects near room temperature is infrared. Humans at normal body temperature radiate mainly at wavelengths around 10 μm. Thermal-infrared imaging is used extensively for military and civilian purposes. Military applications include target acquisition, surveillance, night vision, homing, and tracking. Non-military uses include thermal efficiency analysis, environmental monitoring, industrial facility inspections, remote temperature sensing, short-ranged wireless communication, spectroscopy, and weather forecasting. The infrared spectrum range can be divided into three parts.

(a) *Far-infrared* region extends from 300 GHz to 30 THz (1 mm-10 μm). The lower part of this range may also be called microwaves or terahertz waves. This radiation is typically absorbed by so-called rotational modes in gas-phase molecules, by molecular motions in liquids, and by phonons in solids. The water in Earth's atmosphere absorbs so strongly in this range that it renders the atmosphere as opaque to radiation. However, there are certain wavelength ranges (windows) within the opaque range that allow partial transmission, and can be used in astronomy and in satellite to earth communication. The wavelength range from approximately 200 μm up to a few mm is often referred to as sub-millimeter wave in astronomy, reserving far infrared for wavelengths below 200 μm.

(b) *Mid-infrared* region ranges from 30 to 120 THz (10-2.5 μm). Hot objects (black-body radiators) can radiate strongly in this range. Human skin at normal body temperature radiates strongly at the lower end of this region. This radiation is absorbed by molecular vibrations, where the different atoms in a molecule vibrate around their equilibrium positions. This range is sometimes called the fingerprint region, since the mid-infrared absorption spectrum of a compound is very specific for that compound.

(c) Near-infrared region ranges from 120 to 400 THz (2,500-750 nm). Physical processes that are relevant for this range are similar to those for visible light. The highest frequencies in this region can be detected directly by special types of photographic films, and by many types of solid state image sensors.

2.4.2 Visible region

The visible spectrum is the portion of the electromagnetic spectrum that is visible to the human eye. Electromagnetic radiation in this range of wavelengths is called visible light or simply light. A typical human eye responds to wavelengths from about 390 to 700 nm. In terms of frequency, this corresponds to a band in the vicinity of 430 to 790 THz. The spectrum does not, however, contain all the colours that the human eyes and brain can distinguish. Colours containing only one wavelength are called pure colours or spectral colours.

2.4.3 Ultraviolet region

Next in frequency spectrum is the ultraviolet (UV). The wavelength of UV rays is shorter than the violet end of the visible spectrum but longer than the X-ray. UV in the very shortest range (next to X-rays) is even capable of ionizing atoms, and thereby greatly changing their physical behaviour. At the middle range of UV, rays cannot ionize but can break chemical bonds, making molecules unusually reactive.

2.4.4 Terahertz region

Terahertz radiation falls in between infrared radiation and microwave radiation in the electromagnetic spectrum, and it shares some properties with each of these. In the electromagnetic spectrum, radiation at 1 THz has a period of 1 ps, a wavelength of 300 μm, a photon energy of 4.1 meV, and an equivalent temperature of 47.6 degrees Kelvin. Like infrared and microwave radiation, terahertz radiation travels in a line of sight and is non-ionizing. It can penetrate a wide variety of non-conducting materials such as clothing, paper, cardboard, wood, masonry, plastic, and ceramics. The penetration depth is typically less than that of microwave radiation. Terahertz radiation has, however, limited penetration through fog and clouds and cannot penetrate liquid water or metal. Despite some difficulties, the unique nature of THz waves has stimulated researchers to develop instrumentations and systems at this frequency band for various security applications. THz waves have low photon energies and thus cannot lead to photo-ionization in biological tissues as can X-rays. As a result, THz waves are considered safe for both the samples and the operator. Due to extreme water absorption, THz waves cannot penetrate into the human body like microwaves can, and the effect is limited to skin level. At THz frequencies, many molecules exhibit strong absorption and dispersion due to dipole-allowed rotational and vibrational transitions. These transitions are specific to the molecule, enabling spectroscopic fingerprinting in the THz range. Combined with imaging technology, inspection using THz waves provides both profile and composition information of the sample. As the wavelength of the THz waves is sufficiently short to provide sub-millimeter level spatial resolution, higher spatial resolution down to nm could be achieved in near-field photonic techniques.

2.5 Optical imaging systems

Optical imaging is a two- or three-dimensional reproduction of a physical situation or scene. Imaging can be achieved by any of at least these six groups of techniques:

1. *Photography*: This uses a light source, lenses, and film, or another large-area detector. Photography can be used in reflection, in transmission, or with phase-dependence, at various illuminations and at various wavelengths.

2. *Optical microscopy*: This also uses a light source, lenses, and film (or some other large-area detector). If the illumination is through the sample, i.e., in transmission mode, the technique is termed as bright-field microscopy. Similarly, when the illumination is from the side, the technique is termed oblique microscopy. Dark-field microscopy is a technique where the illumination is confined to an outer ring of light. An even more elaborate illumination system is used in phase-contrast microscopy. If a polarised illumination beam is split into two components that pass the sample at close (but not identical) locations, and recombined afterwards, then the technique is termed differential interference contrast microscopy. In fluorescence microscopy, a sample is treated with a fluorescent dye and the illuminating light is filtered out to observe only the fluorescence.

3. *Telescopy*: This type of optical system is used mostly in geodesy and as-

tronomy. Telescopes can also take images at various wavelengths, ranging from radio frequencies, infrared, visible, and ultraviolet to X-rays. Simple telescopes are lens-based; high-performance telescopes are usually mirror-based. Most of the advanced astronomical telescopes can compensate star images for the effects of the turbulence of the atmosphere by using adaptive optics techniques.

4. *Optical scanners*: Scanning techniques construct images point by point through the motion of the detector, the light source, or both. There are numerous scanning techniques used in microscopy, such as confocal laser scanning microscopy, fibre-based near-field scanning optical microscopy, and combinations of these with fluorescence techniques. Many of these scanning microscopy techniques allow resolutions much lower than the wavelength of light.

5. *Optical tomography*: This technique is performed in transmission mode and uses a source and a detector that are rotated together around an object. Effectively a specialized scanning technique, this allows imaging cross sections of physical bodies.

6. *Holography*: Holography is a technique of constructing holograms using lasers and large-area detectors, and allows taking three-dimensional images of objects. Holography is sometimes termed 3D photography and can be done in reflection or in transmission mode.

The subject of optical imaging can be divided into three areas: (a) Geometrical optics or ray optics, where light is described by rays which show the paths of energy transfer. This treatment provides a good understanding of the propagation of light in transparent media and the operation of optical imaging systems such as cameras, telescopes, and microscopes. (b) Physical optics, sometimes termed wave optics, where the wave nature of light is taken into account. This treatment covers polarisation, interference, and diffraction of light. The physical optics approach is necessary for understanding the limits of resolution of optical imaging systems. (c) Quantum optics, where the particle nature of light is taken into account. This description, where light is considered as consisting of photons, is needed to understand fully the interaction of light and matter. Topics which require the quantum optics approach include the photoelectric effect, photodetectors, lasers, and hosts of photonic devices.

2.5.1 Ray optic theory of image formation

Ray optics, sometimes called geometrical optics, is concerned with the light ray. Light is necessarily thought about in terms of rays, which have been defined from both corpuscular and wave-theory perspectives. Without defining ray as a specification of both position and direction, it is possible to establish some physical phenomena associated with the propagation of rays in the context of image formation. Since the wavelength of light (in the order of $10^{-7}m$) is too small in comparison to most of the optical components that are used for image formation and operations on images, the ray theory is treated assuming wavelength as zero. The usual starting point may be with a simple definition of refractive index n, as the ratio of speed of light in vacuum c, divided by the speed of light in the medium v, through which light is propagating. Refractive index varies with wavelength, but this dependence is not made explicit in ray optics treatment, most of which is limited to monochromatic light. The output of a system in polychromatic light is the sum of outputs at the constituent wavelengths. The medium is a homogeneous medium if n is the same everywhere. The medium is an inhomogeneous or heterogeneous medium if the refractive index varies with position. In an isotropic medium, n is constant at each point for the light travelling in all directions and with all polarisations.

In corpuscular theory, rays are defined as the path of a corpuscle or the path of a photon. However, in such descriptions, energy densities can become infinite. In wave theory, rays are defined as quantities both scalar and electromagnetic. Sometimes rays are treated as a descriptor of wave behaviour in short wavelength or high frequency limit [35].

Rays propagate in straight lines in homogeneous media and have curved paths in heterogeneous media. Difficulties of a physical definition can be avoided by treating rays as mathematical entities, though it is virtually impossible to think purely geometrically unless rays are treated as objects of geometry.

Rays have positions, directions, and speeds. Between any pair of points on a given ray there is a geometrical path length and an optical path length. Geometric path length is geometric distance ds measured along a ray between any two points and the optical path length between two points x_1 and x_2 through which a ray passes. In homogeneous media, rays are straight lines and the optical path length $O_l(x_1, x_2)$ is given by

$$O_l(x_1, x_2) = \int_{x_1}^{x_2} n(x)dx = c \int \frac{ds}{v} = c \int dt \qquad (2.55)$$

where $n(x)$ is the refractive index and dx is the geometrical path length. The integral is taken along the ray path in the media (homogeneous and/or inhomogeneous), and may include any number of reflections and refractions. In homogeneous media, rays are straight lines and the optical path length is simply $n \int ds$.

The time required for light to travel between the two points is simply (optical path length) ÷ (speed of light). At smooth interfaces between media with different indices, rays refract and reflect. Path of rays are also reversible; reflection and refraction angles are the same in either direction. Ray paths are reversible and rays carry energy. Ray power per area is approximated by ray density.

2.5.1.1 Fermat's principle

Fermat's principle is a unifying principle of ray optics that helps in deriving laws of reflection and refraction [36]. This principle is also used to find the equations that describe ray paths and geometric wavefronts in heterogeneous and homogeneous media. According to Fermat's principle [36], [37], the optical path between two points through which a ray passes is an extremum. An extremum may be minimum, maximum, or a point of inflections, therefore the light passing through these two points along any other nearby path would take either more or less time. In a homogeneous medium, the extremum is a minimum as light travels in a straight line, hence, according to Fermat's principle, light travels along the path of least time. Mathematically

$$\delta \int n(r)ds = 0 \tag{2.56}$$

where ds is the differential length along the ray trajectory between two points.

By defining a ray vector $r(s)$, which has three components in three axes of a Cartesian coordinate system, the ray trajectory equation is written as

$$\frac{d}{ds}\left(n\frac{dr}{ds}\right) = \nabla n \tag{2.57}$$

where, ∇n is the gradient of n and is a vector with a Cartesian component.

The ray trajectories are often characterized by the surfaces to which they are normal. An equilevel surface everywhere normal to the rays is shown in Figure 2.2, and is described by a constant scalar function $S(r)$. The ray trajectories can then be readily constructed.

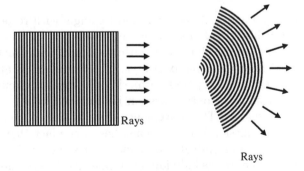

Rays

Rays

FIGURE 2.2: Ray propagation in constant $S(r)$ for plane wave and for spherical wave

The normal to the equilevel surfaces at a position r is in the direction of the gradient vector $\nabla s(r)$. The function $s(r)$ is called the eikonal, and is akin to the potential function where the role of the optical rays is played by the lines of electric field. The eikonal equation is written in vector form as

$$|\nabla s|^2 = n^2 \tag{2.58}$$

Integrating the eikonal equation along a ray trajectory between any two points gives the difference between values of $s(r)$, which is nothing but the optical path length between two points. Therefore, Fermat's principle and the ray equation follows from the eikonal equation. Either the eikonal equation or Fermat's principle is regarded as the principal postulate of ray optics.

2.5.1.2   Law of refraction and reflection

At an interface between two homogeneous and isotropic media, described by indices n_1 and n_2, the incidence angle θ_1 and the outgoing angle θ_2 are related by Snell's law of refraction as

$$n_1 \sin \theta_1 = n_2 \sin \theta_2 \tag{2.59}$$

Therefore, the angles within any two media are related by their indices alone, regardless of the intervening layers. If the refractive indices vary with wavelength, the angles of refraction do likewise. If $\sin \theta_1 > 1$, total internal reflection takes place. Snell's law can also be expressed as $n_1(S \times r_1) = n_2(S \times r_2)$.

The reflection equations can be derived from those for refraction by setting the index of the final medium equal to the negative of that of the incident medium, $n_2 = -n_1$. This gives $\theta_2 = \theta_1$ that is the angles of incidence and reflection are equal. The incident and reflected ray directions are related by $(S \times r_2) = (S \times r_1)$.

2.5.1.3   Matrix formulation of an optical system

A ray is described by its position and its angle with respect to an optical axis. These variables are altered as the ray travels through the system. Assuming that the rays are travelling only within a single plane, matrix formulation can be applied for tracing paraxial rays. The formalism is applicable to systems with planar geometry and to meridional rays in circularly symmetric systems. As a result, the optical system is described by a 2×2 matrix called the ray-transfer matrix. The convenience of using matrix methods, quite akin to electrical transmission line equations, lies in the fact that the ray-transfer matrix of a cascade of optical components (or systems) is a product of the ray-transfer matrices of the individual components (or systems). Matrix optics therefore provides a formal mechanism for describing complex optical systems under the paraxial approximation. The method contains the expressions of the paraxial power and transfer equations, which permit many useful results to be derived mechanically, and are especially useful for lens combinations. With the symbolic manipulation programs now available, matrix methods also provide a means of obtaining useful expressions for optical systems containing a number of elements.

An optical system consisting of several components is placed between two transverse planes at z_1 and z_2, referred to as the input and output planes, respectively. The system is characterized completely by its effect on an incoming ray of arbitrary position and direction (y_1, θ_1), which steers the ray to its new position and direction (y_2, θ_2), at the output plane, as shown in Figure 2.3.

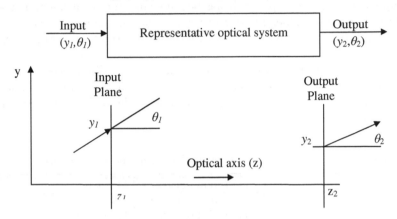

FIGURE 2.3: Ray propagation in an optical system

In the paraxial approximation, at sufficiently small angles, so that $\sin\theta = \theta$, the relation between (y_1, θ_1) and (y_2, θ_2) is linear and can be expressed as

$$y_2 = Ay_1 + B\theta_1 \tag{2.60}$$

$$\theta_2 = Cy_1 + D\theta_1 \tag{2.61}$$

where A, B, C, and D are real numbers. The equations can be written in matrix form, whose elements are A, B, C, and D as

$$\begin{bmatrix} y_2 \\ \theta_2 \end{bmatrix} = \begin{bmatrix} A & B \\ C & D \end{bmatrix} \begin{bmatrix} y_1 \\ \theta_1 \end{bmatrix} \tag{2.62}$$

The matrix M, whose elements are A, B, C, and D, characterizes the optical system completely and is known as the ray-transfer matrix. If a periodic system is composed of a cascade of identical unit systems (stages), each with a ray-transfer matrix (A, B, C, D), and a ray enters the system with initial position y_0, and slope θ_m, then $ABCD$ matrix can be applied m times, to determine the position and slope (y_m, θ_m) of the ray at the exit of the mth stage, as

$$\begin{bmatrix} y_m \\ \theta_m \end{bmatrix} = \begin{bmatrix} A & B \\ C & D \end{bmatrix}^m \begin{bmatrix} y_0 \\ \theta_0 \end{bmatrix} \tag{2.63}$$

2.5.2 Wave optics: diffraction and interference

Wave optics describes two fundamental concepts regarding the wave nature of light: diffraction and interference. Though there is no clear distinction

between the two concepts, diffraction is related to the changing of direction
of a wave as it encounters an obstacle. This also can be explained using an
argument based on wave superposition, or interference. Likewise, interference
between waves implies diffraction, since each wave acts as an obstacle to the
other, redirecting energy in propagation components of opposing phase into
components that are in phase. It has become customary (although not always
appropriate) to consider interference when the phenomenon involves superposition of only a few waves and diffraction when treating a large number of
waves.

At this point, it is necessary to point out the differences between the
wave characteristics of electromagnetic and optical waves (as we understand
in common jargon). Table 2.2 indicates some of the differences between electromagnetic and optical waves.

TABLE 2.2: Differences between electromagnetic and optical waves

Wave characteristics	Electromagnetic	Optical
Principle	Maxwell's equations	Schrodinger's equation
Emission	Classical movement	Quantum transition
Measurement	Electric field	Intensity
Approximations	Uniform field	Uniform medium
Tools	Antenna, waveguide	Lens, mirror, prisms etc

2.5.2.1   Scalar wave equation

An optical wave is mathematically described by a real function of posi-
tion $\mathbf{r} = (x, y, z)$ and time t denoted by $\psi(\mathbf{r}, t)$ known as wave function or
optical disturbance. If the propagation of light is represented as a scalar and
monochromatic wave, the scalar wave equation is given by

$$\nabla^2 \psi(\mathbf{r}, t) - \frac{1}{c^2}\frac{\partial^2}{\partial t^2}\psi(\mathbf{r}, t) = 0 \qquad (2.64)$$

where ∇^2 is the Laplacian operator given by $\nabla^2 = \partial^2/\partial x^2 + \partial^2/\partial y^2 + \partial^2/\partial z^2$.

The real wave function $\psi(\mathbf{r}, t)$ can be replaced by a complex wave function
$\Psi(\mathbf{r}, t)$, as $\psi(\mathbf{r}, t) = Re[\Psi(\mathbf{r}, t)] = \frac{1}{2}[\Psi(\mathbf{r}, t)] + \Psi^*(\mathbf{r}, t)]$. The wave equation is
then given by

$$\nabla^2 \Psi(\mathbf{r}, t) - \frac{1}{c^2}\frac{\partial^2}{\partial t^2}\Psi(\mathbf{r}, t) = 0 \qquad (2.65)$$

In the monochromatic case, $\Psi(\mathbf{r}, t)$ can be separated as

$$\Psi(\mathbf{r}, t) = \Psi(\mathbf{r})e^{i2\pi\nu t} \qquad (2.66)$$

where $\Psi(\mathbf{r})$ is a complex function describing the disturbance's spatial variations in amplitude in phase and ν is the frequency of the wave.

Time dependence is eliminated for a monochromatic disturbance and its spatial component then satisfies the Helmholtz equation given by

$$\left[\nabla^2 + \left(\frac{2\pi\nu}{c}\right)^2\right]\Psi(\mathbf{r}) = (\nabla^2 + k^2)\Psi(\mathbf{r}) = 0 \qquad (2.67)$$

where $k = \dfrac{2\pi\nu}{c} = \dfrac{\omega}{c}$ is the disturbance's wavenumber, and $\lambda = \frac{c}{\nu}$ is the wavelength.

Optical intensity $I(\mathbf{r})$ can be obtained, when averaged over a time longer that optical period $\frac{1}{\nu}$ as

$$I(\mathbf{r}) = |\Psi(\mathbf{r})|^2 \qquad (2.68)$$

The simplest solution of the Helmholtz equation in a homogeneous medium involves the plane wave and spherical wave. Any solution of the wave equation of the form

$$\Psi(\mathbf{r}, t) = \Psi(\mathbf{r} \cdot \hat{s}, t) \qquad (2.69)$$

is called a plane wave solution, since Ψ at any time t is constant over any plane normal to the unit vector $\hat{s}$. For a plane wave

$$\Psi(r, t) = \Psi_1(\mathbf{r} \cdot \hat{s} - ct) + \Psi_2(\mathbf{r} \cdot \hat{s} - ct) \qquad (2.70)$$

replacing $\mathbf{r} \cdot \hat{s} = \varsigma$, $\Psi(\zeta \mp ct)$ represents a plane wave that propagates in the positive ζ direction (upper sign) or in the negative ζ direction (lower sign).

For a spherical wave $\Psi = \Psi(r, t)$, where r is the distance from origin, the solution is given by

$$\Psi(r, t) = \frac{\Psi_1(r - ct)}{r} + \frac{\Psi_2(r + ct)}{r} \qquad (2.71)$$

$\mathbf{r}$ can be expressed as $\mathbf{r} = x\mathbf{i} + y\mathbf{j} + z\mathbf{k}$, and $\mathbf{i}, \mathbf{j}$, and $\mathbf{k}$ are unit vectors along the Cartesian coordinate axis. The spherical wave propagates away from the origin and can be approximated at points near the z axis and sufficiently away from the source, as a paraboloidal wave by Fresnel approximation given by $(x^2 + y^2)^{1/2} \ll z$. At very far points, the spherical wave approaches plane wave.

2.5.2.2 Huygens-Fresnel principle

In 1690, while considering transmission of waves, Huygens enunciated that every point on a propagating wavefront serves as the source of spherical secondary wavelets, such that the wavefront at some later time is the envelope of these wavelets. Further, the secondary wavelets will have the same speed and frequency of the propagating wave while transmitting through a medium. Unfortunately, the Huygens principle failed to resolve the phenomenon of diffraction. The difficulty was resolved by Fresnel with the addition of the concept of interference. This resulted in Huygens-Fresnel principle, which states that each point on a wavefront is a source of secondary disturbance and the secondary wavelets emanating from different points mutually interfere. The amplitude of optical field at any point beyond is the superposition of all these wavelets when their amplitudes and relative phases are considered. As a corollary, if the wavelength is large compared to any aperture through which the wave is passing, the waves will spread out at large angles into the region beyond the obstruction.

2.5.2.3   Diffraction

The phenomenon of diffraction concerns with the interaction of a wavefront with an object in its path and the description of how it propagates following that interaction. In all imaging applications, diffraction's first-order effect is to limit resolution. Diffraction phenomena are usually divided into two classes:

1. Fresnel or near-field diffraction: This is observed when the source of light and the plane of observation (may be a screen) are at a finite distance from the diffracting aperture.

2. Fraunhofer or far-field diffraction: This is observed when the plane of observation is at infinite distance from the aperture.

A representative regime of the diffraction is shown in Figure 2.4. By sufficiently reducing the wavelength of light λ, Fraunhofer diffraction pattern reduces to a Fresnel pattern. If $\lambda \to 0$, the diffraction pattern disappears and the image takes the shape of the aperture as dictated by the geometrical optics. There is another region very near to the source known as Fresnel-Kirchhoff region.

A point source of light that generates plane waves of amplitude A is normally incident on an aperture, as shown in Figure 2.5. The aperture and observation planes are separated in the z-direction and their respective coordinates planes are (ξ, η) and (x, y), respectively.

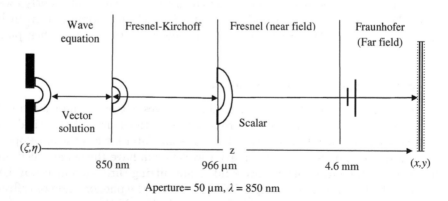

Aperture= 50 μm, $\lambda = 850$ nm

FIGURE 2.4: Representative diffraction regime

To explain the diffraction phenomenon, the optical system is redrawn as a model representation given in Figure 2.5. Considering an infinitesimal area $d\xi d\eta$ on the aperture, the total field at point P due to waves emanating from all infinitesimal areas of the aperture is given by

$$\Psi(P) = \frac{1}{i\lambda} \int \int \frac{Ae^{ikr}}{r} F(\theta) d\xi d\eta \tag{2.72}$$

where $F(\theta)$ is the obliquity factor $\frac{1}{i\lambda}$, is a proportionality constant and integration is over the entire area of the aperture. The quantity r represents the distance between the source having coordinates $(\xi, \eta, 0)$ and the point P having coordinates (x, y, z).

If the amplitude and phase distribution on the plane $z = 0$ is given by $A(\xi, \eta)$, then under small-angle conditions between source and point P, the above integral is modified by replacing obliquity factor $F(\theta) = \cos\theta = \frac{z}{r}$ by

$$\Psi(P) = \frac{z}{i\lambda} \int \int A(\xi, \eta) \frac{e^{ikr}}{r^2} d\xi d\eta \tag{2.73}$$

The geometry between $(\xi, \eta, 0)$ and (x, y, z) is related by

$$\begin{aligned} r &= [(x-\xi)^2 + (y-\eta)^2 + z^2]^{1/2} \\ &= z\sqrt{1+\alpha} \end{aligned} \tag{2.74}$$

where $\alpha = \frac{(x-\xi)^2}{z^2} + \frac{(y-\eta)^2}{z^2}$.

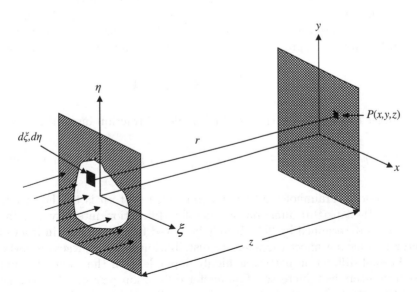

FIGURE 2.5: Diffraction of a plane wave incident normally on an aperture

For $\alpha \ll 1$, the quadratic and higher-order terms of the binomial expansion of $\sqrt{1 + \alpha}$ is neglected to yield the following equation:

$$
\begin{aligned}
r &\simeq z + \frac{(x - \xi)^2}{2z} + \frac{(y - \eta)^2}{2z} \\
&= z + \left(\frac{x^2}{2z} + \frac{y^2}{2z}\right) + \left(\frac{\xi^2}{2z} + \frac{\eta^2}{2z}\right) - \left(\frac{x\xi}{z} + \frac{y\eta}{z}\right)
\end{aligned} \quad (2.75)
$$

and the r^2 in the denominator of the expression for $\Psi(P)$ can be replaced by z^2. Therefore, a general equation may be written as

$$
\Psi(x, y) \approx \frac{e^{ikz}}{i\lambda z} \int \int A(\xi, \eta) \exp\left\{\frac{ik}{2z}[(x - \xi)^2 + (y - \eta)^2]\right\} d\xi d\eta \quad (2.76)
$$

The equation can be rearranged as

$$
\Psi(x, y) \approx \frac{e^{ikz}}{i\lambda z} \exp\left[\frac{ik}{2z}(x^2 + y^2)\right] \int \int A(\xi, \eta) \exp\left[\frac{ik}{2z}(\xi^2 + \eta^2)\right]
$$
$$
\exp\left[\frac{-ik}{z}(x\xi + y\eta)\right] d\xi d\eta \quad (2.77)
$$

These equations are known as the Fresnel diffraction integral and can be used to calculate Fresnel diffraction patterns. In Fresnel approximation, the term proportional to α^2 has been neglected. This is justified when maximum phase change is less than π. The intensity distribution in the diffraction pattern is determined by $|\Psi|^2$. Hence the exponents in front of the integral do not affect the intensity distribution, as their moduli are equal to unity.

In Fraunhofer approximation, z is considered large, and the function

$$
\exp\left[\frac{ik}{2z}(\xi^2 + \eta^2)\right] \to 1 \quad (2.78)
$$

Under this approximation the Fraunhofer diffraction integral is given by

$$
\Psi(x, y) \approx \frac{e^{ikz}}{i\lambda z} \exp\left[\frac{ik}{2z}(x^2 + y^2)\right] \int \int A(\xi, \eta) \exp\left[-\frac{ik}{z}(x\xi + y\eta)\right] d\xi d\eta
$$
$$
(2.79)
$$

Therefore, Fraunhofer diffraction (far field) pattern is the Fourier transform of the aperture function, Fraunhofer diffraction intensity distribution pattern is determined by $|\Psi|^2$. It may be noted that the factor in front of the integral does not affect the intensity distribution in the diffraction pattern.

Fresnel diffraction pattern is highly dependent on the value of z and the pattern alters as z increases. Fraunhofer diffraction pattern of a rectangular aperture can be best studied by passing a collimated bean of coherent source, i.e., laser through a rectangular aperture and seen on a board place perpendicular to be beam path or can be captured on photographic plate. A lens mat be used for focussing the pattern. The Fresnel diffraction pattern of a wire and Fraunhofer diffraction pattern of a rectangular aperture are shown in Figure 2.6.

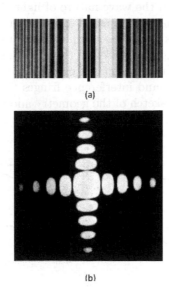

(a)

(b)

FIGURE 2.6: (a) Fresnel diffraction of a wire and (b) Fraunhofer diffraction pattern of a rectangular aperture

2.5.2.4 Interference

When two monochromatic waves of complex amplitudes are superposed, they can constructively add together to create a larger wave, or destructively subtract from each other to create a smaller wave. The occurrence of such waves depend on the relative direction of respective fields of the wavefronts, or in other words, their relative phase. The resultant wave $\Psi(\mathbf{r})$ is given by

$$\Psi(\mathbf{r}) = \Psi_1(\mathbf{r}) + \Psi_2(\mathbf{r}) \tag{2.80}$$

The intensity of the total wave is given as

$$I = |\Psi|^2| = |\Psi_1 + \Psi_2|^2 \tag{2.81}$$

where r has been omitted for convenience and $\Psi_1 = \sqrt{I_1}e^{i\varphi_1}$ and $\Psi_2 = \sqrt{I_2}e^{i\varphi_2}$, φ is the phase. The interference equation is then given by

$$I = I_1 + I_2 + 2\sqrt{I_1 I_2}\cos\varphi \tag{2.82}$$

where $\varphi = (\varphi_2 - \varphi_1)$

Thus the interference between the two waves is not the addition of two intensities, but contains an additional term, which may be positive or negative, corresponding to constructive or destructive interference. Optical power is conserved by spatial redistribution of optical intensity.

In 1801, Thomas Young performed a fundamental experiment for demonstrating interference and the wave nature of light. Monochromatic light from a single pinhole illuminates an opaque screen with two additional pinholes or slits. The light diffracts from these pinholes and illuminates a viewing screen at a distance that is large compared to the pinhole separation. Since the light illuminating the two pinholes comes from a single source, the two diffracted wavefronts are coherent and interference fringes form where the beams overlap. Figure 2.7 shows a sketch of the geometry adopted to derive approximate interference fringe locations observed on a distant screen.

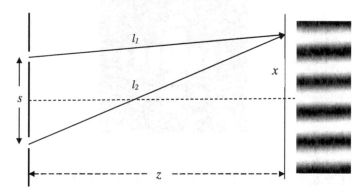

FIGURE 2.7: Young's experiment for the formation of interference pattern

Assuming a time-independent monochromatic wavefront impinges on a screen with two equal-sized slits separated by a distance s that transmits fields of amplitude A. According to the Huygens-Fresnel principle, each slit may be thought of as a source emitting a spherical wave travelling path lengths of (simplifying to one dimension) $l_1(x)$ and $l_2(x)$ to a projection screen at a distance z away. At any point x on the screen the amplitude of the wavefront is found by adding the two components of the waves, as

$$\Psi(x) = \Psi_1(x) + \Psi_2(x) \tag{2.83}$$

Considering s and x to be small relative to z, the intensity pattern on the screen is

$$
\begin{aligned}
I(x) &= I_0\left(2 + 2\cos\left(\frac{2\pi}{\lambda}[l_1(x) - l_2(x)]/\lambda\right)\right) \\
&= 2I_0\left[1 + \cos\left(\frac{2\pi d(x)}{\lambda}\right)\right]
\end{aligned}
\tag{2.84}
$$

where $d(x) = l_1(x) - l_2(x)$ is the optical path difference (OPD) between the two waves at a given point x on the screen. If $s << z$, we have

$$d(x) \approx \frac{xs}{z} \qquad (2.85)$$

Thus, the resulting interference pattern observed on the screen manifests itself as a sinusoidal variation in intensity ranging from 0 to $4I_0$. That the intensity does not vary from 0 to $2I_0$ can be explained through simple conservation of energy arguments.

In interferometry it is convenient to define a metric that quantifies the quality of a fringe pattern. The contrast between destructive and constructive fringes is quantified through the concept of visibility V, where

$$V = \frac{I_{max} - I_{min}}{I_{max} + I_{min}} \qquad (2.86)$$

It is clear from this definition that $0 \leq V \leq 1$, where the closer the visibility is to unity, the higher the contrast of the fringes.

2.5.2.5 Polarisation

Polarisation generally means orientation. Wave polarisation occurs for vector fields. For light (electromagnetic waves) the vectors are the electric and magnetic fields, and the polarisation direction is by convention along the direction of the electric field. Light waves only have two non-vanishing components: the two that are perpendicular to the direction of the wave. In order to satisfy all four of Maxwell's equations, the waves must have the $\mathbf{E}$ and $\mathbf{B}$ fields transverse to the propagation direction. Light is a transverse electromagnetic wave. Thus, if the wave is travelling along the positive z-axis, the electric field $\mathbf{E}$ can be parallel to the $+x$-axis and $\mathbf{B}$ field parallel to the $+y$ axis. The direction the light travels is determined by the direction of the vector cross product $\mathbf{E} \times \mathbf{B}$. Polarisation of light only refers to direction, not sense.

If in the process of wave propagation the vector field $\mathbf{E}$ remains in the same plane parallel to the direction of propagation, the waves are called linearly polarised. The plane in which the vectors $\mathbf{E}$ and $\mathbf{k}$ lie is called the plane of polarisation.

Electric field $\mathbf{E}(z,t)$ of a monochromatic plane wave of frequency ω lies in the x, y plane, and travelling with a velocity c in the z, is given by

$$\mathbf{E}(z,t) = Re\{\mathbf{A} \exp[j\omega(t - \frac{z}{c})]\} \qquad (2.87)$$

where the complex amplitude vector is $\mathbf{A} = A_x \hat{x} + A_y \hat{y}$, with complex components $\hat{x}$ and $\hat{y}$ and the end point of vector $\mathbf{E}(z,t)$ at each point z is a function of time.

$\mathbf{E}(z,t)$ can also be expressed as

$$\mathbf{E}(z,t) = E_x \hat{x} + E_y \hat{y} \qquad (2.88)$$

where the x and y components of the electric field vector are

$$E_x = a_x \cos[\omega(t - \frac{z}{c}) + \varphi_x] \tag{2.89}$$

and

$$E_y = a_y \cos[\omega(t - \frac{z}{c}) + \varphi_y] \tag{2.90}$$

E_x and E_y are the parametric equation of an ellipse given by

$$\left(\frac{E_x^2}{a_x^2}\right)^2 + \left(\frac{E_y^2}{a_y^2}\right)^2 - 2\cos\varphi\left(\frac{E_x E_y}{a_x a_y}\right) = \sin^2\varphi \tag{2.91}$$

where $\varphi = (\varphi_x - \varphi_y)$.

The state of polarisation of the wave is determined by the shape of the ellipse, i.e., the direction of the major axis and the ellipticity, the ratio of the minor to the major axis of the ellipse. The size of the ellipse determines the intensity of the wave $I = (a_x^2 + a_y^2)/2\eta$ where η is the impedance of the medium. At a fixed time t, the locus of the tip of the electric-field vector follows a helical trajectory in space lying on the surface of an elliptical cylinder, and at a fixed value of z. The tip of the electric field vector rotates periodically in the $(x-y)$ plane, tracing out this ellipse. The electric field rotates as the wave advances, repeating its motion periodically for each distance corresponding to a wavelength λ.

If one of the components, either a_x or a_y, is equal to zero, the light is linearly polarised in the direction of the other component (the y direction). The wave is also linearly polarised if the phase difference $\varphi = 0$ or π. Under this condition, the wave is said to be plane polarised.

If, $\varphi = \pm\pi/2$ and $a_x = a_y = a_0$, the equation of ellipse becomes the equation of circle. The elliptical cylinder becomes a circular cylinder and the wave is said to be circularly polarised. In the case of $\varphi = +\pi/2$, the electric field at a fixed position z rotates in a clockwise direction when viewed from the direction toward which the wave is approaching. The light is then said to be right-circularly polarised. The case, $\varphi = -\pi/2$ corresponds to counter-clockwise rotation and the light is left circularly polarised.

A monochromatic plane wave of frequency ω travelling in the z direction is characterized by the complex envelopes $\mathbf{A}_x = a_x \exp(i\varphi_x)$ and $\mathbf{A}_y = a_y \exp(i\varphi_y)$ of the x and y components of the electric field. These complex quantities can be written in a column matrix, known as the Jones vector, as

$$\mathbf{J} = \begin{bmatrix} A_x \\ A_y \end{bmatrix} \tag{2.92}$$

Given the Jones vector, the total intensity of the wave can be determined as $I = (|A_x|^2 + |A_y|^2)/2\eta$ and the phase difference $\varphi = arg[A_y] - arg[A_x]$. The Jones vector can be used to determine orthogonal polarisations of two polarised waves. Polarisation states are orthogonal if the inner product between

$\mathbf{J}_1$ and $\mathbf{J}_2$ is zero. The inner product is given by

$$(\mathbf{J}_1, \mathbf{J}_2) = A_{1x}A_{2x}^* + A_{1y}A_{2y}^* \qquad (2.93)$$

where A_{1x} and A_{1y} are the elements of $\mathbf{J}_1$, and A_{2x} and A_{2y} are the elements of $\mathbf{J}_2$.

Polarisation plays an important role in the interaction of light with matter, related to reflection, refraction, absorption and scattering. These are:

a. The amount of light reflected at the boundary between two materials depends on the polarisation of the incident wave.

b. The amount of light absorbed by certain materials is polarisation dependent.

c. Light scattering from matter is generally polarisation sensitive.

Moreover, the refractive index of anisotropic materials depends on the polarisation. Waves with different polarisations travel at different velocities and undergo different phase shifts, so that the polarisation ellipse is modified as the wave advances (e.g., linearly polarised light can be transformed into circularly polarised light). This property is used in the design of many optical devices. The so-called optically active materials have the natural ability to rotate the polarisation plane of linearly polarised light. In the presence of a magnetic field, most materials rotate the polarisation. When arranged in certain configurations, liquid crystals also act as polarisation rotators.

The transmission of plane wave of certain polarisation through an optical device that maintains the plane-wave nature of the wave, but alters its polarisation can be obtained by using Jones vector. The device is assumed to be linear, so that the principle of superposition of optical fields is obeyed. Two examples of such systems are the reflection of light from a planar boundary between two media, and the transmission of light through a plate with anisotropic optical properties. The complex envelopes of the two electric-field components of the input (incident) wave, A_{1x}, and A_{1y}, and those of the output (transmitted or reflected) wave, A_{2x}, and A_{2y} are related by weighted superposition relations, given by a matrix relation as

$$\begin{bmatrix} A_{2x} \\ A_{2y} \end{bmatrix} = \begin{bmatrix} T_{11} & T_{12} \\ T_{21} & T_{22} \end{bmatrix} \begin{bmatrix} A_{1x} \\ A_{1y} \end{bmatrix} \qquad (2.94)$$

where T_{11}, T_{12}, T_{21}, and T_{22} are constants describing polarisation of the device. If the input and output waves are described by the Jones vectors $\mathbf{J}_1$ and $\mathbf{J}_2$, then the above equation can be written in the compact matrix form as

$$\mathbf{J}_2 = T\mathbf{J}_1 \qquad (2.95)$$

The matrix T is a property of the optical device and determines its effect on the polarisation state and intensity of the incident wave. For example, the

Jones matrix of a linear polariser with a transmission axis making an angle θ with the x axis is

$$T = \begin{bmatrix} \cos^2\theta & \sin\theta\cos\theta \\ \sin\theta\cos\theta & \sin^2\theta \end{bmatrix} \tag{2.96}$$

The reflection and refraction of a monochromatic plane wave of arbitrary polarisation incident at a planar boundary between two dielectric media can also be studied in terms of Jones vectors. If the media are assumed to be linear, homogeneous, isotropic, non-dispersive, non-magnetic, and the refractive indices are n_1 and n_1, then the wavefronts of these waves are matched according to the laws of reflection ($\theta_3 = \theta_1$) and Snell's law of refraction, $n_1\sin\theta_1 = n_2\sin\theta_2$. To relate the amplitudes and polarisations of the the waves, with each wave an $x - y$ coordinate system is associated in a plane normal to the direction of propagation. The electric-field envelopes of these waves as described by the Jones vectors are $\mathbf{J}_1$, $\mathbf{J}_2$, and $\mathbf{J}_3$, and the relations between them are written in matrices form as $\mathbf{J}_2 = \mathbf{t}\mathbf{J}_1$ and $\mathbf{J}_3 = \mathbf{r}\mathbf{J}_1$, where $\mathbf{t}$ and $\mathbf{r}$ are 2×2 Jones matrix for the transmission and reflection of the wave, respectively. Elements of the transmission and reflection matrices may be determined by using the boundary conditions required by electromagnetic theory.

The tangential components of $\mathbf{E}$ and $\mathbf{H}$ and normal components of $\mathbf{D}$ and $\mathbf{B}$ are continuous at the boundary. The magnetic field associated with each wave is orthogonal to the electric field and their magnitudes are related by the characteristic impedances for the incident reflected and transmitted waves.

Two normal modes for this system are linearly polarised waves with polarisation along the x and y directions. The x-polarised mode is called the transverse electric (TE) polarisation or the orthogonal (or s) polarisation, since the electric fields are orthogonal to the plane of incidence. The y-polarised mode is called the transverse magnetic (TM) polarisation, since the magnetic field is orthogonal to the plane of incidence, or the parallel (or s) polarisation, since the electric fields are parallel to the plane of incidence. The independence of the x and y polarisations implies that the Jones matrices $\mathbf{t}$ and $\mathbf{r}$ are diagonal matrices. The coefficients t_x and t_y are the complex amplitude transmittances for the TE and TM polarisations, and the coefficients for the complex amplitude reflectance are r_x and r_y. Applying the boundary conditions to the TE and TM polarisations separately gives the expressions for the reflection and transmission coefficients, known as the Fresnel equations. The Fresnel equations for TE polarisation are

$$r_x = \frac{n_1\cos\theta_1 - n_2\cos\theta_2}{n_1\cos\theta_1 + n_2\cos\theta_2} \tag{2.97}$$

$$t_x = 1 - r_x \tag{2.98}$$

The reflection coefficient r_x is always real and negative for $n_1 < n_2$, corresponding to a phase shift π. The magnitude $\frac{(n_2-n_1)}{(n_2-n_1)}$ at $\theta_1 = 0$ (normal incidence) and increases to unity at $\theta_1 = 90^\circ$ (grazing incidence). When $n_1 > n_2$, for small θ_1 the reflection coefficient is real and positive and its magnitude increases gradually to critical angle $\theta_c = \sin^{-1}(n_2/n_1)$. The total internal reflection is associated with a phase shift $= arg(r_x)$

The Fresnel equations for TM polarisation are

$$r_y = \frac{n_2 \cos \theta_1 - n_1 \cos \theta_2}{n_2 \cos \theta_1 + n_1 \cos \theta_2} \qquad (2.99)$$

$$t_y = \frac{n_1}{n_2}(1 + r_y) \qquad (2.100)$$

The reflection coefficient is real, when $n_1 < n_2$. It decreases from a positive value $\frac{(n_2 - n_1)}{(n_2 - n_1)}$ until it vanishes at an angle $\theta_1 = \theta_B$, known as the Brewster angle where

$$\theta_B = \tan^{-1} \frac{n_2}{n_1} \qquad (2.101)$$

For $\theta_1 > \theta_B$, r_y reverses sign and its magnitude increases gradually, approaching unity at $\theta_1 = 90^\circ$. When $n_1 > n_2$, at $\theta_1 = 0$, r_y is negative. As θ_1 increases beyond θ_B, r_y becomes positive and increases until it reaches unity at the critical angle θ_c. For $\theta_1 > \theta_c$, the wave undergoes total internal reflection accompanied by a phase shift.

2.6 Interaction of light and matter

Light, like any other kind of electromagnetic radiation, interacts with medium in mainly two different ways: absorption, where the photons disappear, and scattering, where the photons change their direction. A third way of interaction also exists, when the light can change its state of polarisation. In the case of absorption it may happen that light is being re-emitted at a different wavelength, i.e., as fluorescence or phosphorescence. Classically, light-matter interactions are a result of an oscillating electromagnetic field resonantly interacting with charged particles. Quantum mechanically, light fields will act to couple quantum states of the matter.

2.6.1 Absorption

Absorption is a transfer of energy from the electromagnetic wave to the atoms or molecules of the medium. Energy transferred to an atom can excite electrons to higher energy states, and energy transferred to a molecule can excite vibrations or rotations. The wavelengths of light that can excite these energy states depend on the energy-level structures and therefore on the types of atoms and molecules contained in the medium. The spectrum of the light after passing through a medium appears to have certain wavelengths removed because they have been absorbed. This is called an absorption spectrum. Selective absorption is also the basis for objects having colour. The colour red is red because it absorbs the other colours of the visible spectrum and reflects only red light.

The intensity of the light beam is weakened by absorption, and the transmitted intensity decreases exponentially with the thickness x of the layer of material through which the light has to pass. The transmitted intensity I is usually written as

$$I = I_0 10^{-\alpha x} \qquad (2.102)$$

where the quantity α is called the absorption, coefficient, and I_0 is the incident light intensity.

2.6.2   Scattering

Scattering is the redirection of light caused by the interaction of light with matter. The scattered electromagnetic radiation may have the same or longer wavelength (lower energy) as the incident radiation, and it may have a different polarisation. Scattering can be subdivided into elastic and inelastic scattering: If the scattered light has exactly the same wavelength as the incident light, meaning that the scattered photons have exactly the same energy as the incident photons, the scattering is called elastic. If the photons come out of the scattering process with a changed energy, the scattering process is termed inelastic. This means that the scattered light will have either a longer or a shorter wavelength than that of the incident light.

If the dimensions of the scatterer are much smaller than the wavelength of light, like a molecule, for example, the scatterer can absorb the incident light and re-emit the light in a different direction. If the re-emitted light has the same wavelength as the incident light, the process is called Rayleigh scattering. Air molecules are Rayleigh scatterers of visible light and are more effective at scattering shorter wavelengths (blue and violet). The Rayleigh scattering intensity per unit volume I_R is given by

$$I_R \approx \frac{K(n-1)^2}{\lambda^4} I_0 \qquad (2.103)$$

where K is a constant, n is the refractive index of the scatterer, and I_o is the mean energy flux density of the incident light. Therefore, the scattering intensity is inversely proportional to the fourth power of the wavelength.

If the re-emitted light has a longer wavelength, the molecule is left in an excited state, and the process is called Raman scattering. In Raman scattering, secondary photons of longer wavelength are emitted when the molecule returns to the ground state. The difference in energy between the excitation and scattered photons corresponds to the energy required to excite a molecule to a higher vibrational mode. The Raman interaction leads to two possible outcomes: (a) The material absorbs energy and the emitted photon has a lower energy than the absorbed photon. This outcome is labelled Stokes Raman scattering. (b) The material loses energy and the emitted photon has a higher energy than the absorbed photon. This outcome is labelled anti-Stokes Raman scattering.

The Raman scattering process takes place spontaneously; i.e., in random time intervals, one of the many incoming photons is scattered by the material. This process is thus called spontaneous Raman scattering. On the other hand, stimulated Raman scattering can take place when some Stokes photons have previously been generated by spontaneous Raman scattering (and somehow forced to remain in the material), or when deliberately injecting Stokes photons (signal light) together with the original light (pump light). In that case, the total Raman scattering rate is increased beyond that of spontaneous Raman scattering. Pump photons are converted more rapidly into additional Stokes photons. The more Stokes photons are already present, the faster more of them are added. Effectively, this amplifies the Stokes light in the presence of the pump light, which is exploited in Raman amplifiers and Raman lasers. Stimulated Raman scattering is a nonlinear-optical effect. It can be described using a third-order nonlinear susceptibility.

Another type of scattering is known as Brillouin scattering. This phenomenon occurs when light, transmitted by a transparent carrier, interacts with that carrier's time-and-space-periodic variations in refractive index. The result of the interaction between the light-wave and the carrier-deformation-wave is that a fraction from the passing-through light-wave changes its momentum (thus its frequency and energy) along grating-like preferential angles. Brillouin scattering denominates the scattering of photons from low-frequency phonons, while for Raman scattering, photons are scattered by interaction with vibrational and rotational transitions in the bonds between first-order neighbouring atoms.

If the scatterer is similar in size to or is much larger than the wavelength of light, matching the energy levels is not important. All wavelengths are equally scattered. This process is called Mie scattering. This type of scattering is observed when water droplets effectively scatter all wavelengths of visible light in all directions.

2.7 Photometric image formation

To produce an image, the scene must be illuminated with one or more light sources. Light sources can generally be divided into point and area light sources. A point light source has an intensity and a colour spectrum and originates at a single location in space, potentially at infinity. In general light from a point source is spread over a larger (spherical) area. A simple area light source is a finite rectangular area emitting light equally in all directions.

2.7.1   Reflectance and shading

When light hits an object, it is scattered and reflected. The interaction can be described by the bidirectional reflectance distribution function (BRDF). Relative to some local coordinate frame on the surface, the BRDF is a four dimensional function that describes how much of each wavelength arriving at an incident direction v_i is emitted in a reflected direction v_r. The function can be written in terms of the angles of the incident and reflected directions relative to the surface frame, as $f_r(\theta_i, \phi_i, \theta_r, \phi_r, \lambda)$. The BRDF is reciprocal, as the direction of incident and reflected light can be interchanged. While light is scattered uniformly in all directions, i.e., the BRDF is constant. Typical BRDFs can often be split into their diffuse and specular components. The diffuse component (also known as Lambertian or matte reflection) scatters light uniformly in all directions and is the phenomenon normally associated with shading, e.g., the smooth (non-shiny) variation of intensity with surface normal that is seen. Diffuse reflection also often imparts a strong body colour to the light since it is caused by selective absorption and re-emission of light inside the object material. This is because of the fact, that the surface area exposed to a given amount of light becomes larger at oblique angles, and completely self-shadowed as the outgoing surface normal points moves away from the light.

The second major component of a typical BRDF is specular (gloss or highlight) reflection, which depends strongly on the direction of the outgoing light. Incident light rays are reflected in a direction that is rotated by 180° around the surface normal. The diffuse and specular components of reflection can be combined with another term, called the ambient illumination. This term accounts for the fact that objects are generally illuminated not only by point light sources but also by a general diffuse illumination corresponding to inter-reflection from other surfaces.

Books for further reading

1. *Fundamentals of Photonics*: B. E. A. Saleh and M. C. Teich, John Wiley, New York, 1991.

2. *Introduction to Information Optics*: F. T. S. Yu, S. Jutamulia and S. Yin (ed.), Academic Press, New York, 2001.

3. *Information Optics and Photonics*: T. Fournel and B. Javidi (eds.), Springer, New York, 2010.

4. *Advances in Information Technologies for Electromagnetics*: L. Tarricone and A. Esposito, Springer, The Netherlands, 2006.

5. *Catching the Light, the Entwined History of Light and Mind*: A. Zajonc, Oxford University Press, New York, 1993.

6. *The Quantum Theory of Light*: R. Loudon, Oxford University Press, New York, 2000.

7. *The Nature of Light — What is a Photon?* C. Roychoudhuri, A. F. Kracklauer and K. Creath, CRC Press, Boca Raton, FL, 2008.

8. *Statistical Optics*: J. W. Goodman. John Wiley Inc., New York, 2000.

9. *Introduction to Statistical Optics*: E. L. ONeill, Addison-Wesley, Reading, 1963.

10. *Coherence of Light*: J. Perina, Van Nostrand-Reinhold, London, 1985.

11. *Optical Coherence Theory*: G. J. Troup, Methuen, London, 1967.

12. *Theory of Partial Coherence*: M. J. Beran and G. B. Parrent, Prentice Hall, New York, 1964.

13. *Electromagnetic Waves*: C. G. Someda, Chapman & Hall, London, 1998.

14. *Electromagnetic Wave Theory*: J. A. Kong, McGraw-Hill, New York, 1984.

15. *Electromagnetics*: E. J. Rothwell and M. J. Cloud, CRC Press, Boca Raton, Florida, 2001.

16. *Electromagnetics*: J. D. Kraus, McGraw-Hill, New York, 1984.

17. *Modern Optics*: R. Guenther, John Wiley, New York, 1990.

18. *Light*: R. W. Ditchburn, Academic Press, New York, 1976.

19. *Geometrical and Physical Optics*: R. S. Longhurst, Longman, Inc., New York, 1973.

20. *Principles of Optics*: M. Born and E. Wolf, Cambridge University Press, New York, 1997.

21. *Optics*: E. Hecht, Addison Wesley, Reading, MA, 1998.

22. *Geometric Optics—The Matrix Theory*: J. W. Blaker, Marcel Dekker, New York, 1971.

23. *Elements of Modern Optical Design*: D. C. O'Shea, John Wiley, New York, 1985.

Bibliography

[1] I. Walmsley and P. Knight. Quantum information science. *Optics and Photonics News*, 43(9), 2002.

[2] European Technology Platform Photonics:21. *Towards 2020: Photonics Driving Economic Growth in Europe: Multiannual Strategic Roadmap 2014 to 2020.* VDI Technologiezentrum GmbH, Dusseldorf, 2014.

[3] Committee on Optical Science and Engineering. *Harnessing Light: Optical Science and Engineering for the 21st Century.* National Research Council, USA, 1998.

[4] M. Watanabe. *Optical Information Technology.* National Institute of Advanced Industrial Science and Technology, 2008.

[5] M. Leis, M. Butter, and M. Sandtke. *The Leverage Effect of Photonics Technologies: the European Perspective.* European Commission, 2011.

[6] European Technology Platform Photonics21. *Lighting the way ahead.* 2010.

[7] S. A. Ramakrishna. Physics of negative refractive index materials. *Progess of Physics*, 68:449–521, 2005.

[8] G. N. Lewis. *Nature*, 118(2):874–875, 1926.

[9] J. Z. Buchwald. *The Rise of the Wave Theory of Light: Optical Theory and Experiment in the Early Nineteenth Century.* University of Chicago Press, Chicago, USA, 1989.

[10] J. C. Maxwell. A dynamical theory of the electromagnetic field. *Trans. Royal Soc.*, 155:459–460, 1865.

[11] I. Bialynicki-Birula. On the wave function of the photon. *Acta Physica Polonica*, 86:97–116, 1994.

[12] J. F. Clauser. Experimental distinction between the quantum and classical field theoretic predictions for the photoelectric effect. *Physical Rev.*, D9(4):853–860, 1974.

[13] A. Zeilinger. Happy centenary, photon. *Nature*, 433:230–238, 2005.

[14] A. Muthukrishnan, M. O. Scully, and M. S. Zubairy. The concept of the photon revisited. *Optics and Photonics News Trends*, 3(1):18, 2003.

[15] R. Kidd, J. Aedini, and A. Anton. Evolution of the modern photon. *Am. J. Physics*, 57:27–35, 1989.

[16] C. Roychoudhuri, A. F. Kracklauer, and K. Creath. *The Nature of Light: What is a Photon?* CRC Press, Boca Raton, FL, 2008.

[17] R. Glauber. *One Hundred Years of Light Quanta*. Physics Nobel Prize Lecture, 2005.

[18] J. E. Sipe. Photon wave functions. *Phy. Rev. (A)*, 52(3):1875–1883, 1995.

[19] I. Bialynicki-Birula. Photon wave function. *Progress in Optics*, 36:245–294, 1996.

[20] V. Gharibyan. Possible observation of photon speed energy dependence. *Phy. Let.*, 611:231–238, 2005.

[21] L. Allen, M. J. Padgett, and M. J. Babiker. The orbital angular momentum of light. *Prog. Opt*, 39:291–372, 1999.

[22] L. B. Okun. Photon: History, mass, charge. *Acta Physica Polonica*, 37(3), 2006.

[23] J. Luo, L. Tu, and G. T. Gilles. The mass of the photon. *Report. Prog. Physics*, 68:77–130, 2005.

[24] S. Eidelman. Review of particle physics. *Physics Letters. B*, 592:1–5, 2004.

[25] D. R. Smith, R. Dalichaouch, N. Kroll, S. Scholtz, S. L. McCall, and P. M. Platzman. Photonic band structure and defects in one and two dimensions. *J. Opt. Soc. Am. (B)*, 10(2):314–321, 1993.

[26] F. Zernike. The concept of degree of coherence and its application to optical problems. *Physica*, 5:785795, 1938.

[27] L. Mandel and E. Wolf. Coherence properties of optical fields. *Rev. Mod. Phy.*, 37:231–287, 1965.

[28] P. Maragos. Tutorial on advances in morphological image processing. *Opt. Eng.*, 26(3):623–630, 1987.

[29] G. J. Troup. *Optical coherence theory*. Methuen, London, 1967.

[30] L. Mandel and E. Wolf. *Selected papers on coherence and fluctuation of light*. Dover, New York, 1970.

[31] B. E. A. Saleh and M. C. Teich. *Fundamentals of Photonics*. John Wiley, USA, 1991.

[32] B. J. Thompson. Image formation with partially coherent light. 1969.

[33] S. A. Schelkunoff. Forty years ago Maxwell's theory invades engineering and grows with it. *IEEE Trans. Education*, E-15:2, 1972.

[34] A. M. Bork. Maxwell and the electromagnetic wave equation. *Am. J. Physics*, 35:844–856, 1967.

[35] M. Born and E. Wolf. *Principles of Optics.* Cambridge University Press, New York, 1997.

[36] W . F . Magie (ed.). *Fermat's Principle published in A Source Book in Physics.* Cambridge University Press, Massachusetts, USA, 1963.

[37] H. H. Hopkins. An extension of Fermat's theorem. *Optica Acta*, 17:223–225, 1970.

Chapter 3

Vision, Visual Perception, and Computer Vision

3.1 Introduction

Human visual perception is an active process of the brain starting with primal processing in the visual system. Visual perception (or vision) can be defined as the process or the result of processes of building an internal representation of an object or a scene in the mind of the viewer. This encompasses entities or relations that are believed to exist in an external reality and that can be derived by processing reflected light rays (or an absence thereof). In other words, visual perception is the human ability to interpret a scene by processing information that is contained in visible light by the eye-brain mechanism.

3.1.1 Human visual system

The human visual system can be regarded as consisting of two parts. In the first part the eyes act as image receptors which capture light and convert it into electrical or neuronal signals, which are then transmitted to image processing centres in the brain. These centres process the signals received from the eyes and build an internal replica of the scene being viewed. While the eyes convert visual signals to electrical impulses via a chemical process. In the second part the brain partly acts by simple image processing and partly by building and manipulating an internal model of the scene.

3.1.1.1 Structure of the human eye

The anatomical structure of the eye can be divided into three structural regions: (a) protective structure of the eye consisting of the orbit, the eyelids, conjunctive, and the sclera; (b) anterior segment consisting of the cornea, the aqueous humour, iris, the crystalline lens, and the ciliary muscle; and (c) posterior segment consisting of the retina and the vitreous humour.

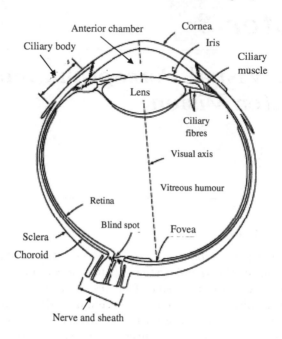

FIGURE 3.1: Basic structure of an eye

Schematic cut-out view of the human eye is shown in Figure 3.1[1] The
protective structures of the eye consist of two sockets (orbits) that are situated
at the front of the skull. Each has a wider opening at the front and narrows
to a small opening at the rear, where the optic nerve connects the visual
pathways and the brain. The structures are protected by the eyelids. The
eyelids contain the glands responsible for maintenance of the tear layer. The
outer surface of the eyeball except the cornea is covered by a transparent
mucous membrane called the conjunctiva. The eyeballs are predominately
formed of and protected by the sclera which extends from the edges of the
clear cornea at the front of the eye (the limbus) to the optic nerve at the back
of the eye. The sclera is a thick, opaque white tissue that covers 95 percent
of the surface area of the eye. At the posterior of the eye, the sclera forms a
net-like structure, through which the optic nerve passes. The sclera also serves
as an anchor tissue for the extra-ocular muscles.

The anterior segment of the eye has most of the structures responsible
for focusing images onto the retina of the eye. The cornea is the primary
focusing structure which provides about 75 percent of the focusing power
of the eye. The crystalline lens further refines the focus, thus allowing the
eye to focus objects at different distances from the eye. The iris controls the

---

[1]Redrawn from the book *Digital Image Processing* by R. C. Gonzalez and R. E. Wood
published by Pearson Prentice Hall, Indian reprint.

aperture or pupil of the eye for different light levels. The iris is an extension of the ciliary body, a structure that has multiple functions in the anterior segment ranging from the production of the fluid that fills the anterior segment (aqueous humour) to the suspension and control of the shape of the crystalline lens of the eye.

The aqueous humour is a kind of fluid that fills the anterior chamber of the eye between the cornea and the front surface of the crystalline lens. The fluid is basically a fortified form of blood plasma. Aqueous humour also provides nutrients to the cornea and is a part of the optical pathway of the eye. As nutrients are drawn from the aqueous into the cornea, the aqueous fluid is circulated out of the eye and replaced by newly produced aqueous. It has an index of refraction of 1.333, which is slightly less than the index of refraction of the cornea (1.376) and less than the index of refraction of the lens (gradient index of 1.406 to 1.386). This difference in indices of refraction between the media coupled with the curvature of the various optical surface results in the bending of light at each interface.

The iris is visible through the cornea and gives the eye its colour. The main purpose of the iris, however, is to block excess light from entering the eye by controlling the aperture or pupil. There are two opposing muscles in the iris; the sphincter muscles that serve to constrict the pupil, and the dilator muscles that serve to dilate the pupil. Most pupil responses are controlled by a complex set of signals sent by the mid-brain in response to the amount of light striking the retina.

Like the cornea, the crystalline lens is a transparent structure; however, unlike the cornea, it has the ability to change its shape in order to increase or decrease the amount of refracting rays of light coming into the eye. The gradient index of refraction of the lens ranges from about 1.406 at the centre to about 1.386 at the peripheral portions of the lens. The lens is surrounded by an elastic extra-cellular matrix of muscles known as the capsule. The capsule not only provides a smooth optical surface, but also acts as an anchor for the suspension of the lens within the eye. A mesh work of non-elastic microfibrils (zonula) anchor into the capsule near the equator of the lens and are connected into the ciliary muscle. These muscles control both the shape and positioning (forward and backwards) of the lens and exert fine control of the light entering into the eye.

The posterior segment of the eye mainly consists of vitreous humour and the retina. Fovea and macula in the retina are also parts of the posterior segment. Initial processing of the image occurs in this highly specialized sensory tissue. Vitreous humour is the clear gel that fills the posterior segment and serves as a medium for light transmission through the eye This also protects the retina. It comprises collagen fibrils in a network of hyaluronic acid. The vitreous body is loosely attached to the retina around the optic nerve head and the macula, and more firmly attached to the retina at the posterior of the ciliary body. The connections at the anterior portion of the vitreous body help to keep the anterior, and posterior chamber fluids separated. The connections

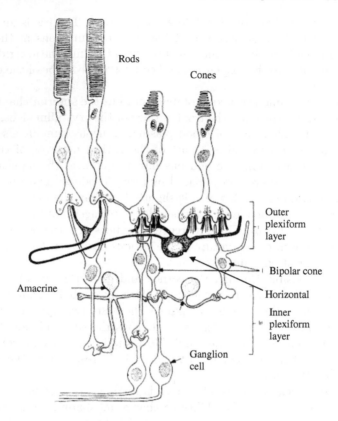

FIGURE 3.2: Basic structure of retina

around the optic nerve and macula help to hold the vitreous body against the retina.

The retina as shown in Figure 3.2 consists of eight layers of mostly transparent thin tissues evolved to capture photons and to initiate processing of the image by the brain. From the surface of the retina to the back of the eye, the layers are:

1. the optic nerve fibre layer (axons of the ganglion cells),

2. the ganglion cell layer,

3. the inner plexiform layer (synapses between ganglion and bipolar or amacrine cells),

4. the inner nuclear layer (amacrine, bipolar, and horizontal cells),

5. the outer plexiform layer (synapses between bipolar, horizontal, and photoreceptor cells),

6. the outer nuclear layer (photoreceptor cells),

7. the receptor layer (outer and inner segments of the photoreceptor cells), and

8. the retinal pigment epithelium, which acts as the final light sink for incoming photons, and that reduces intra-ocular glare.

The fact that the receptor layer is deep within the retina means that photons of light actually must pass through most of the layers of the retina before reaching the receptors. The receptors absorb and convert photons to neuronal signals, which are than processed through the network of bipolar, horizontal, amacrine, and ganglion cells. The output axons of the ganglion cells form the nerve fibre layer that collects neurons at the optic nerve and exit the eye [1].

The intricate interconnections of the various neural cells in the retina complete the initial processing of the visual information being sent to the brain. The signals produced in the retina are propagated backward through the head along the optic fibre tract. The optic nerve of each eye consists of a bundle of approximately a million retinal ganglion cell axons. The nerve connection is referred to as the optic nerve head. Since there are no photoreceptors overlying the optic nerve head, there is a small blind spot (scotoma) which fixes the visual field of each eye. When both eyes are open, the blind spot of each eye is filled in by the visual field of the other eye.

There are several types of neurons and neuronal pathways by which signals are conducted [2], [3]. Types of integrators include horizontal cells, whose communications are entirely within the outer plexiform layer, and which connect photoreceptors with bipolar cells. Bipolar cells can receive signals from the photoreceptors directly, or via mediation of a horizontal cell. Their dendrites are in the outer plexiform layer, and their axons extend inward to the inner plexiform layer, where they synapse with other integrators, specifically amacrine cells and ganglion cells.

The amacrine cells mediate signals between bipolar cells, other amacrine cells, and ganglion cells. The ganglion cells are the final element in the chain, and they receive input from either the bipolar cells or the amacrine cells. The simplest and fastest pathway for transmission of a signal is from cone to ganglion cell via bipolar cell. The ganglion cells' bundled axons make up the beginnings of the optic nerve. The connection for rods is different than it is for cones, and it involves four neurons in the chain from rod to bipolar cell to amacrine cell and ending with the ganglion cell. Again, the synapse with the ganglion cell marks the end of intra-retinal processing and the beginning of transmission of the integrated output into the visual cortex.

The signal routes described above are the simplest possible cases and the actual route of information through the retina is usually a great deal more complicated. Lateral connections between two rods and cones, bipolar cells, etc., are all possible and the number of switches in the system and possible combinations is nearly infinite.

The outer segment of the receptor cells contain two types of light-sensitive visual pigment molecules-the rods and cones, essentially named for their shape [4]. There are approximately 5 million cones and 92 million rods in the retina of an adult human [5]. The tips of rods and cones are embedded in a pigmented layer of cells on the very back of the retina. The cones and rods have the same kind of structure of optical fibre, i.e., they consist of dielectric cylindrical rods surrounded by another dielectric of slightly lower refractive index. The light causes a chemical reaction with photo-pigments, such as iodopsin and rhodopsin.

The distribution of rods and cones is not uniform across the retina. The cones are concentrated towards, and the rods away from the centre as shown in Figure 3.3[2]. The cones and rods from different parts of the retina are not fully identical, but vary in their morphological structure. However, the density of these cells varies and as a result the sensation of the central part of the retina differs from that of the periphery.

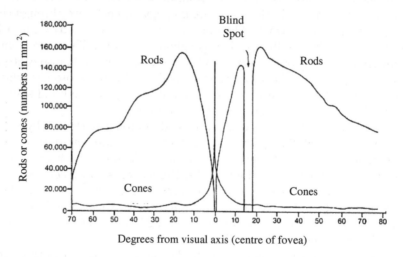

FIGURE 3.3: Distribution of rods and cones in retina

At the most posterior aspect of the retina, where most of the light that the eye receives is focused, is a region called the macula. The macula is an area which has a greater density of pigments that helps to protect the retinal neural cells. Within the macular is an area called the fovea centralis, where the vision is most acute. The average thickness of the retina drops in this foveal pit. In order to maximise photon capture in the central retina, the retinal capillary system does not extend into the fovea centralis, an area known as the foveal vascular zone. In the fovea area there are no rods, and only cones are present.

---

[2]Redrawn from the book *Digital Image Processing* by R. C. Gonzalez and R. E. Wood published by Pearson Prentice Hall, Indian reprint.

3.1.1.2 Visual signal processing in retina and brain

Although there are millions of rods and cones in the vertebrate retina, the transmission of signals is somewhat selective. Some light-sensitive elements are on and others are off at any given time, and the brain has to interpret these on-off patterns. The connections between the primary elements, the rods and cones and the brain, are mediated through the various layers of the retina in such a way that the actual input to the brain is controlled. The information processing pathways are shown in Figure 3.4[3].

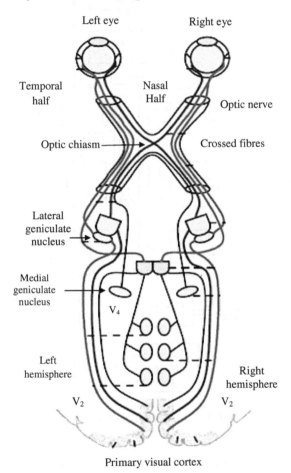

FIGURE 3.4: Visual pathways from eye to brain

The optic nerves of each eye continue posteriorly and then meet at the optic chiasm. It is here that axons of neurons from the nasal retina (temporal visual

---

[3]Figure redrawn from B. Dubuc; The brain from the top to bottom (2011) given in the Website, http://thebrain.mcgill.ca/avance.php

field) cross to the opposite optic tract. Axons of neurons from the temporal retina however, continue along the same side's optic tract. This means that visual signals from the right side of the visual field are travelling to the brain via the left optic tract and signals from the left visual field are travelling via the right optic tract. Signal flow diagram of neural processing of visual signal is shown in Figure 3.5.

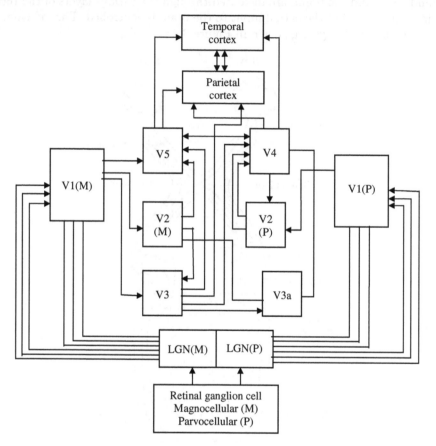

FIGURE 3.5: Signal flow in visual system

Each optic tract terminates at its lateral geniculate nucleus (LGN). The LGN is a paired structure located at the dorsal thalamus. It is here that visual information to the brain, specifically to the visual cortex, appears to be regulated and the first stage of coordinating vision from both eyes begins. Because of the way the retinal ganglion cell axons are distributed through the chiasm and on to the optic tracts, the information processed in any one layer of the LGN represents specific areas of the visual field for one eye. The cell arrangements in LGN are such that they respond to contrast and movement.

Similarly, the paired responses for red-green and yellow-blue are also found in the LGN, suggesting that its main function is to carry out further processing of incoming signals. The LGN then sends forward neurons to the primary visual cortex. The visual cortex is at the most posterior portion of the brain and executes final processing of the neural signals from the retina.

Besides six separate primary areas in the visual cortex, known as the V1, V2, V3, V3a, V4, and V5, there are about thirty further visual areas across parietal and temporal cortices in the brain. The primary visual cortex or V1 is the first structure in the visual cortex where the neuronal signals are interpreted in terms of visual space, including the form, colour, and orientation of objects. V1 dedicates most of its area for mapping and interpreting the information received from the fovea. There is a pronounced functional streaming, or division of labour, for form, colour, and motion processing; some neuroscientists have proposed a fundamental division into two visual systems along lines such as magno and parvo (fast/slow) or even conscious and unconscious vision.

Others designated the visual information from V1 as dividing along two streams: 1) a dorsal *where* stream, which is concerned with the spatial relationship between objects, and 2) a ventral *what* stream, which is concerned with object recognition. The *what* stream sees in detail because it gets most of its input from ganglion cells in the fovea. The *where* stream gets less detailed input from ganglion cells in the peripheral retina and excels at coding the locations of objects. The signals from V1 pass through to V2, where colour perception occurs. As the neuronal signals continue further into other areas of the visual cortex, more associative processes take place. In the temporal visual cortical areas, including the middle temporal (V5) area, recognition of objects through interpretation of complex forms and patterns occurs. The final psychological and perceptual experience of vision includes aspects of memory, expectation/prediction, and interpolation [6].

A question still remains, which simply can be stated as: How is visual inference carried out in the brain? There is a series of computational modules, each performing a relatively encapsulated computational step in image analysis, and postulating that there is a rough correspondence between these modules and the areas in the visual cortex [7]. Here V5 is considered to be the area where the aperture problem is solved by integrating local motion signals into a global whole, while V2 is the area where many gestalt grouping (see the section on visual perception) operations are performed. A central tenet, however, is the decomposition of visual processing into low, intermediate, and high levels, and the belief that the visual cortex could likewise be divided into these functional stages. In this framework, the early visual area is the site of the primal sketch, where local features are detected and grouped together into symbolic tokens and contours. Representations of surfaces and object models are thought to be computed and represented in the cortex areas, such as V4 and V5 [8]. Therefore, at early stages of processing, visual cortex takes part in the computation and representation of perceptual contours, surface shapes, object saliency, and possibly medial axis of forms.

An interesting aspect of the architecture shown in the wiring diagram of Figure 3.5 is the pattern of reciprocating feedback connections. In general, there are pairwise reciprocating connections between visual areas from the deep layers in one area to superficial layers in another area. With the massive feedback projections from primary visual cortex V1 back down to the LGN, where it meets signals from the eyes, these reciprocating projection pathways are perhaps suggestive of a kind of iterative strategy for hypothesis generation, and testing, necessary for understanding the visual environment and the objects.

3.1.1.3   Photopic, mesopic, and scotopic vision

The eye is sensitive to radiation over a range of very high illumination to intensity as low as 5 or 6 photons (when adapted to the dark for at least an hour). The detection capacity of an eye also depends on the intensity of the surrounding illumination. Fluctuations in the intensity of incoming light change the size of the pupils, which controls the radiation input to the retina by a factor of twelve; the remaining adaptation is provided by the cones and rods. Relative performances of vision with respect to light levels are shown in Figure 3.6.

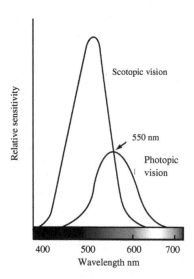

FIGURE 3.6: Comparative performance of photopic and scotopic vision

The high radiation range, called the photopic range, is mediated by the cones, while the low radiation range, what is referred to as the scotopic range, is mediated by the rods. The intermediate range, known as the mesopic range, is mediated by both these elements. The eyes are also not equally sensitive to all wavelengths of visible light. Each human being perceives differently the amount and colour (wavelength) of a given beam of light intensity. The spectral weighting functions for the photopic and scotopic ranges are known as the spectral luminous efficiency. The mesopic range requires a weighting function based upon a gradual transition between the spectral luminous efficiency functions throughout the mesopic region that depends on the visual adaptation conditions. As illumination increases above the mesopic level, the rods become saturated and cease to function and then the cones are mediated. This condition is referred to as photopic vision.

Cones, supporting photopic vision, are present throughout the retina, but are highly concentrated in the fovea and sparsely distributed in the periphery. At the centre of the fovea, cone density is high and such high cone density provides distinct input to the visual centre in the brain. This allows for high-resolution vision of at least 20/20. In the peripheral retina, cone density decreases, and input from cones is pooled, thus limiting visual resolution when visual acuity is about 20/200. In addition to providing high-resolution central vision, the cone system supports colour vision.

Because of their structure, cones capture light most effectively if rays enter straight on, that is, perpendicularly to the retinal surface. Light rays striking at wider angles stimulate the cones less efficiently, a phenomenon known as the Stiles-Crawford effect. Because of this, light rays entering the peripheral pupil appear less bright than rays entering centrally. A benefit of the Stiles-Crawford effect is that light scattered within the eye has little effect on cone vision. Eventually illumination becomes so high that the cones become saturated.

The rods are absent in the fovea but present in the rest of the retina. Rods support vision in low light (scotopic vision) and the structure of the rods maximises photon capture. The rod cell is capable of responding to very few photons of light though perceptual awareness requires simultaneous absorption of 5 to 14 photons. Since there are no rods in the fovea, under scotopic conditions, such as at night, the central 2° of the visual field becomes a tiny blind spot. The scotopic system operates from nearly complete darkness up to luminance values of about 1 candela $/m^2$. Moreover, rods do not contribute to colour perception. Because of the distribution of rods and their supporting neurons, scotopic visual acuity is poorer than cone-mediated acuity. On the other hand, the rod system is better at integrating light from a larger area of the retina, and it therefore provides vision in low light, below the threshold for cones. As illumination increases and approaches the upper limit for the rods, the cones begin to work. For intermediate light levels (mesopic vision), both rods and cones are working [9].

Both rods and cones are sensitive to a wide range of wavelengths, but sensitivity varies for different wavelengths. Both rod and cone sensitivities peak near the middle of their respective ranges. In terms of absolute sensitivity, the rods are more sensitive than the cones. The scotopic (rod) sensitivity spectrum peaks at shorter wavelengths than the photopic (cone) sensitivity spectrum. Because of this, as illumination decreases, vision transits from photopic to scotopic vision. Moreover, shorter wavelength hues become relatively brighter while longer wavelength hues become less bright, a phenomenon known as the Purkinje shift.

Within their working ranges, rods and cones must adapt to changes in light level. When illumination increases (light adaptation), the photoreceptors become less sensitive to light. When illumination decreases (dark adaptation), they become more sensitive. Cones have capability to dark adapt more quickly than rods and reach their maximum sensitivity after about 15 minutes. Rods, on the other hand, require about 40 minutes to fully dark adapt and reach maximum sensitivity. After complete dark adaptation, exposure to any light begins the process of light adaptation and visual sensitivity declines. If a dark-adapted observer needs to use light, yet hopes to preserve dark adaptation, the loss of sensitivity can be minimised by using a long-wavelength red light. Long wavelengths are relatively poorly absorbed by rods, so red light has minimal impact on rod dark adaptation. Meanwhile, cones are about equally sensitive to rods at long wavelengths, so they can contribute to vision in low light.

3.1.1.4   The CIE V(λ) function

The CIE luminosity function $\overline{y}(\lambda)$ or V(λ), is a standard function established by the Commission Internationale de l'Éclairage (CIE) and is used to convert radiant energy into luminous (i.e., visible) energy [10]. Figure 3.7 shows the relation between relative sensitivity of an eye with wavelength λ in visible region as adapted by CIE.

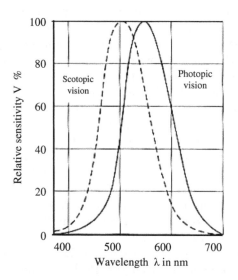

FIGURE 3.7: CIE V$-\lambda$ curve

The luminosity function or luminous efficiency function describes the average spectral sensitivity of human visual perception to brightness. It is based on subjective judgements of brightness of lights of different colour and indicates relative sensitivity to light of different wavelengths. It is a very good representation of visual sensitivity of human eye and it is valuable as a baseline for experimental purposes.

The luminous flux (or visible energy) of a light source is defined by the photopic luminosity function. The following equation calculates the total luminous flux of a source of light:

$$F = 683 \ lm/W \times \int_0^\infty V(\lambda)J(\lambda)d\lambda \qquad (3.1)$$

where F is the luminous flux in Lumen, $V(\lambda)$ is the standard luminosity function, and $J(\lambda)$ is the spectral power distribution. The standard luminosity function is normalised to a peak value of unity at 555 nm wave length of light. The value of the constant in front of the integral is usually rounded off to 683 lm/W. The number 683 is connected to the modern definition of the candela, the unit of luminous intensity.

Colour is a critical feature of vision since it helps in better discrimination of objects. Colour generates a sensation in the visual system primarily based on the wavelength absorbed by the cones. CIE $V(\lambda)$ function is a colour-matching function. There are two luminosity functions for common use. At everyday light levels, the photopic luminosity function best approximates the response of the human eye. For low light levels, the response of the human eye changes, and the scotopic curve applies.

3.2 Visual perception

Hermann von Helmholtz is often credited with the first study of visual perception in modern times. He concluded that vision could only be the result of some form of unconscious inferences, based on previous experiences and guided by the capability of making assumptions and conclusions from incomplete visual data collected by the eye-brain mechanism. Researchers from various disciplines (including, among others, psychology, cognitive science, and neurology) have proposed many hypotheses of visual perception. Besides visual characteristics of the objects, the process is shaped by individual anatomy, prior experiences, current task and context, expectations, aims, and self-regulatory strategies [11], [12]. One of them is a perception theory called the unconscious inference hypothesis, which suggests that the visual system performs some form of Bayesian inference to derive a perception from sensory data [13].

3.2.1   Perception laws

The way the information is decoded from scenes (particularly from graphs and pictures) is of great interest and has been the subject of study and speculation. The basic tasks, those helps in formulating the perception laws, include recognising shapes, discerning colour, judging sizes and distances, and tracing and extrapolating motion in three dimensions. In the course of these studies, a number of laws of perception have been formulated. These laws are mostly based on empirical study, but gave some useful information on the perception mechanism.

3.2.1.1   Weber's psychophysical law

Weber's law is one of the important laws of human perception that is concerned with the judgement of visual intensity discrimination. Let u denote the mean strength of the background field, and δu be the difference or contrast between the background signal and a target signal embedded within. Let $T(u)$ be the yes-no binary decision of an average human observer for detecting the target signal under a fixed background level u. Then psychological evidence shows that $T(u)$ behaves like a Heaviside function $H(\delta u - \delta_T u)$ for some threshold $\delta_T u$ depending on u, such that $T_u|_u = H(\delta u - \delta_T u) = 1$, when $\delta u \geq \delta_T u$, and 0 otherwise.

For each fixed background level u this threshold, $\delta_T u$ is often called the just-noticeable-difference (JND) in psychology. Weber's law says that the JND is linearly proportional to the mean of the background field. Further, there exists a constant W, so that for a large range of mean-of-background field u,

$$\frac{\delta_T u}{u} \equiv W \qquad (3.2)$$

However, in a quantitative or statistical model, decisions are not necessarily binary and a human subject's responses can lead to a fuzzy inference. Therefore, an ideal perceptual response function shall allow T to be a continuous function that takes any confidence value between 0 and 1.

Instead of mathematically arriving at the proof of Weber's law a simple example may serve to explain the simplicity of the law. Suppose that $w_p(x)$ is a positive number such that a line of length $[x + w_p(x)]$ is detected as longer than a line of length x with probability p. Using the statement of Weber's law, for a fixed value of p, $w_p(x) = k_p x$, where the value of k_p does not depend on x. In simple terms, Weber's law states that it is relative rather than absolute differences which are perceived when lengths are compared.

3.2.1.2   Fechner's law

Fechner's law (better referred to as Fechner's scale) provides an explanation for Weber's law [13]. Fechner's explanation has two parts. The first part is that two stimuli will be discriminable, if they generate a visual response that exceeds some threshold. The second part is that the visual response R

to an intensity I varies logarithmically, i.e., the relationship between stimulus and perception is logarithmic. This logarithmic relationship means that if a stimulus varies as a geometric progression (multiplied by a fixed factor), the corresponding perception is altered in an arithmetic progression [14].

3.2.1.3 Steven's power law

Steven's power law [15] [16], states that for a given person and for a given perceptual task, the perceived values follow a general law given by

$$p(x) = \alpha x^\beta \tag{3.3}$$

where α and β depend on the particular person and the perceptual task and different types of stimuli result in different power functions. Extending the law for two encoded values x_1 and x_2 in a graph is perceived as $p(x_1)$ and $p(x_2)$. Then, comparing the relative perceived values, the following equation holds good:

$$\frac{p(x_1)}{p(x_2)} = \left(\frac{x_1}{x_2}\right)^\beta \tag{3.4}$$

The power law is generalized to the broad sensory domain [17], [18], where sensation magnitude S grows as a power function of stimulus magnitude I minus a threshold constant I_0. The rate of the growth, characterised by the exponent β of the function, varies greatly from one continuum to another, according to the equation

$$S = \alpha(I - I_0)^\beta \tag{3.5}$$

Steven's law provides a description of how dimensional values are perceived through a number of encodings, including length, area, and volume. Because of the importance of β, many experiments have been conducted to try to estimate its value for particular ways of encoding graphical information. For an area, the value of β is taken between 0.6 and 0.9, and for volume, the value is taken from 0.5 and 0.8. For length, the value of β is considered to be between 0.9 and 1.1. Because the value of β is close to 1 for length judgements, the ratio of the perceived values are closer to the ratio of the actual values. However, if an area of $1/2$ is considered instead of 1, the perceived value is 0.62, which is rather large. Since the value of β for volume judgement is even further removed from 1, representing very small values may lead to even greater distortions.

3.2.1.4 Grouping laws and gestalt principles

The term *gestalt* means shape or form. In the domain of perception, gestalt psychologists stipulate that perceptions are the products of complex interactions among various stimuli. Contrary to the behaviourist approach to understand the elements of cognitive processes, gestalt psychologists sought to understand their organization. The central principle of gestalt psychology is that the mind forms a global whole with self-organizing tendencies. This principle maintains that when the human mind (perceptual system) forms a percept or gestalt, the whole has a reality of its own, independent of the parts.

Gestalt theory starts with the assumption of active grouping laws in visual perception. These groups are identifiable with subsets of the retina. Whenever points in an object (or previously formed groups) have one or several characteristics in common, they get grouped and form a new larger visual object, a *gestalt*. The list of elementary grouping laws considered are vicinity, similarity, continuity of direction, amodal completion, closure, constant width, tendency to convexity, symmetry, common motion, and past experience. For example, the colour constancy law states that connected regions where luminance (or colour) does not vary strongly are unified (seen as a whole, with no inside parts). The amodal completion law applies when a curve stops on another curve, thus creating a T-junction. In such a case our perception tends to interpret the interrupted curve as the boundary of some object undergoing occlusion. Another example is the constant width law which is applied to group the two parallel curves, perceived as the boundaries of a constant width object. This law is constantly in action since it is involved in the perception of writing and drawing.

An aspect of perceiving scenes as meaningful wholes, as opposed to an atomistic, literal, or elemental description in terms of individual features, is the grouping of features into a 3D model. A classic illustration of this idea of vision as model-building is the Necker cube (Figure 3.8(a)). This is a set of 12 planar line segments that are always seen as a 3D solid (a cube), yet having two conflicting (bistable) visual interpretations. Such bistable percepts are examples of perceptual rivalry: two or more alternative ways to interpret the same visual stimulus. Some other examples are given in Figure 3.8 (b) and (c), where illusory contours are perceived in places where no actual contours exist. The key notion is that percepts are hypotheses: They are top-down interpretations that depend greatly on contexts, expectations, and other extraneous factors that go beyond the actual stimulus.

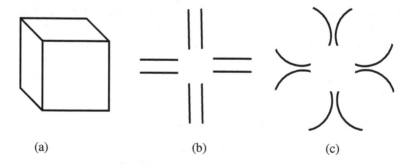

(a)   (b)   (c)

FIGURE 3.8: Some examples of illusion

3.2.1.5 Helmholtz principle

The Helmholtz principle states that whenever some large deviation from randomness occurs, a structure is perceived. Although it is difficult to formulate a mathematical statement or formulation of the principle, it may be addressed by introducing a universal variable expressed as the number of false alarms (NFA). The NFA of an event is the expectation of the number of occurrences of this event. Events whose NFA is less than a preselected ε are called ε-meaningful events [19], [20]. When $\varepsilon \leq 1$, the event is meaningful. An event is ε-meaningful if the expectation of the number of occurrences of this event is less than ε under the constraints of uniform random assumption. The philosophy can be illustrated in a dot alignment example. In the visual field many dots are distributed randomly following uniform (Poisson) spatial distribution. Some dots in sufficient numbers will be called aligned if they fall into a thin strip. The precision of the alignment is measured by the width of the strip. The meaningful and visible alignment is detected as a large deviation from randomness. If more dots are again added randomly, the alignment is no longer meaningful or visible.

A word of caution is necessary here. Visual perception is not based on some global field-like process as suggested in the laws presented. Instead, a perception involves the interaction of multiple processes, representations, and sources of information from many sensory organs and memory. Theoretical framework for a unified model is difficult to evolve in terms of interacting processing streams because of conflicting constraints of biological vision. Decomposition of vision into its component processes and how the visual system integrates information from different processes remains largely an unsolved problem.

3.2.2   Seeing in three dimensions

Despite the fact that humans have strong abilities of spatial perception , the basic visual sensing system which converts illumination into electrical impulses is two dimensional, i.e., the retina. While the presence of two eyes and its associated geometrical and biological mechanism gives the perception of depth to a certain extent, perception also is related to the previously stored experience of seeing an object. Ultimately, the process gives very strong depth cues which enable to perceive depth in a two-dimensional representation of a three-dimensional scene [21]. Unfortunately, perception of truly three-dimensional scenes can be incorrect when those scenes contain structures which can be interpreted as the result of perspective.

From the pair of two-dimensional images formed on the retinas, the brain is capable of synthesizing a rich three-dimensional representation of our visual surroundings. The horizontal separation of the two eyes gives rise to small positional differences, (called binocular disparities), between corresponding features in the two retinal images. These disparities provide a powerful source of information about three dimensional scene structure, and alone can be sufficient in most cases for depth perception. Although neurons selective for binocular disparity have been found in several visual areas, the brain circuits that give rise to stereoscopic vision are not very well understood. Recent electrophysiological studies address four issues: the encoding of disparity at the first stages of binocular processing followed by the organization of disparity-selective neurons into topographic maps. Then the specific visual areas contribute to different stereoscopic tasks, and finally the integration of binocular disparity and information of viewing distance yield egocentric distance.

The perception of depth and three-dimensional structure is, however, possible with information visible from one eye alone, such as differences in object size and motion parallax, which come from the differences in the image of an object over time with observer movement. The impression of depth in these cases is often not as vivid as that obtained from binocular disparities. Therefore, it has been suggested that the impression of separation in depth is linked to the precision with which depth is derived, and that a conscious awareness of this precision is also necessary.

3.2.3 Bayesian interpretation of visual perception

Most of the tasks, whose solution requires intelligence involve degrees of uncertainty. Decision-making under uncertainty is a required characteristic in computer vision. The framework to adopt uncertainty in computer vision is to work with a certain amount of computing probabilities to arrive at a decision based on prior knowledge. A highly influential formalism for integrating prior knowledge with new incoming data from an image sequence is mathematically expressed in terms of Bayesian inference. Bayesianism interprets probability as degree-of-belief, rather than as frequency of occurrence, and argues for weighing all evidence with all possible (imaginable) interpretations and their associated (estimated) probabilities. The rule is a formalism for combining prior knowledge or beliefs with empirical observations. It is a procedure for the integration of evidence, and a protocol for decision-making. Some aspects of Bayesian interpretation in vision are evident in the way the following text is read, in which the same letter stimulus is read in completely different ways depending on local context (see Figure 3.9).

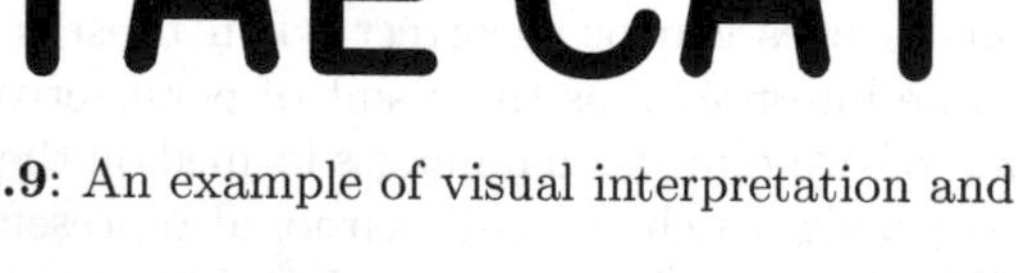

THE CAT

FIGURE 3.9: An example of visual interpretation and inference

An informal statement of Bayesian rule for drawing inferences from data is presented here. If H represents an hypothesis about the object in a scene and D represents the available image data, then the explanatory conditional probabilities $P(H|D)$ and $P(D|H)$ are related to each other and to their unconditional likelihoods $P(H)$ and $P(D)$ is

$$P(H|D) = \frac{P(D|H)P(H|D)}{P(D)} \tag{3.6}$$

Given a state H, there is a corresponding conditional probability $P(D|H)$ of observing certain image data D. However, typically the goal of computer vision with the capability of recognising an object in a scene is to calculate just the inverse of that conditional probability; that is, given image data D, what is the probability $P(H|D)$ that the hypothesis H is true?! Bayesian rule specifies the formal procedure for calculating such inferences $P(H|D)$, given the observations, the unconditional probabilities, and the prior expert knowledge $P(D|H)$. It thereby offers a simple interface between a knowledge base and visual data.

A key feature of Bayesian theorem is that it provides a mechanism for repeatedly updating an assessment of a visual hypothesis as more data arrives incrementally, so that a rule can be applied recursively, using the latest posterior as the new prior for interpreting the next set of data. However, in many vision tasks, there may not be any useful (or strong) priors and the recognition problems need to be solved purely on the basis of some vector of acquired features from a given object or image. Then the task is to decide whether or not this feature vector is consistent with membership in a particular class or object category.

3.3 Computer vision

Vision and visual perception are an effortless and immediate faculty for humans; however, they have proven exceedingly difficult to automate using computing machines and algorithms. Computer vision is concerned with automatic integration of a wide range of processes and representations used in visual perception and cognitive processing. It seeks to generate intelligent and useful descriptions of visual scenes and sequences, and of the objects that are present in the scene, by performing operations on the signals received from video cameras and other sensing devices. The goal is to simplify images to the end of solving a given vision problem, such as recognizing the presence of a certain object or identifying selected points of a three-dimensional (3D) scene. The challenge of computer vision can be described in terms of building a signal-to-symbol converter which is expected to have the capability to process information required for intelligent understanding of a scene and its environment. These signals must ultimately be converted into symbolic representations suitable for computer interpretation and manipulation, thus allowing the machine to interact intelligently with the world.

Some difficulties arise when mimicking the activities of the vision system and its processing within the brain by hardware and algorithms. Firstly, an image is a two-dimensional (2D) optical projection, but the visual system senses three-dimensional (3D) scenes. In this respect, vision is inverse optics where a computer vision system needs to invert the 3D to a 2D projection in order to recover world properties (object properties in space, although the task of inversion of such a projection is, strictly, mathematically impossible). Computer graphics, however, begin with a 3D world description, in terms of object and illuminant properties, viewpoint, etc., and simply computes the resulting 2D image, with its occluded surfaces, shading and shadows, gradients, perspective, and perhaps some other properties. The vision system perform exactly the inverse of this process. In terms of recognition of objects and scenes by the visual system under various constraints, humans perform this task effortlessly, rapidly, reliably, and unconsciously through formidable neural resources. Secondly, very few visual tasks can be successfully performed in a purely data-driven hardware-based system by so-called bottom-up image analysis irrespective of occlusion, perspective, illumination variations, and an object's background. Therefore, most of the problems in computer vision are ill-posed in Hadamard's sense that a typical problem does not qualify the criteria that (a) a solutions exists (b) a solution is impractical and (c) its solution depends continuously on the data [7]. Clearly, few of the tasks at hand in computer vision are well-posed problems in Hadamard's sense.

One of the notable developments in computer vision during this decade was the increased interaction with computer graphics, especially in the cross-disciplinary area of image-based modelling and rendering for automatically creating realistic 3D models from collections of images. Simultaneously, the past decade has seen the emergence of feature-based techniques, combined with learning for object recognition. Some of the notable developments in computer vision during this period include the constellation model [22] and the pictorial structures [23]. Feature-based techniques also dominate other recognition tasks, such as scene recognition, and panorama and location recognition [24]. The current dominating trend is the application of sophisticated machine learning techniques to computer vision problems. This trend coincides with the increased availability of immense quantities of partially labelled data on the Internet, which makes it more feasible to learn object categories without the use of human supervision.

3.3.1 Image formats

There are many different image formats used for storing and transmitting images in compressed form, since raw images are large data structures that contain much redundancy (e.g., correlations between nearby pixels) and thus are highly compressible. There are two types of image-file compression algorithms: lossless and lossy. Lossless compression algorithms reduce file size while preserving a perfect copy of the original uncompressed image. Lossless compression generally, but not always, results in larger files than lossy compression. Lossless compression should be used to avoid accumulating stages of re-compression when editing images, and the algorithms preserve a representation of the original uncompressed image that may appear to be a perfect copy, but it is not a perfect copy. Often lossy compression is able to achieve smaller file sizes than lossless compression. Most lossy compression algorithms allow for variable compression that trades image quality for file size. Different formats are specialized for compressibility, manipulability, or the properties of printers and browsers. Some extensively used formats are:

- .jpeg (*joint photographic experts group*): This format is ideal for variable compression of continuous colour images, with a quality factor (typically 75) that can be specified. The useful range of compression goes from 100:1 (lossy) to about 10:1 (almost lossless). Individual image frames are in general .jpeg compressed, but an equal amount of redundancy is removed temporally by inter-frame predictive coding and interpolation.

- .mpeg (*moving picture experts group*): This format was formed by the International Standard Organization to set standards for audio and video compression and transmission. This is a stream-oriented, compressive encoding scheme used mainly for video and multimedia works.

- .gif (*graphics interchange format*): This format is ideal for sparse binarised images as it is very compressive. Thus it is favoured for web-browsers and other bandwidth-limited media.

- .tiff (*tagged image file format*): This is a complex class of tagged image file formats with randomly embedded tags and can take up to 24-bit colour images. It is non-compressive.

- .bmp (*bit mapped*): This is another non-compressive format in which individual pixel values can easily be extracted.

- .png (*portable network graphics*): This file format was created as a free, open-source alternative to gif. The png file format supports 8 bit paletted images (with optional transparency for all palette colours) and 24 bit true-colour (16 million colours) or 48-bit true-colour with and without alpha channel, while gif supports only 256 colours and a single transparent colour.

In addition to these formats, there are varieties of other formats, where colour coordinates are used for colour separation, such as RGB (red, green, blue) and HSI (hue, saturation, intensity), etc.

It is typical for a monochromatic (black and white) image to have resolution of of 8 bits/pixel. This creates 256 different possible intensity values for each pixel, from black (=0) to white (=255), with all shades of grey in between. A full-colour image may be quantized to this depth in each of the three colour planes, requiring a total of 24 bits per pixel. However, it is common to represent colour more coarsely or even to combine luminance and chrominance information in such a way that their total information is only 8 or 12 bits/pixel.

Because quantized image information is fundamentally discrete, the image array imposes an upper limit on the amount of information contained. One way to describe this is by the total bit count, but this does not relate to the optical properties of image information. A better way is through Nyquist's theorem, which describes how the highest spatial frequency component of information contained within the image is equal to one-half the sampling density of the pixel array. Thus, a pixel array containing 640 columns can represent spatial frequency components of image structure no higher than 320 cycles/image. For the same reason, if image frames are sampled in time at the rate of 30 per second, then the highest temporal frequency component of information contained in the image sequence is 15 Hertz. The total number of independent pixels in an image array determines the spatial resolution of the image. Independent of this is the grey-scale (or colour) resolution of the image, which is determined by the number of bits of information specified for each pixel.

3.3.2   Image acquisition

Before any video or image processing can commence an image must be captured by a camera and converted into a manageable entity. This process is known as image acquisition. The image acquisition process consists of three steps: energy reflected from the object of interest, an optical system which focuses the energy, and finally a sensor which measures or stores the amount of energy with the help of some physical process. In the most elementary system, the three steps are accomplished by a pinhole camera, with the sun as the energy source and a photographic film as the sensor. Focussing is accomplished by a lens system.

A frame-grabber in a digital camera or an additional attachment card in a computer contains a high-speed analogue-to-digital converter which discretises this video signal into a byte stream. Conventional video formats include NTSC (North American standard), which is of 640×480 pixels, at 30 frames/second (actually there is an interlace of alternate lines scanned out at 60 frame per second). The other is a PAL (European, UK standard), which is 768×576 pixels, at 25 frames/second standard.

3.4   Geometric primitives and transformations

In the processing of image information, particularly in graphics applications, geometric objects are defined in terms of a number of building blocks called geometrical primitives. These primitives may correspond to points, lines, curves, surfaces, and are used to describe two- or three-dimensional shapes. For example, a rectangle can be defined by its four sides (or four vertices) and each side is constructed from a line segment by applying numbers of geometric operations, called transformations. The transformations position, orient, or scale the line primitives.

3.4.1 Geometrical primitives

Two dimensional (2D) points or pixel coordinates of an image are denoted using a pair of values, $\mathbf{x} = (x, y) \in \Re^2$, and can also be represented using homogeneous coordinates, $\tilde{\mathbf{x}} = (\tilde{x}, \tilde{y}, \tilde{w}) \in P^2$, where $\Re$ and P are universal and projected space, respectively. Instead of representing a 2D point with two parameters, homogeneous coordinates represent it with three parameters, so that a point in the plane is now represented by the column vector $\tilde{\mathbf{x}} = (\tilde{x}, \tilde{y}, \tilde{w})^T$. The vectors that differ only by scale are considered to be equivalent, as $P^2 = \Re^2 - (0, 0, 0)$. A homogeneous vector $\tilde{\mathbf{x}}$ can be converted back into an inhomogeneous vector $\mathbf{x}$ by dividing by the last element $\tilde{w}$ to yield $\tilde{x} = \tilde{w}\bar{x}$. Homogeneous points whose last element is $\tilde{w} = 0$ are called ideal points or points at infinity and do not have an equivalent inhomogeneous representation.

Another advantage of the homogeneous notation is that it permits simple formulae to calculate a line $\mathbf{I}$ through two points, x_1 and x_2. The condition for a point lying on a line may now be represented in a homogeneous plane as, $\mathbf{I}\tilde{x} = 0$. The line vector must be orthogonal to the point vector, if the point lies on the line. Hence, given two points, the line passing through both must be represented by a vector that is orthogonal to both. Such a vector can be obtained by using the cross product, so that, $\mathbf{I} = x_1 \times x_2$.

2D lines can be represented using homogeneous coordinates $\bar{\mathbf{l}} = (a, b, c)$, and the corresponding line equation is $\bar{x}.\bar{\mathbf{l}} = ax + by + c = 0, a \neq 0, b \neq 0$, and can be rewritten in matrix form as

$$\begin{bmatrix} x \\ y \end{bmatrix} = -\frac{c}{\sqrt{a^2 + b^2}} \tag{3.7}$$

The unit vector $\mathbf{n}$, called a normal vector, is perpendicular to the line. It points toward the line, when $c < 0$, and away from the line, when $c > 0$. The vector (a, b), like $\mathbf{n}$, is perpendicular to the line while the orthogonal vector $(-b, a)$ is parallel to the line.

Similarly, point coordinates in three dimensions (3D) can be written using inhomogeneous coordinates $\mathbf{x} = (x, y, z) \in \Re^3$ or homogeneous coordinates $\tilde{x} = (\tilde{x}, \tilde{y}, \tilde{z}, \tilde{w}) \in P^3$. It is useful to denote a point in 3D space using the augmented vector $\bar{\mathbf{x}} = (x, y, z, 1)$ with $\tilde{\mathbf{x}} = \tilde{w}\bar{x}$.

There are other algebraic curves, such as 2D and 2D conics, that can be expressed with simple polynomial homogeneous equations. For example, the 2D conic sections (so called because they arise as the intersection of a plane and a 3D cone) can be written by using a quadric equation, $\tilde{\mathbf{x}}^T Q\tilde{x} = 0$.

As in the case of 2D planes, 3D planes can also be represented in homogeneous coordinates $\tilde{m} = (a, b, c, d)$ vector with the plane equation given as $\bar{x} \cdot \tilde{m} = ax + by + cz + d = 0$. It is possible to normalise the plane with a normal vector n perpendicular to the plane and d its distance from the origin. n can also be expressed as a function of two angles θ and ϕ of spherical coordinates, as $n = (\cos\theta\cos\phi, \sin\theta\cos\phi, \sin\phi)$. 2D line and 2D representations are shown in Figure 3.10.

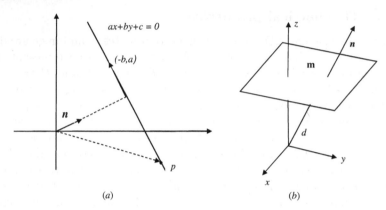

(a)   (b)

FIGURE 3.10: (a) 2D line representation and (b) 3D plane representation

3.4.2 Geometrical transformation

Five types of geometric transformations are particularly relevant in many applications. They are: translation, similarity, rigid (Euclidean), affine, and projective. Each transformation preserves an invariant property. The translation operation preserves orientation, similarity preserves angles, rigid preserves lengths, affine preserves the parallelism, and projective preserves straight lines. Combinations of one or two or several transformations can be used.

Linear transformation of a plane (also called an affine mapping), is a mapping $L : \Re^2 \to \Re^2$ from the plane to itself such that

$$\begin{bmatrix} x \\ y \end{bmatrix} \mapsto A \begin{bmatrix} x \\ y \end{bmatrix} + b \tag{3.8}$$

where $A = \begin{bmatrix} a_{11} & a_{12} \\ a_{21} & a_{22} \end{bmatrix}$ and $b = \begin{bmatrix} b_1 \\ b_2 \end{bmatrix}$

The transformation maps the line $cx + dy + e = 0$, where $c \neq 0$ or $d \neq 0$, to the line $(a_{22}c - a_{21}d)x + (a_{11}d - a_{12}c)y = (a_{12}b_2 - a_{22}b_1)c - (a_{11}b_2 - a_{21}b_1)d = (a_{11}a_{22} - a_{11}a_{22} - a_{12}a_{21})e = 0$.

A translation, $Trans(b_1, b_2)$, is also an affine mapping with the matrix $A = I$. That is, a transformation maps every point to a new point by adding a constant vector b. It has the effect of moving the point in the direction of the x-axis by b_1 units, and in the direction of the y-axis by b_2 units. The inverse translation is denoted as $T^{-1} = Trans(-b_1, -b_2)$.

Similarity transformation of a scaling operation ($Scale(s_x, s_y)$ about the origin is an affine mapping, where the matrix $A = \text{diag}\ (s_x, s_y)$ with $s_x \neq 0$, $s_y \neq 0$ and $b = 0$. This transformation maps a point by multiplying its x and y coordinates by factors s_x and s_y, respectively. In matrix notation the scaling is given by

$$\begin{bmatrix} x' \\ y' \end{bmatrix} = \begin{bmatrix} s_x & 0 \\ 0 & s_y \end{bmatrix} \begin{bmatrix} x \\ y \end{bmatrix} \tag{3.9}$$

The scaling is said to be an enlargement if $s > 1$, and a contraction, if $s < 1$. It is said to be uniform if $s_x = s_y$.

All transformations operations of a rectangular image positioned at origin is illustrated in Figure 3.11.

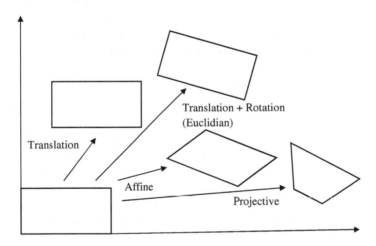

FIGURE 3.11: Geometric 2D transformations

A rotation about the origin through an angle θ maps every point P to a point P', such that P and P' are at the same distance from the origin and the angle from the vector P to the vector P' is θ. The coordinates of P' can be obtained by using a matrix multiplication given by

$$\begin{bmatrix} x' \\ y' \end{bmatrix} = \begin{bmatrix} \cos\theta & -\sin\theta \\ \sin\theta & \cos\theta \end{bmatrix} \begin{bmatrix} x \\ y \end{bmatrix} \tag{3.10}$$

Two-dimensional rotation T_{rot} in homogeneous coordinates is given by the matrix

$$T_{rot} = \begin{bmatrix} \cos\theta & -\sin\theta & 0 \\ \sin\theta & \cos\theta & 0 \\ 0 & 0 & 1 \end{bmatrix} \tag{3.11}$$

Euclidean transformation, which preserves Euclidean distance, is composed of rotation and translation operation done simultaneously.

The affine transformation is written as $y = A\bar{x}$, where A is an arbitrary 2×3 matrix, given by

$$y = \begin{bmatrix} a_{00} & a_{01} & a_{02} \\ a_{10} & a_{11} & a_{12} \end{bmatrix} \bar{x} \tag{3.12}$$

Parallel lines remain parallel under affine transformations.

Projective transformation, also known as a perspective transform or homography, operates on homogeneous coordinates. Perspective transformations preserve straight lines (i.e., they remain straight after the transformation). Homogeneous coordinates must be normalised in order to obtain an inhomogeneous result.

The set of three-dimensional coordinate transformations is very similar to that available for 2D transformations, as discussed. The main difference between 2D and 3D coordinate transformations is related to the parameterisation of the 3D rotation. However, a linear 3D to 2D projection is needed, when 3D primitives are projected onto the image plane.

3.4.2.1 Geometric model of camera

Simply put, a computational model for a camera gives information about the projection of 3D entities (points, lines, etc.,) onto the image, and vice versa. A model also indicates the method of back-projection from the image to 3D. The models may be classified into (a) a global model, which gives a set of parameters such that changing the value of any parameter affects the projection function of all across the field of view, and (b) a local camera model, which gives a set of parameters, each of which influences the projection function only over a subset of the field of view. The perspective projection of a pinhole system is shown in Figure 3.12.

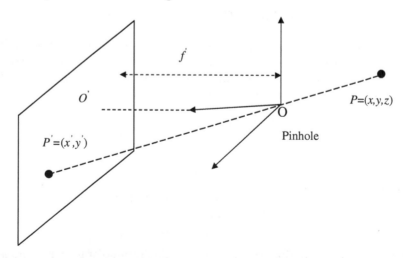

FIGURE 3.12: Perspective projection model of a pinhole camera

The simplest model is a pinhole camera, which is a mathematical approximation of the real-world camera. It can be thought of as a closed box having a single pinhole in the centre of one of the sides. Light passes through the pinhole and forms an inverted image inside the box on the opposite side. If the pinhole is infinitely small then this image will be focused. In reality cameras come equipped with lenses, and gather light from a much larger area. Although an image in a real camera would be inverted, it is common to consider a virtual image, which would be associated with an image plane lying in front of the pinhole. The image on this plane has the same orientation as the object.

Real cameras do not use a pinhole, as (i) it is sensible to collect more light and (ii) a very small pinhole may produce diffraction effects. In practice lenses are used to collect light from a larger area, and focus it on the sensor plane. It is important to note that the system may introduce several known aberrations unless corrective designs are used. Normally, the following aberrations and other difficulties are of concern:

1. *Spherical aberration*: The lens models which predict a perfect sharp image are only approximations of a real lens or lens system, and degrade the image by a blurring of every point.

2. *Chromatic aberration*: Different wavelengths of light are refracted by different amounts, so it is only possible to focus one wavelength on the imaging plane. Presence of other wavelengths will make the image blurred.

3. *Vignetting*: Peripheral parts of the scene may be partially occluded internally in the imaging system when multiple lenses of finite size are used. This results in images which are darker around the edges.

4. *Radial distortion*: In real camera systems, the linear perspective model is not strictly accurate. That is, the world point, image point, and the centre of camera are not exactly collinear. The most important deviation is a radial distortion, which is manifested by a barrelling effect of the image.

Another interesting aspect of the lens system in a camera is that the size of the object in the image increases as focal length increases. This is known as optical zoom. In practice, the effective focal length is changed by rearranging the optics, e.g., the distance between one or more lenses inside the optical system. However, the optical zoom is not related to digital zoom, which is done through the use of image precessing techniques and software. The observable area of a camera is denoted as the field-of-view of the camera. The field-of-view depends, besides the focal length, on the physical size of the image sensor. Often the sensor is rectangular rather than square and from this it follows that a camera has a field-of-view in both the horizontal and vertical directions.

The depth-of-field is influenced by the aperture, which is a flat circular object with a hole in the centre with adjustable radius. The aperture corresponds to the human iris, and controls the amount of incoming light entering the camera. In the extreme case, the aperture only allows rays through the optical centre, resulting in an infinite depth-of-field. The downside is that more light is blocked by the aperture. As a result lower shutter speed is required in order to ensure enough light to create an image. It follows that objects in motion can result in blurry images.

The process of image formation relating image plane and camera plane, in terms of geometry of a camera system, is shown in Figure 3.13. A 3D point $P^w = [x^w, y^w, z^w]^T$ in the world coordinate system is mapped into a camera coordinate system from the world coordinate system, and then mapped to the physical retina or camera film, i.e., the physical image plane, at the image coordinates $[u; v]^T$. For convenience, a normalised image plane located at the focal length $f = 1$, is considered.

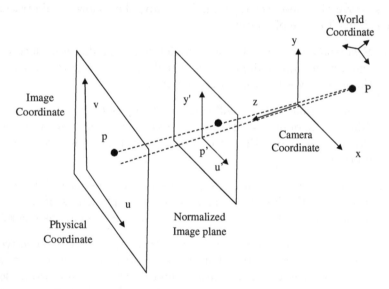

FIGURE 3.13: Coordinate system

In such a normalised image plane, the pinhole is mapped to the origin of the image plane $\tilde{c}$, and p is mapped to $\tilde{P} = [\tilde{u}; \tilde{v}]^T$, where

$$\tilde{P} = \begin{bmatrix} \tilde{u} \\ \tilde{v} \\ 1 \end{bmatrix} = \frac{1}{z_c} \begin{bmatrix} \mathbf{I} & 0 \end{bmatrix} \begin{bmatrix} x^c \\ y^c \\ z^c \\ 1 \end{bmatrix} \tag{3.13}$$

and

$$
\begin{bmatrix} u \\ v \\ 1 \end{bmatrix} = \frac{1}{z_c} \begin{bmatrix} kf & 0 & u_0 \\ 0 & lf & v_0 \\ 0 & 0 & 1 \end{bmatrix} \begin{bmatrix} x^c \\ y^c \\ z^c \end{bmatrix} = \frac{1}{z_c} \begin{bmatrix} kf & 0 & u_0 \\ 0 & lf & v_0 \\ 0 & 0 & 1 \end{bmatrix} \begin{bmatrix} \mathbf{I} & 0 \end{bmatrix} \begin{bmatrix} x^c \\ y^c \\ z^c \\ 1 \end{bmatrix} \quad (3.14)
$$

Converting the equation in terms of intrinsic camera imaging parameters, α, β, u_0, and v_0, where $\alpha = kf$ and $\beta = lf$, the following matrix equation is developed:

$$
\begin{bmatrix} u \\ v \\ 1 \end{bmatrix} = \frac{1}{z_c} \begin{bmatrix} \alpha & 0 & u_0 & 0 \\ 0 & \beta & v_0 & 0 \\ 0 & 0 & 1 & 0 \end{bmatrix} \begin{bmatrix} x^c \\ y^c \\ z^c \\ 1 \end{bmatrix} = \frac{1}{z_c} \begin{bmatrix} \alpha & 0 & u_0 & 0 \\ 0 & \beta & v_0 & 0 \\ 0 & 0 & 1 & 0 \end{bmatrix} \begin{bmatrix} \mathbf{R} & \mathbf{t} \\ 0^T & 1 \end{bmatrix} \begin{bmatrix} x^w \\ y^w \\ z^w \\ 1 \end{bmatrix}
$$
$$(3.15)$$

The parameters $\mathbf{R}$ and $\mathbf{t}$ are extrinsic parameters representing the coordinate transformation between the camera coordinate system and the world coordinate system. If M is the projection matrix, then,

$$
[uv1] = \frac{1}{z^c} M P^w \quad (3.16)
$$

It may be noted that the extrinsic camera parameters are those parameters that define the location and orientation of the camera reference frame with respect to a known world reference frame. Similarly, the intrinsic camera parameters are those parameters which are necessary to link the pixel coordinates of an image point with the corresponding coordinates in the camera reference frame. Typically, determining the parameters helps in finding the translation vector between the relative positions of the origins of the two reference frames, and also determines the rotation matrix that brings the corresponding axes of the two frames into alignment.

The process of estimating the intrinsic, extrinsic, and radial distortion parameters for a given camera is known as calibration. When these parameters are unknown, the camera is termed uncalibrated. Different algorithms can operate in calibrated or uncalibrated settings, but in general any algorithm that produces measurements of the parameters in meaningful units (e.g., mm) will require calibration. Calibration requires a real world object containing some distinctive visual features, such as a chequerboard of known dimensions.

Books for further reading

1. *Visual Prosthetics: Physiology, Bioengineering, Rehabilitation*: G. Dagnelie (ed.). Springer, New York, 2011.

2. *Adler's Physiology of the Eye*: P. L. Kaufman and A. Alm, St. Louis, Mosby, 2003.

3. *Introduction to the Optics of the Eye*: D. Goss and R. West, Butterworth-Heinemann, Boston, 2002.

4. *Wolff's Anatomy of the Eye and Orbit*: R. Tripathi, A. Bron, and B. Tripathi, Chapman and Hall, London, 1997.

5. *Optics of the Human Eye*: D. Atchison and G. Smith, Elseviermedical, London, 2000.

6. *Treatise on Physiological Optics*: H. Von Helmholtz, Dover, New York, 1999.

7. *Perceptual Organization and Visual Recognition*: D. Lowe, Kluwer Academic Publishers, Boston, 1985.

8. *Organization in Vision*: G. Kanizsa, H. Rinehart and C. Winston, 1979.

9. *Computer Vision: A Modern Approach*: D. A. Forsyth and J. Ponce, Pearson Education, New Delhi, 2015.

10. *Multiple View Geometry in Computer Vision*: R. Hartley and A. Zisserman, Cambridge University Press, Cambridge, 2000.

11. *Introductory Techniques for 3D Computer Vision*: E. Trucco and A. Verri, Prentice Hall, 1998.

12. *Multiple View Geometry*: R. I. Hartley and A. Zisserman, Cambridge University Press, Cambridge, UK, 2004.

Bibliography

[1] H. Kolb. How the retina works. *Am. Scientist*, 91:28–35, 2003.

[2] J. M. Alonso, W. M. Usrey, and R. C. Reid. Rules of connectivity between geniculate cells and simple cells in cat primary visual cortex. *J. Neurosc.*, 21(7):4001–15, 2002.

[3] A. Angelucci, J. B. Levitt, and E. J. Walton. Circuits for local and global signal integration in primary visual cortex. *J. Neurosc.*, 22(19):8633–46, 2002.

[4] A. Roorda and D. Williams. The arrangement of the three cone classes in the living human eye. *Nature*, 397:520–522, 1999.

[5] C. M. Cicerone and J. L. Nerger. The relative numbers of long-wavelength-sensitive to middle-wavelength-sensitive cones in the human fovea centralis. *Vis. Res.*, 29(1):115–128, 1989.

[6] J. Bullier. Integrated model of visual processing. *Brain Res. Rev.*, 36(2,3):96–107, 2001.

[7] J. Marroquin, S. Mitter, and T. Poggio. Probabilistic solution of ill-posed problems in computational vision. *J. Am. Stat. Asso.*, 82(397):76–89, 1987.

[8] T. S. Lee. Computations in the early visual cortex. *J. Physio.*, 97:121–139, 2003.

[9] H. Hofer, J. Carroll, J. Neitz, M. Neitz, and D. R. Williams. Organization of the human trichromatic cone mosaic. *J. Neurosc.*, 25(42):9669–9679, 2005.

[10] K. McLaren. The development of the CIE 1976 uniform colour-space and colour-difference formula. *J. Soc. of Dyers and Colourists*, 92:338–341, 1976.

[11] I. Biederman. Recognition-by-components: A theory of human image understanding. *Psycho. Rev.*, 94(2):115–147, 1987.

[12] J. Feldman. How does the cerebral cortex work? development, learning, and attention. *Trends in Cognitive Sc.*, 7(6):252–256, 2003.

[13] D. M. McKay. Psychophysics of perceived intensity: A theoretical basis for Fechner's and Stevens' laws. *Science*, 139:1211–1216, 1963.

[14] S. C. Masin, V. Zudini, and M. Antonelli. Early alternative derivations of Fechner's law. *J.of Hist. of the Behavioral Sc.*, 45:56–65, 2009.

[15] S. S. Stevens. The relation of saturation to the size of the retinal image. *Am. J. of Psycho.*, 46:70–79, 1934.

[16] On the psychophysical law. *The Psychological Review*, 64(3):153–181, 1957.

[17] G. A. V. Borg and L. E. Marks. Twelve meanings of the measure constant in psychophysical power functions. *Bull. Psychonomic Soc.*, 21:73–75, 1983.

[18] G. Lindzey (ed.). *A history of psychology in autobiography*. Prentice Hall, New Jersy, 1974.

[19] A. Desolneux, L. Moisan, and J. M. Morel. Meaningful alignments. *Int. J. Comp. Vis.*, 40(1):7–23, 2000.

[20] A. Desolneux, L. Moisan, and J. M. Morel. Computational gestalts and perception thresholds. *J. Physio.*, 97)2-3):311–324, 2003.

[21] S. Ullman. The visual recognition of three-dimensional objects. In *Attention and performance*, MIT Press, Cambridge, MA, 1993.

[22] R. Fergus, P. Perona, and A. Zisserman. Weakly supervised scale-invariant learning of models for visual recognition. *Int. J. of Comp. Vis.*, 71(3):273–303, 2007.

[23] P. F. Felzenszwalb and D. P. Huttenlocher. Pictorial structures for object recognition. *Int. J. Comp. Vis.*, 51(1):55–59, 2005.

[24] M. Brown and D. Lowe. Automatic panoramic image stitching using invariant features. *Int. J. Comp. Vis.*, 74(1):59–67, 2007.

Chapter 4

Photonic Sources and Detectors for Information Processing

4.1 Introduction

Systems used in information processing at optical frequencies require a wide variety of photonic devices for generation and detection of photonic signals, for switching, modulation, and storing of information. Photonic devices are thus components for creating, manipulating, and detecting light. Most devices are of low power and therefore attempts are made to integrate them into photonic integrated circuits. Almost all photonic devices are manufactured and have evolved with semiconductor materials, therefore, it is necessary to introduce elements of semiconductor physics as the staring platform.

4.2 Elements of semiconductor physics

In a semiconductor, the allowed states of the electrons of its constituent atoms form continuous energy bands rather than discrete levels. The photonic processes associated with such electrons are a strong function of the characteristics of the energy bands. For a solid material in thermal equilibrium at a temperature T, the probability of any electronic state at an energy E being occupied by an electron is given by the Fermi-Dirac distribution function, as

$$f(E) = \frac{1}{1 + e^{(E - E_F)/k_B T}} \qquad (4.1)$$

where E_F is the Fermi level of the material and k_B is the Boltzmann constant.

At temperature $0^o K$, all states below the Fermi level are occupied by electrons, while those above it are empty. If the Fermi level lies between two separate energy bands, all energy bands are either completely filled or completely empty at $T \to 0^o K$. The filled bands are known as the valence bands, whereas the empty bands are known as the conduction bands. The lowest conduction band and the highest valence band are separated by an energy gap known as the bandgap, E_g. The minimum of the conduction bands is called the conduction-band edge, E_c, and the maximum of the valence bands is called the valence-band edge, E_v. The bandgap therefore is $E_g = E_c - E_v$. In semiconductors, the band gap is, in general, less than 4.0 eV and decreases with temperature. For example, in Ge, the energy gap is 1.12 eV and in Si it is 0.72 eV.

A material is an insulator if all bands are either completely filled or completely empty, as electrons in a completely filled energy band cannot move under an electric field. A solid with one or more partially filled bands is a metal because electrons in a partially filled band can move under an electric field to conduct electric current. At a nonzero temperature, however, electrons in higher valence bands have a probability of being thermally excited to lower conduction bands. The probability of this thermal excitation is a function of $\frac{E_g}{k_B T}$. The electrons that are excited to conduction bands become carriers of negative charges. The removal of electrons from valence bands results in electron deficiencies, known as holes, in the form of unoccupied electron states. Holes in a valence band behave like carriers of positive charges. Both contribute to the electrical conductivity of a semiconductor.

The energy of an electron is a function of its quantum-mechanical wave vector, $\mathbf{k}$, and depends on electron energy which forms the band structure of the semiconductor. When the minimum of conduction band and the maximum of valence band do not occur at the same $\mathbf{k}$ value, the semiconductor is called an indirect bandgap semiconductor. In contrast, a semiconductor like GaAs is a direct bandgap semiconductor because the minimum of the conduction bands and the maximum of the valence bands of this semiconductor occur at the same value of $\mathbf{k}$. The free-space optical wavelength λ_g of a photon that has an energy equal to the bandgap of a given material is given by

$$\lambda_g = \frac{hc}{E_g} \tag{4.2}$$

The electron concentration in a semiconductor is the number of conduction electrons in the conduction bands per unit volume of the semiconductor, and the hole concentration is the number of holes in the valence bands per unit volume of the semiconductor. The number of electrons in a particular energy band is determined by both the number of available states in that band and the probability of occupancy for each state. In a bulk semiconductor, the density of states is the number of states per unit material volume. The density of electron states within the energy range between E and $E + dE$ for $E \geq E_c$ near the conduction band edge is given by $\rho_c(E)dE$ and is:

$$\rho_c(E)dE = A_c(E - E_c)^{1/2}dE \tag{4.3}$$

Similarly, the density of states for the same energy difference for $E \leq E_v$, near the valence-band edge, is given by $\rho_v(E)dE$ as

$$\rho_v(E)dE = A_v(E_v - E)^{1/2}dE \tag{4.4}$$

The constants A_c and A_v are related to the effective mass of electron and hole.

The concentration of electrons in the conduction band (negatively charged carriers) whose energies fall between E and $(E + dE)$ is given by,

$$n_0(E)dE = f(E)\rho_c(E)dE \tag{4.5}$$

The probability of finding a hole at an energy E is $[1 - f(E)]$ as a hole is an unoccupied electron state. A similar expression for hole concentration between the same energy difference in the valence band is

$$p_0(E)dE = [1 - f(E)]\rho_v(E)dE \tag{4.6}$$

The total concentrations of electrons and holes in thermal equilibrium are given by

$$n_0 = \int_{E_c}^{\infty} f(E)\rho_c(E)dE \tag{4.7}$$

and

$$p_0 = \int_{-\infty}^{E_v} [1 - f(E)]\rho_v(E)dE \tag{4.8}$$

Approximate relations for carrier concentrations in terms of the effective densities of states for conduction band N_c and that of the valence band is given by,

$$n_0 = N_c(T)e^{-(E_c - E_F/k_B T)}$$
$$p_0 = N_v(T)e^{-(E_F - E_v/k_B T)} \tag{4.9}$$

where N_c and N_v are the effective densities of states for conduction and valence bands, respectively.

In an intrinsic semiconductor, impurities contribute negligibly to the electron and hole concentrations. Consequently, there are as many holes as electrons so that $n_0 = p_0 = n_i$, and the Fermi level lies very close to the middle of the bandgap. In an extrinsic or doped semiconductor, however, n_0 and p_0 are different, because of the contribution of carriers from the impurities in a semiconductor. An impurity atom that contributes a conduction electron is a donor, and one that can contribute a hole to the valence bands is an acceptor. The requirement for charge neutrality in a semiconductor leads to the following relation,

$$n_0 + N_a^- = p_0 + N_d^+ \tag{4.10}$$

where N_a^- is the concentration of the immobile negatively ionized acceptors and N_d^+ is that of the immobile positively ionized donors.

When $N_d^+ > N_a^-$, the semiconductor is an n-type semiconductor with $n_0 > p_0$. In an n-type semiconductor, electrons are the majority carriers, and holes are the minority carriers. When $N_a^- > N_d^+$, the semiconductor is a p-type semiconductor with $p_0 > n_0$. In a p-type semiconductor, holes are the majority carriers, and electrons are the minority carriers. In an n-type semiconductor, the Fermi level moves toward the conduction band edge and in a p-type semiconductor, it moves toward the valence band edge. In a heavily p-doped semiconductor, the Fermi level can move into the valence band, and similarly, the Fermi level can move into the conduction band in a heavily n-doped semiconductor. Such a heavily doped semiconductor is called a degenerate semiconductor.

An electron in a conduction band and a hole in a valence band can experience a recombination process. When a semiconductor is in thermal equilibrium with its environment, recombination of the carriers is exactly balanced by thermal generation of the carriers so that the electron and hole concentrations are maintained at their respective equilibrium values of n_0 and p_0. When the electron and hole concentrations in a semiconductor are higher than their respective equilibrium concentrations, due to current injection or optical excitation, the excess carriers will return toward their respective thermal equilibrium concentrations through recombination processes. If the external excitation persists, the semiconductor can reach a quasi-equilibrium state in which electrons and holes are not characterised by a common Fermi level but are characterised by two separate quasi-Fermi levels E_{Fc} and E_{Fv} for both conduction and valence bands. The probability of occupancy in the conduction bands and the valence bands are then described by two separate Fermi-Dirac distribution functions. In such quasi-equilibrium, the electron and hole concentrations as a function of energy are

$$n = N_c(T)e^{-(E_c - E_{Fc}/k_B T)}$$

$$p = N_v(T)e^{-(E_{Fv} - E_v/k_B T)} \tag{4.11}$$

As a consequence, the product of n and p is given by,

$$np = n_0 p_0 e^{(E_{Fc} - E_{pv})/k_B T} \tag{4.12}$$

The relaxation time constant for excess electrons or holes, known as the electron lifetime and hole lifetime, respectively, are given by

$$\tau_e = \frac{n - n_0}{R}$$

and

$$\tau_h = \frac{p - p_0}{R} \tag{4.13}$$

where R is the recombination rate.

The lifetime of the minority carriers in a semiconductor is called the minority carrier lifetime, and that of the majority carriers is called the majority carrier lifetime. A sufficient condition for electrons and holes to have the same lifetime is that the electron and hole concentrations are both very large compared to the density of the recombination centres.

4.2.1 Current density

The current density flowing in a semiconductor is the current flowing through a cross-sectional area of semiconductor in amperes per square meter. There are two mechanisms that can cause the flow of electrons and holes: drift, in the presence of an electric field, and diffusion, in the presence of a spatial gradient in the carrier concentration. The electron current density, $\mathbf{J}_e$, and the hole current density, $\mathbf{J}_h$, can be expressed in terms of excess electron and hole concentrations $\nabla n = (n - n_0)$ and $\nabla p = (p - p_0)$ as

$$\mathbf{J}_e = e\mu_e n \mathbf{E}_e + eD_e \nabla n$$

$$\mathbf{J}_h = e\mu_h n \mathbf{E}_h + eD_h \nabla p \tag{4.14}$$

where e is the electronic charge, μ_e and μ_h are the electron and hole mobilities respectively, E_e and E_h are the electric fields seen by electrons and holes, and D_e, and D_h are the diffusion coefficients of electrons and holes.

The electron and hole mobilities strongly depend on temperature, the type of semiconductor, and the impurities and defects in the semiconductor. They generally decrease with increasing concentration of impurities and defects. For non-degenerate semiconductors, the diffusion coefficients are related to the mobilities by the Einstein relations given as

$$D_e = \frac{k_B T}{e} \mu_e$$

$$D_h = \frac{k_B T}{e} \mu_h \tag{4.15}$$

The electric fields seen by electrons and holes can be expressed in terms of the gradients ∇E_c and ∇E_v in the conduction and valence band edges. In a homogeneous semiconductor where the conduction- and valence-band edges are parallel to each other, the gradients are generally the same.

The drift components $\mathbf{J}_e^{dri}$ and $\mathbf{J}_h^{dri}$ of the electron and hole current densities respectively can be expressed as

$$\mathbf{J}_e^{dri} = \mu_e n \nabla E_c$$

$$\mathbf{J}_h^{dri} = e\mu_h p \nabla E_v \tag{4.16}$$

Similarly, the diffusion components J_e^{dif} and J_h^{dif} of the electron and hole current densities can be expressed in terms of quasi-Fermi levels E_{Fc} and E_{Fv} as

$$J_e^{dif} = \mu_e n \nabla E_{Fc} - \mu_e n \nabla E_c$$

$$J_h^{dif} = \mu_h p \nabla E_{Fv} - \mu_h p \nabla E_v \tag{4.17}$$

By combining the drift and diffusion components, the total current density $\mathbf{J}$ can therefore be expressed as

$$\mathbf{J} = \mathbf{J}_e + \mathbf{J}_h = \mu_e n \nabla E_{Fc} + \mu_h p \nabla E_{Fv} \tag{4.18}$$

A semiconductor in thermal equilibrium carries no net electric current and thus $\mathbf{J} = 0$. On the other hand, when a semiconductor carries an electric current, it is in a quasi-equilibrium state with separate quasi-Fermi levels $E_{Fc} \neq E_{Fv}$.

The electric conductivity σ of a material is the proportional constant between the current density and the electric field. In a semiconductor, the conductivity is contributed by both electrons and holes. Only the drift current contributes to the conductivity because the diffusion current is not generated by an electric field. The current $\mathbf{J}$ is given by

$$\mathbf{J} = e(\mu_e n + \mu_h p)\mathbf{E} = \sigma \mathbf{E} \qquad (4.19)$$

where $\sigma = e(\mu_e n + \mu_h p)$ and $\mathbf{E}$ is the electric field.

Evidently, for intrinsic semiconductors, $\sigma_i = e(\mu_e + \mu_h)n_i$. Because $\mu_e > \mu_h$ for most semiconductors, an n-type semiconductor has a higher conductivity than a p-type one of the same impurity concentration. The conductivity of a given semiconductor is a strong function of temperature as the carrier concentrations and carrier mobilities are both sensitive to temperature.

4.2.2 Semiconductor materials for photonic devices

The elemental semiconductor materials are the group IV elements Si and Ge. Though C is not a semiconductor, Si and C can form IV-IV compound semiconductor (SiC), which has many different structural forms of different bandgaps. By moving down a particular column of the periodic table, the bandgap decreases, as the atomic weight of an element in a semiconductor increases. Further, the refractive index at a given optical wavelength corresponding to a photon energy below the bandgap increases while moving down the column in the periodic table.

The most useful semiconductor materials for photonic devices are the III-V compound semiconductors. These compound materials are formed by combining group III elements (such as Al, Ga, and In), with group V elements (such as N, P, As, and Sb). Si and Ge can be mixed to form the IV-IV alloy semiconductor $Si_x Ge_{1-x}$. These group IV crystals and IV-IV compounds are indirect-gap materials. There are more than ten binary III-V semiconductors, such as GaAs, InP, AlAs, and InSb. Different binary III-V compounds can be alloyed with varying compositions to form mixed crystals of ternary compound alloys and quaternary compound alloys.

A ternary III-V compound consists of three elements, two groups of III elements and one group of V elements, such as $Al_x Ga_{1-x}$ As, or one group III element and two group V elements, such as $GaAs_{1-x} P_x$. A quaternary III-V compound consists of two group III elements and two group V elements, such as $In_{1-x} Ga_x As_{1-y} P_y$. A III-V compound can be either a direct bandgap or an indirect bandgap material. A III-V compound can, however, be a direct bandgap material or indirect bandgap material depending on bandgap.

Among the III-V compounds, the nitrides are very useful as materials for photonic devices. These compounds arre frequently used in fabricating in lasers, light-emitting diodes, and semiconductor photodetectors in the blue, violet, and ultraviolet spectral regions. The nitride compounds and their alloys cover almost the entire visible spectrum and extend to the ultraviolet region. The binary nitride semiconductors AlN, GaN, and InN, and their ternary alloys such as InGaN, are all direct bandgap semiconductors.

Zn, Cd, and Hg are group II elements, which can also be combined with group VI elements, such as S, Se, and Te, to form binary II-VI semiconductors. Among such compounds, the Zn and Cd compounds, ZnS, ZnSe, ZnTe, CdS, CdSe, and CdTe, are direct bandgap semiconductors. The II-VI compounds can be further mixed to form mixed II-VI compound alloys, such as the ternary alloys $Hg_x Cd_{1-x}Te$, and $Hg_x Cd_{1-x}Se$. The photonic devices made with these materials can have a wide range of bandgaps covering the visible to the mid-infrared spectral regions.

Lattice-matched compounds have also found applications in evolving photonic devices. By definition, if crystals have the same lattice structure, and the same lattice constant, they are called lattice-matched. The bandgap of such ternary or quaternary compounds varies with the composition of the compound when the lattice constant of the compound is, for the purpose of lattice matching, kept at a fixed value as the composition varies. As the bandgap varies, the refractive index of the compound also varies in opposite directions. The quaternary compound alloys such as $Al_x Ga_{1-x}As$ and $In_{1-x}Ga_xAs_yP_{1-y}$ normally have a large flexibility for lattice matching by varying x and y in the ranges $0 \leq x \leq 1$ and $0 \leq y \leq 1$. Moreover, in the range $0 \leq x \leq 0.45$, $Al_x Ga_{1-x}As$ is closely lattice matched with GaAs and behaves as a direct bandgap semiconductor, however for larger values of x, it becomes an indirect bandgap semiconductor.

4.2.3 Semiconductor junction

Inhomogeneous semiconductors that have either spatially nonuniform doping distribution, or spatially varying bandgaps, or both, have found wide applications in photonic devices.

A semiconductor junction can be either an abrupt junction or graded junction. An abrupt junction has a sudden change of doping and/or bandgap from one region to the other, and in a graded junction, the change of doping and/or bandgap is gradual. There are two categories of semiconductor junctions: homojunctions and heterojunctions. A homojunction is formed by different doping in the same semiconductor, whereas a heterojunction is formed between two different semiconductors. A p-n homojunction is formed between a p-type region and an n-type region with different doping in the same semiconductor. A homojunction can also be formed as a p-i or i-n junction between between a p-type region and an undoped intrinsic region of the same semiconductor, or an undoped intrinsic region and an n-type region of the same semiconductor.

In addition, a metal-semiconductor junction can also be formed between a metal and a semiconductor.

In the case of homojunction structure, bandgap at thermal equilibrium, remains constant and therefore, at any given location the conduction and valence band edges have the same gradient. This spatially varying band-edge gradient creates a spatially varying built-in electric field that is seen by both electrons and holes as $E_e = E_h$ at any given location. The built-in electric field results in a built-in electrostatic potential of height V_0 across the p-n junction and this potential is called the contact potential of the junction. The contact potential is higher on the n side than on the p side. Energy band diagram showing bandgaps in a typical abrupt p-n junction is shown in Figure 4.1.

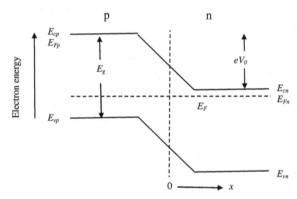

FIGURE 4.1: Energy band diagram of an abrupt p-n junction

Denoting E_{cp} and E_{cn} as the conduction-band edges in the homogeneous p and n regions, and E_{vp} and E_{vn} as the valence-band edges in the homogeneous p and n regions, the following equation can be written:

$$E_{cp} - E_{cn} = E_{vp} - E_{vn} = eV_0 \tag{4.20}$$

where eV_0 is the energy barrier for an electron on the n side to move to the p side and vice versa for a hole on the p side to move to the n side. The relation for carrier concentrations can be written as

$$\frac{p_{p0}}{p_{n0}} = \frac{n_{n0}}{n_{p0}} = e^{eV_0/k_BT} \tag{4.21}$$

Since $p_{p0}n_{p0} = p_{n0}n_{n0} = n_i^2$ according to the law of mass action, and under the conditions, $p_{p0} \approx N_a \gg n_{p0}$ and $n_{n0} \approx N_d \gg p_{n0}$, the contact potential is given by

$$V_0 = \frac{k_BT}{e} \ln \frac{N_aN_d}{n_i^2} \tag{4.22}$$

4.2.3.1 Homojunction under bias voltage

In the case of a p-n homojunction, a bias voltage V changes the electrostatic potential between the p and n regions. A junction is under forward bias if $V > 0$. A forward bias voltage raises the potential on the p side with respect to that on the n side, resulting in a lower potential barrier of $(V_0 - V)$. As a result the potential barrier between the p and n regions is reduced by the amount of eV. Similarly, a junction under reverse bias, $V < 0$, lowers the potential on the p side with respect to that on the n side, thus raising the potential barrier to $(V_0 - V) = V_0 + |V|$, which results in an increase in the energy barrier between the homogeneous p and n regions by the amount of $e|V|$. Energy band diagrams and potentials for an abrupt p-n homojunction is shown in Figure 4.2.

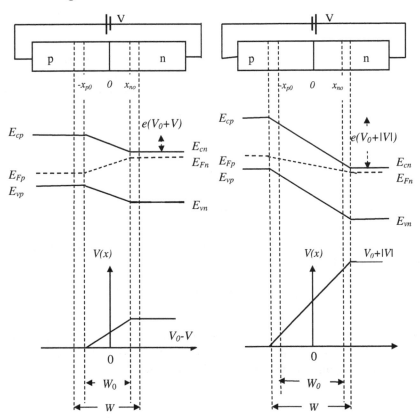

FIGURE 4.2: Energy band and potentials in (a) forward biased and (b) reverse biased abrupt p-n homojunction

A bias voltage changes the difference between E_{cp} and E_{cn} and that between E_{vp} and E_{vn} in p and n regions. A bias voltage causes an electric current to flow in a semiconductor by splitting the Fermi level into separate quasi-Fermi levels, E_{Fc} and E_{Fv} for electrons and holes, and creates spatial gradients in them to support an electric current in the semiconductor. When an electric current flows in a semiconductor, the gradients in E_{Fc} and E_{Fv} existing throughout the semiconductor vary from one location to another, in conjunction with the variations in the local concentrations of electrons and holes. The largest splitting of E_{Fc}, and E_{Fv} occurs in the depletion layer, and the largest gradients in E_{Fc} and E_{Fv} exist in the diffusion regions just outside the depletion layer. The quasi-Fermi levels, however gradually merge into E_{Fp} in the homogeneous p region and into E_{Fn} in the homogeneous n region. Therefore, the difference in Fermi levels in p and n sides of the junction can be written as $E_{Fn} - E_{Fp} = eV$. In the case of forward bias $E_{Fn} > E_{Fp}$, and $E_{Fc} > E_{Fv}$. In the case of reverse bias $E_{Fn} < E_{Fp}$, and $E_{Fc} > E_{Fv}$.

The depletion layer or the space-charge region is created by the diffusion of holes from the p side, (where the hole concentration is high), to the n side, (where the hole concentration is low). Diffusion of electrons from the n side, where the electron concentration is high, also occurs to the p side, where the electron concentration is low. The depletion layer has a width $W = x_p + x_n$ where x_p and x_n are the penetration depths of the depletion layer into the p and n regions. x_p and x_n are related to the donor and acceptor densities as

$$x_p = W\frac{N_d}{N_a + N_d}$$

$$x_n = W\frac{N_a}{N_a + N_d} \tag{4.23}$$

When a junction is in thermal equilibrium without bias, $W = W_0$. For $V > 0$, $W < W_0$, and for $V > 0$, $W > W_0$. Therefore the depletion layer narrows with forward bias, and broadens with reverse bias.

The width of the depletion layer can also be expressed as a function of the applied bias voltage V, and permittivity ϵ as

$$W = \left[\frac{2}{\epsilon}e\left(\frac{N_a + N_d}{N_a N_d}\right)(V_0 - V)\right]^{1/2} \tag{4.24}$$

The depletion layer acts as a capacitor by holding negative space charges on the p side and positive space charges on the n side. For a cross-sectional area A_c of the junction the charge Q is given by

$$Q = eN_a x_p A_c = eN_d x_n A_c = eWA_c\frac{N_a N_d}{N_a + N_d} \tag{4.25}$$

Therefore the junction capacitance C_j associated with the depletion layer is given by

$$C_j = |\frac{dQ}{dV}| = \frac{\epsilon A_c}{W} \tag{4.26}$$

A bias voltage can cause substantial changes in the minority carrier concentrations at $x = -x_p$ and $x = x_n$, where the edges of the depletion layer are located. The changes in the minority carrier concentrations from their equilibrium values are given as

$$\triangle n_p = n_p|_{-x_p} - n_{p0} = n_{p0}(e^{eV/k_BT} - 1)$$

$$\triangle p_n = p_n|_{x_n} - p_{n0} = p_{n0}(e^{eV/k_BT} - 1)$$

The diffusion of minority carriers caused by a bias voltage is, however, not localized at the edges of the depletion layer. Instead the minority carrier concentrations have the spatially dependent variations across the diffusion region given as

$$n_p(x) - n_{p0} = n_{p0}(e^{eV/k_BT} - 1)e^{(x+x_p)/L_e} \text{ for } x < -x_p$$

$$p_n(x) - p_{n0} = p_{n0}(e^{eV/k_BT} - 1)e^{-(x-x_n)/L_h} \text{ for } x > -x_n \qquad (4.27)$$

where $L_e = \sqrt{D_e \tau_e}$ is electron diffusion length in the p region, and $L_h = \sqrt{D_h \tau_h}$ is the hole diffusion length in the n region and τ is the lifetime of respective the minority carriers.

Due to charge neutrality in the diffusion regions the concentration of majority carriers also varies in space, and is given by

$$p_p(x) - p_{p0} = n_p(x) - n_{p0} \text{ for } x < -x_p$$

$$n_n(x) - n_{n0} = p_n(x) - p_{n0} \text{ for } x > -x_n \qquad (4.28)$$

4.2.3.2   Current flow in homojunction under bias

The electric current flowing in a semiconductor under bias consists of an electron current and a hole current, each having both drift and diffusion components. The total current is constant throughout the semiconductor under a constant bias voltage. In the depletion layer, currents are due to drift and diffusion for both electrons and holes. In this layer a large electric field exists, and the carrier concentration gradients for both electrons, and holes are large. There are negligible generation and recombination of carriers in the depletion layer, as the large electric field in the depletion layer swiftly sweeps the carriers across this layer. Total electron current density J_e, and total hole current density, $J_h(x)$, are constant across the depletion layer. Assuming $x = 0$ at the physical junction (see Figure 4.2), the minority carrier currents in the depletion regions are purely diffusive, and can be expressed as

$$J_e(-x_p) = \frac{eD_e}{L_e} n_{p0}(e^{eV/k_BT} - 1)$$

$$J_h(x_n) = \frac{eD_h}{L_h} p_{n0}(e^{eV/k_BT} - 1) \qquad (4.29)$$

Consequently the total current density J which varies with the bias voltage V is given as

$$J = J_e(-x_p) + J_h(x_n) = J_s(e^{eV/k_BT} - 1) \qquad (4.30)$$

where J_s is the saturation current density given by

$$J_s = \frac{eD_e}{L_e} n_{p0} + \frac{eD_h}{L_h} p_{n0} \qquad (4.31)$$

Because of higher mobility of electrons than holes $D_e > D_h$. Unless the p side is much more heavily doped than the n side, the injection current through a homojunction is predominantly carried by the electrons injected from the n side into the p side.

A junction that has cross-sectional area A_c, and biased with voltage V, the total current is given by

$$I = I_s(e^{eV/k_BT} - 1) \tag{4.32}$$

where I_s is the saturation current. The voltage-current characteristics is shown in Figure 4.3.

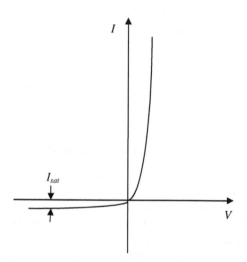

FIGURE 4.3: V-I characteristics of a pn junction

4.2.3.3   Semiconductor heterojunction

A heterojunction is normally formed between two lattice-matched semiconductors of different bandgaps. Nomenclature followed naming a heterojunction by a lowercase letter, n, p, or i, for small bandgap, and a large-gap semiconductor by an uppercase letter, N, P, or I. Heterojunction can be isotype or anisotype depending on junction (p-P or n-N) formed between two dissimilar semiconductors of same conductivity or dissimilar semiconductors of different conductivity (p-N and P-n). The differences in p-n homojunctions and p-N and P-n heterojunctions, however, are their energy band structures.

For a p-N heterojunction, the semiconductor on the n side has larger bandgap than that on the p side, that is, $E_{gn} > E_{gp}$. The energy barrier for an electron on the n side is lowered from eV_0 by ΔE_c due to the conduction-band offset, but that for a hole on the p side is raised by ΔE_v due to the valence-band offset. Therefore

$$E_{cp} - E_{cn} = eV_0 - \Delta E_c$$

$$E_{vp} - E_{vn} = eV_0 + \Delta E_v \tag{4.33}$$

The contact potential V_0 for a non-degenerate p-N junction can therefore also be modified. A similar case occurs in the case of P-n heterojunction. The relations that describe the carrier distributions are valid for p-N and P-n heterojunctions as well as for p-n homojunctions. In the case of a heterojunction, the exponential dependence on bandgap difference can be significant if $\Delta E_g > k_B T$. This dependence dominates in most practical heterojunctions. Consequently, $J_e >> J_h$ for a p-N junction where $E_{gn} > E_{gp}$. For a P-n junction where $E_{gn} < E_{gp}$, $J_e << J_h$. Therefore, in the case of a p-N junction the diffusion current is mainly contributed by the injection of electrons from the wide-gap n type semiconductor to the narrow-gap p type semiconductor, whereas in the case of a P-n junction it is mainly contributed by the injection of holes from the widegap p type semiconductor to the narrow-gap n type semiconductor.

4.2.4 Quantum well structure

A quantum well is a very thin (about 50 nm) double heterostructure (DH). Quantum wells are formed in semiconductors by having a material like GaAs sandwiched between two layers of a material with a wider bandgap, like AlAs. Another example is a layer of InGaN sandwiched between two layers of GaN. These structures can be grown by molecular beam epitaxy or chemical vapour deposition with control of the layer thickness down to monolayers. As the active layer of a semiconductor DH gets thinner, the effect of quantum confinement for the electrons and holes in the thin active layer starts to appear in the direction perpendicular to the junction plane.

A semiconductor quantum well is not an infinite potential well because the heights of the energy steps at the DH junctions are finite. The confinement leads to quantisation of momentum in the perpendicular direction, resulting in discrete energy levels associated with the motion of electrons and holes in this direction. In the horizontal dimensions, however, the electrons and holes remain free and form energy bands. Since both conduction and valence bands are split into a number of subbands corresponding to the quantised levels, the effective bandgap for a quantum well is no longer E_g of the semiconductor material in the active layer. As an effect, the energy band is the separation between the lowest subband of the conduction band and the highest subband of the valence band. They are both associated with the so-called $q = 1$ quantised levels.

An important property of the quantum well is that its density of states is different from that of the bulk semiconductor. Because of quantisation in the perpendicular direction, the density of states for each subband is that of a

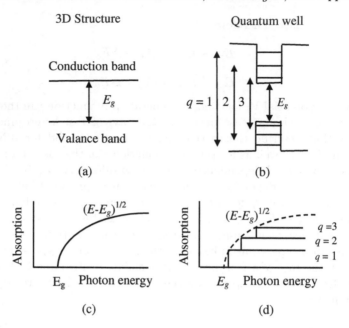

FIGURE 4.4: Relative differences in bandgaps in 3D and quantum well structure are shown in (a) and (b), quantisations of absorption bands are shown in (c) and (d)

two-dimensional system divided by the thickness of the well. Every subband in the conduction band has the same density of states, and every subband in the valence band also has the same density of states as shown in Figure 4.4.

In quantum well, solution of Schrodinger's equation arrives at the formation of series of discrete energy levels instead of the continuous energy bands of the bulk material. With the approximation that the well is infinitely deep and the allowed energy levels are given by,

$$E_n = \frac{(n\pi\hbar)^2}{2m_c L_x^2} \tag{4.34}$$

where $n = 1, 2, 3...$, and m_c is the effective mass of the particle in the well, and L_x is the quantum well thickness.

Setting the energy at the top of the valence band equal to zero, the allowed energies for an electron in the conduction band of a semiconductor QW become $E = E_g + E_{mcc}$ where E_{mc} is the energy for effective mass m_c. The allowed energies for a hole in the valence band are then $E = -E_{mcv}$. The allowed transition energies are then limited to

$$E = E_g + E_{mcc} + E_{mcv} + \frac{\hbar^2 k^2}{2m_c} \tag{4.35}$$

where k is the wave vector.

The selection for optical transitions of electrons between quantised conduction subbands and quantised valence subbands requires that only transitions between a conduction subband and a valence subband of the same quantum number q are allowed, that is, optical transitions take place only between the conduction subband and the valence subband of the same quantum numbers. The lowest photon energy required for a transition between the conduction band and the valence band of a quantum well is that with $q = 1$. In general, the photon energy required for transition between the conduction subband of quantum number q and the valence subband of quantum number q is

$$h\nu > E_g + \frac{q^2 h^2}{8m^* d_{qw}^2} \qquad (4.36)$$

where m^* is the reduced effective mass. Therefore, a quantum well has an effective bandgap which is larger than that of the bulk semiconductor by an amount that is inversely proportional to the square of the well width.

Quantum wells have several advantages over bulk semiconductor media. The injected carriers are more concentrated in a quantised subband of a quantum well than in the entire band of a bulk semiconductor. Because the density of states for each subband of a quantum well is a constant that does not vary with energy, there are already a large number of electrons of the same energy near the edge of a conduction subband and a large number of holes of the same energy near the edge of a valence subband. Therefore, a quantum well has a much larger gain cross section than a bulk semiconductor. The gain bandwidth of a typical quantum well is in the range of 20-40 THz, which is almost twice that of a typical bulk semiconductor.

4.3   Photonic sources

Various sources of photonic signal useful for information processing are named either after their structures or by the processes of light emission [1]. Lasers are the most prominent member of photonic sources and make use of processes that either produce or amplify light signals. The other types of photonic sources are light-emitting diodes and injection laser diodes. These devices are based on the phenomenon of injection luminescence in semiconductors.

Laser is an acronym for light amplification by stimulated emission of radiation and is a source that emits light of various wavelengths. The term laser generally refers to a laser oscillator, which generates laser light without an input light wave. A device that amplifies a laser beam by stimulated emission is the other type of laser, generally called a laser amplifier.

Lasers differ from other sources of light because of their coherence property, i.e., they maintain a high degree of both spatial as well as temporal coherence. Spatial coherence allows a laser to be focused to a tight spot, whereas temporal coherence results in a very narrow line spectrum. The spatial coherence and directional property allows the laser beam to stay narrow over long distances. Apart from the property of coherence, lasers are also highly monochromatic, intense, and directional sources.

Nearly all lasers are produced as a result of electron transitions from an excited energy level, within a radiating species to a lower energy level and, in the process, radiate light that contributes to the laser beam. The radiating materials can be

(a) Atoms such as in the helium-neon (HeNe) laser, argon ion, helium-cadmium (HeCd) lasers, and copper vapour lasers (CVL).

(b) Molecules such as those in the carbon dioxide (CO_2) laser, the excimer lasers, such as ArF and KrF, and the pulsed nitrogen laser.

(c) Liquids such as those involving various organic dye molecules dilutely dissolved in various solvent solutions.

(d) Dielectric solids such as those involving neodymium atoms doped in YAG or glass to make the crystalline Nd:YAG or Nd:glass lasers.

There are a wide variety of lasers, covering a spectral range from the soft X-ray to the far infrared, delivering output powers from microwatts to terawatts, operating from continuous wave to femtosecond pulses, and having spectral linewidths from just a few hertz to many terahertz. The sizes range from microscopic, on the order of $10 \mu m^3$, to huge dimensions, to stellar, of astronomical dimensions.

In its simplest form, a laser consists of a gain or amplifying medium. The gain media utilised include plasma, free electrons, ions, atoms, molecules, gases, liquids, solids, and so on. In a practical laser device, however, it is generally necessary to have certain positive optical feedback in addition to optical amplification provided by a gain medium. This requirement can be met by placing the gain medium in an optical resonator. The optical resonator provides selective feedback to the amplified optical field.

On the other hand, semiconductor light sources such as light-emitting diodes and injection laser diodes are based on electroluminescence, which results from the radiative recombination of electrons and holes in a semiconductor. Semiconductor materials such as gallium arsenide or indium phosphide crystals, or various mixtures of impurities blended with those and other semiconductor material are used in evolving these lasers. These sources are rugged and reliable and have long operating lifetimes, because of their small integrated structures. They have also very high efficiencies and consume very little power in comparison with other light sources of similar brightness. Usually these sources are called cold light sources, operating at temperatures that are much lower than the equilibrium temperatures of their emission spectra. The sources are either monochromatic or wide-band, depending on the spectral distribution. The main characteristics of such sources are their wavelength, the spectral line-width (or band-width), and the optical output power.

4.3.1   Laser amplifiers

Atoms and molecules can exist only in discrete energy states, and except for the lowest energy state (ground state), all other states are excited states. Under equilibrium conditions, almost all atoms and molecules are in their ground states. Three types of processes or transitions occur in a two-level atomic system, as shown in Figure 4.5. At first, an incoming photon excites the atoms or molecules from a lower energy state into a higher energy state through a process called *absorption*. This process reduces the lower level population N_1 having energy E_1 and increases the upper level population N_2 having energy E_2. *Spontaneous emission* occurs when electrons in the higher

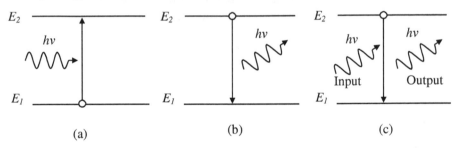

(a)   (b)   (c)

FIGURE 4.5: Transition processes in a two-level atomic system: (a) absorption, (b) spontaneous emission, and (c) stimulated emission

energy state drop down to the lower energy state, thus emitting photons. This reduces the population of E_2 and increases the population of E_1. *Stimulated emission* occurs when a photon with energy equal to the difference of energy of two states interacts with an electron in the higher energy level. Spontaneous emission is necessary to induce stimulated emission, since a photon of energy $h\nu$ induces stimulated emission, resulting in the emission of a photon of energy $h\nu$ in the same direction and having the same phase, polarisation, and frequency as the photon inducing it. Therefore, due to the process of stimulated emission, an optical wave amplified by a laser amplifier preserves most of the characteristics of the input wave.

The resonance frequency, ν_{21}, of the transition is determined by the separation between the energy levels and is given by

$$\nu_{21} = \frac{E_2 - E_1}{h} \tag{4.37}$$

The probability per unit time for a resonant optical process to occur is measured by the transition rate of the process. The transition rate W_{12} per second of stimulated upward transition from level 1 to 2 associated with absorption in the frequency range between ν and $\nu + d\nu$ is

$$W_{12}(\nu)d\nu = B_{12}u(\nu)g(\nu)d\nu \qquad (4.38)$$

where $g(\nu)$ is the normalised Gaussian line-shape, which is a characteristic of transition, and B is the Einstein coefficient.

Similarly, the downward transition rate W_{21} per second from level 2 to 1, associated with stimulated emission in the frequency range between ν and $\nu + d\nu$, is

$$W_{21}(\nu)d\nu = B_{21}u(\nu)g(\nu)d\nu \qquad (4.39)$$

The spontaneous emission rate W_{sp} per second, is independent of the energy density of the radiation and for a particular resonant transition, it is determined solely by the line-shape function of the transition, as given by

$$W_{sp}(\nu)d\nu = A_{21}g(\nu)d\nu \qquad (4.40)$$

The constants A and B are known as the Einstein A and B coefficients. The ratio of A_{21} and B_{21} is related to physical constants, as

$$\frac{A_{21}}{B_{21}} = 8\pi \left(\frac{nh\nu}{c}\right)^3 \qquad (4.41)$$

and

$$g_1 B_{12} = g_2 B_{21} \qquad (4.42)$$

where g_1 and g_2 are degeneracy factors.

In an atomic or molecular system, however, a given energy level usually consists of a number of degenerate quantum-mechanical states having the same energy. In thermal equilibrium at temperature T, the population ratio of the atoms in the upper and the lower levels follows the Boltzmann distribution. Taking into account the degeneracy factors, g_2 and g_1, of these energy levels, the ratio of N_2 and N_1 for the population densities associated with a transition energy of $h\nu$ is given by

$$\frac{N_2}{N_1} = \frac{g_2}{g_1}e^{-h\nu/k_BT} \qquad (4.43)$$

The steady-state population distribution in thermal equilibrium satisfies the relation

$$\frac{N_2}{N_1} = \frac{W_{12}}{W_{21} + W_{sp}} = \frac{B_{12}u(\nu)}{B_{21}u(\nu) + A_{21}} \qquad (4.44)$$

Further, it can be proved that the transition rates of both of the induced processes of absorption and stimulated emission are directly proportional to the spontaneous emission rate.

The finite spectral width of a resonant transition is dictated by finite relaxation time constants. The radiative relaxation rate of the transition from level 2 to level 1 is characterised by Einstein coefficient A_{21}, or a time constant $\tau_{sp} = 1/A_{21}$, known as the spontaneous radiative lifetime between levels 2 and 1.

4.3.1.1   Population inversion

Population inversion is the basic condition for the presence of an optical gain, and thus for lasing action. The spectral characteristic of a resonant transition is never infinitely sharp. At thermal equilibrium, a low-energy state is always more populated than a high-energy state, and hence population inversion is not possible. Population inversion in a system can only be accomplished through a process called pumping, by actively exciting the atoms in a low-energy state to a high-energy state. Since population inversion is a non-equilibrium state, to maintain a constant optical gain, continuous pumping is required to keep the population inversion at a constant level. However, this is possible only if there is a source supplying the energy that will channel the energy to the gain medium for the amplification of an optical wave. The use of a particular pumping technique depends on the properties of the gain medium being pumped. Many pumping techniques include electric discharge, current injection, optical excitation, chemical reaction, and excitation with particle beams. Semiconductor gain media is optically pumped, or pumped with electric current injection. The most commonly used pumping technique is optical pumping, either with incoherent light sources, such as flash-lamps and light-emitting diodes, or with coherent light sources from other lasers. The lasers and optical amplifiers of particular interest in photonic systems are made of either dielectric solid-state media doped with active ions, such as Nd:YAG and Er:glass fibre, or direct-gap semiconductors, such as GaAs and InP.

4.3.1.2   Rate equation

Rate equation indicates the net rate of increase of population density in a given energy level. For simplicity, rate equations for two energy levels, 2 and 1, with the population densities N_2 and N_1 is considered. The rates of increase or decrease of the population densities of levels 2 and 1 arising from pumping and decay are given by

$$\frac{dN_2}{dt} = P_{r2} - \frac{N_2}{\tau_2}$$

$$\frac{dN_1}{dt} = -P_{r1} - \frac{N_1}{\tau_1} + \frac{N_2}{\tau_{21}} \tag{4.45}$$

where P_{r1} and P_{r2} are the total rates of pumping into energy levels 1 and 2, τ_1 and τ_2 are overall lifetimes, respectively, permitting transitions to lower level 1 and 2, and τ_{21} is the lifetime of population decay including radiative and non-radiative spontaneous relaxation, from level 2 to level 1.

Under steady-state conditions $\frac{dN_1}{dt} = \frac{dN_2}{dt} = 0$ and the population difference $N_0 = (N_2 - N_1)$ can be found as

$$N_0 = P_{r2}\tau_2 \left(1 - \frac{\tau_1}{\tau_{21}}\right) + P_{r1}\tau_1 \tag{4.46}$$

A large gain coefficient clearly requires a large positive value of population difference N_0, which can be achieved by long τ_2, large P_{r1}, P_{r2}. Therefore the lifetime τ_2 of the upper laser level is an important parameter that determines the effectiveness of a gain medium. Generally speaking, the upper laser level has to be a metastable state with a relatively large τ_2 for a gain medium to be useful.

In the presence of a monochromatic, coherent optical wave of intensity I at a frequency ν, the rate equations are modified by considering optical amplification and absorption. While stimulated emission leads to amplification of an optical field, optical absorption results in attenuation of the same. The transition rates per second between the levels at frequency ν are then related as

$$W_{21} = \frac{I}{h\nu}\sigma_e$$

$$W_{12} = \frac{I}{h\nu}\sigma_a \tag{4.47}$$

where σ_e and σ_a are emission and absorption cross section at frequency ν.

Net power transfer W_p from optical field to the material is

$$W_p = I[N_1\sigma_a - N_2\sigma_e] \tag{4.48}$$

When $W_p < 0$, net power flows from the medium to the optical field, resulting in an amplification to the optical field with a gain coefficient g at frequency ν given by

$$g = \sigma_a \left(N_2 - \frac{g_2}{g_1}N_2 \right) \tag{4.49}$$

The rate equations then can be modified when coherent optical wave is present, as

$$\frac{dN_2}{dt} = P_{r2} - \frac{N_2}{\tau_2} - \frac{I}{h\nu}(N_2\sigma_e - N_1\sigma_a)$$

$$\frac{dN_1}{dt} = -P_{r1} - \frac{N_1}{\tau_1} + \frac{N_2}{\tau_{21}} + \frac{I}{h\nu}(N_2\sigma_e - N_1\sigma_a) \tag{4.50}$$

Population inversion is guaranteed when optical gain at frequency ν fulfils the condition

$$N_2\sigma_e - N_1\sigma_a > 0 \tag{4.51}$$

The pumping requirement for the condition to be satisfied is according to the properties of a medium. However, no matter how a true two-level system is pumped, it is not possible to achieve population inversion for an optical gain in the steady state, since the pump for a two-level system has to be in resonance with the transition between the two levels, thus inducing downward transitions as well as upward transitions. Thus, at steady state, a two-level system would reach a thermal equilibrium with the pump inhibiting population inversion. However, if a system is not a true two-level system, but is a quasi-two-level system, either or both of the two levels involved are split into closely spaced bands in such a manner that by pumping properly, it is possible to reach the

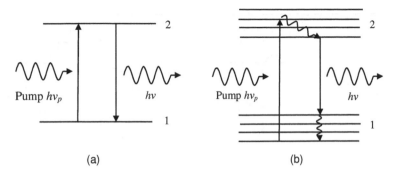

FIGURE 4.6: Two-level pump involving band

needed population inversion at steady state for an optical gain at a particular laser frequency ν. The situation is indicated in Figure 4.6.

Population inversion in steady state is possible for a system that has three energy levels involved in the process. The lower laser level 1 is the ground state, or is very close to the ground state, so that it is initially populated. The atoms are pumped to an energy level 3 above the upper laser level 2. In an effective three-level system, population relaxation time or lifetime from level 3 to level 2 is very fast such that $\tau_2 \gg \tau_{32} \approx \tau_3$. This helps the atoms excited by the pump to quickly end up in level 2. Moreover, level 3 lies sufficiently high above level 2 that the population in level 2 cannot be thermally excited back to level 3, and the lower laser level 1 is the ground state, or its population relaxes very slowly. The pumping condition for a constant optical gain under steady-state population inversion is obtained as

$$W_p > \frac{\sigma_a}{\sigma_e \tau_2} \tag{4.52}$$

where W_p is the effective pumping transition probability rate for exciting an atom in the ground state to eventually reach the upper laser level, and is proportional to the power of the pump.

A four-level system differs from a three-level system in that the lower laser level 1 lies sufficiently high above the ground level, termed the 0 level. At thermal equilibrium, the population in 1 is negligibly small compared to that in 0 and pumping takes place from level 0 to level 3. In addition to the conditions of a three-level system, a four-level system has to satisfy the condition that the population in level 1 relaxes very quickly back to the ground level, so that level 1 remains relatively unpopulated in comparison with the level 2 when the system is pumped. A four-level system is still much more efficient than a three-level system, as no minimum pumping requirement for population inversion is needed, since level 1 is initially empty. To compare, the transitions from different levels in a three-level system indicates a nonradiative transition form level 3 to level 2, whereas in a four-level system there are two

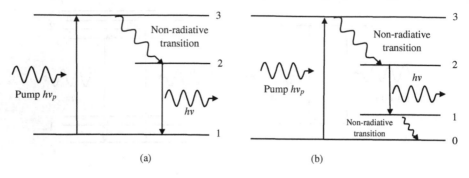

FIGURE 4.7: Three-level and four-level transitions

nonradiative transitions, one from level 3 to level 2 and another from level 1 to ground level. The transitions are shown in Figure 4.7.

4.3.1.3 Characteristics of laser amplifier

Amplifier gain, efficiency, bandwidth, and noise are the four important characteristics of a laser amplifier. The amplification of the intensity, I_s, of an optical signal propagating through a laser amplifier can be expressed in terms of unsaturated gain coefficient $g_0 z$, which is spatially varying in the longitudinal z direction, and saturation intensity I_{sat} of the gain medium, and is given by

$$g = \frac{g_0(z)}{1 + I_s/I_{sat}} \qquad (4.53)$$

The unsaturated gain coefficient $g_0(z)$ of an optically pumped laser amplifier depends on the pump intensity and the geometry of the medium. If the contribution of noise is neglected and the beam is collimated, then currents in the above equation can be replaced by power. Integrating the equation, it can be established that the power of the optical signal grows exponentially with distance. As the power approaches the value of saturation power, the growth slows down. Eventually, the signal grows only linearly with distance. The power gain G of signal is given by

$$G = \frac{P_{out}^s}{P_{in}^s} \qquad (4.54)$$

where P_{in}^s and P_{out}^s are input signal power and output signal power. For a weak optical signal, the power gain is simply the unsaturated power gain sometimes called small signal power gain. If the signal power approaches or even exceeds the saturation power of the amplifier, the gain drops down.

The efficiency of a laser amplifier is measured either as power conversion efficiency or as quantum efficiency. The power conversion efficiency, η_c, of a laser amplifier is defined as

$$\eta_c = \frac{P_{out} - P_{in}}{P_p} \qquad (4.55)$$

where P_p is the pump power.

The quantum efficiency, η_q, of an optically pumped laser amplifier is defined as the number of signal photons generated per pump photon. In the case of electrical pumping, the quantum efficiency is the number of photons generated per pump electron that is absorbed by the gain medium. Because the maximum value of quantum efficiency is unity, the maximum possible power conversion efficiency of an optically pumped laser amplifier is $\frac{\lambda_p}{\lambda_s}$, where λ_s and λ_p are the free-space signal and pump wavelengths.

The optical bandwidth, B, of a laser amplifier is determined by the spectral width of the gain coefficient $g(\nu)$ and an optical filter that might be incorporated into the device. The optical bandwidth of a laser amplifier is generally a function of the pumping rate. Because of the resonant nature of the laser transition that is responsible for the gain of a laser amplifier, the optical bandwidth of a laser amplifier is quite small.

Noise sources in a laser amplifier are quantum noise due to spontaneous emission and thermal noise associated with black-body radiation. Thermal noise dominates at long wavelengths, whereas quantum noise dominates at short wavelengths. Therefore, thermal noise in a laser amplifier that operates in the optical region at room temperature is negligible in the presence of quantum noise caused by spontaneous emission. Because of the spontaneous emission noise, the signal-to-noise ratio (SNR) of an optical signal always degrades after the optical signal passes through an amplifier.

4.3.2   Laser oscillators

The laser oscillator comprises a resonant optical amplifier whose output is fed back into its input with matching phase. A small amount of input noise within the amplifier bandwidth initiates the oscillation process. The input is amplified and the output is fed back to the input, where it undergoes further amplification. The process continues until the saturation of the amplifier gain limits further growth of the signal. The system thus reaches a steady state in which an output signal is created at the frequency of the resonant amplifier. Two conditions must be satisfied for oscillation to occur: (a) The amplifier gain must be greater than the loss in the feedback system so that a net gain is achieved in a round trip through the feedback loop, and (b) The total phase shift in a single round trip must be a multiple of 2π for matching with the phase of the original input. The system becomes unstable once the oscillation starts and continues till saturation limits the oscillation power and the system reaches a stable state.

Thus the laser oscillator is a spatially and temporally coherent and collimated features of laser light. The laser oscillation takes place only along a longitudinal axis of a straight or folded optical resonator. The gain medium emits spontaneous photons in all directions, but only the radiation that propagates along the longitudinal axis within a small divergence angle obtains sufficient regenerative amplification to reach the threshold for oscillation. Therefore, a laser oscillator is necessarily an open cavity with optical feedback only along the longitudinal axis.

A laser cavity can take a variety of forms with a gain medium in the cavity. A linear optical resonator cavity, also known as a Fabry-Perot cavity, is formed by a pair of precisely aligned mirrors facing each other. The resonator cavity provides frequency-selective optical feedback, thus amplifying the optical field for the gain medium. Laser output is obtained from one end of the cavity where the mirror is partially reflecting. To analyse the optical cavity, the properties of the optical fields or modes inside the cavity need to be examined. A mode is the field distribution due to an oscillating wave. It occurs due to constructive interference of the to and fro traversing waves within the cavity, thereby generating standing waves. Since modes are self-reproducing oscillations, if one mode is excited, it continues to oscillate in the cavity without inducing any other mode. Two consecutive longitudinal modes traversing a cavity of length L are separated by a frequency difference $\triangle\nu$

$$\triangle\nu = \frac{c}{2nL} \tag{4.56}$$

where c is the velocity of light and n is the refractive index of the medium.

In the Fabry-Perot laser cavity, as shown in Figure 4.8, the radii of curvature of the left and right mirrors are R_{c1} and R_{c2}, respectively. The sign of the radius of curvature is taken to be positive for a concave mirror and negative for a convex mirror. For the cavity to be a stable cavity in which a Gaussian mode can be sustained, the radii of curvature of both end mirrors have to match the wavefront curvatures of the Gaussian mode at the surfaces of the mirrors, that is, $R(z_1) = -R_{c1}$ and $R(z_2) = R_{c2}$, where z_1 and z_2 are the coordinates of the left and right mirrors measured from the location of the

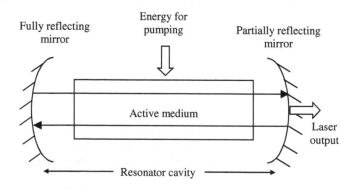

FIGURE 4.8: Schematic diagram of a laser cavity

Gaussian beam waist. For the length of the cavity $L = (z_2 - z_1)$ the following two relations can be established as

$$z_1 + \frac{z_R^2}{z_1} = -R_{c1}$$

$$z_2 + \frac{z_R^2}{z_2} = R_{c2} \qquad (4.57)$$

Given the values of R_{c1}, R_{c2}, and L, stable Gaussian modes exist for the cavity if both the above relations are satisfied with a positive, real parameter $z_R > 0$ for a finite, positive spot size. This condition yields the stability criterion for a Fabry-Perot cavity as

$$0 \le \left(1 - \frac{L}{R_{c1}}\right)\left(1 - \frac{L}{R_{c2}}\right) \le 1 \qquad (4.58)$$

The same analysis can be used to find the stability criterion of a cavity with multiple mirrors, such as a folded Fabry-Perot cavity or a ring cavity.

A laser oscillator must have a gain medium in a resonant laser cavity. The gain medium provides amplification of the intra-cavity optical field while the resonator provides optical feedback. If no external optical field is injected into the optical cavity for laser oscillation, then the intra-cavity optical field has to grow from the field generated by spontaneous emission from the intra-cavity gain medium. The gain medium may fill up the entire length of the cavity, or it may occupy a fraction of the cavity length. It is assumed that the length of the gain medium is equal to the length of the cavity L, i.e., the gain filling factor is negligible. When an inter-cavity field completes a round trip back to position z, it is modified by a complex factor a (which can be amplification or attenuation), expressed as

$$a = Ge^{i\varphi_{RT}} \qquad (4.59)$$

where G is the round-trip gain factor for the field amplitude, equivalent to the power gain in a single pass through the linear Fabry-Perot cavity, and φ_{RT} is the round-trip phase shift for the field. Both G and φ_{RT} have real values. If $G > 1$, the intra-cavity field is amplified, otherwise the intra-cavity field is attenuated.

When both the optical path length L_{RT} and the localized phase shifts are fixed, as in a laser resonator, the resonance condition of $\varphi_{RT} = 2q\pi$ (for $q = 1, 2, ...$) is satisfied only if the optical frequency satisfies the following relation:

$$\nu_q = \frac{c}{L_{Rt}}\left(q - \frac{\varphi_L}{2\pi}\right) \qquad (4.60)$$

where, ν_q is the optical frequency, and φ_L is the local fixed phase shift. L_{Rt} takes values $2nL$ for linear cavity and nL for ring cavity.

The discrete resonance frequencies are the longitudinal mode frequencies of the optical resonator because they are defined by the resonance condition of the round-trip phase shift along the longitudinal axis of the cavity. Thus when the optical spacing between the mirrors is L, the resonance condition along the axis of the cavity is given by

$$L = \frac{\lambda}{q} 2n \tag{4.61}$$

Alternatively, discrete emission frequencies ν are given by

$$\nu = \frac{qc}{2nL} \tag{4.62}$$

The resonant optical field inside an optical cavity cannot be a plane wave because of a finite transverse cross-sectional area. Therefore, there exist certain normal modes for the transverse field distribution and such transverse field patterns are known as the transverse modes of a cavity. When an optical cavity supports multiple transverse modes, the round-trip phase shift is generally a function of the transverse mode indices and the phase shift equation is modified accordingly.

Considering a cavity which contains an isotropic gain medium (with a gain filling factor $\simeq 1$), the reflection coefficient r_1 and r_2 at the two end mirrors can be expressed by taking into account the phase change at the reflections as

$$r_1 = \sqrt{R_1} e^{i\varphi_1}$$
$$r_2 = \sqrt{R_2} e^{i\varphi_2} \tag{4.63}$$

where R_1 and R_2 are the intensity reflectivities of the left and right mirrors, and φ_1 and φ_2 are the phase changes on reflections. When steady-state oscillation is reached, the coherent laser field at any given location inside the cavity becomes constant with time in both phase and magnitude. The gain and the phase conditions for steady-state laser oscillation are given by

$$a = G e^{i\varphi_{RT}} = 1 \tag{4.64}$$

where $|G| = 1$ and $\varphi_{RT} = 2q\pi, \quad q = 1, 2, ...$ The condition implies that there are a threshold gain and a corresponding threshold pumping level for laser oscillation. The threshold gain coefficient, g^{th}, can be obtained as,

$$g^{th} = [\alpha - \frac{1}{L} \ln \sqrt{R_1 R_2}] \tag{4.65}$$

where α is the mode dependent loss and can be neglected.

The power required to pump a laser to reach its threshold is called the threshold pump power. Because the threshold gain coefficient is mode and frequency dependent, the threshold pump power is also mode and frequency dependent. The threshold pump power of a laser mode can be found by calculating the power required for the gain medium to have a gain coefficient equal to the threshold gain coefficient of the mode.

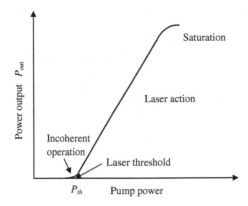

FIGURE 4.9: Output characteristics of a laser oscillator

The output power of a laser can also be estimated. A representative relation between pump power and output power of a laser oscillator is shown in Figure 4.9. For a continuous wave laser oscillating in a single longitudinal and transverse mode, the threshold intra-cavity energy growth rate for laser oscillation is the cavity decay rate. The intra-cavity average photon density S is given by

$$S = \frac{n}{c} \frac{I}{h\nu} \qquad (4.66)$$

where n is the average refractive index of gain medium and I is the spatially averaged intensity inside the laser cavity, and $h\nu$ is the photon energy of the oscillating laser mode. However, average photon density S is limited by a saturation photon density.

4.3.3 Light-emitting diodes

A light-emitting diode (LED) is a p-n junction device which, under forward bias, promotes radiative recombination of minority carriers. When the majority carriers (holes in p-type and electrons in n-type) are injected across such a forward-biased p-n junction, the electrons and holes recombine and emit incoherent light, using the property of electro-luminescence—a process of spontaneous photon emission. Under this condition, electrons re-occupy the energy states of the holes, emitting photons with energy equal to the difference between the electron and hole states involved. The injected carriers have to find proper conditions before the recombination process can take place and therefore the recombination is not instantaneous, and both energy and momentum conservation have to be met. Certain impurities in a semiconductor, however, can form neutral iso-electronic centres which introduce local potential that can trap an electron or a hole. Conservation of momentum can easily be satisfied in the radiative recombination process through an iso-electronic centre no matter whether the host semiconductor is a direct bandgap or an in-

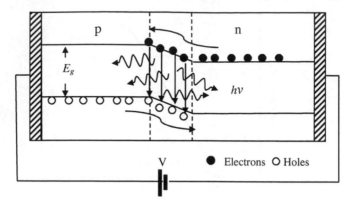

FIGURE 4.10: Basic operation of LED

direct bandgap material [2]. A schematic diagram of LED operation is shown in Figure 4.10 which shows a forward biased p-n homojunction of bandgap energy E_g. Photons of energy $h\nu$ are usually generated in the depletion region.

For maximising of the radiative recombination process, internal quantum efficiency η_i is made high. η_i is given by

$$\eta_i = \frac{\tau_n}{\tau_n + \tau_r} \tag{4.67}$$

where τ_n, and τ_r are the lifetime of minority carriers for non-radiative and radiative recombination, respectively. Excellent material quality is required to obtain large τ_n and efficient radiative recombination conditions for small τ_r. Under such conditions, the internal quantum efficiency may approach nearly 100 percent.

The wavelength of the light emitted, and thus its colour, depends on the band gap energy E_g of the material forming the p-n junction. The bandgap energy E_g and emitted wavelength λ are approximately related by

$$\lambda \simeq \frac{hc}{E_g} \simeq \frac{1.24}{E_g} \tag{4.68}$$

where h is Plank's constant, c is the speed of the light, and E_g is the energy bandgap in electron-volt.

For an LED that emits in the visible spectral region, a photometric efficiency, or luminous efficiency, is also considered to account for the spectral response of the human eye. The luminous efficiency is defined as the luminous flux emitted by an LED per watt of electric input power. A green LED therefore may appear much brighter than a blue or red LED of the same efficiency.

Bare uncoated semiconductors such as silicon exhibit a very high refractive index relative to air, which prevents passage of photons arriving at sharp angles relative to the air-contacting surface of the semiconductor. Therefore, the light extraction process after internal generation of photons affects the light-emission efficiency of LEDs. The efficiency of LEDs is thus measured by the external quantum efficiency η_e. It quantifies the efficiency of the conversion of electrical energy into emitted optical energy.

In general, a flat-surface uncoated LED semiconductor chip emits light only perpendicular to the semiconductor's surface, and a few degrees to the side, in a cone shape referred to as the light cone, or the escape cone. The maximum angle of incidence is referred to as the critical angle. At angles above critical, photons no longer escape the semiconductor but are instead reflected internally inside the semiconductor crystal. Internal reflections can escape through other crystalline faces, if the incidence angle is low enough and the crystal is sufficiently transparent to discourage re-absorption of the emitted photons.

Most of the commercial LEDs are realised using a highly doped p and an n junction. LED development began with GaAs-based LEDs emitting infrared ($\lambda = 900$nm) with bandgap energy at $E_g = 1.4$ eV and red ($\lambda = 635$ nm) for bandgap energy at $E_g = 1.9$ eV. Bandgap can be tailored to any value between 1.4 and 2.23 by adding GaP, which results in ternary alloy $GaAs_{1-y}P_y$ [3]. Current bright blue LEDs are based on the wide band gap compound semiconductors such as ZnSe or the nitride system GaN, AlGaN, or AlGaInN and InGaN. Blue LEDs have an active region consisting of one or more InGaN quantum wells sandwiched between thicker layers of GaN, called cladding layers. By varying the relative InGa fraction in the InGaN quantum wells, the light emission can be varied from violet to amber. Green LEDs manufactured from the InGaN/GaN compound are far more efficient and brighter than green LEDs produced with non-nitride material systems. With nitrides containing Al, most often AlGaN and AlGaInN, even shorter wavelengths are achievable. Blue LEDs can be added to existing red and green LEDs to produce the impression of white light.

LEDs are usually built on an n-type substrate, with an electrode attached to the p-type layer deposited on its surface. Substrates of p-type, while less common, occur as well. The process of growth is called epitaxial growth. Commonly used epitaxial structures are grown or diffused homojunctions and single or doubled confinement heterojunctions. A homostructure has two major deficiencies. The excess carriers are neither confined nor concentrated but are spread by diffusion. For this reason, the thickness of the active layer in a homostructure is normally on the order of one to a few μ m. Moreover, the structure is not suitable for optical confinement. Basic homostructure is shown in Figure 4.11(a), where the junction is formed on highly doped GaAs substrate. Au-Ge and Al are used to form electrical contacts [4].

The distribution of the excess carriers in a homostructure is largely determined by the diffusion of the minority electrons in the p region. To counter this effect, it is possible to place a P-p heterojunction in the p region to

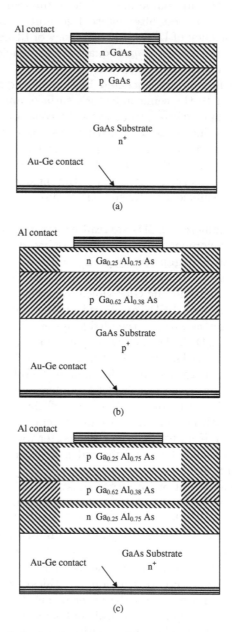

FIGURE 4.11: Basic homostructure and heterostructure of a representative LED

restrict the diffusion of excess electrons that are injected into the p region across the p-n junction. This additional P-p heterojunction results in a P-p-n junction. A heterojunction can be fabricated with lattice-matched layers of compound semiconductors that have different compositions. Many variations are possible, though ternary compound of GaAlAs is used as base material. Some typical heterojunctions are shown in Figure 4.11(b) and (c) [5].

Another LED geometry which is providing significant benefits for communication applications is the super-luminescent diode or SLD [6]. This device type offers advantages of: (a) a high output power; (b) a directional output beam; and (c) a narrow spectral linewidth.

4.3.3.1   Light-current characteristics

When a p-n junction is forward biased by applying an external voltage, electric current begins to flow as a result of carrier diffusion. The current I increases exponentially with the applied voltage V (as shown in Figure 4.12) according to the well-known relation given by

$$I = I_s(e^{eV/k_BT} - 1) \tag{4.69}$$

where I_s is the saturation current.

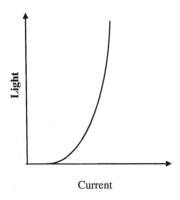

Current

FIGURE 4.12: Optical power-current relationship of LED

However, major interest in photonic sources is related to the amount of optical power generated with the passage of current. At a given current I, the carrier-injection rate is I/e, where e is the electronic charge. At steady state, the rate of recombination of electron-hole pairs through radiative and nonradiative processes is equal to the carrier injection rate. Since the internal quantum efficiency η_i determines the fraction of electron-hole pairs that recombine through spontaneous emission, the rate of photon generation is $\eta_i I/e$. The internal optical power is thus given by

$$P_i = \eta_i(\hbar\nu/e)I \tag{4.70}$$

where $\hbar\nu$ is the photon energy. If η_e is the fraction of photons escaping from the device, the emitted power P_e is given by

$$P_e = \eta_e P_i = \eta_e \eta_i (\hbar\nu/e) I \qquad (4.71)$$

External quantum efficiency can be calculated by taking into account internal absorption and the total internal reflection at the semiconductor and air interface. Only light emitted within a cone of angle $\theta_c = \sin^{-1}(1/n)$ escapes from the LED surface, where n is the refractive index of the semiconductor material. Therefore, the external quantum efficiency can then be written, under the assumption that the radiation is emitted uniformly in all directions over a solid angle of 4π, as

$$\eta_e = \frac{1}{4\pi} \int_0^{\theta_c} F_t(\theta)(2\pi \sin\theta) d\theta \qquad (4.72)$$

where F_t is the Fresnel transmissivity which depends on the incidence angle θ. Considering that the LED material has refractive index n and that of the ambient material is 1, the above equation can be simplified as

$$\eta_e = \frac{1}{n(n+1)^2} \qquad (4.73)$$

A measure of the LED performance is the total quantum efficiency η_t, defined as the ratio of the emitted optical power P_e to the applied electrical power, $P_{ele} = V_o I$, where V_0 is the voltage drop across the device. η_t is given by

$$\eta_t = \eta_e \eta_i (\hbar\nu/eV_o) \qquad (4.74)$$

As $\hbar\nu \approx eV_0$, $\eta_t \approx \eta_e \eta_i$, the total quantum efficiency, η_t, is called the power-conversion efficiency and is a measure of the overall performance of the device. Another quantity sometimes used to characterize the LED performance is the responsivity, defined as the ratio $R_{LED} = P_e I = \eta_t V_0$. The responsivity remains constant as long as the linear relation between P_e and I holds. In practice, this linear relationship holds only over a limited range of current.

4.3.3.2 Spectral width and modulation bandwidth

The spectral linewidth of the emission can be derived by differentiating the wavelength generation equation $\lambda = \frac{hc}{E_g}$, and is given by

$$\triangle\lambda = -\frac{hc}{E_g^2}\triangle E_g \qquad (4.75)$$

Therefore, the fractional spectral linewidth is given by

$$\gamma = \left|\frac{\triangle\lambda}{\lambda}\right| = \frac{\triangle E_g}{E_g} \simeq \frac{2.4kT}{E_g} \qquad (4.76)$$

In most of the high-speed photonic communication and photonic signal processing where LED is used as the source, modulation bandwidth becomes a limiting factor. The factors contributing to the modulation bandwidth of the LED are its response to the changes in the injected current as well as the junction capacitance of the device. If the DC power input to the device is P_o, the relative output at frequency ω is given by

$$\frac{P_\omega}{P_o} = [1 + (\omega\tau)^2]^{-1/2} \tag{4.77}$$

Thus, the inherent modulation speed is limited by the carrier lifetime τ. At an injection current I, the output optical power and the small-signal power-bandwidth product of an LED is given by

$$P_{out}f_{3dB} = \eta_e \frac{h\nu}{e} \frac{I}{2\pi\tau} \tag{4.78}$$

Therefore, at a given current level, the modulation bandwidth of an LED is inversely proportional to its output power.

4.3.4   Semiconductor lasers: Injection laser diodes

An injection laser diode is an electrically pumped semiconductor laser in which the active medium is a p-n junction semiconductor diode. Forward electrical bias across the p-n junction causes the majority charge carriers, i.e., holes and electrons, to be injected from opposite sides of the junction into the depletion region. Lasing in a semiconductor laser is made possible by the existence of a gain mechanism plus a resonant cavity. Due to charge injection mechanism in laser diode, this class of lasers is sometimes termed injection lasers or injection laser diode (ILD).

An ILD has several advantages over other semiconductor lasers:

1. Injection laser diodes have a sharper linewidth around 1 nm than the line width of an LED, which is around 50 μm.

2. Injection laser diodes provide quite high radiance due to stimulated emission.

3. Temporal and spatial coherence of injection laser diodes is relatively good than an LED.

The differences between an ILD and an LED are that when operated below threshold, laser diodes produce spontaneous emission and behave as light-emitting diodes. The laser operates on the basis of stimulated emission, which causes the laser light to be concentrated in particular modes. Up to a threshold current ILD operation is equivalent to an LED operation. Above the threshold current, laser action starts. It is evident from the current-optical power characteristics of a representative ILD as shown in Figure 4.13.

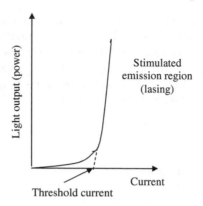

FIGURE 4.13: Optical power-current relationship of ILD

In a semiconductor laser the gain mechanism is provided by light generation from the recombination of holes and electrons. In order for the light generation to be efficient enough to result in lasing, the active region of a semiconductor laser, where the carrier recombination occurs, must be a direct bandgap semiconductor. The surrounding carrier injection layers, which are called cladding layers, can be indirect bandgap semiconductors. When the injected carrier density in the active layer exceeds a certain value, known as the transparency value, population inversion is realised and the active region exhibits optical gain. An input signal propagating inside the active layer would then amplify as e^{gz}, where g is the gain coefficient. As carrier density N increases, g becomes positive over a spectral range. The peak value of the gain, g_p, also increases with N, with a shift of the peak toward higher photon energies.

The earlier semiconductor lasers were homojunction, which used photoemissive properties of a single p-n junction and did not operate at room temperature. The radiative properties of a junction diode may be improved by the use of heterojunctions which may be isotype or anisotype. The isotype heterojunction provides a potential barrier within the structure for confining minority carriers in a small active region. On the other hand anisotype heterojunctions improve the injection efficiency. For a practical laser, the cladding layers have a wider bandgap and a lower index of refraction than the active region. This type of semiconductor laser is called a double heterostructure (DH) laser, since both cladding layers are made of a different material than the active region. The narrow bandgap of the active region confines carrier recombination to a narrow optical gain region. The sandwich of the larger refractive index active region surrounded by cladding layers forms a waveguide, which concentrates the optical modes generated by lasing in the active region [7].

For efficient carrier recombination, the active region must be fairly thin, so that a significant fraction of the optical mode spreads into the cladding layers. In order to completely confine the optical mode in the semiconductor structure, the cladding layers are made fairly thick. The resonant cavity of a simple semiconductor laser is formed by cleaving the ends of the structure. Lasers are fabricated with their lasing cavity oriented perpendicular to a natural cleavage plane. The reflectivities of the end facets can be modified by applying dielectric coatings to them.

Unpumped semiconductor material absorbs light of energy greater than or equal to its bandgap. When the semiconductor material is pumped optically or electrically, it reaches a point at which it stops being absorbant. This point is called transparency. If it is pumped beyond this point it will have optical gain, which is the opposite of absorption. A semiconductor laser is subject to both internal and external losses. For lasing to begin, that is to reach threshold, the gain must be equal to these optical losses. The threshold gain per unit length is given by

$$g_{th} = \alpha_i + \frac{1}{2L} \ln \left(\frac{1}{R_f R_r} \right) \tag{4.79}$$

where α_i is the internal loss per unit length, L is the laser cavity length, and R_r and R_f are the front and rear end reflectivity.

Double heterostructure (DH) semiconductor lasers can be fabricated from a variety of lattice-matched semiconductor materials. The two material systems most frequently used are GaAs / $Al_x Ga_{1-x}As$ and $In_{1-x}Ga_x As_y P_{1-y}/$ InP. x and y must be chosen appropriately to achieve both lattice match and the desired lasing wavelength. All of these semiconductors are III-V alloys. For GaAs-based lasers, the active region is usually GaAs. There is a great deal of interest in developing true visible lasers for optical data storage applications where $(Al_x Ga_{1-x})_{0.5}In_{0.5}P$ materials are used. In order to fabricate a blue semiconductor laser, other material systems with II-VI semiconductors are required. A basic DH structure is shown in Figure 4.14.

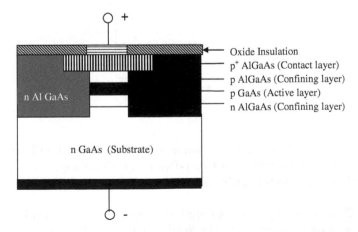

FIGURE 4.14: Schematic diagram of a double heterostructure structure used in an ILD

While double heterostructure design can confine carriers and act as waveguides in the transverse direction, confinement in the lateral direction is also possible. Three possible lateral confinement methods are gain guiding, positive index guiding, and negative index guiding. In the gain guided or stripe geometry laser, electrical contact to the upper semiconductor surface is made by a metal strip, the purpose of which is to confine the injected current into an area determined by the dimensions of the strip. In positive and negative index guided lasers physical structures provide real dielectric lateral index waveguiding. Strong index guiding along the junction plane by using buried heterostructure (BH) can provide improved transverse mode control. In Bh the active volume is completely buried in a material of wider bandgap and lower refractive index.

A traditional Fabry-Perot cavity can also be formed in a diode laser by cleaving the ends of the laser in the longitudinal dimension. Because of the high refractive index change between the cleaved ends and the outside, a high relectivity coefficient can be obtained. Diode lasers have flexibility in that a Fabry-Perot cavity is not the only feedback structure available. A distributed feedback (DFB) configuration creates a laser that is frequency selective and for certain feedback types surface emitting as well. Schematic diagram of a DFB laser and an elementary Fabry-Perot laser is shown in Figure 4.15.

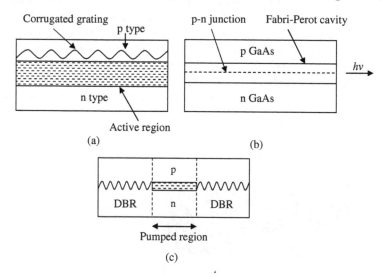

FIGURE 4.15: Schematic diagram of structure of: (a) distributed feedback (DFB) laser, (b) conventional Fabry-Perot injection diode laser, and (c) distributed Bragg reflector (DBF) of multi-section laser

In DFB, a corrugation (grating) is etched in the waveguide section of the large optical cavity laser. The grating acts as a Bragg diffraction grating which provides feedback for the light and hence determines the wavelength of longitudinal emission. The distributed feedback laser benefits from a lower sensitivity to current and temperature variations. Another advantage of DFB structure is that the laser can be made tunable. Several other designs of tunable DFB lasers have been developed in recent years. In one scheme, the built-in grating inside a DBR laser is chirped by varying the grating period. In another scheme, a superstructure grating is used for the distributed Bragg reflector (DBR) section of a multi-section laser as shown in Figure 4.15 (c). A superstructure grating consists of an array of gratings (uniform or chirped) separated by a constant distance. As a result, its reflectivity peaks at several wavelengths whose interval is determined by the spacing between the individual gratings forming the array [8].

Another class of semiconductor lasers, known as vertical-cavity surface-emitting lasers (VCSELs), operate in a single longitudinal mode by virtue of an extremely small cavity length, for which the mode spacing exceeds the gain bandwidth [9], [10]. They emit light in a direction normal to the active-layer plane. Moreover, the emitted light is in the form of a circular beam, which is advantageous in many photonic information processing applications. Fabrication of VCSELs requires growth of multiple thin layers on a substrate. The active region, in the form of one or several quantum wells, is surrounded by two high-reflectivity DBR mirrors that are grown epitaxially on both sides of the active region. Each DBR mirror is made by growing many pairs of alternating GaAs and AlAs layers, each $\lambda/4$ thick, where λ is the wavelength emitted by the VCSEL [11].

The other variety of recent interest in semiconductor lasers is a quantum well and quantum dot laser. The active region in conventional double heterostructure (DH) semiconductor lasers is wide, acts as bulk material, and no quantum effects are possible. Figure 4.16 shows a structure of a quantum well laser and its lasing direction [12].

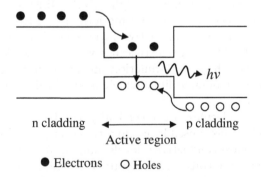

n cladding $\longleftrightarrow$ p cladding

Active region

● Electrons   ○ Holes

FIGURE 4.16: Schematic structure of a quantum well laser

In such lasers the conduction band and valence band are continuous. If the active region of a semiconductor laser is very thin in the order of the De Broglie wavelength of an electron, quantum effects become important. If the active region is still narrower the structure is a quantum well. Both single-quantum-well (SQW), corresponding to a single active region, and multi-quantum well (MQW), corresponding to multiple active regions, are used.

Lasers have also been developed in which the device contains a single discrete atomic structure or so-called quantum dot (QD). Quantum dots are small elements that contain a vary small region of free electrons forming a quantum well structure. They are fabricated using semiconductor crystalline materials and have typical dimensions between nanometers and a few microns [13].

The quantisation of energy levels will change the density of states and the difference in density of energy states directly affects the modal optical gain generated by injection carriers. The optical confinement factor, defined as the ratio of light intensity of the lasing mode within the active region to the total intensity in the overall space, is strikingly small. Therefore the loss of optical confinement may pose difficulties. A guiding layer is needed between the well and cladding layer.

The use of quantum well active regions has significantly improved the performance of semiconductor lasers with respect to the dramatic reductions in threshold current density, improvements in the realisation of high quantum efficiency, operation at high temperature better modulation speed, and narrower spectral linewidth.

4.3.5 Organic light-emitting diode (OLED)

An organic light-emitting diode (OLED) is a photonic source composed of electroluminescent organic thin films in between two conductors or electrodes. Typically, at least one of these electrodes is transparent. When electric current passes through it, the electrons and holes recombine within the electroluminescent organic layer, thus emitting light. Early OLEDs had a single organic layer between the electrodes, whereas modern-day OLEDs are bi-layered, i.e., they have an emissive and conductive layer sandwiched in between the two electrodes [14]. OLED requires a structure with a low work-function cathode, to inject electrons, a high work-function anode, to inject holes, and between them, a luminescent organic film, which is specified as a polymer, such as polyvinylcarbazole.

The organic molecules are electrically conductive as a result of delocalisation of electrons caused by conjugation over part or all of the molecule. These materials have conductivity levels ranging from insulators to conductors, and are therefore considered organic semiconductors. The highest occupied and lowest unoccupied molecular orbitals (HOMO and LUMO) of organic semiconductors are analogous to the valence and conduction bands of inorganic semiconductors. Electrons flows through the device from cathode to anode, as electrons are injected into the LUMO of the organic layer at the cathode and withdrawn from the HOMO at the anode. This latter process may also be described as the injection of electron holes into the HOMO. Electrostatic forces bring the electrons and the holes towards each other and they recombine forming an exciton, a bound state of the electron and hole. This happens closer to the emissive layer, because in organic semiconductors holes are generally more mobile than electrons. The decay of this excited state results in a relaxation of the energy levels of the electron, accompanied by emission of radiation whose frequency is in the visible region. The frequency of this radiation depends on the band gap of the material, in this case, the difference in energy between the HOMO and LUMO. The device structure is shown in Figure 4.17.

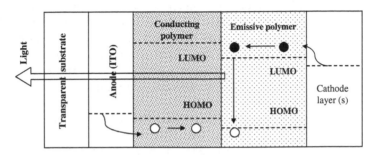

FIGURE 4.17: Representative OLED structure

The device is made on a glass substrate coated with indium tin oxide (ITO). ITO has a high work function, and acts as the anode, injecting holes. The next layer transports holes (hole transport layer, HTL), but also inhibit the passage of electrons. Next to this layer is the electroluminor, where the holes and electrons will recombine. There is then an electron transport layer (ETL), inhibiting hole passage. The final metal cathode is a low-work-function metal, such as Mg:Ag. Electron emitted from Mg:Ag cathode is allowed into a layer of Alq3, tris(8-hydroxyquinolinato) aluminium, with a layer of diamine. Alq3 is a complex molecule that is can adopt a number of crystalline phases and is usually deposited by evaporation, resulting in an amorphous film. It was later found that a better luminor would be a polymer, and, in particular, a main-chain conjugated polymer (p-phenylene vinylene) which could be prepared by spin coating [15], [16], [17]. This resulted in polymer light-emitting diodes (PLEDs). PLEDs, also called light-emitting polymers (LEP), emit light when connected to an external voltage. They are used as a thin film for full-spectrum colour displays. Polymer OLEDs are quite efficient and require a relatively small amount of power for the amount of light produced. Attempots were also made to evolve OLED emitting white light [18], [19]

4.3.6   Typical lasers used in photonics

4.3.6.1   Gas lasers

The gas lasers are the most frequently encountered type of laser oscillator. The red-orange, green, and blue beams of the HeNe, Ar^+, and HeCd gas lasers, respectively, are by now widely used in photonic instrumentation. In He-Ne, helium is excited by electron impact and the energy is then transferred to Neon by collisions. The transition between two excited levels of the neon atom emits infrared radiation at 1.15 m. Lasing can be achieved at many wavelengths 632.8 nm, 543.5 nm, 593.9 nm, 611.8 nm, 1.1523 μm, 1.52 μm, and 3.3913 μm. Pumping is achieved by electrical discharge.

Similar to the HeNe laser, the argon ion gas laser is pumped by electric discharge and emits light at wavelengths 488.0 nm, 514.5 nm, 351 nm, 465.8 nm, 472.7 nm, and 528.7 nm. The krypton ion gas laser is analogous to the argon gas laser with wave lengths of 416 nm, 530.9 nm, 568.2 nm, 647.1 nm, 676.4 nm, 752.5 nm, and 799.3 nm. Pumped by electrical discharge, the Kr+ laser readily produces hundreds of milliwatts of optical power at wavelengths ranging from 350 nm in the ultraviolet to 647 nm in the red and beyond. These lasers can be operated simultaneously on a number of lines to produce so called white laser light.

Molecular gas lasers such as CO_2, and CO, which operate in the middle-infrared region of the spectrum, are highly efficient and can produce large amounts of power. The laser transistions are related to vibrational-rotational excitations. The main emission wavelengths of CO_2 laser are 10.6 m and 9.4 m. They are pumped by transverse (high power) or longitudinal (low power) electrical discharge. These lasers are widely used in the material processing industry for cutting and welding of steels and in medical surgery. Carbon monoxide (CO) gas lasers produce wavelengths of 2.6 to 4 μm and 4.8 to 8.3 μm and are pumped by electrical discharge. Output powers as high as 100 kW have been demonstrated.

A gas laser in the ultraviolet region is the excimer laser. An excimer laser typically uses a combination of a noble gas (argon, krypton, or xenon) and a reactive gas (fluorine or chlorine). Under the appropriate conditions of electrical stimulation and high pressure, a pseudo-molecule called an excimer (or in the case of noble gas halides) is created, which can only exist in an energized state and can give rise to laser light in the ultraviolet range. Excimer (KrF) exist only in the form of excited electronic states since the constituents are repulsive in the ground state. The lower laser level is therefore always empty and unstable and thus provides a built-in population inversion. Wavelengths that can be generated are 193 nm (ArF), 248 nm (KrF), 308 nm (XeCl), and 353 nm (XeF).

4.3.6.2 Chemical lasers

A chemical laser is a laser that obtains its energy from a chemical reaction. Chemical lasers can reach continuous wave output with power reaching to megawatt levels. They are used in industry for cutting and drilling. Common examples of chemical lasers are the chemical oxygen iodine laser (COIL), all gas-phase iodine laser (AGIL), and the hydrogen fluoride (HF) and deuterium fluoride (DF) lasers, both operating in the mid-infrared region. There is also a DFCO2 laser (deuterium fluoridecarbon dioxide), which, like COIL, is a transfer laser.

4.3.6.3   Liquid dye lasers

The importance of liquid dye lasers stems principally from their medium. The laser gain media are organic dyes in solution of ethyl, methyl alcohol, glycerol, or water. These dyes can be excited optically with argon lasers, for example, and emit at 390-435 nm, 460-515 nm, and 570-640 nm (rhodamine 6G), and many other wavelengths. These lasers are tunable. Unfortunately, dyes are carcinogenic and therefore dye lasers became extinct.

4.3.6.4   Solid-state lasers

The term solid-state laser indicates a laser using a dielectric crystal (or a glass) doped with a small quantity of impurities as the active medium. Lasers using ruby emit at 694.3 nm. Ruby consists of the naturally formed crystal of aluminium oxide called corundum. In this crystal some of the Al^{3+} ions are replaced by Cr^{3+} ions. The chromium ions give ruby its fluorescence, which is related to the laser transitions. The active medium is a cylindrical monocrystal with a length of a few centimeters and diameter of a few millimeters. The ruby laser is a three-level laser: level 0 is the ground level, level 1 is the first excited level, level 2 is represented by the two absorption bands. The pump power comes from a lamp, emitting light on a broad spectrum that includes the green and violet wavelengths required to excite the chromium ions to level 2. Level 1 is populated by spontaneous decay processes from level 2. The lifetime of level 2 is of the order of picoseconds, while that of level 1 is of the order of milliseconds.

Nd^{3+}:YAG and Nd^{3+}:glass are the other lasing materials used in solid state lasers. The main difference with respect to the ruby laser is that the laser transition involves two excited levels, that is, one deals with a 4 level scheme. The threshold is about an order of magnitude lower than that of ruby by virtue of its being a four-level system. Because it can be optically pumped to its upper laser level by light from other semiconductor laser diodes, Nd^{3+}:YAG serves as an efficient compact source at 1.064 μm. Initially, Nd:YAG was flash-lamp pumped and today, efficient pumping is accomplished with laser diodes and diode arrays. Diode-pumped versions, which can be very compact and efficient have become useful photonic source in materials processing, range finding, and surgery. Furthermore, neodymium laser light can be passed through a second-harmonic generating crystal which doubles its frequency, thereby providing a strong source of radiation at 532 nm in the green. Neodymium can also be doped in a host of other crystals such as yttrium lithium fluoride (Nd:YLF) emitting at 1047 μm and glass (Nd:glass) at 1.062m (silicate glasses), 1.054 μm (phosphate glasses). Yttrium scandium gallium garnet (YSGG) glass lasers have been used to build extremely high power in terawatt range.

Another variety of solid-state laser is based on YAG. It is a quasi three-level laser emitting at 1.030 μm. The lower laser level is only 60 meV above the ground state and therefore at room temperature the medium is heavily thermally populated. The laser is pumped at 941 or 968 nm with laser diodes to provide the high brightness pumping needed to achieve gain. Solid-state laser amplifiers and oscillators also include alexandrite ($Cr^{3+}:Al^2 BeO$). These lasers offer tunable output in the wavelength range between 700 and 800 nm.

In contrast to neodymium, which is a rare earth element, titanium is a transition metal. The Ti^{3+} ions replace a certain fraction of the Al^{3+} ions in sapphire (Al_2O_3). The Ti:sapphire laser is tunable over an even broader range, from 660 to 1180 nm, and Er^{3+}:YAG, is often operated at 1.66 μm.

4.4 Photodetectors

Photodetectors are devices that convert the incident light into electrical voltages/current, and whose response is related to the intensity of the incident light. All photodetectors are square-law detectors and respond to the power or intensity, rather than on the amplitude of optical signal. Photodetectors can be classified into (a) photon detectors and (b) thermal detectors. Photon detectors can again be classified into two groups depending on the process involved. One group is based on the external photoelectric effect and another is based on the internal photoelectric effect. Three steps are involved in the internal photoelectric effect devices: (a) absorption of incident photons resulting in carrier generation, (b) movement of generated carriers across the active/absorption region, and, finally, (c) collection of carriers, generating photocurrent in the device. Photodetectors based on the external photoelectric effect are photoemissive devices, such as vacuum photodiodes and the photomultiplier tubes, in which photoelectrons are ejected from the surface of a photocathode. Thermal detectors, on the other hand, convert optical energy into heat and thus respond to the optical energy, rather than the number of photons absorbed by the detector [20].

The choice of a photodetector for a particular application depends on the wavelength range of operation, speed of response or response time, and quantum efficiency. The wavelength range over which a photodetector operates is determined by the material structure. Various photodetectors used in information communication are the photoconductors, p-n junction photodiodes, p-i-n photodetectors, avalanche photodiodes, and charge-coupled devices (CCDs). Other photodetectors such as, photon detectors are extremely sensitive and can have a high speed of response. In contrast, thermal detectors are relatively slow in response because of the slow process of heat diffusion. For these reasons, photon detectors are suitable for detecting low level optical signals in photonic systems, whereas thermal detectors are most often used for optical power measurement or infrared imaging.

The characteristic properties of a photodetector are its responsivity or dynamic wavelength range of detection, quantum efficiency, damage threshold or maximum tolerable power, response time, and dark current [21]. Some of these properties are explained below:

1. *Quantum efficiency:* Quantum efficiency η is defined as the ratio of the number of photo-electrons produced for every photon incident on the photosensitive surface. In other words, quantum efficiency is the probability of generating a charge carrier for each incident photon. Thus, with n_{ph} incident photons, if n_e is the total number of photoelectrons generated, then quantum efficiency is

$$\eta = \frac{n_e}{n_{ph}} \times 100\% \qquad (4.80)$$

   Equation 4.80 can be rewritten as

$$\eta = \frac{i_p h\nu}{P_i e} \qquad (4.81)$$

   where i_p is the generated photocurrent, $h\nu$ is the incident photon energy, e is the charge, and P_i is the incident optical power at the wavelength corresponding to photon energy. The quantum efficiency of a photodetector is a function of the wavelength of the incident photons because of the spectral response of the detector.

2. *Responsivity*: The responsivity of a photodetector is defined as the ratio of the output current or voltage signal to the power of the input optical signal. For a photodetector that has an output current without any internal gain, the responsivity is defined as

$$\mathcal{R} = \frac{i_p}{P_i} = \eta \frac{e}{h\nu} \qquad (4.82)$$

   The spectral response of a photodetector is usually characterised by the responsivity of the detector as a function of optical wavelength, $\mathcal{R}(\nu)$, and is known as the spectral responsivity.

3. *Response speed and frequency response*: The response speed of a photodetector is determined by the transit time of carriers in the active region of the device and the time-constant of the device. The carrier transit time t is defined as the time taken by the carriers to travel across the active region, and is given by

$$t = \frac{d}{v_{sat}} \qquad (4.83)$$

   where d is the thickness of the active region and v_{sat} is the saturation velocity of electrons/holes. To minimise the response time, the transit time is reduced to minimum, thereby maximising the speed. This is achieved by reducing the thickness of the active region. But making the active region thin increases the time constant of the device, and so the speed is slowed. Hence the response time and time constant need to be optimised to get the fastest photodetector response. The frequency response is characterised by the frequency dependence of the responsivity $\mathcal{R}(\nu)$ at a given optical wavelength.

4. *Dark current*: In the absence of any incident illumination, the leakage current in the device under reverse bias condition is known as dark current. For better performance of photodetectors, the dark current should be as low as possible. The dark current I_d can be written as the summation of three contributions,

$$I_d = I_d^{gr} + I_d^{Dif} + I_d^{sl} \qquad (4.84)$$

where the term I_d^{gr} refers to the dark current due to carrier generation and recombination, the term I_d^{Dif} is due to carrier diffusion, and the term I_d^{sl} is contributed by surface leakage. Dark current can be measured by probing the device, keeping it in the dark under reverse-bias condition. Dark current is usually of the order of nano-amperes.

5. *Detectivity*: The detectivity is the ability of a photodetector to detect a small optical signal. It is defined as the inverse of the noise equivalent power (NEP) of the detector. The background radiation current, i_b, and the dark current, I_d, are the dominant sources of noise for a photodetector, and are often proportional to the surface area A, of a photodetector. For a dark-current-limited photodetector without an internal gain, the specific detectivity D^* is given by

$$D^* = \frac{\sqrt{A}\mathcal{R}}{\sqrt{2ei_d}} \qquad (4.85)$$

4.4.1 Photoconductors

Photoconductors are photodetectors based on the phenomenon of photoconductivity. With photoexcitation, the conductivity of the detector increases, since electron-hole pairs are generated. A suitable bias is applied across the device to collect the electrons and holes. The photoconductivity is the enhanced conductivity $\triangle\sigma$ contributed by these photo-generated excess carriers, and is given by

$$\triangle\sigma = e(\mu_e \triangle n + \mu_h \triangle p) \qquad (4.86)$$

where $\triangle n = (n - n_0)$ and $\triangle p = (p - p_0)$ are the photo-generated excess electron and hole concentrations, and μ_e and μ_h are the electron and hole mobilities, respectively. To reduce dark current, it is necessary to minimise the thermal equilibrium concentrations, n_0 and p_0, of free electrons and free holes in a photoconductor.

Photoconductivity has a threshold photon energy, E_{th}, and a corresponding threshold wavelength, λ_{th}, that are characteristic of a given photoconductive material. The threshold photon energy of intrinsic photoconductivity is clearly the bandgap energy of the photoconductor, which is $E_{th} = E_g$. The extrinsic photoconductivity is contributed by carriers that are generated by optical transitions associated with impurity levels within the bandgap of an extrinsic semiconductor. Photoconductors cover a broad spectral range, from the ultraviolet to the far-infrared. Along with E_{th} and λ_{th}, the absorption coefficient at the wavelength of radiation determine the spectral response of a photoconductor.

Both direct-gap and indirect-gap semiconductors can be used for photoconductors. The group IV semiconductors, the III-V and II-VI compounds, and the IV-VI compounds can be used for intrinsic photoconductors. Intrinsic silicon photoconductors are widely used photoconductive detectors in the visible and near-infrared spectral regions. The active region of the photoconductor is formed by epitaxial growth or ion-implantation over a highly resistive substrate. The thickness of the active region plays a vital role in absorbing more photons, but it is optimised to minimise the noise current.

A photoconductor has a photoconductive gain that depends on many parameters of the photoconductor and on the properties of the electrical contacts. The operation requires that voltage be applied to the device. Photoconductive gain of the detector is defined as the ratio of the time-rate of flow of electrons through the device to the generation of electron-hole pairs within it. A simple photoconductor of a length l between its electrodes, a width w, and a thickness d, is considered. The generation of electron-hole pairs in the volume lbw of detector is given by g_{eh} as

$$g_{eh} = \frac{\eta}{lbw}\phi \tag{4.87}$$

where η is the quantum efficiency, and ϕ is the incident photon flux.

For an n type photoconductor, at the steady-state condition, the rate of generation is equal to the rate of electron-hole recombination. When illuminated, the conductivity increases, as given by

$$\triangle\sigma = e\triangle n(\mu_e + \mu_h) = e(\tau g_{eh})(\mu_e + \mu_h) = e\tau\frac{\eta}{lbw}(\mu_e + \mu_h)\phi \tag{4.88}$$

where μ_e and μ_h are the electron and hole mobilities, respectively, $\triangle n$ is the excess electron density, and τ is the lifetime of minority carriers.

The generated photocurrent i_s due to applied voltage V is

$$i_s = V\frac{\triangle\sigma wd}{l} \tag{4.89}$$

Considering the electron velocity to be more than that of the holes in an n-type extrinsic photoconductor, the generated electrons flow faster to the end of the device in the external circuit, whereas other electron flow from the external circuit takes place into the device, to take its place. Thus, with one incident photon, several electrons move across the detector until recombination occurs. The photoconductive gain is thus expressed as

$$G = \frac{\tau}{\tau_e} \tag{4.90}$$

where τ is the minority carrier lifetime and τ_e is the carrier transit time.

When the external quantum efficiency and the gain of a photoconductor are known, its responsivity is given by,

$$\mathcal{R} = G\eta \frac{e}{h\nu} \tag{4.91}$$

Since the gain G varies with the applied voltage V, the responsivity of photoconductor is also a function of V, in addition to being a function of the optical wavelength and the device parameters.

4.4.2 Junction photodiodes

Junction photodiodes are the most commonly used photodetectors in information photonics technology [22]. They can be made from semiconductor homojunctions, heterojunctions, and metal-semiconductor junctions. The photo-response of a photodiode results from the photogeneration of electron-hole pairs as in the case of photoconductors. In contrast to photoconductors, which can be of either intrinsic or extrinsic type, a photodiode is normally of intrinsic type. The threshold photon energy of a semiconductor photodiode is the bandgap energy of its active region given by $E_{th} = E_g$.

In a semiconductor photodiode, generation of electron-hole pairs by optical absorption can take place in any of the different regions: the depletion layer, the diffusion regions, and the homogeneous regions. When an electron-hole pair is generated in the depletion layer by photo-excitation, the internal field sweeps the electron to the n side and the hole to the p side. This process results in a drift current that flows in the reverse direction from the cathode on the n side to the anode on the p side. If a photo-excited electron-hole pair is generated within one of the diffusion regions at the edges of the depletion layer, the minority carrier can reach the depletion layer by diffusion and then be swept to the other side by the internal field. This process results in a diffusion current that also flows in the reverse direction. However, photons absorbed in the homogeneous regions do not generate any photocurrent, and therefore the active region of a photodiode consists of only the depletion layer and the diffusion regions. The homogeneous regions on the two ends of the diode act like blocking layers for the photogenerated carriers because carriers neither drift nor diffuse through these regions.

A photodiode has a unity gain, $G = 1$, with the external signal current simply being equal to the photocurrent, and is given by

$$i_s = i_{ph} = \eta \frac{eP_0}{h\nu} \tag{4.92}$$

This photocurrent is a reverse current that depends only on the power of the optical signal. When a bias voltage is applied to the photodiode, the total current of the photodiode is the combination of the diode current given in and the photocurrent given as,

$$i(V, P_0) = I_s(e^{eV/K_B T} - 1) - i_s \tag{4.93}$$

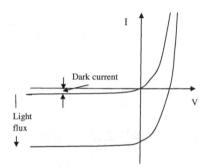

FIGURE 4.18: Current-voltage characteristics of a photodiode at two different power levels of optical illumination

Figure 4.18 shows the current-voltage characteristics of a junction photodiode at various power levels of optical illumination. As seen from the figure, there are two modes of operation for a junction photodiode. The device functions in photoconductive mode in the third quadrant of its current-voltage characteristics, including the short-circuit condition on the vertical axis. It functions in photovoltaic mode in the fourth quadrant, including the open-circuit condition on the horizontal axis. The mode of operation is determined by the external circuitry and the bias condition.

4.4.2.1   p-i-n photodiode

A p-i-n photodiode consists of an intrinsic region sandwiched between heavily doped p+ and n+ regions. A reverse bias voltage applied to the device drops almost entirely across the intrinsic region and as a result, the depletion layer is almost completely defined by the intrinsic region. The electric field in the depletion layer is also uniform across the intrinsic region. The depletion-layer width does not vary significantly with bias voltage.

Both the quantum efficiency and the frequency response of a p-i-n photodiode can be optimised by the geometric design of the device. One major limitation of p-i-n photodiodes that are made of indirect-gap semiconductors, such as Si and Ge, is the small absorption coefficients of these semiconductors in the spectral regions, where only indirect absorption takes place in such semiconductors.

A p-i-n photodetector can be of the heterojunction type, thus offering additional flexibility in optimising the performance. In a heterojunction photodiode, the active region normally has a bandgap that is smaller than one or both of the homogeneous regions. A large-gap homogeneous region, which can be either the top p^+ region or the substrate n region, serves as a window for the optical signal to enter. An intrinsic layer (lightly doped) is grown on a highly doped N^+ substrate and a p^+ layer is grown over the intrinsic layer [23]. The detector structure can be grown on a P^+ substrate over which an intrinsic layer and n^+ layer are grown. A representative structure of diffused p-i-n photodiode is shown in Figure 4.19. Ohmic contacts are taken out of the device for electrical connections. In most of the devices an anti-reflection coating is applied.

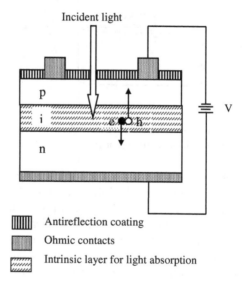

FIGURE 4.19: Structure of a diffused p-i-n photodetector

4.4.2.2 Avalanche photodiode

An internal gain is built into an avalanche photodiode (APD) to multiply the photo-generated electrons and holes [24], [25]. The physical process responsible for the internal gain in an APD is avalanche multiplication of charge carriers through impact ionization. When an incoming photon is absorbed, an electron-hole pair is generated, with the electron residing in the conduction band and hole in the valence band. Under strong reverse-bias, the electrons are accelerated towards the n-side and holes towards the p-side. As the electron gets accelerated, it collides with semiconductor lattice, thereby losing some of its energy. An electron gaining energy more than the bandgap energy E can generate another electron-hole pair, forcing the electron to move up to the conduction band and leaving the hole behind at the valence band. If the two electrons now gain more energy than E, they can further produce more electron-hole pairs and thus the process can continue. Similarly, the holes can also create more electron-hole pairs by impact ionization.

Under strong reverse-bias condition across the p$^+$ and n$^+$ layers, the minority carriers undergo impact ionization, thereby multiplying the number of minority carriers in avalanche photodiodes (APDs). If the impact ionization coefficients for electrons and holes are α_e and α_h, respectively, then the ionization ratio κ in APD is given by

$$\kappa = \frac{\alpha_h}{\alpha_e} \qquad (4.94)$$

The value of κ is made as low as possible to achieve maximum gain in APD.

There are many different structures developed for APDs. In principle, a p-n or p-i-n diode biased near its breakdown voltage can have an avalanche multiplication gain, thus functioning as an APD. Figure 4.20 shows a basic structure and its electric field distribution diagram for an avalanche photodiode. As can be seen high electric field is generated in the p type layer and almost constant electric field in intrinsic layer. Electric fields drops in depletion regions of both p$^+$ and p regions.

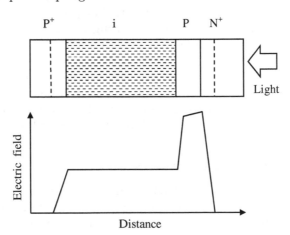

FIGURE 4.20: Structure of an APD and its electric field distribution

The total current gain G of an APD is denoted by the avalanche multiplication factor M of photo-generated carriers. It depends on the thickness and the structure of the avalanche region in the APD, as well as on the reverse voltage applied to the APD. In ideal situation, the avalanche multiplication gain for electron or hole injection into the avalanche region is given by

$$M = \frac{1 - \kappa}{e^{-(1-\kappa)\alpha_e d} - \kappa} \tag{4.95}$$

where d is the width of the region where carrier multiplication occurs. M can also be expressed in terms of α_h by replacing κ by $1/\kappa$ in the above relation. The gain of an APD is very sensitive to both reverse-bias voltage and temperature. The multiplication gain M increases nonlinearly with an increase in the value of $\alpha_e d$ and $\alpha_h d$, for any given value of κ. The value of M may approach infinity leading to avalanche breakdown. Because of the internal gain, the responsivity of an APD is $\mathcal{R} = M\mathcal{R}_0$, where $\mathcal{R}_0$ is the intrinsic responsivity of an equivalent photodiode without an internal gain.

In practice, the empirical gain of an APD is sometimes expressed as

$$M = \frac{1}{1 - (V_r/V_{br})^n} \tag{4.96}$$

where V_r is the reverse voltage on the APD, V_{br} is the avalanche breakdown voltage, and n is a constant in the range of 3 to 6.

The quantum efficiency of APDs can be increased by increasing the thickness of the absorption layer, but again the thickness has to be optimised to achieve stable controlled avalanche multiplication. Si and Ge having indirect bangaps, are materials used for APDs, since there is less band-to-band tunnelling. Generally, Si is used for wavelengths between 0.8 μm to 1.0 μm and Ge is used for wavelengths between 1.0 μm to 1.55 μm. Even group III-V materials are used for APDs in this wavelength range. A concept for optimising both photogeneration and avalanche multiplication in an APD is to use a separate absorption and multiplication (SAM) structure. This structure has separate regions for the two functions of photogeneration and avalanche multiplication. Photogeneration takes place in a relatively thick region of a moderately high field to reduce the carrier transit time, and the ionizing carriers are injected into a thin region of a very high field for avalanche multiplication.

4.4.3 Charge-coupled devices

A charge-coupled device (CCD) is a light-sensitive integrated circuit that stores and displays the data for an image in such a way that each pixel in the image is converted into an electrical charge. The movement of electrical charge, usually from within the device to an area where the charge can be manipulated into a digital value, is achieved by shifting the signals between stages within the device one at a time [26], [27]. CCD has found wide application in digital cameras for high-resolution imaging.

A CCD contains a photoactive region (an epitaxial layer of silicon), and a transmission region made out of a shift register. The photoactive region consists of a two-dimensional array of pixels that converts the incoming photons into electron charges. The pixels are represented by p-doped metal oxide semiconductor (MOS) capacitors. These capacitors are biased above the threshold for inversion when image acquisition begins, allowing the conversion of incoming photons into charges at the semiconductor-oxide interface. In n-channel CCDs, the silicon under the bias gate is slightly p-doped. An image is projected through a lens onto the capacitor array, causing each capacitor to accumulate an electric charge proportional to the light intensity at that location. The charges are accumulated in the capacitor bins or potential wells. The CCD is then used to read out by means of a control circuit, which allows each capacitor bin to transfer its contents to its neighbouring bin (operating as a shift register). The last capacitor in the array dumps its charge into a charge amplifier, which converts the charge into a voltage. By repeating this process, the controlling circuit converts the entire contents of the array in the semiconductor to a sequence of voltages. In a digital device, these voltages are then sampled, digitized, and usually stored in memory. Thus, CCD can detect spatially-varying image information as well as display the same [28].

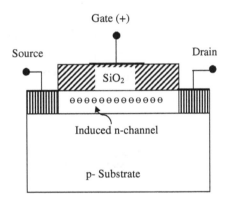

FIGURE 4.21: Depletion-type MOSFET CCD

Figure 4.21 shows the semiconductor layers of a MOSFET type CCD. The gate, the substrate, and the SiO_2 layer in between form a parallel plate capacitor. The gate is then biased at a positive potential, above the threshold for strong inversion, which will eventually result in the creation of an n-channel below the gate as in a MOSFET. Due to positive bias on the gate, the holes are pushed far into the substrate, and no mobile positive carriers are present at or near the surface. The CCD thus operates in a non-equilibrium state called deep depletion. When light is incident on the CCD sensor, then electron-hole pairs are generated in the depletion region. The pairs are then separated by the electric field, the electrons move toward the surface, and the holes move toward the substrate. Thus the free electrons generated in the vicinity of the capacitor are stored and integrated into the potential well.

The quantity of electrons stored in the well is a measure of the incident light intensity and is detected at the edge of the CCD, by transferring the charge package. Figure 4.22 shows the transfer of charge in CCD arrays by propagation of potential wells due to the sequential clocking of the gates. There are three main mechanisms that enable charge to be transferred from one CCD element to the next: self-induced drift, thermal diffusion, and fringe field drift. Self induced drift or carrier diffusion is a repulsion action between the charges that is responsible for majority of transfer. Charge transfer efficiency η_{cte} is a measure of the ability of the array to transfer charge, and is defined as the fraction of charge transferred from one element to the next. If a single charge packet of total integrated charge Q_0 is transferred down a register n times, the net charge remaining is $Q_n = Q_0(1 - n\epsilon_{cte})$, where ϵ_{cte} is the transfer loss of charge.

The sequential clocking of gates and the subsequent charge transfer is explained through Figure 4.22. During Phase I, when gates G_2 and G_5 are turned on and other gates are off, electrons are collected in potential wells W_2 and W_5. During Phase II of the next clock cycle, G_2, G_3, G_5, and G_6 are all on, whereas the rest are off, then W_2 and W_3 merge, forming a wider well and so also W_5 and W_6.

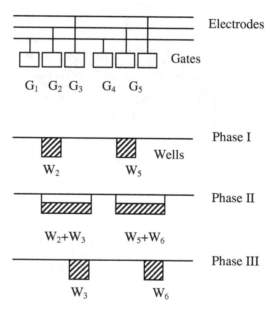

FIGURE 4.22: Shift of potential wells with sequential clocking of gates

During Phase III of the next clock cycle, W_2 and W_5 are turned off, whereas W_3 and W_6, are left on; then electrons previously stored in W_2 and W_5 are shifted and stored in W_3 and W_6, respectively. With the subsequent clocking of gates, all charge packages will be transferred to the edge and then picked up by the charge amplifier, which finally converts it into voltage.

4.4.4 Silicon photonic devices

Silicon photonics is evolving as an important device technology for information communication, realised by integrating photonic and electronic components on a silicon-based platform. Silicon photonics by definition is the study and application of photonic devices and systems which use silicon as an optical medium. The rationale behind the interest in silicon photonics technology is to use a single manufacturing process, the CMOS process. In this way, the level of integration of photonic devices can be increased, which in turn is reflected in an increase in the performance of such devices. Although the base material is silicon other materials, mainly Ge and IIIV semiconductors, have been used to evolve functionally better building blocks.

The silicon is usually patterned with sub-micrometre precision, into microphotonic components. Most of these devices operate in the infrared at the 1.55 μm wavelength. The wavelength range over which so-called Si-based photonic integrated circuit (PIC) can operate depends upon whether the photonic waveguide material is silicon or another material. Intrinsic silicon is transparent from its 1.1 μm indirect bandgap wavelength out to about 100 μm, allowing a wide scope for silicon waveguide operation. However, individual Si or SiGe/Si components function well at the mid-wave, long-wave, and very long-wave infrared (THz) ranges.

The silicon typically lies on top of a layer of silica and is known as silicon on insulator (SOI). The components are built upon a layer of insulator in order to reduce parasitic capacitance to give improved performance. The propagation of light through silicon devices is governed by a range of nonlinear optical phenomena including the Kerr effect, the Raman effect, two photon absorption and interactions. The presence of nonlinearity is of fundamental importance, as it enables light to interact with light, thus permitting applications such as wavelength conversion and all-optical signal routing, in addition to the passive transmission of light. The benefits provided by silicon photonic devices are:

1. wide band infrared transparency;

2. low-noise, high-speed integrated circuits;

3. high heat conductance; and

4. rugged 3D platforms for packaging.

Silicon is transparent to infrared light with wavelengths above about 1.1 μm. Silicon also has a very high refractive index, of about 3.5. The tight optical confinement provided by this high index allows for microscopic optical waveguides, which may have cross-sectional dimensions of only a few hundred nanometres. This is substantially less than the wavelength of the light itself, and is analogous to a sub-wavelength diameter optical fibre. Single-mode propagation can be achieved (as with single-mode optical fibre), thus eliminating the problem of modal dispersion.

The strong dielectric boundary effects that result from this tight confinement substantially alter the optical dispersion relation. By selecting the waveguide geometry, it is possible to tailor the dispersion to have desired properties, which is of crucial importance in applications requiring ultrashort pulses. In particular, the group velocity dispersion (that is, the extent to which group velocity varies with wavelength) can be closely controlled. In bulk silicon at 1.55 μm, the group velocity dispersion (GVD) is normal in that pulses with longer wavelengths travel with higher group velocity than those with shorter wavelength. By selecting a suitable waveguide geometry, however, it is possible to reverse this, and achieve anomalous GVD, in which pulses with shorter wavelengths travel faster.

In order for the silicon photonic components to remain optically independent from the bulk silicon of the wafer on which they are fabricated, it is necessary to have a layer of intervening material. This is usually silica, which has a much lower refractive index (of about 1.44 in the wavelength region of interest), and thus light at the silicon-silica interface will (like light at the silicon-air interface) undergo total internal reflection, and remain in the silicon. While silicon photonics allows realising compact photonic integrated circuits in the near-infrared, the possibility is still in research stages, to realise efficient current injection based light emitters and amplifiers, due to the indirect band gap of silicon. Some studies have recommended the integration of III-V semiconductor layer stacks on the silicon-based waveguide platform. This heterogeneous integration can be realised using semiconductor wafer bonding technology. This III-V layer can also be used for the active optical functionality of optical modulators, photodetectors, and switches.

4.4.4.1 Silicon photonic technologies for light sources

A silicon-based light-emitting device with high efficiency is a missing part in the evolution of light sources. In direct bandgap semiconductors, internal quantum efficiency may approach 100% because the radiative lifetime is shorter than the nonradiative lifetime. However, in Si, which is an indirect bandgap semiconductor, the radiative lifetime is long, and quantum efficiency drops by several orders of magnitude [29]. Therefore, it is necessary to suppress the nonradiative pathways to improve upon the internal quantum efficiency. In this context light emitting properties of bulk SiGe alloys are studied. Both crystalline silicon (c-Si) and crystalline germanium (c-Ge) are indirect bandgap semiconductors, and therefore SiGe crystalline alloys exhibit properties of an indirect band-gap semiconductor. SiGe alloys have crystalline material but with strong compositional disorder. As a result carrier recombination selection rules in SiGe alloys are partially relaxed enhancing the possibility of a generation of photo-luminescence spectra. This type of luminescence in SiGe bulk crystalline alloys has little practical use due to thermal quenching at room temperature [30]. However, $Si_{1-x}Ge_x$ multilayers with $0.1 < x < 0.2$ were intensively studied as a possible method to confine electron hole pairs in a double heterojunction or quantum well configuration, and to reduce the luminescence thermal quenching. An intense luminescence, but still restricted to low temperature, has been demonstrated [31].

Doping with rare earth ions offers another possibility of evolving the photonic source, whose emissions are characterised by sharp, atomic-like spectra with predictable and temperature-independent wavelengths. Semiconductor hosts doped with rare earth ions offer an important advantage, that is, dopants can be excited not only by a direct absorption of energy but also indirectly, by energy transfer from the host. This can be triggered by optical band-to-band excitation, giving rise to photo-luminescence, or by electrical carrier injection, resulting in electro-luminescence [32]. Among many possible rare earth-doped devices research interest has been mostly concentrated on Yb in InP and Er in Si. When Er^{3+} ions are embedded in SiO_2, the device is excited optically [33]. This is done by pumping resonantly with a laser beam directly to one of the excited states. Er-doped crystalline Si (c-Si:Er) emerged as an important source, where the most advanced and successful Si technology could be used whose emission coincides with the 1.5 μm communication wavelength [34], [35].

Band-engineered Ge-on-Si materials are also used as lasers. Epitaxial Ge-on-Si is a promising candidate for efficient monolithic light emitters and lasers due to its pseudo-direct gap behaviour and compatibility with Si CMOS technology. However, two critical aspects related to band-engineered Ge-on-Si light emitters need attention: (1) introduction of tensile strain [11], [36], [37] and (2) N-type doping [38]. The experimental demonstration of room temperature optical gain, and lasing under optical and electrical pumping, is a promising future for the applications of monolithic Ge-on-Si lasers. Developing electrically pumped Ge-on-Si lasers is the ultimate goal for practical applications.

4.4.4.2   Silicon photonic technologies for light modulators

One of the key challenges for the development of silicon photonics are the development of high speed, low loss, and compact optical modulators that provide modulated optical signal at the output driven by an electrical command when a continuous input beam is provided at the input [39], [40]. Some of the most successful demonstrations of optical modulators formed in silicon have been based upon the plasma dispersion effect, which relates changes in free electron and hole concentrations to changes in refractive index and absorption [41]. The change in refractive index produces phase modulation, which is later converted to intensity modulation in an interferometric or resonant structure. Various electrical structures have been demonstrated, however, generally they can be categorized into one of three types: carrier injection [42], [43], carrier accumulation, and carrier depletion [44], [45]. Carrier injection structures are generally based upon a PIN diode where the waveguide is formed within the intrinsic region. When the device is forward biased, electrons and holes are injected into the waveguide region of the device, causing a reduction in refractive index and a change in the phase of the light propagating through the device. Although the high-speed performance of carrier injection devices had progressively improved, the modulation bandwidth of this type of device is limited by the long minority carrier lifetime in silicon. Structures of carrier accumulation type use a thin insulating layer (or barrier) positioned in the waveguide. When the device is biased, free carriers accumulate on either side of the barrier much like around the dielectric layer in a capacitor. The accumulation of free carriers again reduces the refractive index of the silicon and causes phase modulation. Carrier depletion is an alternative technique where the structures employ a reversed biased PN junctions. The junction of the PN diode is positioned in interaction with the propagating light and as the device is reverse biased, the depletion region widens reducing the density of free carriers within the waveguide. The phase efficiency of the device is highly dependant on the doping concentrations of the p and n regions, as well as the junction position with respect to the waveguide.

Another category of modulators is based on electro-absorption that employs an electric field-dependent change in absorption to modulate the intensity of light. There are two main mechanisms used in electro-absorption modulators: the Franz-Keldysh effect (FKE) [46], and the quantum-confined Stark effect (QCSE) [47]. FKE is a bulk material mechanism while QCSE is the quantum-confined version of FKE. These effects are mainly seen in direct band gap semiconductors, such as GaAs, InP, Ge, and SiGe. FKE is a bulk semiconductor material mechanism, and can be viewed as the tunnelling of the electron and hole wave functions into the band gap with the application of an electric field, allowing for absorption below the band edge. Similar to FKE, QCSE is a shift toward lower energy absorption (longer wavelengths) with the application of the electric field. When two direct band gap semiconductor materials (well and barrier) form a heterostructure, the electron and holes become confined within the well material.

Among the different possibilities to get high-performance modulators, silicon depletion modulator and Ge-based electro-absorption devices are today probably the most promising solutions [48], [49].

4.4.4.3   Silicon photonic technologies for light detection

The state-of-the-art photodetectors in photonic communication are based on IIIV based materials. At a wavelength 850 nm, GaAs components are typically utilised. However, at 1300 nm and 1550 nm wavelengths, InGaAs detectors are universally used. Similarly, infrared imagers operating in the eye-safe regime ($\lambda > 1300$ nm) utilize InGaAs. To overcome the limitation of using Si platform, Ge photodetectors that can be integrated onto Si substrates are used. Ge is a nearly ideal material for infrared optical detection. It has an absorption edge of roughly wavelength 1800 nm, and shows strong absorption for wavelength lower than 1550 nm, a wavelength that corresponds to the direct transition energy in Ge. However, integration of Ge onto Si represents a significant materials challenge due to the large lattice mismatch between Si and Ge. Photodetectors grown on relaxed $Si_{1-x}Ge_x$ buffer layers, Ge grown directly on Si, Ge-on-insulator [50], and Ge-on-silicon-on-insulator (SOI) detectors are some of the recently evolved photodetectors. Ge-on-Si bulk photodetectors can be configured in a number of geometries, including vertical p-i-n, lateral p-i-n, and metal-semiconductor-metal structures. Recently Ge photodetectors have utilised vertical p-i-n geometries [51]. Vertical p-i-n geometries allow for a uniform field, created through the absorbing region, and therefore carrier transport to the contacts is primarily dominated by drift, not diffusion. This feature improves the speed responsivity trade-off as it allows the absorption layer to be made thicker while maintaining good carrier collection properties [52].

Ge-on-SOI and Ge-on-insulator free-space detectors are other alternative approaches to evolve photodetectors, where a buried insulating layer can eliminate a collection of carriers generated in the Si. This allows the devices to use thinner Ge absorption layers at wavelengths where Si is absorbing. The buried insulator can also act as a back-side reflector allowing improved efficiencies. The Ge/Si heterostructure system offers an opportunity to create APDs with improved performance, by creating separate absorption and multiplication regions. In Ge/Si APDs, the Ge is used for absorption, while the avalanche multiplication takes place in the Si [53].

Books for further reading

1. *Photonic Devices*: Jia-Ming Liu, Cambridge University Press, 2005.

2. *Fundamentals of Photonics*: B. E. A. Saleh and M. C. Teich, Wiley, 1991.

3. *Lasers and Electro-Optics: Fundamentals and Engineering*: C. C. Davis, Cambridge University Press, 1996.

4. *Elements of Photonics in Free Space and Special Media* (Vol-1): K. Iizuka, Wiley, 2002.

5. *Lasers*: P. W. Milonni and J. H. Eberly, Wiley, 1988.

6. *Fundamentals of Optoelectronics*: C. R. Pollock, Irwin, 1995.

7. *Lasers*: A. E. Siegman, University Science Books, 1986.

8. *Laser Fundamentals*: W. T. Silfvest, Cambridge University Press, 1996.

9. *Laser Electronics*: J. T. Verdeyen, Prentice-Hall, 1995.

10. *Applied Photonics*: C. Yeh, Academic Press, 1994.

11. *Solid State Lasers: New Development and Applications*: M. Inguscio and R. Wallenstein, Plenum Press, 1993.

12. *Solid-State Laser Engineering*: W. Koechner, Springer-Verlag, 1988.

13. *Lasers*: P. W. Milonni and J. H. Eberly, Wiley, 1988.

14. *Laser Fundamentals*: W. T. Silfvest, Cambridge University Press, 1996.

15. *Semiconductor Lasers*: G. P. Agrawal and N. K. Dutta, Van Nostrand Reinhold, 1993.

16. *Semiconductor Optoelectronic Devices*: P. Bhattacharya, Prentice-Hall, 1997.

17. *Physics of Optoelectronic Devices*: S. L. Chuang, Wiley, 1995.

18. *Optoelectronics*: B. E. Rosencher and B. Vinter: Cambridge University Press, 2002.

19. *Physics of Semiconductor Devices*: S. M. Sze, Wiley, 1981.

20. *Physical Properties of Semiconductors*: C. M. Wolfe, N. Holonyak, and G. E. Stillman, Prentice-Hall, 1989.

21. *Light Emitting Diodes*: K. Gillessen and W. Schairer, Prentice-Hall, 1987.

22. *Photodetectors: Devices, Circuits, and Applications*: S. Donati, Prentice-Hall, 2000.

Bibliography

[1] A.V. Mitrofanov and I. I. Zasavitskii. Optical sources and detectors. *Physical methods, instruments and measurements*, II, 2005.

[2] E. F. Schubert. *Light-Emitting Diodes*. Cambridge University Press, New York, 2006.

[3] K. Streubel, N. Linder, R. Wirth, and A. Jaeger. High brightness AlGaInP light-emitting diodes. *IEEE J. Sel. Top. Quant. Elect.*, 8(2):321–332, 2002.

[4] J. Kovac, L. Peternai, and O. Lengyel. Advanced light emitting diodes structures for optoelectronic applications. *Thin Solid Films*, 433:22–26, 2003.

[5] G. F. Neumark, I. L. Kuskovsky, and H. Jiang. *Wide Bandgap Light Emitting Materials and Devices*. Wiley-VCH, 2007.

[6] Y. Kashima, M. Kobayashi, and T. Takano. High output power GaInAsP/InP superluminescent diode at 1.3 m. *Elect. Let.*, 24(24):1507–1508, 1988.

[7] H. Kressel and M. Ettenburg. Low-threshold double heterojunction AlGaAs laser diodes: theory and experiment. *J. App. Phy.*, 47(8):3533–3537, 1976.

[8] K. Saito and R. Ito. Buried-heterostructure AlGaAs lasers. *IEEE J. Quantum Elect.*, 16(2):205–215, 1980.

[9] S. F. Yu. *Analysis and Design of Vertical Cavity Surface Emitting Lasers*. John Wiley, New York, 2003.

[10] C. W. Wilmsen, H. Temkin, and L. A. Coldren. *Vertical-Cavity Surface-Emitting Lasers: Design, Fabrication, Characterization, and Applications*. Cambridge University Press, Cambridge, 2001.

[11] C. J. Chang-Hasnain. Tunable VCSEL. *IEEE J. Sel. Top. Quant. Elect.*, 6(6):978–987, 2000.

[12] V. M. Ustinov, A. E. Zhukov, A. Y. Egorov, and N. A. Maleev. *Quantum Dot Lasers*. Oxford University Press, Oxford, 2003.

[13] X. Huang, A. Stintz, H. Li, J. Cheng, and K. J. Malloy. Modeling of long wavelength quantumdot lasers with dots-in-a-well structure. *Conf. Proc. CLEO 02. Tech. Dig*, 1:551–557, 2002.

[14] F. J. Duarte, L. S. Liao, and K. M. Vaeth. Coherence characteristics of electrically excited tandem organic light-emitting diodes. *Opt. Let.*, 30(22):30724, 2005.

[15] J. H. Burroughes, D. D. C. Bradley, A. R. Brown, R. N. Marks, K. MacKay, R. H. Friend, P. L. Burns, and A. B. Holmes. Light-emitting diodes based on conjugated polymers. *Nature*, 347 (6293):539541, 1990.

[16] R. Holmes, N. Erickson, B. Lussem, and K. Leo. Highly efficient, single-layer organic light-emitting devices based on a graded-composition emissive layer. *App. Phy. Let.*, 97:083308, 2010.

[17] S. A. Carter, M. Angelopoulos, S. Karg, P. J. Brock, and J. C. Scott. Polymeric anodes for improved polymer light-emitting diode performance. *App. Phy. Let.*, 70 (16):2067, 1997.

[18] J. N. Bardsley. International OLED technology roadmap. *IEEE Journal of Selected Topics in Quantum Electronics*, 10:3–4, 2004.

[19] K. T. Kamtekar, A. P. Monkman, and M. R. Bryce. Recent advances in white organic light-emitting materials and devices (WOLEDs). *Adv. Materials*, 22 (5):572–582, 2010.

[20] F. Stockmann. Photodetectors: their performance and their limitations. *App. Phy.*, 7(1):1–5, 1975.

[21] K. L. Anderson and B. J. McMurty. High speed photodetectors. *Proc. IEEE*, 54:1335, 1966.

[22] Melchior, M. B. Fisher, and F. R. Arams. Photodectors for optical communication systems. *Proc. IEEE*, 58:1466, 1970.

[23] R. G. Hunsperger (ed). *Photonic devices and systems*. Marcel Dekker, 1994.

[24] T. Kaneda. *Silicon and germanium abalenche photodiode*. Academic Press, New York, 1985.

[25] R. J. Mcintyre. Multiplication noise in avalanche photodiode. *Trans. Elect. Devices*, 13:164–168, 1966.

[26] W. S. Boyle and G. E. Smith. Charge coupled semiconductor devices. *Bell Syst. Tech. J.*, 49 (4):587–593, 1970.

[27] G. F. Amelio, M. F. Tompsett, and G. E. Smith. Experimental verification of the charge coupled device concep. *Bell Syst. Tech. J.*, 49 (4):593–600, 1970.

[28] J. P. Albert. *Solid-State Imaging With Charge-Coupled Devices.* Springer, 1995.

[29] INSPEC. *Properties of crystalline silicon.* IEE, 1999.

[30] L. C. Lenchyshyn, M. L. W. Thewalt, J. C. Sturm et al. High quantum efficiency photoluminescence from localized excitons in Si-Ge. *App. Phy. Let.*, 60(25):3174–3176, 1992.

[31] Y. H. Peng, C. H. Hsu et al. The evolution of electroluminescence in Ge quantum-dot diodes with the fold number. *App. Phy. Let*, 85(25):6107–6109, 2004.

[32] H. Vrielinck, I. Izeddin, and V. Y. Ivanov. Erbium doped silicon single and multilayer structures for LED and laser applications. *MRS Proc.*, 866(13), 2005.

[33] N. Q. Vinh, N. N. Ha, and T. Gregorkiewicz. Photonic properties of Er-doped crystalline silicon. *Proc. IEE.*, 97(7):1269–1283, 2009.

[34] M. A. Loureno, R. M. Gwilliam, and K. P. Homewood. Extraordinary optical gain from silicon implanted with erbium. *App. Phy. Let.*, 91:141122, 2007.

[35] A. Karim, G. V. Hansson, and M. K. Linnarson. Influence of Er and O concentrations on the microstructure and luminescence of SiEr LEDs. *J. Phy. Conf. Series*, 100:042010, 2007.

[36] M. J. Lee and E. A. Fitzgerald. Strained Si SiGe and Ge channels for high-mobility metal-oxide-semiconductor field-effect transistors. *J. App. Phy.*, 97:011101, 2005.

[37] X. Sun, J. F. Liu, L. C. Kimerling, and J. Michel. Room-temperature direct bandgap electroluminesence from Ge-on-Si light-emitting diodes. *Opt. Let.*, 34:1198–1200, 2009.

[38] J. M. Hartmann, J. P. Barnes, M. Veillerot, and J. M. Fedeli. Selective epitaxial growth of intrinsic and in-situ phosphorous-doped Ge for optoelectronics. *Euro. Mat. Res. Soc. (E-MRS) Spring Meet. Nice, France*, 14(3), 2011.

[39] D. A. B. Miller. Device requirements for optical interconnects to silicon chips. *Proc. IEEE*, 97(7):1166–1185, 2009.

[40] A. Liu, R. Jones, L. Liao, D. Samara-Rubio, D. Rubin, O. Cohen, R. Nicolaescu, and M. Paniccia. A high speed silicon optical modulator based on a metal-oxide-semiconductor capacitor. *Nature*, 427:615–618, 2004.

[41] L. Friedman, R. A. Soref, and J. P. Lorenzo. Silicon double-injection electro-optic modulator with junction gate control. *J. App. Phys.*, 63:1831–1839, 1988.

[42] P. D. Hewitt and G. T. Reed. Improving the response of optical phase modulators in SOI by computer simulation. *J. Lightwave Tech.*, 18:443–450, 2000.

[43] C. E. Png, S. P. Chan, S. T. Lim, and G. T. Reed. Optical phase modulators for MHz and GHz modulation in silicon-on-insulator. *J. Lightwave Tech.*, 22:1573–1582, 2004.

[44] F. Y. Gardes, G. T. Reed, N. G. Emerson, and C. E. Png. A submicron depletion-type photonic modulator in silicon on insulator. *Opt. Exp*, 13:8845–8854, 2005.

[45] C. Gunn. CMOS photonics for high-speed interconnects. *Micro. IEEE*, 26:58–66, 2006.

[46] W. Franz. Influence of an electric field on an optical absorption edge. *Z. Naturforsch*, 13a:484–490, 1858.

[47] D. A. B. Miller, D. S. Chemla, and S. Schmitt-Rink. Relation between electroabsorption in bulk semiconductors and in quantum wells: The quantum-confined Franz-Kedlysh effect. *Phy. Rev. B*, 33(10):6876–6982, 1986.

[48] A. Frova and P. Handler. Shift of optical absorption edge by an electric field: modulation of light in the space-charge region of a Ge p-n junction. *App. Phy. Let.*, 5(1):11–13, 64.

[49] R. M. Audet, E. H. Edwards, P. Wahl, and D. A. B. Miller. Investigation of limits to the optical performance of asymmetric Fabry-Perot electroabsorption modulators. *IEEE J. Quant. Elec.*, 48(2):198–209, 2012.

[50] L. Colace, G. Masini, and G. Assanto. Ge-on-Si approaches to the detection of near-infrared light. *IEEE J. Quant. Elec.*, 35:1843–1852, 1999.

[51] H. Y. Yu, S. Ren, W. S. Jung, D. A. B. Okyay, A. K.and Miller, and K. C. Saraswat. High-efficiency p-i-n photodetectors on selective-area-grown Ge for monolithic integration. *IEEE Elect. Device Let.*, 30:1161–1163, 2009.

[52] M. Jutzi, M. Berroth, G. Wohl, M. Oehme, and E. Kasper. Zero biased Ge-on-Si photodetector on a thin buffer with a bandwidth of 3.2 Ghz at 1300 nm. *Mat. Sci. in Semiconductor Processing*, 8:423–427, 2005.

[53] O. I. Dosunmu, D. D. Cannon, M. K. Emsley, L. C. Kimerling, and M. S. Unlu. High-speed resonant cavity enhanced Ge photodetectors on reflecting Si substrates for 1550 nm operation. *IEEE Photonics Tech. Let.*, 17:175–177, 2005.

Chapter 5

Photonic Devices for Modulation, Storage, and Display

5.1 Photonic devices for light-beam modulation

Photonic devices for light beam modulation are known as modulators. These devices are used to modulate a beam of light travelling through a medium or free space, or through a waveguide. Depending on the properties of light, photonic modulators are categorized into amplitude modulators, phase modulators, polarisation modulators, and the like. Modulators often use the electro-optic effect, acousto-optic effect, or magneto-optic effect, or they take advantage of polarisation changes in liquid crystals. There are also modulators which use Faraday effects. For low- and medium-speed applications, liquid crystal-based modulators are used [1].

5.1.1 Electro-optic effect

The electro-optic effect is related to the change in the index of refraction due to an applied electric field [2]. If the refractive index change is proportional to the applied field, then the effect is known as linear electro-optic effect or the Pockels effect. If the index change is quadratically dependent on the applied field, then the effect is known as the quadratic electro-optic effect or the Kerr effect. The effect is caused by forces that distort the positions, orientations, or shapes of the molecules of the material, which is reflected into the changes of refractive index. The change is typically very small yet its effect on an optical wave propagating a distance much greater than the wavelength of light in the medium can be significant [3].

The optical properties of anisotropic media as a function of electric field E can be described by electric impermeability η since $\eta = \epsilon_0/\epsilon = 1/n^2$. The incremental change of η is given by

$$\Delta\eta = (d\eta/dn)\Delta n = (-2/n^3)(-\frac{1}{2}\zeta n^3 E - \frac{1}{2}\varrho n^3 E^2) = \zeta E + \varrho E^2 \qquad (5.1)$$

Therefore,

$$\eta(E) \approx \eta(0) + \zeta E + \varrho E^2 \qquad (5.2)$$

where ζ and ϱ are the electro-optic coefficients and $\eta(0)$ is the impermeability at $E = 0$. The values of the coefficients ζ and ϱ depend on the direction of the applied electric field and the polarisation of the light.

The refractive index $n(E)$ of an electro-optic medium is a function of the applied electric field E and can be expressed about $E = 0$ as

$$n(E) = n(0) + a_1 E + \frac{1}{2} a_2 E^2 + \ldots\ldots, \tag{5.3}$$

where $a_1 = dn/dE$ and $a_2 = d^2n/dE^2$. Terms higher than the third can safely be neglected. $n(E)$ can also be expressed in terms of electro-optic coefficients as

$$n(E) \approx n - \frac{1}{2}\zeta n^3 E - \frac{1}{2}\varrho n^3 E^2 + \ldots \tag{5.4}$$

In many materials, the term associated with E^2 is negligible in comparison with the second term. Such a medium is known as a Pockels medium (or a Pockels cell) and the coefficient ζ is called the Pockels coefficient or the linear electro-optic coefficient. The most common crystals used as Pockels cells include $NH_4H_2PO_4$, (ADP), KH_2PO_4 (KDP), $LiNbO_3$, $LiTaO_3$, and CdTe.

If the material is centro-symmetric, $n(E)$ is an even symmetric function, as it is invariant to the reversal of E, and therefore the first derivative of E vanishes to yield the relation

$$n(E) \approx n - \frac{1}{2}\varrho n^3 E^2 \tag{5.5}$$

The material is then known as a Kerr medium (or a Kerr cell) and the parameter ϱ is called the Kerr coefficient or the quadratic electro-optic coefficient. The Kerr effect is seen in some gases, liquids, and in certain crystals [4].

5.1.1.1 Electro-optic modulators

Phase modulation: In a phase modulator, an electric field E is applied across a Pockels cell of length L. If a beam of polarised light traverses the cell, it undergoes a phase shift of $\varphi = 2\pi n(E)L/\lambda_0$, where λ_0, is free-space wavelength of incident light and n is the refractive index when the field is 0. In terms of Pockels constant ζ, the phase shift φ is given by

$$\varphi = \frac{2\pi n L}{\lambda_0} - \frac{\pi}{\lambda_0}\zeta n^3 E L \tag{5.6}$$

The electric field E is related to voltage V by the relation $E = V/d$ where V is the applied voltage across two faces of the cell separated by distance d. The phase change can then be related in terms of a half-wave voltage V_{hwv}

$$\varphi = \frac{2\pi n L}{\lambda_0} - \pi \frac{V}{V_{hwv}} \tag{5.7}$$

where the parameter V_{hwv} is the half-wave voltage, which is the applied voltage at which the phase shift changes by π. The half-wave voltage depends on the material properties n and ζ and is given by

$$V_{hwv} = \frac{d}{L} \frac{\lambda_0}{\zeta n^3} \tag{5.8}$$

Therefore, it is possible to modulate the phase of an optical wave linearly by varying the voltage V that is applied across a material through which the light passes. The electric field may be applied in a direction perpendicular to the direction of light propagation (transverse modulators) or parallel to that direction (longitudinal modulators). A schematic diagram of phase modulation using Pockels cell is shown in Figure 5.1.

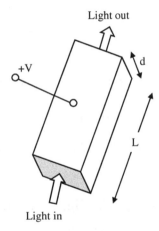

FIGURE 5.1: Phase modulation using a Pockels cell

The speed at which an electro-optic modulator operates is limited by electrical capacitive effects and by the transit time of the light through the material. However, electro-optic modulators can be constructed as integrated-optic devices which operate at higher speed and at lower voltage than the bulk devices.

Intensity modulation: Phase modulation alone does not affect the intensity of a light beam. However, a phase modulator placed in one branch of an interferometer is common to obtain intensity modulation and can function as an intensity modulator. In the Mach-Zehnder interferometer configuration a phase modulator is placed in one branch of an interferometer. If the beamsplitters divide the optical power equally, the transmitted intensity I_0 is related to the incident intensity I_i by

$$I_0 = I_i \cos^2 \frac{\varphi}{2} \tag{5.9}$$

where φ is the difference between the phase shifts encountered by light as it travels through the two branches of the interferometer. Phase difference φ is controlled by the applied voltage V in accordance with the linear relation as discussed earlier. A schematic diagram of a representative intensity modulator is shown in the Mach-Zehnder interferometer configuration is shown in Figure 5.2.

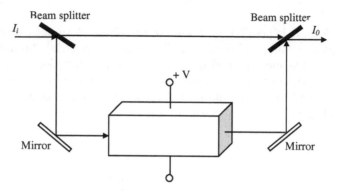

FIGURE 5.2: Intensity modulation using a Mach-Zehender interferometer

An optical intensity modulator using a Pockels cell can also be evolved by placing it between two crossed polarisers. The device modulates light by rotating the plane of polarisation of the output polariser when the applied voltage is half-wave voltage V_{hwv}.

5.1.2 Liquid crystal spatial light modulators

To process information and introduce it into the image processing system as spatially modulated optical signal through an electronic interface, a large number of photonic devices are explored. Such devices are broadly called spatial light modulators (SLMs) [5]. The SLM is a category of device that spatially transforms an input signal (either electrical or optical) as encoded information in the intensity or phase or both of an optical beam. Thus, by definition, a spatial light modulator (SLM) is an object that imposes some form of spatially varying modulation of the intensity of a beam of light. It is also possible to produce devices that modulate the phase of the beam or both the intensity and the phase simultaneously.

SLMs are generally fabricated with liquid crystal (LC) as the control element. Liquid crystal is an intermediate state between liquid and solid that exhibits the properties of both the crystalline solid and the isotropic liquid. Molecules in liquid crystals tend to align themselves in some specific direction. Most liquid crystal compounds exhibit polymorphism, or a condition where more than one phase is observed in the liquid crystalline state. The term mesophase is used to describe the sub-phases of liquid crystal materials. Mesophases are formed by changing the amount of order in the sample, either by imposing order in only one or two dimensions, or by allowing the molecules to have a degree of translational motion.

A characteristic of liquid crystals arises from the rod-like shape of their molecules. The molecules in liquid crystal, however, do not exhibit any positional order. Positional order refers to the extent to which an average molecule or group of molecules shows translational symmetry although LC molecules do possess a certain degree of orientational order. Orientational order represents a measure of the tendency of the molecules to align on a long-term basis. The molecules, however, do not all point in the same direction all the time. They merely tend to point more in one direction over time than in other directions. This direction is referred to as the *director* of the liquid crystal. The amount of order is measured by the order parameter of the liquid crystal, which can be found by averaging the orientation angle θ for all the molecules in the sample, where θ is the angle the long axis of the molecule makes with a particular direction. This order parameter is highly dependent on the temperature of the sample.

Depending on the temperature and nature of the crystal, liquid crystals can exist in one of several distinct phases known as nematic, smectic, or cholesteric, as shown in Figure 5.3.

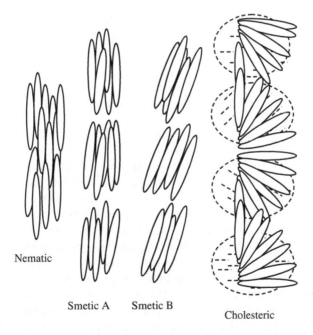

Nematic

Smetic A   Smetic B

Cholesteric

FIGURE 5.3: Phases of liquid crystals: (a) nematic phase, (b) smectic A, (c) smectic B, and (d) cholesteric phase

The classes are differentiated by different molecular order. Liquid crystals in a nematic phase are parallely oriented in the same direction, but randomly located centres in the volume. In the smectic phase, the crystals are oriented in the same direction, but their centres lie in parallel layers with randomness of location only within the layer. Depending on the degree of orientation smectic liquid crystals are called smectic A or smectic B as shown in Figure 5.3. The cholesteric phase can be describes as a special type of smectic LC in which the thin layers of crystals have their longitudinal axes rotated in adjacent layers at certain angles. The direction changes gradually and regularly with distance, so that the director follows a helical path.

A class of smectic liquid crystals are better known as ferroelectric liquid crystals (FLCs) where the angle of the molecules within a single layer are constrained to lie at an angle θ with layer normal, as shown in Figure 5.4. The FLC molecules possess a permanent dipole moment and hence retain the molecular orientation even after the electric field is removed, and thus possess memory, too. Applying an electric field to a liquid crystal molecule

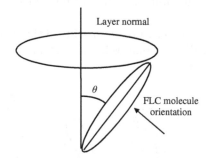

FIGURE 5.4: Molecular orientation of an FLC molecule

with a permanent electric dipole will cause the dipole to align with the field. If the molecule did not originally have a dipole, then it is induced when a field is applied. Either of these situations has the ultimate effect of aligning the director of the liquid crystal with the electric field being applied. Since the electric dipole across liquid crystal molecules varies in degree along the length and the width of the molecule, some kinds of liquid crystal require less electric field and some require much more in order to align the director. The ratio of electric dipole per unit volume of crystal to the field strength is called the electric susceptibility and is a measure of electrically polarising the material. To refresh the FLC molecular state, DC fields of opposite polarity are applied across the FLC layers. Because of the permanent dipole moment of FLC molecules, the electrical response time or switching time of FLC is much faster ($\sim 50\mu$ s) than that of nematic type (~ 20 ms) molecules.

Liquid crystal is birefringent. It possesses two different indices of refraction. One index of refraction corresponds to light polarised along the director of the liquid crystal, and the other is for light polarised perpendicular to the director. Light propagating along a certain direction has electric and magnetic field components perpendicular to that direction. In case the director lies perpendicular to the direction of propagation of the light, each component can be broken into two components, one parallel to the director of the crystal and one perpendicular. Essentially, these two components of either the electric or the magnetic field propagate through the liquid crystal at different speeds, and therefore, most likely will be out of phase when they exit the crystal. If the light is linearly polarised before it enters the liquid crystal, meaning that the electric and magnetic field components of the light oscillate in one direction, then the light will not be linearly polarised when it leaves the liquid crystal, unless the angle between the polarisation direction and the director of the crystal is 0 or 90 degrees. Depending on how far out of phase the two waves are when they emerge (which depends on the thickness of the crystal), the resulting light might again be linearly polarised, or more likely, the light will be elliptically polarised, meaning that the polarisation of the light will rotate around the direction of propagation in some fashion.

A type of liquid crystal, known as chiral nematic liquid crystal, is bire-fringent in a different way. Considering that the helical structure is aligned with the direction of propagation of the light, circularly polarised light will travel through the crystal at different speeds depending on whether it is right-circularly polarised or left-circularly polarised Thus, circular birefringence is exhibited in chiral nematic liquid crystal because of the helical structure the molecules form. The circular birefringence is highly wavelength dependent, so light of different colours gets modified by different amounts.

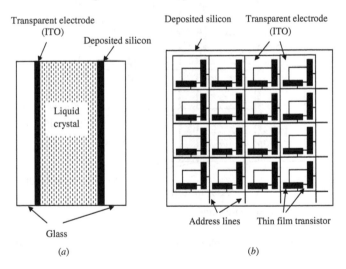

FIGURE 5.5: (a) A generic electrically addressed liquid crystal cell, and (b) its addressing mode

Liquid crystal-based spatial light modulators are broadly categorized as electrically addressed or optically addressed SLM modulators depending on the mode of transmission. In the former category, the electrical signal representing the input information drives the device in such a way that the spatial distribution of light is controlled through absorption, transmission, or phase shift. In the latter category, the SLM functions in such a way that an incoherent image may be converted to a coherent image; thus it can cause image amplification from a weak image to an intense image. Such a situation can also provide wavelength conversion. A generic transmissive type SLM and its addressing mode is shown in Figure 5.5(a) and (b). A common approach to fabricating an EASLM is to apply matrix-addressing electrodes to a substrate. The pixels are formed at the intersection of the electrodes, which modulate the voltage dropped across the sandwiched light-modulating medium.

5.1.2.1 Electrically addressed liquid crystal-based spatial light modulator

Electrically addressed spatial light modulators (EASLMs) are generally based on nematic liquid crystals or the ferroelectric liquid crystals. The layer of liquid crystal is contained in between two glass plates. Boundary conditions are imposed on the alignment of liquid crystals by polishing soft alignment layers coated on the glass plates with strokes in the desired direction. The inside surface of the glass plates is coated with a transparent conductive layer, usually indium tin oxide films, as shown in Figure 5.6.

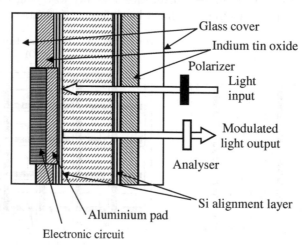

FIGURE 5.6: An electrically addressed liquid crystal cell

An EASLM may either be of the transmissive or reflective type. In a reflective EASLM, the reflective surface is the functional area. The input image is read by light modulation through an electrical field. With the FLC director parallel to the input polariser, the light transmitted through the cell remains linearly polarised and is blocked by the output polariser. If the state of the FLC is then switched by means of an applied voltage, the FLC director rotates through an angle 2ϕ, where ϕ is the angle between the director and the normal to the layer planes. The polarisation vector of light transmitted through the cell is rotated by twice the angle through which the director has shifted.

To make it possible to address a large number of rows sequentially, a transistor at each pixel transfers the electrical information for that pixel. Therefore, the addressing of the device can be carried out rapidly while the reading can be done more slowly in the remaining frame period. EASLMs are also fabricated by incorporating an integrated circuit (IC) silicon substrate addressing matrix, and a liquid crystal light-modulating medium. The pixel size in these active backplane SLMs is limited to that which can be formed by very large scale integration (VLSI) IC fabrication. While using an EASLM, the factors such as the maximum number of pixels, frame rate, fill factor (fraction of an area that is optically usable), number of distinct gray levels, and contrast ratio need to be addressed. EASLM drive circuits now utilize conventional interface such as VGA or DVI input for addressing pixels.

5.1.2.2   Optically addressed spatial light modulator (OASLM)

An optically addressed spatial light modulator (OASLM) consists of a photosensor (amorphous silicon (α Si:H) or photo transistor) followed by a layer of liquid crystals. Schematic diagram of an OASLM is shown in Figure 5.7.

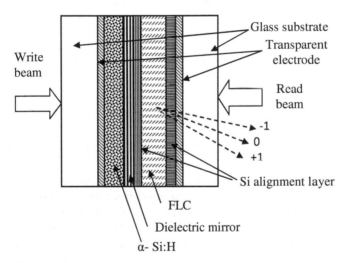

FIGURE 5.7: Schematic structure of a reflective OASLM

Light incident on the photosensor side produces a spatial light variation at the liquid crystal layer. The image on an optically addressed spatial light modulator is created and changed by shining light encoded with an image on its front or back surface. A photosensor allows the OASLM to sense the brightness of each pixel and replicate the image using liquid crystals. As long as the OASLM is powered, the image is retained even after the light is extinguished. An electrical signal is used to clear the whole OASLM at once. OASLMs can be of a transmissive or reflective type. The structure of a reflective type

OASLM is similar to that of an EASLM, except for the introduction of a photosensor layer, $\alpha Si : H$, which replaces the electrodes of control electronics matrix. Read beam and write beam are also shown in Figure 5.7. Diffracted beams are shown as $-1, 0, +1$. The optical write-image is directly sensed by the photosensor and the image is transferred to the modulating layer.

A typical application of OASLM is wavelength conversion or conversion from incoherent light to coherent light as shown in Figure 5.8. Images illuminated by incoherent light are allowed to fall as write beam on the OASLM and a coherent light source is used as read beam via a beam-splitter. The image of coherent light is available in the read beam. A smart OASLM can be built into this OASLM in the electronic circuit between the photosensor and the pixels, implementing some specific optical image processing functions such as edge detection.

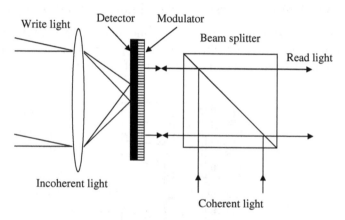

FIGURE 5.8: OASLM in incoherent to coherent image conversion

OASLMs are often used as the second stage of a very-high-resolution display, such as one for a computer-generated holographic display. In a process called active tiling, images displayed on an EASLM are sequentially transferred to different parts on an OASLM, before the whole image on the OASLM is presented to the viewer. As EASLMs can run at fast frame rate, it is possible to tile many copies of the image on the EASLM onto an OASLM while still displaying full-motion video on the OASLM.

5.1.3 Magneto-optic spatial light modulator

The EASLMs are based on electro-optic effect. However, SLMs that are based on Faraday effect or operate through rotation of polarisation direction with the application of magnetic field are magneto-optic spatial light modulators (MOSLMs). This magnetically induced rotation of the plane of polarisation of a linearly polarised optical wave is called Faraday rotation. However, the sense of Faraday rotation is independent of the direction of wave propagation. In a paramagnetic or diamagnetic material, which has no internal magnetisation, the Faraday rotation for a linearly polarised wave propagating over a distance l is linearly proportional to the externally applied magnetic field. The Faraday rotation angle θ_F in this case is generally expressed as

$$\theta_F = V_c H l \tag{5.10}$$

where, V_c is called the Verdet constant. The Verdet constant has positive values for diamagnetic materials and negative values for paramagnetic materials. The Verdet constant of a given material is a function of both optical wavelength and temperature. In the optical spectral region, its absolute value usually increases when the optical wavelength or the temperature decreases. In a ferromagnetic or ferrimagnetic material, which has an internal magnetisation, Faraday rotation is determined by the magnetisation rather than by the applied magnetic field, but is determined only by the direction of the external magnetic field, or that of the magnetisation if the material is ferromagnetic or ferrimagnetic.

A magneto optic spatial light modulator (MOSLM), as shown in Figure 5.9, consists of a 1D or 2D spatial array of independently addressable Faraday rotators placed between a polariser and an analyser [6]. A single-crystal magnetic thin film, commonly a bismuth-substituted iron garnet film of high specific Faraday rotation coefficient, is grown on a lattice-matched, transparent, non-magnetic garnet substrate.

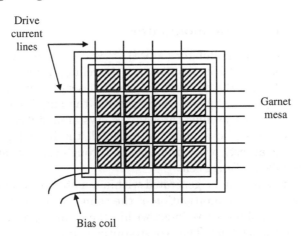

FIGURE 5.9: Structure of MOSLM

The magnetic film is structured into a one- or two-dimensional array of isolated mesas with X-Y drive lines or metallic electrodes deposited alongside each mesa for current switching of the individual mesas. The garnet mesas are transparent to the incident light, but when magnetized, the axis of polarisation of incident light transmitted through the film is rotated by the Faraday effect in a direction determined by the direction of magnetisation of the mesa. A bias coil surrounds the pixelated garnet film through which current can be driven in both directions, thereby establishing magnetic field parallel to the direction of light propagation as well as anti-parallel to it.

The bias coil is driven by current pulses injected across the X-Y matrix of metallic electrodes, such that a strong magnetic field is induced in the direction of the desired magnetisation. The state of an individual mesa changes only when the magnetic field is strong enough to nucleate such a change. Each mesa is written with one cycle of current pulse across the bias coil.

The Faraday effect causes a rotation of the polarisation plane of the incident light, but this rotation angle θ is generally quite small, much less than 90°. The Faraday rotation results from the difference in refractive indices of the left-circularly polarised and right-circularly polarised light component of a propagating wave. The rotation angle θ is given by

$$\theta = \frac{\pi(n_2 - n_1)d}{\lambda} \qquad (5.11)$$

where λ is the wavelength of light, d is the film thickness, and n_1 and n_2 are the refractive index experienced by the left-circularly polarised and right-circularly polarised light components respectively. A polarisation analyser converts this effect into the modulation of image brightness. To use the device as an intensity modulator, the output analyser should be perpendicular to the direction of polarisation of light in one of the two states of rotation.

5.1.4   Acousto-optic modulator

The acousto-optic modulator (AOM) [7] is used to modulate an incident optical wavefront by the interaction of a travelling acoustic wave. The frequency, intensity, and the direction of a light (laser) beam can be modulated by an acousto-optic modulator where varying the amplitude and frequency of the acoustic waves travelling through the crystal modulate the light beam. Within the acousto-optic modulating material, the incoming light is Bragg diffracted by the acoustic wave propagating through the crystal and therefore the device is called a Bragg cell.

An acoustic wave can be generated within a material by a vibrating piezoelectric transducer. The propagation of the sound wave through the medium causes a variation of refractive index within the medium, resulting in a phase grating within the medium. The acousto-optic modulator operates in two different regimes, namely the Raman-Nath regime and the Bragg regime, and exhibits different properties in the two regimes. The incoming light incident on such a medium of varying refractive index scatters off the light, and interference occurs similar to Bragg diffraction. A diffracted beam emerges at an angle θ that depends on the wavelength of the light λ and the wavelength of the sound Λ and and are related by

$$2\Lambda \sin \theta = m\lambda \qquad (5.12)$$

where the incident light is normal to the direction of propagation of sound waves, and $m = ..., -2, -1, 0, 1, 2, ...$ are the diffraction orders.

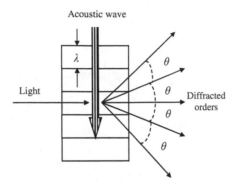

FIGURE 5.10: Acousto-optic diffraction

A schematic diagram of acousto-optic modulator and its diffracted orders are shown shown in Figure 5.10. The incident light may be modulated in its intensity, frequency, or phase. Since sound wave propagates through the medium, Bragg diffraction occurs due to scattering from moving planes. As a consequence, the frequency f of the diffracted beam in diffraction order m will be Doppler-shifted by an amount equal to the frequency F of the sound wave, to conserve energy and momentum. That is, $f \rightarrow f + mF$, where m is an integer representing the order of the shift. The phase of the diffracted light beam will also be shifted by the phase of the sound wave.

AOMs are used in lasers for Q-switching, mode-locking, in telecommunications for signal modulation, and in spectroscopy for frequency control. The time an AOM takes to modulate the incident light beam is roughly limited by the transit time of the sound wave across the beam (typically 5 to 100 ns). This is fast enough to create active mode-locking in an ultrafast laser.

5.1.5 Deformable mirror devices for light modulation

Deformable mirror devices (DMD) use electrically induced mechanical deformation of membranes or mirrors to modulate a reflected light wave [8], [9]. Although classically, these devices used membranes which deformed under an electrical field exerted by the driving electrodes, now DMDs have evolved as an array of cantilevered mirrors, which are individually addressed by voltages applied across floating MOS (metal-oxide semiconductor) sources. DMDs also use mirror elements mounted on flexible supports such that they can twist under an electrical field.

The membrane DMD consists of a metal-coated polymer membrane stretched between spacers, such that the membrane is separated from the underlying address electrode by the spacer. When a positive voltage with respect the ground substrate is applied across the address electrodes, the membrane bends towards the underlying electrode, but returns to its original state when the voltage across electrodes is removed. Thus a phase modulation of the incident light occurs, and this can be appropriately converted to intensity modulation by the use of proper optics.

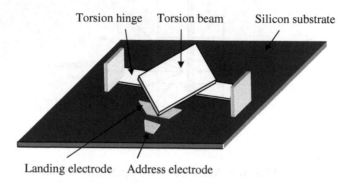

FIGURE 5.11: Deformable mirror device

Figure 5.11 shows a cantilevered mirror DMD. This type of cantilever beam DMD consists of a hinged metallised beam over a spacer support, which is biased to a negative voltage. When a positive voltage with respect the ground substrate is applied across the address electrodes, the metal beam deflects downwards, thereby deflecting the incident light beam, and resulting in intensity modulation. The most advanced DMDs use a torsion beam connected at two points, rather than hinged at a single point support. The torsion rod connects the metallised mirror to supports at the ends of the diagonal of a square pixel, as shown in Figure 5.12.

The mirror has a negative bias voltage across it. Two address electrodes and two landing electrodes exist for each pixel, one on either side of the rotating axis, as shown in Figure 5.12(a). When one of the address electrodes is given a positive voltage, the mirror twists in one direction, whereas it twists in the opposite direction when the other address electrode is made positive. Thus when the address electrodes are activated, the incident light is deflected in either directions. For example, five pixels are shown in Figure 5.12(b), which shows the deflection of light on two pixels, where the pixels are at off state.

5.2   Photonic switching devices

The functioning of electro-optic, magneto-optic, and acousto-optic devices discussed in previous sections are based on the fact that the optical properties of a material depend on the strength of an electric, magnetic, or acoustic field that is present in an optical medium. At a sufficiently high optical intensity,

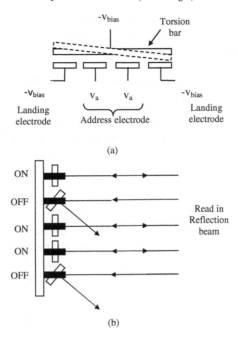

FIGURE 5.12: (a) Torsion type deformable mirror device (side view) and (b) deflection in OFF pixels

the optical properties of a material also become a function of the optical field. Such nonlinear response to the strength of the optical field results in various nonlinear optical effects. Optical switches are one of the attractive applications of optical nonlinearity in which optical bistable devices are a major subset. Switching devices are evaluated on the basis of these parameters:

1. *Switching Time τ*: This is a measure of the time in which the output intensity changes from 10% to 90% of the maximum intensity. It is related to the -3dB bandwidth $\Delta\nu$ as

$$\tau = \frac{0.35}{\Delta\nu} \tag{5.13}$$

2. *Contrast ratio R_c*: This is also termed as the on-off ratio. It is the ratio of maximum I_{max} to minimum I_{min} of transmitted light intensity and is expressed in decibels as

$$R_c = 10\log\frac{I_{max}}{I_{min}} \tag{5.14}$$

3. *Insertion loss R_l*: It is the fraction of power loss in the system and is expressed in decibels as

$$R_l = 10\log\frac{P_{out}}{P_{in}} \tag{5.15}$$

where P_{out} is the transmitted power when the switch is not placed in the system and P_{in} is the transmitted power when the switch is placed in the system.

4. *Power Consumption*: It measures the power consumed by the switching device when in operation.

5. *Cross-talk*: In the case of a routing switch having one input and two output channels, cross-talk describes how much of the input signal appears on the unselected output channel. Ideally, the whole signal should appear on the selected channel, and no signal in the unselected channel.

A bistable device has two stable output states under one input condition. The necessary conditions for optical bistability are optical nonlinearity and positive feedback. Depending on the optical nonlinearity responsible for the bistable function from the real or the imaginary part of a nonlinear susceptibility, two types of nonlinear optical elements can be evolved. These may be a dispersive nonlinear element, for which the refractive index n is a function of the optical intensity, or a dissipative nonlinear element, for which the absorption coefficient α is a function of the optical intensity. Depending on the type of feedback as well, a bistable optical device can also be classified as either intrinsic or hybrid. In an intrinsic bistable device, both the interaction and the feedback are all optical. In a hybrid bistable device, electrical feedback is used to modify the optical interaction, thereby creating an artificial optical nonlinearity.

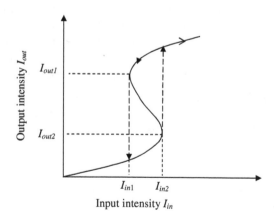

FIGURE 5.13: Characteristics of intensity bistability

Figure 5.13 shows a generic characteristic for intensity bistability of a bistable optical device. For each input intensity I_1 within the range between I_{in1} and I_{in2}, there are three values for the output intensity I_{out}. Only the two values I_{out1} and I_{out2} that lie on the upper and the lower branches of the curve are stable output values. The one that lies on the middle branch is unstable because the middle branch has a negative slope. When the input intensity is gradually increased from zero, the output intensity traces the lower branch of the curve until the input intensity reaches the up-transition point I_{in1} where the output makes a sudden jump to the upper branch. Once the system is in a state that lies on the upper branch, it can be brought back to the lower branch only when the input intensity is lowered to the down-transition point at I_{in2}. If the input intensity is set at a value within the bistable region, the output can be in either stable state, depending on the history of the system. With a proper external excitation, it can be switched from one of the stable states to the other, otherwise, it stays in one state indefinitely. Because both I_{in} and I_{out} are real and positive quantities, only two stable transition points are possible at (I_{in1}, I_{out1}) and (I_{in2}, I_{out2}).

Because of the binary feature, bistable devices can be used for many digital operations, such as switches, memories, registers, and flip-flops. Thus bistable optical devices are used as optical logic gates, memories, and analog-to-digital converters in photonic signal processing systems.

5.2.1   Etalon switching devices

Dispersive nonlinear elements have been devised whose transmittance is a nonmonotonic function of an intensity-dependent refractive index $n = n_0 + n_1 I_0$, where the refractive indices n_0 and n_1 are constants and I_o is an intensity value. Examples of dispersive elements are interferometers, such as the Mach-Zehnder, and the Fabry-Perot etalon, with a medium exhibiting optical Kerr effect.

In the Mach-Zehnder interferometer, the nonlinear medium is placed in one branch of the path. The power transmittance T of the system is

$$T(I_0) = \frac{1}{2} + \frac{1}{2} \cos\left(\frac{2\pi d n_1}{\lambda_0}\right) I_0 + \varphi \tag{5.16}$$

where the phase φ is a constant, d is the length of the active medium of refractive index n_1, and λ_0 is the free-space wavelength.

Another simple bistable optical device, known as the Fabry-Perot etalon can be constructed by placing a nonlinear optical medium inside a cavity.

The mirrors of the cavity provide the needed optical feedback to the non-linear optical interaction. Optical wave in a Fabry-Perot cavity travels through the nonlinear medium twice in each round trip and forms a standing wave pattern. Since the intracavity intensity is much greater than the input intensity due to repeated to-and-fro reflections within the cavity. Self-switching of the device is also possible. As shown in Figure 5.14, two mirrors of the etalon have high relectivities R_1 and R_2 and are separated by distance L with a medium of refractive index n. The maximum transmittance is 1 when $R_1 = R_2$ and the minimum transmittance approaches 0 when reflectances approach unity. For simplicity, the mirrors of the Fabry-Perot cavity are assumed identical and lossless having reflectances R. Under this condition, if I_{in} and I_{out} are

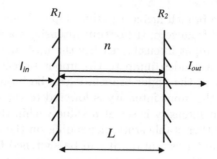

FIGURE 5.14: A Fabry-Perot etalon

the input and output intensities through the etalon, then the transmittance T is given by

$$T = \frac{I_{out}}{I_{in}} = \frac{(1 - R)^2 e^{-\alpha L}}{(1 - Re^{-\alpha L})^2} + 4Re^{-\alpha L} \sin^2 kL \qquad (5.17)$$

where k and α are the propagation constant and the absorption coefficient, respectively. The expression given above is modified for dispersive bistability in a Fabry-Perot cavity when the nonlinear medium has an intensity-dependent index of refraction due to the optical Kerr effect.

5.2.2 Self-electro-optic effect device

The self-electro-optic effect device or SEED [10] utilizes the optical absorption and current generation associated with multiple quantum well p-i-n structures of alternating thin layers of GaAs and AlGaAs. The basic SEED structure, also called the resistor-biased SEED, is shown in Figure 5.15(a).

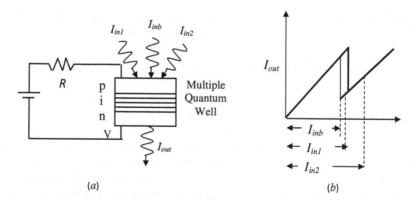

FIGURE 5.15: (a) Schematic structure of a SEED and (b) input/output characteristics of a SEED

Here, the p-i-n diode acts both as a modulator and as a detector. Because the bandgap of AlGaAs is greater than that of GaAs, quantum potential wells are formed which confine the electrons to the GaAs layers. An electric field is applied to the material using an external voltage source. The absorption coefficient is a nonlinear function of the applied voltage V at the wells and on the optical intensity I since the light absorbed by the material creates charge carriers which alter the conductance. In the absence of incident light on the diode, no current flows through the circuit and the applied voltage drop occurs across the multiple quantum well in the intrinsic region. When optical radiation is incident on the device, some of it gets absorbed and the p-i-n diode acting as a photodetector generates current. Hence there occurs a greater voltage drop across the resistor and less across the diode. The lower voltage drop across the diode increases the absorption in the device, and creates greater current generation, resulting in even more drop across the resistor and even less drop across the diode. The reverse mechanism occurs when the incident optical power is decreased. So, pixels on which light is incident become absorbing, whereas the rest of the pixels are transparent to light. The state of increasing absorption creates the nonlinearity in the output signal, I_{out} as shown in Figure 5.15(b). An optical logic gate can be formed by biasing the SEED close to nonlinearity I_{inb} and then applying low-level and high-level signals I_{in1} and I_{in2} to the device.

Amongst the various other SEED structures, another structure is the symmetric SEED or S-SEED, as shown in Figure 5.16. The two diodes are interconnected and constitute a single pixel.

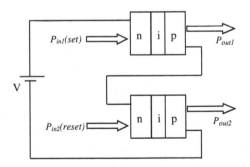

FIGURE 5.16: Structure of a symmetric SEED (s-SEED)

The device operates with a complementary set of inputs P_{in1} and P_{in2}, thereby producing a complementary set of outputs P_{out1} and P_{out2}. In the absence of incident light, the voltage gets equally divided between the two diodes. Now, when two complementary beams are incident on the two diodes, i.e., light across P_{in1} and no light across P_{in2}, the top diode becomes absorbant, resulting in current generation. This manifests in high voltage drop across the top diode and low drop across the bottom diode. The low voltage drop across the diode makes the top diode more absorbant, whereas the bottom diode which is not illuminated remains transmissive. But if the top diode is not illuminated and the bottom is illuminated, then the top diode becomes transmissive and the bottom absorbing. The device is bistable in the relation of the two input beams, thus it is insensitive to input power fluctuations, as long as the two beams are derived from the same source.

5.3 Photonic storage in three dimensions (3D)

Information in digital form to be stored and to be recalled later is a string of bits, and thus information is one-dimensional (1D). In computer and communication systems, this 1D information is formatted as two-dimensional (2D) or three-dimensional (3D) format to achieve maximum storage density. Optical disks, commonly known as CDs or DVDs (digital versatile disks) of various forms are used to store information, mostly in 2D. Multi-layer optical disks are also available which can record and store information in 3D. The details of operations of such storage devices can be obtained in many books, and thus are not repeated here. We only mention here some of the unconventional photonic techniques which can store information in 3D.

5.3.0.1 Storage based on photo-refractive material

Photo-refractive effect is a non-linear effect seen in certain materials in which the refractive index of the material changes when exposed to light. In regions where light is incident, electrons absorb the light and are photo-excited from an impurity level into the conduction band of the material, leaving a hole behind. Impurity levels have an energy intermediate between the energies of the valence band and conduction band of the material. The electrons in the conduction band are free to move and diffuse throughout the material, or there is also some probability that the electrons may recombine with the holes and return to the impurity levels. The rate of recombination is determined by the distance of electron diffusion, which in turn determines the overall strength of the photo-refractive effect in that material. Once back in the impurity level, the electrons are trapped and can no longer move unless re-excited by light.

Photo-refractive crystal materials [11] include the $LiNbO_3$, $BaTiO_3$, organic photo-refractive materials, and certain photo-polymers. Such materials utilize the photo-refractive effect for the storage of information [12] and for optical display of information. The image stored inside the photo-refractive crystal persists until the pattern is erased. This can be done by flooding the crystal with uniform illumination which will excite the electrons back into the conduction band and allow them to be distributed more uniformly. The fundamental limits on writing time, access time, and storage density are related to the physical properties of the photo-refractive materials involved.

5.3.0.2 Storage based on photochromic materials

Photochromism is the reversible transformation of a chemical species be-tween two forms by the absorption of light, where the two forms have differ-ent absorption spectra. Photochromatic effect can be described as a reversible change of colour upon exposure to light. Photochromic materials thus have two distinct stable states, called colour centres, because they are associated with the absorption bands of two distinct wavelengths of light. When illuminated by one wavelength of light, the material undergoes a transition from one stable state to other and remains insensitive to that wavelength after the transition. The absorption band is then shifted to the other wavelength. Thus optical data can be written with one wavelength and can be read by the other wave-length. Photochromism requires that the two states of the molecule should be thermally stable under ambient conditions for a reasonable time interval. Photochromatic materials can dramatically increase memory densities. Thus it is possible to use photochromic materials for 3D data storage of terabytes of capacity.

Photochromic materials can also undergo photon-mode recording [13], which is based on a photochemical reaction within the medium. In photon-mode recording, light characteristics such as wavelength, polarisation, and phase can be multiplexed to enable data storage and thus have the poten-tial to increase the achievable memory density. A category of photochromic storage material is used in rewritable digital video discs (DVDs). The ratio of transmissivity before and after the write-in process is 3:1 at λ of 450 nm (blue) wavelength and 2:7 at λ at 650 nm (red) wavelength. The transition time achievable is higher than 2 to 3 ns.

5.3.0.3 Switching using two-photon absorption material

The two-photon absorption phenomenon is generally known as photon gat-ing, which means using a photon to control the behaviour of other photon. Two-photon absorption materials require a write beam to record data and a read beam to stimulate the excited parts of the material to radiate lumi-nescence. One reason two-photon excitation is preferable for recording in 3D optical memory systems is the minimal crosstalk between two adjacent layers of the 3D memory. Two-photon thermal excitation also reduces multiple scat-tering. This reduction occurs because the illumination beam that is utilised has an infrared wavelength. Advances in two-photon absorbing spirobenzopy-ram molecules, as potential materials for 3D optical memory/storage, have been reported in [14], [15], [16].

Although two-photon excitation is the most desirable method for bit-data recording in 3D optical memories, single-photon recording also is an acceptable method since it gives good separation between the recorded planes. Single-photon recording does not require ultrashort-pulse lasers, and conventional semiconductor lasers can be used.

5.3.0.4 Storage based on bacteriorhodopsin

Bacteriorhodopsin belongs to a family of bacterial proteins related to rhodopsin which is a pigment that senses light in the retina. It is an integral membrane protein and is usually found in two-dimensional crystalline patches, which can occupy almost half the surface area of the archaeal cell. The bacteriorhodopsin molecule is purple and hence also known as purple membrane. The molecules absorb green light of wavelength in the range 500 to 650 nm, with the absorption maximum at 568 nm. With the absorption of photons by the membrane protein, the light energy is used to move photons across the membrane out of the cell. Optical films containing the genetic type of bacteriorhodopsin (BR-D96N) were experimentally studied to find out their properties as a medium for storage [17], [18].

Halobacterial photosynthesis of bacteriorhodopsin is a light-driven pump converting light energy into chemical energy. The initial state of bacteriorhodopsin is called B-state. After absorption of photons at a wavelength (570 nm), the B-state changes to another state called J-state followed by relaxation of the other states. Finally, the relaxed state transforms to another stable state called M-state. The stable M-state can be driven back to B-state by absorption of a photon at another wavelength (412 nm). Thus M-state and B-state are two stable states which can be used as flip-flops of bistable memory elements.

5.3.1 Holographic storage devices

Unlike a photographic film, which records a 2D image of a 3D object, a hologram can record and reconstruct a 3D image. A photographic film records only the intensity of the incident light beam, whereas a hologram records not only the intensity but also the phase of the incident beam, thus providing a means of recording 3D information of an object. Though the hologram does not resemble the object, it records and reconstructs all the information about the amplitude and phase of the object.

Holography is a two-step process: 1) *writing the hologram*, which: this involves recording the amplitude and phase information on a special film, and 2) *reading the hologram*, where the hologram is illuminated with a reference coherent beam similar to that in the writing process. Recording the phase in addition to the amplitude of the incident light is made possible by recording an interference pattern of the object wave $O(x, y) = |O(x, y)|e^{i\phi(x,y)}$ and a reference wave $R(x, y) = |R(x, y)|e^{i\psi(x,y)}$ [19]. The interference of the object beam $O(x, y)$ and reference beam $R(x, y)$ waves can be expressed as

$$I(x, y) = |O(x, y)|^2 + |R(x, y)|^2 + 2|O(x, y)||R(x, y)| \cos[\psi(x, y) - \phi(x, y)] \tag{5.18}$$

The first two terms of the expression depend on the intensities of the object and reference beams, while the third term depends on their relative phase. Thus the information of phase and amplitude is generated and recorded on the photographic plate. Usually, special-purpose photographic plates or films are used for recording holograms. The developed film, if assumed to have a linear response, has an amplitude transmittance $T(x,y)$ proportional to the exposure E and is given by

$$T(x,y) = T_0 + \beta[\{|O(x,y)|^2 + |R(x,y)|^2 + O(x,y)R^*(x,y) + O^*(x,y)R(x,y)\}]$$
(5.19)

where T_0 is the bias transmittance and β is a constant which depends on the type of film used. β is a negative number for a negative transparency and a positive number for positive transparency. $T(x,y)$ expresses amplitude and phase information about the object. The recording and readout optical set up is shown in Figure 5.17 (a) and (b), respectively.

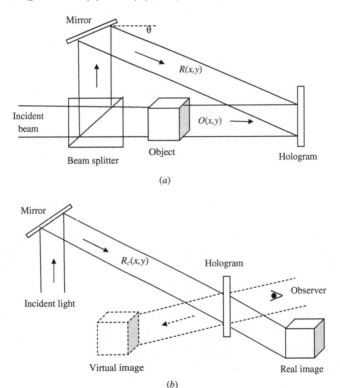

FIGURE 5.17: Hologram (a) recording and (b) read-out

For reconstruction, the developed transparency is illuminated by a coherent reconstruction wave, $R_c(x,y)$. The light transmitted by the transparency is

$$R_cT = T_0R_c + \beta OO^*R_c + \beta R^*R_cO + \beta RR_cO^*$$
(5.20)

It may be noted that the positional term (x,y) is not written explicitly to shorten the equation.

If R_c is derived from the same source of the reference beam $R(x, y)$, then the third term of the equation is simply $\beta |R(x, y)|^2 O(x, y)$, that is, a reconstructed wave is an exact duplication of the original wavefront $O(x, y)$ with a multiplication constant. Similarly, if $R_c(x, y)$ is made equal to the conjugate of the reference beam, then the fourth term of the reconstructed beam is $\beta |R(x, y)|^2 O^*(x, y)$. This wave is proportional to the conjugate of the original wavefront.

However, for the case when $R(x, y) = R_c(x, y)$ and when $R_c(x, y) = R^*(x, y)$, the field component is masked by three additional field components which need to be separated out. In general, during reconstruction twin images of the object are obtained, when illuminated by a plane wave of the same wavelength. To avoid the formation of inseparable twin images along the axis, the reference beam is made to incident at an angle θ with respect to the object beam, so that the real image and the virtual image of the object can be viewed separately during reconstruction for off-axis holography.

Holograms are recorded on a photosensitive optical material or plate. Thin films or plates record plane holograms, whereas thick photosensitive materials record volume holograms. By adjusting the reference beam angle, wavelength, or the recording media, a multitude of holograms can be stored on a single volume. The hologram can be read using the same reference beam in the same set-up.

(a) Digital holography: Advancement in the theory of information and computing opened an era in the field of holography by either writing or reading the hologram digitally rather than optically. In digital holography, the recording schemes are the same as that of conventional, but the recording material is replaced by an electronic device, such as a charge-coupled device (CCD). It is also possible to produce a digital hologram without any optical interference, more popularly known as a computer-generated hologram. The hologram can then be sent to a display device for optical reconstruction. However, in all cases of digital holography, interference between the object and the hologram in the front-end is simulated. Then diffraction between the hologram and the diffraction plane or the observation plane in the back end is also simulated for the conclusion of of the overall process. Thus, creating the digital signal and digital calculations of diffraction are the core of digital holography. This processing offers a number of significant advantages such as the ability to acquire the images rapidly, the availability of both the amplitude and the phase information of the optical field, and the versatility of the processing techniques that can be applied to the complex field data.

(b) Plane holographic storage: Holographic data storage records information using an optical interference pattern of two beams, the object beam reflected from the object or transmitted through the object, and a reference beam. Information to be recorded and stored are in general available as a bit string. This string is first arranged in a 2D format called a page memory. In an optical system, the page memory is displayed on a spatial light modulator (SLM) which acts as a page composer. A modulated beam, gen-

erated from a collimated coherent laser source, is Fourier transformed by a lens. It is advantageous to record the Fourier transform hologram on the page type memory because of the minimum requirement of space bandwidth. A holographic medium records the interference pattern of a reference beam and the Fourier transform records the page memory on the focal plane. The page memory can be reconstructed by passing its Fourier transform through another lens. The reconstructed image of the page memory needs to be read by an array detector. Since only the same reference beam can be employed during recording and reconstruction, a number of holograms can be recorded with reference beams at different incident angles, thus providing multiplexing capability. Several plane holograms can be stacked together in a layered structure to form a 3D optical storage medium. An electrical signal can select and activate a particular stack layer.

(c) Volume hologram for 3D optical storage: In contrast to conventional imaging that stores light pattern on a plane, the holographic technique transforms a 2D pattern into a 3D interference pattern for recording. The recorded 3D pattern is then known as a volume hologram. A volume hologram can be transformed back to the 2D object in the reconstruction process. Using a volume hologram provides manifold increase of storage capacity. It may be noted that a plane hologram can be used to produce 3D display of an object, while a volume hologram is used to record and display a 2D page memory.

5.4   Flat panel displays

Display devices are used to interface a scene containing information by stimulating human vision. Information may be pictures, animation, movies, or any other objects. The basic function of a display device is to produce or reproduce colours and images in two or three dimensions. However, as the name implies, all flat panel displays (FLDs), though the term is actually a misnomer, have a relatively thin profile, i.e., several centimetres or millimetres. FLDs have a large orthogonal array of display elements, called pixels, which form a flat screen. Most television sets and computer monitors currently employ FLDs, replacing yesteryear's cathode-ray tubes.

Flat panel display technologies must have: (1) full colour, (2) full gray scale, (3) high efficiency and brightness, (4) the ability to display full-motion video, (5) a wide viewing angle, and (6) a wide range of operating conditions. Flat-panel displays should also provide the following benefits: (1) thinness and light weight, (2) good linearity, (3) insensitivity to magnetic fields, and (4) no x-ray generation. These four attributes are not possible in a cathode-ray tube.

Any flat panel display is characterised by the following features:

204 Information Photonics: Fundamentals, Technologies, and Applications

1. *Size and aspect ratio*: The size of a display is typically described by diagonal length, generally in units of inches. A 15-inch display means the diagonal of the viewable area of the display is 38.1 cm. The display format can be landscape, equal, or portrait, corresponding to the display width being larger than, equal to, and smaller than its length. The width-to-length ratio of a display is called the aspect ratio, typically 4:3, 16:9, or 16:10.

2. *Resolution*: An FPD typically consists of a dot matrix which can display images and characters. Resolution is related to the total number of pixels in a display, which affects the quality of the image. More resolution means more pixels, which results in better picture quality. Besides, for a full-colour display, at least three primary colours are needed to compose a colour pixel. Hence, each colour pixel is divided into three sub-pixels (RGB) sharing the area.

3. *Pitch*: This is the centre-to-centre distance (usually in millimetres) between adjacent pixels. The smaller the pitch, the higher the resolution. One can define the fill factor as the ratio of the display area in a pixel over the whole pixel size.

4. Brightness and colour: Luminance and colour are two important optical characteristics of an FPD. Typically, the luminance of an FPD should be as bright as (or slightly brighter than) the real object. A display with high luminance looks dazzling in a dark room. On the other hand, a display with insufficient brightness appears washed out under high ambience. Uniformity of the luminance and colour change over a display area is of concern as human eyes are sensitive to luminance and colour differences.

5. *Viewing angle and viewing distance*: There are several ways to define the viewing angle of an FPD, for example, to find the viewing cone with a luminance threshold or with minimum contrast ratio, or with maximum value of colour shift. The viewing distance between the display panel and the viewers is an important factor in determining the suitability of a display for a particular application. Longer distances require less resolution and shorter distances require higher resolutions.

6. *Power consumption*: Power consumption is a key parameter, especially for displays used in mobile handsets, as it affects battery life. For displays with wall-plug electrical input, lower power consumption implies lower heat generation, which means heat dissipation is less serious. For notebooks and TVs, high optical efficiency also translates into less heat dissipation. Thermal management in a small-chassis notebook is an important issue.

Flat panel displays can be divided into three types: transmissive, emissive, and reflective. A transmissive display has a backlight, with the image being formed by a spatial light modulator. A transmissive display is typically low in power efficiency; the user sees only a small fraction of the light from the backlight. An emissive display generates light only at pixels that are turned on. Emissive displays should be more efficient than transmissive displays, but due to low efficiency in the light-generation process most emissive and transmissive flat panel displays have comparable efficiency. Reflective displays, which reflect ambient light, are most efficient. They are particularly good where ambient light is very bright, such as direct sunlight. They do not work well in low-light environments. [20].

Currently, most commercially manufactured flat panel display devices are liquid crystal displays (LCDs). The benchmark for flat panel display performance is the active matrix liquid crystal display (AMLCD). Most portable computers use AMLCDs. Competing flat panel display technologies include electroluminescent displays, plasma display panels, vacuum fluorescent displays, and field-emission displays. Electroluminescent displays are often used in industrial and medical applications because of their ruggedness and wide range of operating temperatures. Plasma display panels are most often seen as large flat televisions, while vacuum fluorescent displays are used in applications where the information content is fairly low, such as the displays on appliances or in automobiles. Field-emission displays are the most recent of these flat-panel technologies.

Basic requirements for any flat panel display capable of showing a high-resolution colour picture are (a) an electro-optical effect, (b) a transparent conductor, and (c) a nonlinear addressing element. For many years, the cathode ray tube (CRT) was one of the main electronic device where images in gray or in colour were displayed. Because of its large shape and high operating voltage, a need was felt for evolving flat panel displays of high image quality, working at a low voltage and having a fast response. Today, FPDs exploiting electro-optical effects with an addressing system provide a solution by converting an electrical signal into a visible change. Two methods are generally utilised for this purpose. The first one gives emission of light, and the second is the modulation of ambient illumination, sometimes called subtractive displays. Apart from the most useful light-emitting media, the modulating effects that have been studied mainly by controlling absorption, reflection, or scattering phenomena. Subtractive displays are energy efficient, but in the absence of ambient light, a secondary source is needed. Emissive displays are attractive to view at low or moderate light levels, but are difficult to read when bright light falls on the panel.

The preferred solution for delivering data to a million or so picture elements (PELs) across the display area is an amorphous silicon thin-film transistor deposited at each cross-point in an X-Y matrix. The general method of matrix addressing consists of sequentially addressing one side of the matrix, for example by selecting the rows one by one and applying the picture information on the other side at the columns row by row. This sequential addressing or multiplexing running down the X rows one by one, applying the half-voltage to the appropriate Y columns at the correct time, solves the ambiguity problem but chances of cross-talk becomes high. The solution is to add to each pixel a circuit element that does have a sharp threshold, such as a semiconductor diode or a transistor, converting the passive matrix into an active matrix as shown in Figure 5.18, where the data lines and scan lines are shown. The elements need to be integrated on the panel, in the form of a thin film. The thin film transistors (TFT) maintain the state of each pixel between scans while improving response times. TFTs are micro-switching transistors (and associated storage capacitors) that are arranged in a matrix on a glass substrate to control each picture element (or pixel).

The display medium also needs two electrodes. At least one of these must be transparent, so that the optical path from the pixel to the eye will suffer little absorption. For most practical effects, the pixel electrodes are orthogonal to this path, and run the full width of the display. The conductivity of the electrodes must then be high enough to give a uniform voltage over the display area. The most preferred medium for flat panel display has been indium oxide doped with tin (ITO), prepared by a sputtering technique. A possible alternative is ZnO doped with Al also prepared by sputtering.

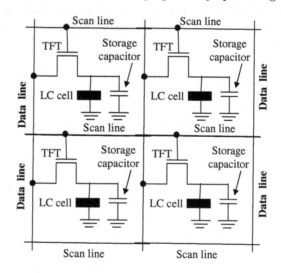

FIGURE 5.18: Active matrix addressing

5.4.1 Liquid crystal flat panel displays

A liquid crystal display (LCD) [21], [22] is a flat panel display, electronic visual display, or video display that uses the light modulating properties of liquid crystals. Each pixel of an LCD typically consists of a layer of molecules aligned between two transparent electrodes, and two polarising filters (parallel and perpendicular), the axes of transmission of which are (in most of the cases) perpendicular to each other. By controlling the voltage applied across the liquid crystal layer in each pixel, light can be allowed to pass through in varying amounts, thus constituting different levels of gray.

Physical parameters of LCs that control the electro-optical behaviour important for display are the elastic constants and the rotational viscosity. The threshold voltage for switching, V_t , is given by

$$V_t = \pi \left[\frac{k}{\varepsilon_0(\varepsilon_\parallel - \varepsilon_\perp)} \right]^{1/2} \qquad (5.21)$$

where k is a constant related to twist and bend, while $\varepsilon_\parallel$ and $\varepsilon_\perp$ are the dielectric constants along and across the long molecules of LC.

By carefully adjusting the amount of voltage applied in very small increments, it is possible to create a gray-scale effect. Most of today's LCD displays support a minimum of 256 levels of brightness per pixel although high-end LCD panels used in high definition LCD televisions support up to 1024 different levels of brightness [23].

The visual appearance of a twisted nematic cell depends strongly on the angle of view, and both the viewing angle and the contrast ratio. In the off state, the molecules adopt a more or less homeotropic structure. When voltage is applied, within each domain the molecules align orthogonally to the raised structures, which are actually pyramids, so the light is channelled into a cone, giving a wide angle of view. The wide-view (WV) film is an optical compensation film, which remarkably enhances the field of view of twisted nematic (TN) thin-film transistor liquid crystal displays. Triacetate film (TAC film) polariser is used in LCD displays. Different layers are shown in Figure 5.19.

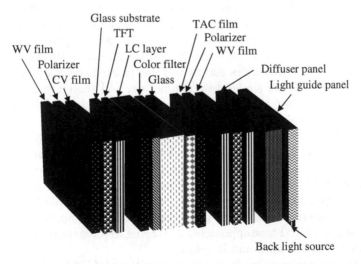

FIGURE 5.19: Layers of LC flat panel display

LCD displays may be either the backlit or reflective type. Unlike CRTs, LCD displays require an external light source to display the picture. The least expensive LCD displays make use of a reflective process to reflect ambient light so that it displays the information. However, computer monitor and LCD TV displays are lit with an external light source, which typically takes the form of built-in micro-fluorescent tubes that are often just a few millimetres in diameter. These are placed above, beside, and sometimes behind the LCD. A white diffusion panel is used behind the LCD to redirect and scatter the light evenly, to ensure uniform display brightness. The latest developments in LCD backlight have also brought about the use of LED-based backlight systems. LED-based LCD displays can also be either edge-lit or full array with local dimming; the latter are capable of exceptional picture quality while making use of less energy.

Displays for a small number of individual digits and/or fixed symbols (as in digital watches and pocket calculators) is implemented with independent electrodes for each segment. In contrast, full alphanumeric and/or variable graphics displays are usually implemented with pixels arranged as a matrix consisting of electrically connected rows on one side of the LC layer and columns on the other side, which makes it possible to address each pixel at the intersections.

LCD display-addressing systems can again be effected through passive or active matrix. In displays using passive matrix, panels rely on the display persistence to maintain the state of each display element (pixel) between refresh scans. The use of such displays is very much limited to a certain extent by the ratio between the time to set a pixel and the time it takes to fade. To operate, passive-matrix LCDs use a simple grid to supply the charge to a particular pixel on the display. The point of intersection of the row and column represents the designated pixel on the LCD panel to which a voltage is applied to untwist the liquid crystals at that pixel for the control of light. Once the voltage between the respective electrodes addressing a pixel is removed, the pixel behaves similarly to a discharging capacitor, slowly turning off as charge dissipates and the molecules return to their twisted orientation. Passive matrix LCD displays are simple to manufacture, and therefore cheap, but they have a slow response time on the order of a few hundred milliseconds, and a relatively imprecise voltage control.

To overcome the sunlight readability issue while maintaining high image quality, a hybrid display called a transflective liquid crystal display (TR-LCD) has been developed. In a TR-LCD, each pixel is divided into two subpixels: transmissive (T) and reflective (R). The area ratio between T and R can be adjusted depending on the application. Within this TR-LCD family, there are still some varieties: double cell gap versus single cell gap, and double TFTs versus single TFT. These approaches are trying to solve the optical path length disparity between the T and R subpixels.

In the transmissive mode, the light from the backlight unit passes through the LC layer once, but in the reflective mode the ambient light traverses the LC medium twice. To balance the optical path length, we could make the cell gap of the T subpixels twice as thick as that of the R subpixels. This is the so-called dual cell gap approach. The single cell gap approach has a uniform cell gap throughout the T and R regions. To balance the different optical path lengths, several approaches have been developed, e.g., dual TFTs, dual fields (stronger field for T region and weaker field for R region) and dual alignments. Presently, the majority of TR-LCDs adopt the double cell gap approach for two reasons: (1) both T and R modes can achieve maximum light efficiency, and (2) the gamma curve matching between the voltage-dependent transmittance (VT) and reflectance (VR) is almost perfect. However, the double cell gap approach has two shortcomings: First, the T region has a slower response time than the R region because its cell gap is about twice as thick as that of the R region; second, the viewing angle is relatively narrow, especially when homogeneous cells are employed. To widen the viewing angle, a special rod-like LC polymeric compensation film must be used.

5.4.2   LED and OLED display panels

LED panel displays are now universal, with a full range of colours and huge sizes. LED displays are also very popular in the market because they are available with a combination of light, colour, motion, and graphics that get noticed. A LED display is a two-lead semiconductor light source that resembles a basic p-n junction diode, except that an LED also emits light. This form of display is the most prevalent in use today and is still a liquid crystal display that uses a LED back-light. LED-backlit LCD displays are commonly marketed as LED televisions.

Televisions with true LED-displays have now been developed, in which LEDs are used to produce actual images rather than acting as back-lighting for LCD display. Called the crystal LED display, it uses ultra-fine LEDs in each of the red-green-blue (RGB) colours, equivalent to the number of pixels. The RGB LED light sources are mounted directly on the front of the display, dramatically improving the light use efficiency. This results in images with strikingly higher contrast (in both light and dark environments), wider vibrant colour gamut, superb video image response time, and wider viewing angles when compared to existing LCD and plasma displays.

An OLED is also an electroluminescent device, like an LED, except that its materials are organic thin films with amorphous structures. Details of OLED operation is given in Chapter 4. However, due to the amorphous character-istics, fabrication with large size (greater than 40 inches) is possible. Two advantages of OLEDs are: (1) low process temperature, and (2) non-selective to the substrate material, which is suitable for flexible displays. Due to the possibility of large-size fabrication, the potential manufacture cost of OLEDs is lower than that of LEDs.

OLEDs, like regular LEDs, operate on the principle of electro-luminescence, where injected charge carriers recombine and generate light.

All OLEDs have four basic components: substrate, anode, organic layers, and cathode. Flexible substrate materials are usually plastic, thin glass, or metal foils. The anode is a transparent layer of metal of low work function which serves to remove electrons when a current flows through the device [24]. The cathode is a metal layer of high work function which injects electrons when a current flows through the device. In between the cathode and the anode are the organic layer(s) where transport and recombination of the electrons and holes occur. Depending on the device, the OLED could have one, two, or multiple organic layers.

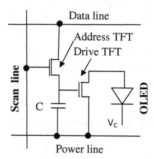

FIGURE 5.20: Active matrix addressing of OLED panel (one individual cell is shown)

OLED can also be addressed like LED or LCD active matrix display as shown in Figure 5.20. Active matrix OLED (AMOLED) display consists of OLED pixels that have been deposited or integrated onto a TFT array to form a matrix of pixels that illuminate light upon electrical activation. The active-matrix TFT backplane acts as an array of switches that control the amount of current flowing through each OLED pixel. The TFT array continuously controls the current that flows to the pixels, signaling to each pixel how brightly to shine. Typically, this continuous current flow is controlled by at least two TFTs at each pixel, one to start and stop the charging of a storage capacitor and the second to provide a voltage source at the level needed to create a constant current to the pixel.

A major part of manufacturing OLEDs is applying the organic layers to the substrate. This can be economically done in two ways: organic vapour phase deposition and inkjet printing. Organic vapour phase deposition involves a carrier gas and a low- pressure, hot-walled reactor chamber. The carrier gas transports evaporated organic molecules onto cooled substrates, where they condense into thin films. Using a carrier gas increases the efficiency and reduces the cost of making OLEDs. With inkjet technology, the organic layers are sprayed onto substrates just like inks are sprayed onto paper during printing. Inkjet printing greatly reduces the cost of OLED manufacturing by enabling roll-to-roll processing. In addition, it allows OLEDs to be printed onto very large films for large displays, like electronic billboards. The OLED flat panel structure is shown in Figure 5.21.

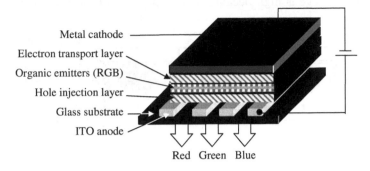

Metal cathode

Electron transport layer

Organic emitters (RGB)

Hole injection layer

Glass substrate

ITO anode

Red   Green   Blue

FIGURE 5.21: Flat panel display using OLED

5.4.3   Plasma display panels

A plasma is a distinct state of matter containing a significant number of electrically charged particles. A plasma display panel (PDP) is an emissive display which can be thought of as very many miniature fluorescent lamps on a panel. PDPs use a similar operation mechanism to fluorescent lamps but the gases commonly used in PDPs are neon and xenon instead of the argon and mercury used in fluorescent lamps. As an emissive display, it typically has a better display performance, such as good colour saturation and wide viewing angle. Due to the limitation of fabrication, the pixel size of a PDP cannot be too small. For a finite pixel size, the video content is increased by enlarging the panel size. A rare gas is sealed inside a tube and high voltage (several hundred volts) is applied across the electrodes at the ends of the tube. A plasma discharge is first induced by the positive period of an AC field and a layer of carriers is shortly thereafter formed on top of the dielectric medium. This stops the discharge, but is induced again when the voltage changes polarity. In this way, a sustained discharge is achieved. The discharge creates a plasma of ions and electrons which gain kinetic energy from the electric field. These particles collide at high speed and are excited to higher-energy states. After a while, the excited atoms return to their original state and energy is dissipated in the form of ultraviolet radiation. This radiation, in turn, excites the phosphors, which glow with red, green, and blue (RGB) colours, respectively. The I-V characteristic of a gas discharge is an S-shaped negative resistance, and the working point is determined by the resistive load line. It is possible to energize the pixel by a certain voltage, and sustain it by a lower voltage.

Although the PDP structure is similar to a fluorescent lamp, which is composed of two electrodes, phosphor and gases, an additional barrier rib structure is needed in PDPs to sustain the space between upper plate and lower plate, as shown in Figure 5.22. Because of the structure of the barrier rib, the unit cell size of PDPs cannot be made too small. In addition, PDP operation voltage is high because a typical plasma generation is needed.

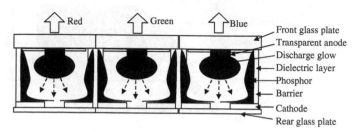

FIGURE 5.22: Plasma flat panel display

A PDP is essentially a matrix of such tiny fluorescent tubes as described above, which are individually controlled. Since each discharge cell can be individually addressed, it is possible to switch picture elements (PELs) on and off. In panels the column electrodes are deposited as metal stripes on a glass substrate, and above them are a series of parallel ridges, dividing the display area into a number of narrow troughs. The sides of the troughs are coated with the phosphors. The row electrodes are ITO stripes on the glass sheet that forms the top of the panel.

The advantages of plasma displays over LCD are:

1. Provides superior contrast ratio.

2. Has wider viewing angles than LCD; images do not suffer from degradation at high angles like LCDs.

3. Less visible motion blur, because of high refresh rates and faster response time.

4. Bigger screen sizes available than with LCDs.

PDPs exhibit a wider view angle, faster response time, and wider temperature range than LCDs. In other words, PDPs remain good candidates for large-panel displays spanning from static pictures to motion pictures, from cold ambient to hot ambient, and from personal use to public use. But the disadvantage of a plasma panel is that it uses more electrical power than an LCD TV. .

5.4.4 Flexible flat panel display

A flexible flat panel display is constructed of thin (flexible) substrates that can be bent, flexed, conformed, or rolled to a radius of curvature of a few centimetres without losing functionality. The development of flexible display technology promises to reshape the global flat panel display landscape and open up compelling new applications for displays. At first glance flexible flat panel displays may appear to be modified renditions of current flat panel display technology, but they will differ in many significant ways. The recent convergence of many technologies (substrates, conducting layers, barrier layers, electro-optic materials, thin film transistor technologies, and manufacturing processes) is accelerating the flexible flat panel display concept closer to commercial reality. To enable a flexible flat panel display, a flexible substrate must be used to replace conventional glass substrates, which can be either plastic or thin glass. Flexible flat panel display technologies offer many potential advantages, such as very thin profiles, light-weight and robust display systems, the ability to flex, curve, conform, roll, and fold a display for extreme portability, high-throughput manufacturing, wearable displays integrated in garments and textiles, and ultimate engineering design freedom (e.g., odd-shaped displays).

Since flexible displays utilize polymer materials, a barrier layer is essential in protecting and enclosing the functional materials and layers from oxygen and water degradation. Since organic materials tend to oxidise and hydrolyse, oxygen and water permeation through a flexible substrate is of particular importance in flexible electronics. Although single-layer barrier layers do provide the packaged materials with some protection, it appears that multiple layers are necessary for organic light-emitting diode applications, for long-term stability. Indium tin oxide is the typical conducting layer used in display technology because of its excellent sheet resistance and optical clarity. Alternatively, conducting polymers are also being considered for flexible display applications. For emissive applications, small molecules and polymers are being used for OLED applications. In order to have a truly low power display, a reflection mode of operation will have to be implemented on flexible substrates. Polymer-dispersed liquid crystals, encapsulated electrophoretics, gyricon, and bichromic ball composites all operate in the reflective mode. For many electro-optic materials, such as OLEDs, polymer-dispersed liquid crystals, electrophoretics, and gyricon materials, an active matrix backplane are required for high resolution. The success of TFTs for plastic substrates has been an enabler for flexible flat panel displays. Currently, poly and amorphous silicon are the standards for flexible displays. However, organic thin film transistors on polymeric substrates are also being considered as a candidate for flexible, lightweight, and inexpensive switching devices.

5.4.5 Electrophoresis and e-book displays

The motion of solid particles in a liquid medium under the action of an electric field, known as electrophoresis, has been studied for many years. The system becomes a display when the particles have a colour that contrasts with that of the liquid. Typically, white TiO_2 particles in blue ink give most impressive high-contrast image. Micro-encapsulating the pigment particles in tiny spheres within the carrier fluid, or by dividing the panel volume into segments by barriers between the surfaces, would solve many of the problems of large flat panel displays, such as gravitational straining and coagulation of particles.

An electronic book, now generally known as an e-book or reader, is the digital equivalent of a printed book, in the simplest version [25]. It is capable of presenting monochrome text one page at a time, with page changing taking a second or two. It is simplest to think of an e-book as a reader in which the visual effect is provided by a flat or flexible electronic display that rivals the effect of print on paper, and so is called electronic paper. Today electro-phoretic (EP) display technology is regarded as a leading candidate for electronic paper and e-book. This technology uses encapsulation and two sets of pigmented particles, one set black and negatively charged, the other white and positive. The fluid within the sphere is clear. In principle, the positive particles could be coloured, or there could be micrometre-sized colour filters on the front surface. Operation of an EP display requires long positive and negative pulses, and this complicates addressing. The response time is set by the transit time of the pigment particles [26]. One requirement of visual effects for electronic paper is bistability, since power is then required only when the display needs updating. If the bistable states are separated by a significant energy barrier, so that transition requires a short pulse with voltage above a threshold, passive matrix addressing is possible, with much simplification in back plane design, and a considerable reduction in processing cost.

Books for further reading

1. *Fundamentals of Liquid Crystal Devices*: D. K. Yang and S. T. Wu, John Wiley, 1986.

2. *Liquid Crystal Displays: Addressing Schemes and Electro-Optical Effects*: E. Lueder, John Wiley, 1997.

3. *Lasers and Electro-Optics*: C. C. Davis, Fundamentals and Engineering, Cambridge University Press, 1996.

4. *Optical Electronics in Modern Communications*: A. Yariv, Oxford University Press, 1997.

5. *Optical Waves in Crystals: Propagation and Control of Laser Radiation*: A. Yariv and P. Yeh, Wiley, 1984.

6. *Magneto-Optics*: S. Sugano and N. Kojima (eds.), Springer, 2000.

7. *Introduction to Flat Panel Displays*: J. H. Lee, D. N. Liu, and S. T. Wu, John Wiley, 2008.

Bibliography

[1] M. Gottlieb, C. L. M. Ireland, and J. M. Ley. *Electro-optic and Acousto-optic Scanning and Deflection*. Marcel Dekker, 1983.

[2] I. P. Kaminow and E. H. Turner. Electro-optic light modulators. *Proc. IEEE*, 54 (10):13741390, 1966.

[3] T. A. Maldonado and T. K. Gaylord. Electro-optic effect calculations: Simplified procedure for arbitrary cases. *App. Opt.*, 27:5051–5066, 1988.

[4] M. Melnichuk and L. T. Wood. Direct Kerr electro-optic effect in non-centrosymmetric materials. *Phys. Rev. A*, 82:013821, 2010.

[5] U. Efron. *Spatial Light Modulator Technology: Materials, Devices, and Applications*. Marcel Dekker, 1995.

[6] W. E. Ross, D. Psaltis, and R. H. Anderson. Two-dimensional magneto-optic spatial light modulator for signal processing. *Opt. Eng.*, 22(4):224485, 1983.

[7] N. J. Berg and J. N. Lee. *Acousto-optic signal processing: Theory and Applications*. Marcel Dekker, 1983.

[8] R. K. Mali, T. G. Bifano, N. Vandelli, and M. N. Horenstein. Development of microelectromechanical deformable mirrors for phase modulation of light. *Opt. Eng.*, 36:542, 1997.

[9] L.J. Hornbeck. Deformable-mirror spatial light modulators. In *Proc. SPIE 1150*, 1990.

[10] D. A. B. Miller. Quantum well self electro-optic effect devices. *Opt. and Quant. Elect.*, 22:S61–S98, 1990.

[11] W. E. Moerner, A. Grunnet-Jepsen, and C. L. Thompson. Photorefractive polymers. *Annual Rev. Materials Sc.*, 27:585–623, 1997.

[12] T. A. Rabson, F. K. Tittel, and D. M. Kim. Optical data storage in photorefractive materials. In *Proc. SPIE 0128*, 1977.

[13] J. Zmija and M. J. Malachowski. New organic photochromic materials and selected applications. *J. Achievement in materials and manufacturing Eng.*, 41:97–100, 2010.

[14] S. Kawata and Y. Kawata. Three-dimensional optical data storage using photochromic materials. *Chem. Rev.*, 100:1777–1–71788, 2000.

[15] I. Polyzos, G. Tsigaridas, M. Fakis, V. Giannetas, P. Persephonis, and J. Mikroyannidis. Three-dimensional data storage in photochromic materials based on pyrylium salt by two-photon-induced photobleaching. In *Proc. Int. Conference on New Laser Technologies and Applications*, volume 5131, page 177. SPIE, 2003.

[16] K. Ogawa. Two-photon absorbing molecules as potential materials for 3d optical memory. *App. Sc.*, 4:1–18, 2014.

[17] B. Yao, Z. Ren, N. Menke, Y. Wang, Y. Zheng, M. Lei, G. Chen, and N. Hampp. Polarization holographic high-density optical data storage in bacteriorhodopsin film. *App. Opt.*, 44:7344–8, 2005.

[18] A. S. Bablumian, T. F. Krile, D. J. Mehrl, and J. F. Walkup. Recording shift-selective volume holograms in bacteriorhodopsin. In *Proc. SPIE*, pages 22–30, 1997.

[19] P. Hariharan. *Basics of Holography*. Cambridge University Press, 2002.

[20] C. Hilsum. Flat-panel electronic displays: a triumph of physics, chemistry and engineering. *Phil. Trans. R. Soc. A*, 368:1027–1082, 2010.

[21] R.H. Chen. *Liquid Crystal Displays: Fundamental Physics and Technology*. John Wiley Sons, 2011.

[22] D.K. Yang. *Fundamentals of Liquid Crystal Devices, 2nd Edition*. John Wiley Sons, 2014.

[23] P. Kirsch. 100 years of liquid crystals at Merck. *Proc. 20th Int. Liquid Crystal Conf*, London, UK, 2004.

[24] C. Adachi, S. Tokito, T. Tsutsui, and S. Saito. Electroluminescence in organic films with three-layer structure. *Jpn. J. App. Phy.*, 27:269–271, 1988.

[25] A. Henzen et al. Development of active-matrix electronic-ink displays for handheld devices. *J. Soc. Inf. Display*, 12:17–22, 2004.

[26] R. Sakurai, S. Ohno, S.I. Kita, Y. Masuda, and R. Hattori. Color and flexible electronic paper display using QR-LPD technology. *Soc. Inf. Display Symp. Dig*, 37:1922–1925, 2006.

Chapter 6

Photonics in Transform Domain Information Processing

6.1 Introduction

All optical systems can perform mapping, or transformation of images from the input plane to the output plane. However, the processing and analysing of information of two dimensional (2D) images are usually done by converting the image into electronic domain by photonic hardware and then by using image processing software, where such an image is described by a 2D function $I = f(x, y)$, x and y being the spatial coordinates. Amplitude of f at any pair of coordinates (x, y) is the intensity I and is expressed by gray values for gray images. When spatial coordinates and amplitude values are all finite and discrete, the image is called the digital image. Common processing of such images is carried out mostly in the spatial domain by various methods. The term spatial domain refers to the aggregate of pixels composing an image, and spatial domain methods operate directly on these pixels to yield processed output $g(x, y) = \mathtt{T}[f(x, y)]$, where $\mathtt{T}$ is an operator on the image f defined over a neighbourhood square area centred around (x, y).

However, the processing can also be done by converting images from the time or space domain to the frequency domain, usually through transforms. The analysis of signals in the frequency domain is related to the analysis of transform properties in the frequency spectrum. In addition to frequency information, phase information is often needed in most of the frequency domain processing. In this chapter, a few useful transforms are introduced which are helpful in frequency domain processing using photonics hardwares and systems.

6.2 Fourier transform

The genesis of the Fourier transform lies in the Fourier series [1]. A Fourier series with time-period T is an infinite sum of sinusoidal functions (cosine and sine), each with a frequency that is an integer multiple of $\frac{1}{T}$. The Fourier series is expressed in exponential form as

$$g(t) = \sum_{-\infty}^{+\infty} c_k^i e^{j2\pi k f_0 t} \tag{6.1}$$

where the k-th Fourier series coefficient c_k^i can be evaluated from the signal as

$$c_k^i = \frac{1}{T} \int_{\langle T \rangle} g(t) e^{-j2\pi k f_0 t} dt \tag{6.2}$$

where $k = \pm 1, \pm 2, ..$ and T indicates any contiguous interval of time.

For an amplitude A, using the Euler's relation, $g(t) = \frac{A}{2}[e^{j2\pi f_0 t} + e^{-j2\pi f_0 t}]$, $c_1^i = A/2$, and $c_k^i = 0$ for $k \neq \pm 1$. The equation also states that a periodic signal is a weighted sum of complex exponentials at frequencies $k f_0 = kT$, where k is an integer. The complex exponential for $k = 0$ is a constant, and is called the DC term and $k = 1$ corresponds to the fundamental frequency.

When the time-period of the periodic function T is increased to infinity, the signal becomes non-periodic and the spacing between harmonics goes to zero, causing the frequency domain characterisation to become continuous. Under this condition, the Fourier series can be expressed as a Fourier transform given by

$$G(f) = \int_{-\infty}^{+\infty} g(t) e^{-j2\pi f t} dt \tag{6.3}$$

The Fourier transform represents the frequency spectrum of the original time-domain signal with infinite time period, or repetition after an infinite period of time. It is possible to find the signal $g(t)$ from its FT, $G(f)$ through inverse Fourier transform (IFT), as

$$g(t) = \int_{-\infty}^{+\infty} G(f) e^{j2\pi f t} df \tag{6.4}$$

Occasionally, the radial frequency $\omega = 2\pi f$ is used rather than the natural frequency f.

The Fourier transform of a two-dimensional (2D) function $f(x, y)$ (which may be an image) in the spatial coordinates x and y can be expressed as

$$FT[g(x, y)] = F(u, v) = \int_{-\infty}^{+\infty} f(x, y) e^{-j2\pi(ux+vy)} dx dy \tag{6.5}$$

where $F(u, v)$ is a complex-valued function of two independent spatial frequencies u and v.

Similarly, the inverse Fourier transform of $F(u, v)$ can be represented as

$$FT^{-1}[F(u, v)] = f(x, y) = \int_{-\infty}^{+\infty} F(u, v) e^{j2\pi(ux+vy)} du dv \tag{6.6}$$

If the signal is separable, that is, $g(x, y) = g(x).g(y)$, then its 2D FT is also separable and can be computed using only two 1D transforms.

6.2.1   Fourier transform properties

The properties of Fourier transform along with their significance are indicated below:

1. Linearity theorem: The weighted sum of two or more functions when Fourier transformed is identical to the weighted sum of their individual transforms. Hence

$$FT[\alpha f_1(x,y) + \beta f_2(x,y)] = \alpha F_1(u,v) + \beta F_2(u,v) \qquad (6.7)$$

where f_1 and f_2 are the two functions and F_1 and F_2 denote Fourier transform of the functions respectively.

2. Shift theorem: Translation of a function $f_1(x,y)$ through quantities a and b in the space domain is equivalent to a linear phase shift of the transform in the frequency domain.

$$FT[(f(x-a, y-b)] = F(u,v)e^{-j2\pi(ua+vb)} \qquad (6.8)$$

3. Similarity theorem: Fourier transform of a scaled-up space domain function produces an amplitude-modulated Fourier transform having condensed frequency coordinates, that is,

$$FT[f(ax, by)] = \frac{1}{|ab|} FT \left[\frac{u}{a}, \frac{v}{b} \right] \qquad (6.9)$$

where a, b represent the scaling values.

4. Parseval's theorem: The energy contained in the signal $f(x,y)$ in the space domain is equivalent to the energy density of its transform in the frequency domain, that is

$$\int_{-\infty}^{\infty} \int |f|(x,y)|^2 dx dy = \int_{-\infty}^{\infty} \int |F|(u,v)|^2 du dv \qquad (6.10)$$

5. Fourier integral theorem: A Fourier transform followed by the inverse Fourier transform on a function yields the same function, except at the points of discontinuity.

$$FT[FT^{-1}]f(x,y) = FT^{-1}[FT\{f(x,y)\}] = f(x,y) \qquad (6.11)$$

6. Convolution theorem: The convolution of two functions in the space domain is equivalent to the multiplication of their Fourier transforms in the frequency domain. Thus

$$f_1(x,y)*f_2(x,y) = FT \left[\int \int_{-\infty}^{\infty} f_1(x,y) f_2(x-x', y-y') dx dy \right] = F_1(u,v) F_2(u,v) \qquad (6.12)$$

7. Correlation theorem:

$$f_1(x,y) \odot f_2(x,y) = FT \left[\int_{-\infty}^{\infty} f_1(x,y) f_2(x-x', y-y') dx dy \right] = F_1(u,v) F_2^*(u,v) \qquad (6.13)$$

8. Modulation property: A function is modulated by another function if they are multiplied in the spatial domain, that is, their Fourier transforms get convolved in the frequency domain.

$$FT[f_1(x,y)f_2(x,y)] = F_1(u,v) * F_2(u,v) \tag{6.14}$$

If $f_1(x,y) = e^{-j2\pi u_0 x} f_2(x,y)$, then

$$FT[f_1(x,y)] = F_1(u - u_0, v) \tag{6.15}$$

9. Conjugation property: If $f_1(x,y) = \overline{f_2(x,y)}$, then

$$FT[f_1(x,y)] = F_1(u,v) = F_2(-u,-v) \tag{6.16}$$

6.2.2   Discrete Fourier transform

While analog processors can handle continuous time signals and images, digital computers can accommodate only discrete time signals. However, discreteness in one domain implies periodicity in the other domain. Since digital computers represent both space and time domain signals and their FTs by discrete arrays, both the discrete time signals and their FTs are periodic. The N-point discrete Fourier transform (DFT) of a sequence $i[n], 0 \leq n \leq (N-1)$ results in an array $I[k], 0 \leq k \leq (N-1)$ related to the discrete time signal as

$$I[k] = \sum_{n=0}^{(N-1)} i[n]e^{-j2\pi \frac{nk}{N}} \tag{6.17}$$

where $k = 0, 1, 2 .... (N-1)$.

Evidently, an N-point DFT produces an N-point discrete sequence $I[k]$ in the transform domain from an N-point sequence $i[N]$ in the time domain. Also $I[k]$ is periodic with period N. Since an increase in N leads to denser sampling in the frequency domain, a DFT size larger than the number of points in the DT signal can be used by simply concatenating zeros (called zero-padding) to the original DT signal. While computing DFT from FT, it is observed that the N-point DFT basically samples one period of the continuous time FT at uniform intervals of $\frac{1}{N}$, that is, $I[k] = I_c\left(\frac{k}{N}\right)$, where I_c is the FT in continuous domain. By increasing N, denser sampling in the frequency domain can be obtained.

Inverse DFT of DT signal can be obtained from $I[k]$ by using the N-point inverse DFT (IDFT) defined as

$$i[n] = \frac{1}{N} \sum_{k=0}^{(N-1)} I[k]e^{j2\pi \frac{nk}{N}} \tag{6.18}$$

Similar expressions for 2D-DFT and 2D IDFT can be worked out for a 2D image or data structure where 2D DFT is given as

$$I[k,l] = \sum_{n=0}^{(N-1)} \sum_{m=0}^{(M-1)} i[n,m] e^{-j2\pi(\frac{nk}{N}+\frac{ml}{M})} \qquad (6.19)$$

where $k = 0, 1, 2.....(N-1)$ and $l = 0, 1, 2.....(M-1)$.

Similarly, 2D IDFT is given by

$$i[n,m] = \frac{1}{MN} \sum_{k=0}^{(N-1)} \sum_{l=0}^{(M-1)} I[k,l] e^{j2\pi(\frac{nk}{N}+\frac{ml}{M})} \qquad (6.20)$$

where $n = 0, 1, 2.....(N-1)$ and $m = 0, 1, 2.....(M-1)$.

2D DFT can be carried out using 1D DFTs, by first performing N-point 1D DFTs along each of the M columns with N points. The results of these 1D DFTs then replace the original columns. Secondly, M-point 1D DFTs are performed along each of the N rows with M points. The resultant DFTs replace the rows. The resulting array is the (N, M)-point 2D DFT $I[k,l]$. Therefore, the (N, M)-point 2D DFT is done using N, M-point and M, N-point 1D DFTs. 2D DFT of an $N \times N$ image can also be carried using N, N-point DFTs along rows and N, N-point DFTs along the column. By assuming N is an integer power of 2, a 2D DFT requires $2N^2 \log_2 N$ multiplications.

6.2.2.1   Fast Fourier transform

The fast Fourier transform (FFT) is an efficient algorithm to compute DFT, thus the DFT is a transform while the FFT is an algorithm based on the *divide and conquer* principle to implement the DFT. For an N-point DFT, the FFT provides a computational advantage by a factor of $(N/N \log_2 N)$. As N increases, the FFT efficiency ratio increases. Many algorithms are now available to compute FFT. Most of them are designed to satisfy different requirements, such as having identical stages so that the same hardware can be used for multiple stages, allowing the generation of the required complex coefficients on line and minimising the memory access needed.

Discrete time (DT) correlation and convolution operations can be performed efficiently via FFT. The correlation of the two signals $i[n]$ and $h[n]$ can be obtained as $IDFT\{[DFT(i[n])]\} \cdot [DFT(h[n])]^*$. The convolution between the DT signals is also related to its DFTs as IDFTs. The correlation and convolution operations in terms of IDFTs are given by

$$IDFT : \{I[k] \cdot H[k]\} = i[n] * h[n] = \sum_{m=0}^{N-1} i[m]h[(n-m)_N] \qquad (6.21)$$

$$IDFT : \{I[k] \cdot H^*[k]\} = i[n] \otimes h[n] = \sum_{m=0}^{N-1} i[m]h[(m+n)_N] \qquad (6.22)$$

where $i[n]$ and $h[n]$ are DT sequences which are zero outside the interval $0 \le n \le (N-1)$, and $I[k], H[k]$ denote their N-point DFTs.

Thus the operations require three N-point DFTs. Assuming that N is an integer power of 2, the three FFT operations need about $3N \cdot log_2 N$ operations. A direct correlation operation, however, will require about N^2 operations. Thus computing convolutions and correlations via FFT is more efficient compared to direct correlation operation by a factor of about $N/(3log_2 N)$. There are also a few other ways to make the convolution and correlation operation using the FFT even more efficient.

6.2.3 Three dimensional Fourier transform

The Fourier transform in higher dimensions (in n-th space) can theoretically be very well written as

$$F(u_1, u_2, ...u_n) = \int_{R^n} f(x_1, x_2, ...x_n)e^{[-j2\pi(u_1x_1+u_2x_2+...u_nx_n)]}dx_1 dx_2..dx_n$$

(6.23)

Generally, the 2D Fourier transform is used for various image processing applications. But it has been seen that 3D Fourier transform is needed for applications such as in shape-descriptors used for image retrieval [2]. A 3D shape-descriptor would possess much more 3D feature information of a 3D object than a 2D shape descriptor and hence would be much more useful during image retrieval.

A 3D Fourier transform is written as

$$F(u, v, w) = \int \int \int f(x, y, z)e^{-j2\pi(ux+vy+wz)}dxdydz \qquad (6.24)$$

6.2.4 Fourier transform in log-polar coordinate system

The log-polar transform is a space-invariant image representation, that is, the effects of scale and rotation of an image become equivalent to the shift of a column or row in the log-polar transformed image. The image mapped from the Cartesian space into the log-polar space is called log-polar mapping. The mapping is done by changing the (x, y) coordinates as

$$x = r \cos \theta$$
$$y = r \sin \theta$$

(6.25)

where

$$r = \sqrt{x^2 + y^2}$$
$$\theta = \arctan(\frac{y}{x})$$

(6.26)

Considering z as the Cartesian plane with coordinates x and y and w as the polar plane with coordinates u and v, the complex logarithmic mapping can be expressed as

$$w = \log z = \log e^{(j\theta)}$$
$$z = x + jy$$

(6.27)

Therefore,

$$w = \log r + j\theta = u + jv \qquad (6.28)$$

where $u = \log r$ and $v = \theta$

Thus, if a horizontal line is drawn from the centre of the image to its right edge, then log-polar transform would be to open out the image radially from this line in the clockwise direction, i.e., to read the values of the pixels radially outward starting from the horizontal line in a clockwise manner. Therefore, a spatially varying image after log-polar transform looks like the one shown on the right of Figure 6.1.

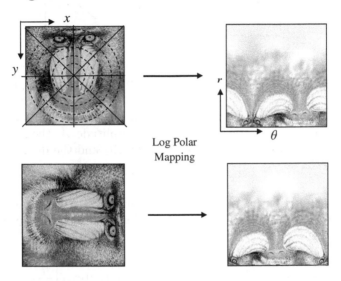

Log Polar
Mapping

FIGURE 6.1: Log-polar transformation of images

Although log-polar mapping offers in-plane rotation and scale-invariance, it is not shift invariant. Moreover, rotation and scale-invariance properties hold only if the rotation and scale changes of the image occur with respect to the origin of the Cartesian image space. The log-polar transformed image can be Fourier transformed like any other image.

6.2.5 Fourier transform properties of lens

Fourier optics makes use of the spatial frequency domain (u, v) as the conjugate of the spatial (x, y) domain. A converging lens can achieve two-dimensional Fourier transform easily. If a transmissive object is placed at the focal length in front of a lens, then its Fourier transform will be formed one focal length behind the lens. A thin lens can produce a phase transformation.

A remarkable property of a converging lens is its ability to perform Fourier transform. The input image is introduced into the parallel beam of light incident on the lens, by a photographic plate or through an electrically addressed spatial light modulator. The lens is placed at a distance f form the input image and FT is obtained at the back focal plane of the lens. The schematic diagram is shown in Figure 6.2.

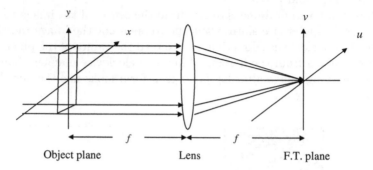

FIGURE 6.2: Lens producing an FT at the back focal plane

If the input has an amplitude transmittance t_i and is illuminated by a collimated monochromatic plane wave of amplitude A, then the amplitude distribution incident on the lens is given by At_i and the distribution behind the lens becomes

$$\xi_t(x,y) = At_i e^{[\frac{-jk}{2f}(x^2+y^2)]} \tag{6.29}$$

The field distribution at a distance f behind the lens in the (u,v) plane can be written in terms of the Fresnel distribution equation (described in Chapter 2):

$$\xi_f(u,v) = \frac{\exp(jkf)\exp\{\frac{jk(u^2+v^2)}{2f}\}}{j\lambda f} \int\int_{-\infty}^{\infty} \xi_t(x,y)\exp\left[\frac{jk}{2f}(x^2+y^2)\right] \times$$
$$\exp\left[\frac{-j2\pi}{\lambda f}(xu+yv)\right]dxdy$$

$$= (At_i)\left[\frac{\exp(jkf)\exp(\frac{jk(u^2+v^2)}{2z})}{j\lambda f}\right]\int\int_{-\infty}^{\infty}\exp[\frac{-j2\pi}{\lambda f}(xu+yv)]dxdy \tag{6.30}$$

The first exponential $\exp(jkf)$ which represents a constant factor can be neglected and will henceforth be dropped. Thus the field distribution $\xi_f(u,v)$ at a distance f behind the lens is proportional to the two-dimensional Fourier transform of the field at the lens aperture. Thus, Fourier components of the input is obtained at frequencies $(\frac{u}{\lambda f}, \frac{v}{\lambda f})$.

The intensity distribution at the Fourier plane or the power spectrum is given by

$$I_f(u,v) = \frac{A}{\lambda^2 f^2}\left[\int\int_{-\infty}^{\infty}t_i\exp\{\frac{-j2\pi}{\lambda f}(xu+yv)dxdy\}\right]^2 \tag{6.31}$$

Optical spatial filtering is based on the Fourier transform property of a lens. It is possible to display the two-dimensional spatial frequency spectrum of an object in such a way that individual spatial frequencies can be filtered out. The input object is described by a space function $t_i(x, y)$. In the infinite lens limit, the field distribution in the back focal plane of the lens is given by Equation 6.30.

6.2.6 Fractional Fourier transform

The fractional Fourier transform (FRFT) [3],[4],[5] is a family of linear transformation that generalizes the Fourier transform with an order parameter. In all the time-frequency representations, a plane with two orthogonal axes corresponding to time and frequency is used. A signal $x(t)$ is represented along the time axis and its ordinary Fourier transform $X(f)$ is represented along the frequency axis. FRFT can transform a function to any intermediate domain between time and frequency. The Fourier transform can be visualized as a change in representation of the signal corresponding to a counterclockwise rotation of the axis by an angle $\pi/2$. The FRFT is a linear operator that corresponds to the rotation of the signal through an angle which is not a multiple of $\pi/2$. It is the representation of the signal along the axis u making an angle α with the time axis.

Mathematically, an α-order FRFT, $F_\alpha(u)$, is the Fourier transform (FT) raised to power α, where α need not be an integer. The first-order FRFT is the ordinary Fourier transform and all properties of the Fourier transform are special cases of the FRFT.

The FRFT $F_\alpha(u)$ of order α of $f(t)$ is defined using a transformation kernel $K_\alpha(t, u)$ and is given by

$$F_\alpha(u) = \int_{-\infty}^{\infty} f(t) K_\alpha(t, u) dt \tag{6.32}$$

The kernel is defined as

$$K_\alpha(t, u) = \begin{cases} \delta(t - u) & \text{if } \alpha = 2\pi N \\ \delta(t + u) & \text{if } (\alpha + \pi) = 2\pi N \\ \sqrt{\frac{-je^{j\alpha}}{2\pi \sin \alpha}} e^{[j(t^2 + u^2)/2] \cot \alpha - jut \csc \alpha} & \text{if } \alpha \neq N\pi \end{cases} \tag{6.33}$$

where N=1,2,3,...

The inverse of FRFT can be written as

$$f(t) = \int_{-\infty}^{\infty} F_\alpha(u) K_{-\alpha}(u, t) du \tag{6.34}$$

Thus the signal $f(t)$ and its FRFT (of order α) $F_\alpha(u)$ form a transform pair and are related to each other.

6.2.6.1 Properties of fractional Fourier transform

The FRFT of a signal $f(t)$ exists under the same conditions in which its Fourier transform exists. The kernel K_α also has the same properties as the FRFT. The properties of the kernel as well as the FRFT are:

1. **Identity:** FRFT of order $\alpha = 0$ gives the input signal itself, and the FRFT of order $\alpha = 2\pi$ corresponds to the successive application of the ordinary Fourier transform four times. That is $F_0 = F_{\pi/2} = I$.

2. **Inverse:** The FRFT of order $-\alpha$ is the inverse of the FRFT of order α since $F_{-\alpha}F_\alpha = F_{\alpha-\alpha} = F_0 = I$.

3. **Additivity:** Successive applications of FRFT are equivalent to a single transform whose order is equal to the sum of the individual orders. That is, $F_\alpha F_\beta = F_{\alpha+\beta}$.

Apparently, fractional Fourier transforms can transform a signal (either in the time domain or frequency domain) into the domain between time and frequency: it is a rotation in the time-frequency domain by angle α. This is generalized by the linear canonical transformation, which generalises the fractional Fourier transform and allows linear transforms of the time-frequency domain other than rotation [6].

The FRFT belongs to the class of time-frequency representations that have been extensively used for signal processing. In all the time-frequency representations, two orthogonal axes correspond to time and frequency. If a signal $f(t)$ is represented along the time axis and its ordinary Fourier transform $F(u)$ is represented along the frequency axis, then the Fourier transform operator (denoted by FT) can be visualized as a change in representation of the signal corresponding to a counterclockwise rotation of the axis by an angle $\pi/2$. This is consistent with some of the observed properties of the Fourier transform. For example, two successive rotations of the signal through $\pi/2$ will result in an inversion of the time axis.

Moreover, four successive rotations or a rotation of the signal through angle 2π will leave the signal unaltered. The FRFT is a linear operator that corresponds to the rotation of the signal through an angle which is not a multiple of $\pi/2$, i.e., it is the representation of the signal along the axis u making an angle n with the time axis. From the above explanations, it is evident that for $\alpha = \theta$, there will be no change after applying fractional Fourier transform, and for $\alpha = \pi/2$, fractional Fourier transform is nothing but a Fourier transform, which rotates the time frequency distribution with $\pi/2$. For other value of α, fractional Fourier transform rotates the time frequency distribution according to α.

6.2.6.2 Discrete fractional Fourier transform

Many attempts have been made to find the discrete version of the fractional Fourier transform [7]. Though quite a few different algorithms were suggested by various researchers, the most consistent definition that agrees with the various properties of the FRFT, such as (a) unitarian, (b) index additivity, and (c) reduction to DFT when $\alpha = \pi/2$ [8]. The analytical method to calculate the DFRFT and the theoretical framework is beyond the scope of this book.

6.3 Hartley transform

A transform related to Fourier transform is the Hartley transform (HT). This transform of signal $i(x)$ is expressed in terms of cos and sin functions as

$$HT[i(x)] = \int_{-\infty}^{+\infty} i(x)[\cos(2\pi fx) + \sin(2\pi fx)]dx \qquad (6.35)$$

In contrast to the expression of FT as $\int_{-\infty}^{+\infty} i(x)[\cos(2\pi fx) - j\sin(2\pi fx)]dx$, the HT of a real signal is real, while the FT of a real signal can be complex. The applications of HT, however appear to be limited because of this restriction [9].

6.4 Wavelet analysis and wavelet transform

A wavelet or a small-wave is a wave-like oscillation of limited duration. Sets of wavelets are generally needed to analyse data fully. To define the wavelet transform, it is necessary to introduce a generating wavelet or mother wavelet $\psi(t)$. Such a wave is mainly concentrated near zero and is characterised by a rapid decrease when $|t|$ increases. $\psi(t)$ is oscillatory and well localised in the frequency domain [10]. The generating wavelet or mother wavelet allows the introduction of smaller wavelets called daughter wavelets, $\psi_{s,\tau}(t)$, depending on two parameters: a scale factor s and a translation or shift factor τ. The factor $s^{-1/2}$ is used for energy normalization across different scales. The expansion of wavelet function $\psi_{s,\tau}(t)$ is given by

$$\psi_{s,\tau}(t) = \frac{1}{\sqrt{s}}\psi\left(\frac{t-\tau}{s}\right) \qquad (6.36)$$

where $0 < s < 1$ is a compression, $s > 1$ is a dilation, and τ is a translation. The wavelet scale and notion of frequency are inversely related.

The wavelets generated from the mother wavelet have different scales s and locations τ, but all have the identical shape. Manipulation of wavelets by translation (change of position) and scaling (change in levels) are shown in

Figure 6.3(a) and (b). During translation the central position of the wavelet is changed along the time axis, and during scaling the width of the wavelet is changed.

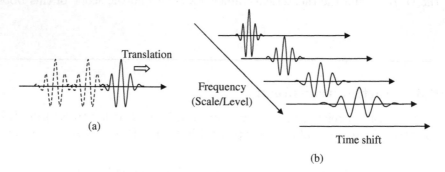

FIGURE 6.3: (a) Translation of wavelet and (b) scaling of wavelet

Two examples of continuous time wavelets, known as Morlet wavelet and Maxican hat wavelet, are shown in Figure 6.4(a) and (b), respectively. The Morlet wavelet corresponds to a signal related to the second derivative $(1 - t^2)e^{-\frac{t^2}{2}}$ of Gaussian function $e^{-\frac{t^2}{2}}$. It turned out to be the modulated Gaussian function.

The Mexican hat wavelet is derived from the second derivative of a Gaussian function, $e^{-t^2/2\sigma^2}$, and is expressed as

$$\psi(t) = \frac{2}{\pi^{1/4}\sqrt{3\sigma}} \left(\frac{t^2}{\sigma^2} - 1 \right) e^{-t^2/\sigma^2} \tag{6.37}$$

Due to the fast decay of the Gaussian function, this function drops to zero very fast.

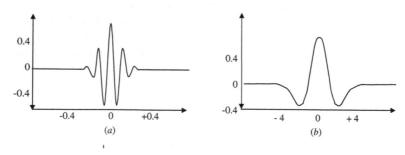

FIGURE 6.4: (a) Morlet wavelet and (b) Mexican hat wavelet

6.4.1 Wavelet transform

Unlike Fourier transform (FT), which converts a time-varying or a spatially varying signal to its frequency variations, a wavelet transform is a time-frequency representation. For a signal whose frequency varies with time or is localised, Fourier transform of such a signal is not able to represent the time-intervals at which such frequencies exist. Moreover, all spatial information is hidden in the *phases* of the expansion coefficients of FT and are therefore not readily available. Thirdly, if the dominant frequency changes in space, only average frequencies are encoded in Fourier coefficients. Wavelet analysis, on the other hand, supplies information about both time and frequency, although both parameters cannot be exactly determined simultaneously due to the Heisenberg uncertainty relation.

In order to estimate the frequency contents of a signal $f(t)$ at an arbitrary time t, one can cut off a piece of f and compute the Fourier transform of this piece only. This method is called short-time Fourier transform (STFT). Since it is a Fourier transform on a short piece of the function, i.e., during a short period of time, the restriction in time may be viewed as a translated window. STFT S_g of a signal $f(t)$ is described mathematically by multiplying the function with a real and symmetric window function $g(t - \tau)$ and then converting in (ω, τ) domain as

$$S_w(\omega, \tau) = \int_{-\infty}^{+\infty} f(t) g^*(t - \tau) e^{-j\omega t} dt \qquad (6.38)$$

where $g(t)$ is a square integrable short-time window, which has a fixed width and is shifted along the time axis by a factor τ and $*$ denotes complex conjugate of the function.

As shown in Figure 6.5, cropping is done by multiplication with the conjugate of window function $g(t-\tau)$, located at τ, and qualifies the local matching of wavelet with the signal. A perfect matching yields high wavelet transform value. The transformed coefficient has two independent parameters. One is

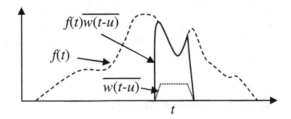

FIGURE 6.5: Time-frequency analysis with STFT and continuous wavelet

the time parameter τ, indicating the instant of concern, and the other is the frequency parameter ω, just like that in the Fourier transform. The transform can be thought of as a measure of the frequency content of f at a time t

and frequency ω. STFT is therefore a compromise between time-based and frequency-based views of a signal, where time and frequency are represented precisely by the size of the window selected, as shown in Figure 6.5. The window function has the time spread $\sigma_t(g)$ and a frequency spread $\sigma_\omega(g)$.

The window can take any form and is designed to extract a small portion of the signal $f(t)$ and then take Fourier transform. A particular type of window of interest is a Gabor window, which yields Gabor transform. The Gabor window is given by $g(t-\tau)e^{j\omega't}$. The Gabor window is generated from a basic window function $g(t)$ by translations τ along the time axis. The phase modulations $e^{j\omega't}$ correspond to translations of the Gabor function spectrum along the frequency axis by ω'. The Fourier transform of the basic Gabor function is expressed as

$$G(\omega - \omega') = \int g(t)e^{j\omega't}e^{j\omega t}dt \qquad (6.39)$$

Another type of windows which is also frequently used is Gaussian window. The Gaussian function has the minimum time-bandwidth product and the Fourier transform of the Gaussian window is also Gaussian. Time and frequency expressions of the Gaussian window is given by

$$g(t) = \frac{1}{\sqrt{2\pi}s}e^{-t^2/2s^2} \qquad (6.40)$$

$$G(\omega) = e^{-s^2\omega^2/2} \qquad (6.41)$$

STFT provides both time and frequency representation with limited precision, where the precision is determined by the size of the window. However, many signals require a more flexible approach to determine both time or frequency more accurately. Wavelet transform provides a perfect solution for such cases.

Wavelet transforms are broadly divided into continuous wavelet transform (CWT) and discrete wavelet transform (DWT). CWT is a function of two parameters and contains a high amount of redundant information when analysing a signal. There are two main differences between the STFT and the CWT: (a) The Fourier transforms of the windowed signals are not taken, and therefore single peak is seen corresponding to a sinusoid, i.e., negative frequencies are not computed. (b) The width of the window is changed as the transform is computed for every single spectral component, which is one of the significant characteristics of the wavelet transform.

6.4.1.1 Continuous wavelet transform (CWT)

Formally, continuous wavelet transform (CWT) is written as

$$W_\psi(s, \tau) = \int f(t)\psi_{s,\tau}^*(t)dt \qquad (6.42)$$

where $*$ denotes complex conjugate. This equation shows how a function $f(t)$ is decomposed into a set of basis wavelet functions, $\psi_{s,\tau}(t)$.

The wavelet transform of $f(t)$ is a correlation between the signal and the scaled wavelets $\psi(t/s)$. The inverse wavelet transform is described as

$$f(t) = \frac{1}{C_\psi} \int \int W_\psi(s,\tau) \frac{\psi_{s,\tau}(t)}{s^2} d\tau ds \qquad (6.43)$$

where C_ψ is expressed in terms of $\psi(u)$, which is the Fourier transform of $\psi(t)$ as

$$C_\psi = \int \frac{|\psi(u)|^2}{|u|} du \qquad (6.44)$$

The last two equations are reversible as long as the admissibility criterion $C_\psi < \infty$. This implies that $\psi(u)$ turns to 0 very fast as $u \to \infty$ and $\psi(0) = 0$.

6.4.1.2 Wavelet transform in frequency domain

The Fourier transform of the wavelet is given by

$$\Psi_{s,\tau}(\omega) = \int \frac{1}{\sqrt{s}} \psi\left(\frac{t-\tau}{s}\right) e^{-j\omega t} dt = \sqrt{s}\Psi(s\omega)e^{-j\omega\tau} \qquad (6.45)$$

where $\Psi(\omega)$ is the Fourier transform of the basic wavelet $\psi(t)$. In the frequency domain, the wavelet is multiplied by a phase factor $e^{-j\omega\tau}$ and by the normalisation factor $\sqrt{s}$. The amplitude of the scaled wavelet is proportional to $\frac{1}{\sqrt{s}}$ in the time domain and is proportional to $\sqrt{s}$ in the frequency domain. When the wavelets are normalised in terms of amplitude, the Fourier transforms of the wavelets with different scales have the same amplitude.

As seen, the continuous wavelet transform is also a function of two parameters: (a) the translation parameter τ, and (b) the scaling parameter s. As the scale parameter is defined in terms of the frequency as $s = 1/\text{frequency}$, the low frequencies (high scales) correspond to global information of a signal, whereas high frequencies (low scales) correspond to detailed information of a certain portion of the signal. Much like a mathematical operation, scaling either dilates or compresses a signal. Larger scales correspond to dilated (or stretched out) signals and small scales correspond to compressed signals.

In the definition of the wavelet transform, the kernel function, (wavelet), is not specified. This is a difference between the wavelet transform and other transforms such as the Fourier transform. The theory of wavelet transform also deals with general properties of transforms, such as the admissibility, regularity, and orthogonality. The wavelet basis is built to satisfy those basic conditions.

6.4.1.3 Time-frequency analysis

The decomposition of a function when a time perspective is preferred in addition to information about frequency content is called time-frequency analysis. The wavelet functions are localised both in time and frequency, yet cannot be an exact localization. The time-frequency joint representation has an intrinsic limitation; the product of the resolutions in time and frequency is limited by the uncertainty principle

$$\sigma_t \sigma_\omega \geq \frac{1}{2} \tag{6.46}$$

where σ_t and σ_ω are the spreads in time and frequency. This limitation is referred to as the Heisenberg inequality, familiar in quantum mechanics as Heisenberg uncertainty and important for time-frequency joint representation.

A signal cannot be represented as a point in the time-frequency space. One can only determine its position in the time-frequency space within a rectangle of $\sigma_t \sigma_\omega$ in the time-frequency plain, as shown in Figure 6.6. These boxes all have the same area, but their sides are stretched and contracted by the same factors, s and s^{-1}. The wavelet transform has higher time resolution at higher frequencies and this makes the wavelet transform useful for analysis of signals that contain both low frequencies and short (high frequency) transients. In

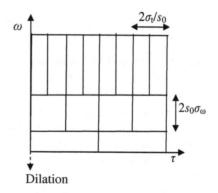

FIGURE 6.6: Heisenberg boxes of (a) wavelet $\psi_{s,\tau}$ and (b) Heisenberg boxes of STFT

contrast STFT localization measures represent the sides of an STFT Heisenberg box. The size of such a box is independent of (t, ω), hence the STFT has the same resolution across the entire time-frequency plane.

6.4.1.4 Discrete wavelet transform (DWT)

The CWT maps a 1D signal to a 2D time-scale joint representation that is highly redundant. Most applications seek a signal description with as few components as possible. To overcome this problem, discrete wavelets have been introduced. Discrete wavelets are not continuously scalable and translatable but can only be scaled and translated in discrete steps. Instead of continuously varying the parameters, the signal is analysed with a small number of scales, with a varying number of translations at each scale. Therefore DWT may be viewed as the discretisation of the CWT through sampling specific wavelet coefficients.

A discrete wavelet transform (DWT) decomposes the signal into mutually orthogonal set of wavelets. The wavelet coefficients of a discrete set of child wavelets is computed from the mother wavelet $\psi(t)$ by shifting and scaling, as given below

$$\psi_{p,k}(t) = \frac{1}{\sqrt{s_0^p}}\psi\left(\frac{t - k\tau_0 s_0^p}{s_0^p}\right) \tag{6.47}$$

where p and k are integers and $s_0 > 1$ is a fixed dilation step. The translation factor τ_0 depends on the dilation step.

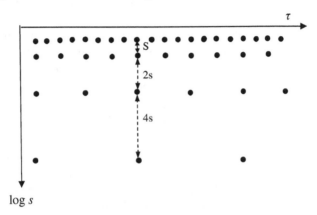

FIGURE 6.7: Localisation of the discrete wavelets in the time-scale space on a dyadic grid

The effect of discretising the wavelet is that the time-scale space is now sampled at discrete intervals, as shown in Figure 6.7. The sampling along the time axis has the interval, $\tau_0 s_0^p$, which is proportional to the scale, s_0^p. The time sampling step is small for small-scale wavelet analysis and is large for large-scale wavelet analysis. The discrete time step is $\tau_0 s_0^p$. By selecting $\tau_0 = 1$, the time sampling step is a function of the scale and is equal to 2^p for the dyadic wavelets. With the varying scale, the wavelet analysis is able to zoom in on singularities of the signal using more concentrated wavelets of very small scale. The behaviour of the discrete wavelets depends on the steps, s_0 and τ_0. When s_0 is close to 1 and τ_0 is small, the discrete wavelets are close to the continuous wavelets. For a fixed scale step s_0, the localization points of the discrete wavelets along the scale axis are logarithmic as $\log s = p \log s_0$. Conversely, CWT can be viewed as a set of transform coefficients $\{W_\psi(s, \tau)\}$ that measure the similarity of $f(t)$ with a set of basis functions, $\{\psi_{s,\tau}(t)\}$. Since, $\psi_{s,\tau}(t)$ is real valued and $\psi_{s,\tau}(t) = \psi_{s,\tau}^*(t)$, each coefficient of $W_\psi(s, r)$ is the integral inner product, $\langle f(t), \psi_{s,\tau}(t) \rangle$, of $f(t)$ and $\psi_{s,\tau}(t)$. For dyadic sampling of frequency, axis s_0 is chosen as 2. The translation factor is also chosen as $\tau_0 = 1$ so that it is also a dyadic sampling of the time axis.

Under dyadic sampling, the child wavelet coefficient of a signal $f(t)$ is the projection of $f(t)$ onto a wavelet. Let $f(t)$ be a signal of length 2^N where N

is an integer, then

$$W_{pk} = \int_{-\infty}^{+\infty} f(t)\frac{1}{\sqrt{2^p}}\psi\left(\frac{t-k2^p}{2^j}\right) dt$$

$$= f(t) * \psi_{p,k}(t)$$

(6.48)

For a particular scale p, W_{pk} is a function of k only. From the above equation, it is seen that W_{pk} can be viewed as a convolution of the signal $f(t)$ with a dilated (scaled), reflected, and normalized version of the mother wavelet, $W(t) = \frac{1}{\sqrt{2^p}}\psi\left(\frac{-t}{2^p}\right)$.

The discrete wavelets can be made orthogonal to their own dilations and translations by special choices of the mother wavelet. An arbitrary signal can be reconstructed by summing the orthogonal wavelet basis functions, weighted by the wavelet transform coefficients as

$$f(t) = \sum_{p,k} W(p,k)\Psi_{p,k}(t)$$

(6.49)

The discrete wavelet transform results in a time-frequency tiling as shown in Figure 6.8, where time allocation of three transforms are shown as (a) Fourier transform, (b) STFT, and (c) wavelet transform. The CWT and DWT resolve all frequencies simultaneously, localised in time to a level proportional to their wavelength.

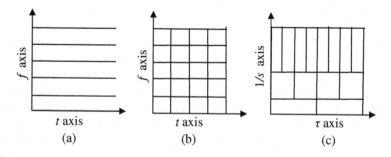

(a)   (b)   (c)

FIGURE 6.8: (a) Time-frequency tiling of FT, (b) time-frequency tiling of STFT, and (c) time-frequency tiling of DWT

6.4.1.5 Haar wavelet

Unlike Fourier series, to represent a signal in wavelets, a single function and its dilations and translations is used to generate a set of orthonormal basis functions. The number of such functions is infinite, and depending on application that suits the purpose, such functions can be selected. Unfortunately, most of the wavelets used in discrete wavelet transform are fractal in nature. They are expressed in terms of a recurrence relation so that several iterations are needed. Fortunately, two special functions called Haar wavelet functions and scaling functions have explicit expression.

A Haar wavelet is the simplest type of wavelet. In discrete form, Haar wavelets are related to a mathematical operation called the Haar transform. The Haar transform serves as a prototype for all other wavelet transforms. To calculate the discrete 1-level Haar wavelet, the mother wavelet is $\psi_1^1 = [\frac{1}{\sqrt{2}}, -\frac{1}{\sqrt{2}}, 0, 0...]$. The 1-level Haar wavelets have a number of interesting properties. First, they each have an energy of 1. Second, they each consist of a rapid fluctuation between just two non-zero values, $\pm 1/\sqrt{2}$, with an average value of 0. Finally, they all are very similar to each other in that they are each a translation in time by an even number of time units of the first Haar wavelet. The second Haar wavelet is a translation forward in time by two units of the first, and the third is a translation forward in time by four units of the first, and so on. The second 1-level Harr wavelet is the dilated, reflected, and normalised version of the first Harr wavelet given as $\psi_2^1 = [0, 0, \frac{1}{\sqrt{2}}, -\frac{1}{\sqrt{2}}, 0, 0, ...0]$ and $(N/2)$-th is $\psi_{N/2}^1 = [0, 0, ..., \frac{1}{\sqrt{2}}, -\frac{1}{\sqrt{2}}]$.

For an input represented by a list of 2^N numbers, the Haar wavelet transform may be considered to simply pair up input values, storing the difference and passing the sum. This process is repeated recursively, pairing up the sums to provide the next scale, finally resulting in $2^{(n-1)}$ differences and one final sum. The Haar DWT can be performed in $O(n)$ operations. It captures not only the frequency content of the input at different scales, but also depicts the times at which these frequencies occur.

6.4.1.6   Daubechies wavelet

The difference between the Haar wavelets and the simplest of Daubechies wavelet (known as Daub4) lies in the way that the scaling signals and wavelets are defined [11]. Let the scaling numbers be $\alpha(1), \alpha(2), \alpha(3)$, and $\alpha(4)$, and they are defined by

$$\alpha(0) = \frac{1 + \sqrt{3}}{4\sqrt{2}}, \quad \alpha(1) = \frac{3 + \sqrt{3}}{4\sqrt{2}}, \quad \alpha(2) = \frac{3 - \sqrt{3}}{4\sqrt{2}}, \quad \alpha(3) = \frac{1 - \sqrt{3}}{4\sqrt{2}}, \quad (6.50)$$

Using these scaling numbers, the 1-level Daub4 scaling signals are

$$\begin{aligned} V_1^1 &= [\alpha(1), \alpha(2), \alpha(3), \alpha(4), 0, 0...0] \\ V_2^1 &= [0, 0, \alpha(1), \alpha(2), \alpha(3), \alpha(4), 0, 0...0] \\ ..... &= ..... \\ V_{n/2}^1 &= [\alpha(3), \alpha(4), 0, 0, ...0, 0, \alpha(1), \alpha(2)] \end{aligned} \quad (6.51)$$

These scaling signals are all very similar to each other. For example, each scaling signal has a support of just four time-units. The Haar scaling signals also have this property of being translations by multiples of two time-units of the first scaling signal.

6.4.1.7   Multiresolution analysis

Although the time and frequency resolution problems are results of a physical phenomenon (the Heisenberg uncertainty principle) and exist regardless of the transform used, it is possible to analyse any signal by using an approach called the multiresolution analysis (MRA). As the name implies, MRA analyses the signal at different frequencies with different resolutions. A scaling factor is used to create a series of approximations of a function (or an image), each differing by a factor of 2 from its nearest neighbour approximation.

The wavelets can be used to achieve multiresolution analysis, which was otherwise achieved through pyramidal coding (in image processing) or subband coding (in signal processing). The main advantage of using wavelet transform in image processing is that it is well suited to manage different image resolutions and allows the image decomposition in different kinds of coefficients, while preserving the image information. The level of detail varies from location to location within an image. Some locations contain significant details, thus requiring analysis at a finer resolution, whereas at other locations, a more coarse resolution is sufficient. Thus MRA facilitates efficient compression of the image by removing the redundancies across the various resolutions.

MRA is designed to give good time resolution but poor frequency resolution at high frequencies and good frequency resolution and poor time resolution at low frequencies. This approach makes sense, especially when the signal at hand has high frequency components for short duration and low frequency components for long duration. Fortunately, the signals that are encountered in practical applications are often of this type.

The MRA is an increasing sequence of closed subspaces $\{V_p\}_{p \in Z}$ which approximate $L^2(R)$, where the space L^2 consists of all square integrable functions. The scaling function $\phi \in V_0$ with a non-vanishing integral exists such that the family $\{\phi(x-k)\}, k \in Z$ forms an orthonormal basis for the reference space V_0. The relations $[... \subset V_{-1} \subset V_0 \subset V_1 \subset ...]$ describe the analysis, where the spaces V_p are nested. The space $L^2(R)$ is a closure of the union of all V_p and the intersection of all V_p is empty. The spaces V_p and V_{p+1} are similar. If the space V_p is spanned by $\phi_{pk}(x), k \in Z$, then the space V_{p+1} is spanned by $\phi_{p+1,k}(x), k \in Z$ and is generated by the function $\phi_{p+1,k}(x) = \sqrt{2}\phi_{pk}(2x)$. Because $V_0 \subset V_1$, any function in V_0 can be written as a linear combination of the basis functions $\sqrt{2}\phi(2x - k)$ from V_1. In particular,

$$\phi(x) = \sum_k h(k)\sqrt{2}\phi(2x - k) \qquad (6.52)$$

where the coefficients $h(k)$ are defined as $\langle \phi(x) \cdot \sqrt{2}\phi(2x - k) \rangle$. Now the orthogonal compliment of wavelet W_p of V_p can be used to transit to V_{p+1} as

$$V_{p+1} = V_p \oplus W_p \qquad (6.53)$$

where $\oplus$ is the union of spaces, like unions of sets.

In other words, each element of V_{p+1} can be written, in a unique way, as the sum of an element of W_p and an element of V_p. The space W_p contains the detailed information needed to go from an approximation at resolution j to an approximation at resolution $(p+1)$.

It can be shown that $\{\sqrt{2}\psi(2x-k)\}$ is an orthonormal basis for W_1, where $\psi(x)$ is given by

$$\psi(x) = \sqrt{2}\sum_k (-1)^k h(-k+1)\phi(2x-k) \tag{6.54}$$

The similarity property of MRA gives that $\{2^{p/2}\psi(2^p x - k)\}$ is a basis for W_p, and the family $\{\psi_{pk}(x)\} = 2^{p/2}\psi(2^p x - k)$ is a basis for $L^2 R$.

For a given function $f \in L^2 R$ one can find N, such that $f_N \in V_N$ approximates f up to preassigned precision in terms of L_2 closeness. If $g_i \in W_i$ and $f_i \in V_i$, then

$$f_N = f_{N-1} + g_{N-1} = \sum_{i=1}^M g_{N-M} + f_{N-M} \tag{6.55}$$

The above equation is the wavelet decomposition of f.

The sequences $h(k)$ and $g(k)$ are quadrature mirror filters in the terminology of signal analysis. The sequence $h(k)$ is known as a low pass or low band filter while $g(k)$ is known as the high pass or high band filter. The connection between them is given by

$$g(n) = (-1)^n h(1-n) \tag{6.56}$$

By using Fourier transforms and orthogonality, it can be proved that $\sum h(k) = \sqrt{2}$ and $\sum g(k) = 0$. The Mallat's algorithm is an appropriate technique used for multiresolution analysis (MRA). A compact way to describe the Mallat's MRA and also to evolve a procedure for image processing is to determine the wavelet coefficients as the operator representation of filters. For a sequence $a = \{a_n\}$, the operators H and G, known as one step wavelet decomposition, are defined by the coordinate-wise relations as

$$(Ha)_k = \sum_n h(n-2k)a_n \tag{6.57}$$

$$(Ga)_k = \sum_n g(n-2k)a_n \tag{6.58}$$

6.4.2   2D wavelet transform and image processing

Any gray image can be approximated by a square matrix A of dimensions $2^n \times 2^n$, in which the entries a_{ij} correspond to intensities of gray in the pixel (i,j). To operate on such an image, two-dimensional (2D) wavelet transform can be used. 2D wavelet transform is defined as,

$$W_\psi(s_x, s_y; u, v) = \frac{1}{\sqrt{s_x, s_y}} \int \int A\psi\left(\frac{x-u}{s_x}; \frac{y-v}{s_y}\right) dxdy \tag{6.59}$$

This is a four-dimensional (4D) function. It is reduced to a set of two-dimension (2D) functions of (u, v) with different scales, when the scale factors are equal.

The process of 2D image wavelet decomposition goes as follows. On the rows of the matrix A the filters H and G are applied. Two resulting matrices are obtained as $H_r A$ and $G_r A$, both of dimension $2^n \times 2^{n-1}$. The subscript r suggests that the filters are applied on rows of the matrix A. Now on the columns of matrices $H_r A$ and $G_r A$, filters H and G are applied again and the four resulting matrices $H_c H_r A, G_c H_r A, H_c G_r A$, and $G_c G_r A$ of dimensions $2^{n-1} \times 2^{n-1}$ are obtained. The matrix $H_c H_r A$ is the average, while the other matrices give the details of the image. The process can be continued with the average matrix $H_c H_r A$ until a single number of the whole original matrix A is obtained. The processing steps are shown in Figure 6.9.

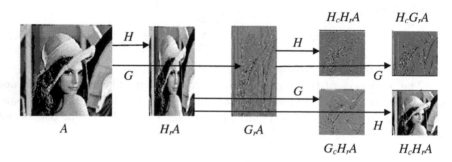

FIGURE 6.9: Wavelet processing of image

As has been seen, using simple notations, the two-step filtering and down-sampling result in four subband images: (LL) for the low-pass image filtered both horizontally and vertically, (HH) for the high-pass image filtered both horizontally and vertically, (LH) for the low-pass image filtered in a horizontal direction and the high-pass image filtered in a vertical direction, and (HL) for the high-pass image filtered in a vertical direction and high-pass image filtered in a horizontal direction. The decomposition continues at different levels, such as 1,2,3 shown in Figure 6.10, where all four images have half the size of the input image. The detail images (LH), (HL), and (HH) are put in three respective quadrants. The image (LL) is the approximation image in both horizontal and vertical directions and is down-sampled in both directions. Then, the whole process of two-step filtering and down-sampling is applied again to the image (LL) in this lower resolution level. The iteration can continue many times until, for instance, the image (LL) has only the smallest size possible.

Some general comments can be made on the application of wavelet transform in image processing. The wavelet transform is a local operation. The wavelet transform of a constant is equal to zero and the wavelet transform of a polynomial function of degree, n, is also equal to zero if the Fourier transform of the wavelet has the zero of order $(n + 1)$ about the zero frequency.

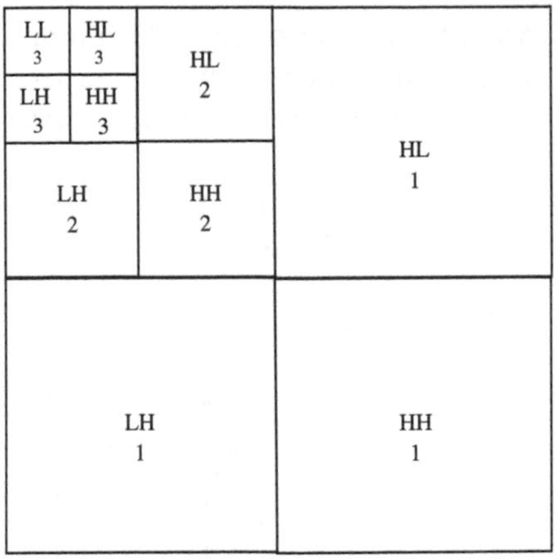

LL 3	HL 3	HL 2	
LH 3	HH 3		
LH 2		HH 2	HL 1
LH 1		HH 1	

FIGURE 6.10: 2D wavelet decomposition of image

Hence, the wavelet transform is useful for detecting singularities of functions and edges of images. The local maxima of the wavelet transform with the first-derivative of a Gaussian wavelet can be extracted as edges. This is an equavalent operation of the Canny edge detector. The zero-crossing of the wavelet transform with the Mexican hat wavelet corresponds to the inflection points of the smoothed image, which can be extracted as edges. This operation is an equivalent to the zero-crossing Laplacian edge detector.

6.5   Radon transform

In general, Radon transform is used for shape detection. The Radon transform maps a function in terms of its (integral) projections [12]. The inverse Radon transform corresponds to the reconstruction of the function from the projections. The original formulation of the Radon transform of a function $f(x, y)$ is as

$$R(\rho, \theta) = \int \int f(x, y)\delta(\rho - x\cos\theta - y\sin\theta)dxdy \qquad (6.60)$$

where δ is the delta function.

The parameters used in the definition of Radon transform specify the position of the line. The parameter ρ is the shortest distance from the origin of the coordinate system to the line, and θ is an angle corresponding to the angular orientation of the line. All lines can be described by choosing $0 \leq \theta \leq 2\pi$ and $\rho \geq 0$, but frequently other limits are used. If negative values of ρ are introduced, the parameter domain is bounded by $0 \leq \theta \leq 2\pi$ and

$-\rho_{max} \leq \rho \leq \rho_{max}$, where ρ_{max} is positive. Both limitations of the parameter domain are valid, and they are very much related to $R(\rho, \theta) = R(-\rho, \theta + \pi)$. The projection and reconstruction is shown in Figure 6.11.

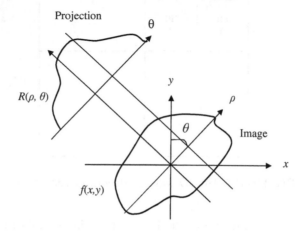

FIGURE 6.11: Two-dimensional projection of image at an angle θ to the vertical axis

The task of reconstruction is to find $f(x,y)$ given the knowledge of $R(\rho, \theta)$ and is obtained as

$$f_{BP}(x,y) = \int_0^\pi R(x\cos\theta + y\sin\theta, \theta)d\theta \qquad (6.61)$$

Geometrically, the back-projection operation simply propagates the measured value of Radon transform back into the image space along the projection paths. The back-projection image is a blurred version of the original image. A point source at the origin, the intensity of the back-projection image rolls off slowly away from the origin because of the convolution process involved [13].

6.5.1 Central slice theorem

The solution to the inverse Radon transform is based on the central slice theorem (CST), which relates the 2D Fourier transform (FT) $F(\nu_x, \nu_y)$ of $f(x,y)$ and the 1D FT of $R(\rho, \theta)$. The CST theorem states that the value of the 2D FT along a line at the inclination angle θ is given by the 1D FT of Radon transform and the projection profile acquired at an angle θ. Hence, with enough projections, 1D FT of Radon transform can fill the (ν_x, ν_y) space to generate $F(\nu_x, \nu_y)$. A flow chart of direct Fourier reconstruction is shown in Figure 6.12.

The 3D Radon transform is defined using 1D projections of a 3D object $f(x,y,z)$ where these projections are obtained by integrating $f(x,y,z)$ on a plane, whose orientation can be described by a unit vector τ. Geometrically,

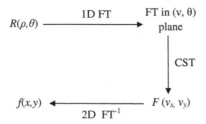

FIGURE 6.12: Flowchart of reconstruction of image

the continuous 3D Radon transform maps a function into the set of its plane integrals. Each point in the (ρ, θ) space corresponds to a plane in the spatial domain (x, y, z). The 3D Radon transform satisfies the 3D Fourier slice theorem, which states that the central slice $f_s(\rho, \theta)$ in the direction θ of the 3D Fourier transform of $f(x, y, z)$ equals to Radon transform $R(\rho, \theta)$. Reconstruction of a 2D image $f(x, y)$ from the projections (Figure 6.13 (a)), taken at angles 0 degree, 45 degree and 90 degree, is shown in Figure 6.13(b).

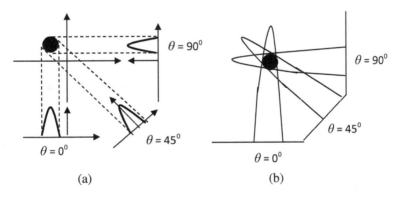

(a)	(b)

FIGURE 6.13: (a) Projection from 2D image $f(x, Y)$ and (b) image reconstruction from projected data

Works exploiting the various properties of Radon transform have also been reported [14],[15], [16]. They generalize the 2D Radon transform to three dimensions.

6.6 Hough transform

The Hough transform (HT) is a powerful feature extraction technique used in image analysis, particularly for detecting edges (straight lines and parametric curves) in an image [17], [18]. Closely related to the Radon transform, HT converts a global feature detection problem in image space into an easily solvable peak detection problem in parameter space. The detection is done easily by looking for accumulation points in the parameter space. Thus this approach maps distributed and disjoint elements of the image into a localised accumulation point,

The Hough transform has several desirable features [19], [20]. In this transform, each point is treated independently and so it is possible to treat all the points simultaneously in parallel making shape detection easy. Secondly, HT can recognize partial or slightly deformed shapes, since the size and spatial localization of a peak is a measure of its shape. Computational load of the method however, increases rapidly with the number of parameters which define the detected shape. Lines have two parameters, circles three, and ellipses (circles viewed at an angle) have five. Hough method has been applied to all of these, but the detection of ellipse probably is most difficult.

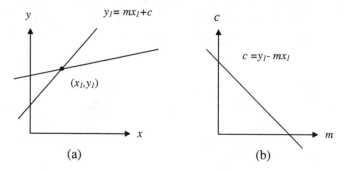

FIGURE 6.14: (a) A straight line in (x_1, y_1) space and (b) a straight line passing through (x_1, y_1) in (m, c) space

6.6.1 Line detection

A straight line as shown in Figure 6.14(a) is represented by the equation

$$y_1 = mx_1 + c \tag{6.62}$$

where (x_1, y_1) are the coordinates through which the straight line passes, m is the slope and c is the intercept. Since m, and c are variables, several lines can pass through the point (x_1, y_1). But only one straight line can pass through (x_1, y_1) if x_1 and y_1 are considered constants and m and c are variable parameters. The above equation can be rewritten as

$$c = y_1 - mx_1 \tag{6.63}$$

Each different line through the point (x_1, y_1) corresponds to one of the points on the line in (m, c) space (also called the Hough space) as shown in Figure 6.14(b).

The steps to detect straight lines in an image begin with the quantization of (m, c) space into a two-dimensional array $A(m, c)$ with appropriate steps of m and c and then initialization of all elements of $A(m, c)$ to zero. Now for each pixel (x_1, y_1) which lies on the edge of the image, all elements of $A(m, c)$ whose indices m and c satisfy $y_1 = mx_1 + c$ are incremented by 1. This step is followed by a search for elements of $A(m, c)$ which have large values, since each value corresponds to a line in the original image.

However, the above-mentioned technique for detecting straight lines by Hough transform fails for vertical lines since slope m becomes infinite. Hence instead of $y_1 = mx_1 + c$, straight lines are more usefully parametrised by distance r and angle θ, as

$$r = x_1 \cos\theta + y_1 \sin\theta \tag{6.64}$$

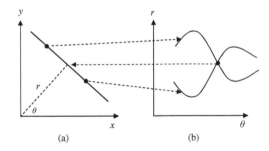

(a)   (b)

FIGURE 6.15: (a) A straight line represented by r, θ in (x, y) space and (b) representing lines passing through point (x, y) in the (r, θ) space

In Figure 6.15, all points lying on the line $x_1 \cos\theta + y_1 \sin\theta = r$ yield sinusoidal curves that intersect at (r_H, θ_H), where θ varies from $-\pi/2$ to $+\pi/2$, in the (r, θ) space.

The HT of a binary image $I(x, y)$ of size $(M \times N)$ can be mathematically represented as

$$HT(r, \theta) = \sum_x^M \sum_y^N I(x, y) * \delta(x \cos\theta + y \sin\theta - r) \tag{6.65}$$

For each data point, say (x_1, y_1), a number of lines can pass through it, all at different angles. For each line passing through (x_1, y_1), a line perpendicular to it is drawn from the origin. The length and angle of each perpendicular line representing r and θ is measured. This is repeated for each straight line passing through (x_1, y_1). A graph of length r against angle θ, known as a Hough space graph, is then created in tabular form. Two different points in (x_1, y_1) space through which one straight line passes will thus be represented by two different curves in (r, θ) space. The intersection of the two curves in (r, θ) space corresponds to the parameters of the straight line in (x_1, y_1) space, formed by joining those two different points.

The algorithm for the Hough transform can be summarized as:

1. Quantize the (r, θ) space into a two-dimensional array $H(r, \theta)$ and initialize to zero.

2. For all values of θ, for the edge pixels,

$$r = x \cos \theta + y \sin \theta + H(r, \theta) \qquad (6.66)$$

3. The values of (r_H, θ_H) are calculated for which $H(r, \theta)$ is maximum. Then, $r_H = x \cos \theta_H + y \sin \theta_H$ is the detected straight line.

6.6.2 Circle detection

The above method can, in principle, be generalized to parametric curves, with the implication that points in the (x, y) space map into surfaces in multidimensional parameter space. The parametrisation of circle, known as the circle HT, is given by

$$f_c(x, y) = (x - a)^2 + (y - b)^2 - r^2 = 0 \qquad (6.67)$$

where (a, b) denotes the centre and r is the radius of the circle. Thus each pixel in the image plane is transformed to a cone in a 3-D parameter space as shown in Figure 6.16.

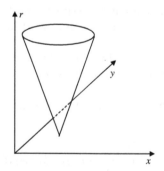

FIGURE 6.16: Circle detection through the creation of a cone

There are many variants of HT, such as probabilistic HT, generalized HT [21],[22],[19], fuzzy HT, etc. In every HT variant, major computational effort is related to scanning the parameter space, in order to determine how many point transforms (lines, curves, or surfaces) intersect each cell. Therefore clusters of intersections, corresponding to prominent image features, are seen. Several methods have been proposed to find these features efficiently [23].

6.7   Photonic implementation of linear transforms

Fast digital algorithms for transform processing have been developed mostly in the electronic domain, but for high-resolution applications, high processing time is a major disadvantage [24]. Mainly for this reason, photonic signal processing techniques are explored. Photonic techniques for the implementation of linear transforms can utilise the parallel processing capability of optical paths. Moreover, the advantage of 2D representation of data in the form of images in a 2D set up can be exploited. In principle, every linear transform in $L^2(\mathbb{R}^2)$ space can be implemented by an optical/photonic technique. The input, a two dimensional image representing coded data, is placed in the parallel beam of coherent or incoherent light. The wavefront is then modified by photonic components and subsystems. The resulting pattern is then decoded and allowed to fall on a recording or detecting device. In such an elementary system the highest possible parallelism is maintained, which results in minimum processing time.

A transmissive or reflective type SLM is used for registering the data arranged in the form of a 2D image which can be addressed electrically or optically. Sometimes an acousto-optic cell is also used. The parallel beam of light illuminates the SLM. The wavefront coming out of the SLM is convolved by optical or photonic components depending on the desired type of transform of interest. Ultimately, the results of convolution operation is captured by a CCD or line scan camera. If needed, some spatial filtering can be incorporated. The convolution operations can be build in an optical system by the well-known $4f$ set up (Figure 6.17), which is based on the convolution theorem of the Fourier transform.

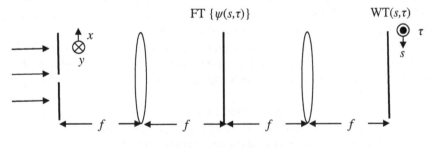

FIGURE 6.17: 4f setup

However, some restrictions for the photonic realisations have to be appreciated. For instance, because of the pixelated nature of SLMs, the discrete arrays exhibit only a finite resolution; therefore, the available spatial frequency range is restricted. Furthermore, only intensities are recorded in a conventional camera and it is difficult to record phase information. If phase information cannot be detected in the SLMs, the information encoded in the amplitude and phase spectrum of the image is lost [25].

6.7.1 Implementation of wavelet transform

Photonic implementation of 2D wavelet transformation can be realised by a conventional 4f system [21], [26], as shown in Figure 6.17. The 2D WT is obtained as correlation of the Fourier transform of the input image with different scaled versions of the wavelet. Mathematically, for a function $f(x)$, the WT transform is given by

$$W_\psi(s,\tau) = \int f(x)\psi^*(s, x - \tau)dx = f(\tau) \otimes \psi(s,\tau) \qquad (6.68)$$

where the symbol $\otimes$ represents a 1-D correlation operation with respect to the x axis.

The WT coefficients may be written in the frequency domain as the inner product between $f(x)$ and $\Psi_{s,\tau}$ and is given by

$$W_\Psi(s,\tau) = \sqrt{s} \int \Psi^*(s\omega)e^{+j\omega\tau}F(\omega)d\omega \qquad (6.69)$$

where $F(\omega)$ is the FT of $f(x)$ and $\Psi_{s,\tau}(\omega)$ is the wavelets expressed in frequency domain, and is given by

$$\Psi_{s,\tau}(\omega) = \sqrt{s}e^{-j\omega\tau}\Psi(s\omega) \qquad (6.70)$$

Thus, the WT may be implemented in an optical correlator using a bank of optical filters $\Psi(s\omega)$ with different scales s. The WT filter is the FT of the dilated wavelet $f(x/s)$ but without translation. Since wavelets and their FT are, in general, complex valued, the WT filter will have phase information. Although a computer-generated hologram is an ideal candidate for the generation of WT filter, considering symmetric wavelets, the WT may be approximated by optical transmittance masks generated on an SLM.

The photonic system for the implementation of WT is shown in Figure 6.18, where a 2D optical correlator performs 1D wavelet transform. The 1D input $f(x)$ is displayed in an acousto-optic modulator, illuminated by a parallel light beam [27].

The 1D FT of $f(x)$ is performed by a cylindrical lens along the x axis. In the Fourier plane, the 1-D signal spectrum $F(\omega)$ is obtained. The filter bank $\Psi(s\omega)$ is introduced as several horizontal strips in the Fourier plane. Each strip represents 1D WT filter $\Psi(s\omega)$ with a different dilation factor s. A spherical and cylindrical lens ensemble performs the inverse FT along the horizontal axis and forms images at the output plane, which is again divided into several horizontal strips. Each strip represents a 1D WT of the input signal with the parameter τ corresponding to the horizontal axis and the parameter s corresponding to the dilation factors along the vertical axis. The WT output is detected by a CCD camera and normalisation is done from the recording by a computer connected with the camera via a frame grabber.

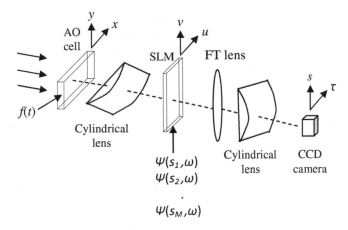

FIGURE 6.18: Schematic of photonic setup for the implementation of WT

6.7.2 Photonic implementation of Hough transform

A simple scheme for the photonic implementation of Hough transform is shown in Figure 6.19. The electrically addressed SLM at object plane is

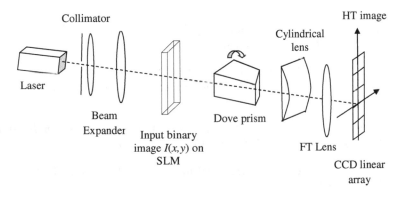

FIGURE 6.19: Arrangement for photonic implementation of Hough transform

illuminated by a parallel beam from a laser source after collimation. The image whose HT is desired is placed on the SLM through a computer. A rotating Dove prism is located on the optical axis between (x, y) plane and the plane of a cylindrical lens. A spherical lens which can perform FT is placed in the beam path. Combination of these two lenses performs imaging in the vertical direction and 1D FT in the horizontal direction.

The linear array of a photodetector (in the form of a 1D CCD array) is oriented vertically and records zero-order components of the FT, which is proportional to the intensity of the line in HT, at an angle θ. Angular scanning of the object is done by rotating the Dove prism. Another cylindrical lens can be added after the first cylindrical lens so that HT in both directions can be obtained. The system, though simple, is prone to errors due to signal-to-noise ratio and space invariance.

Books for further reading

1. *The Fourier Transform and Its Applications*: R. N. Bracewell, McGraw-Hill, 1986.

2. *Introduction to Fourier Optics*: J. W. Goodman, Roberts and Company, 2005.

3. *The Fourier Transform and Its Applications to Optics*: P. M. Duffieux, John Wiley, 1983.

4. *Wavelet Transform*, in *The Transforms and Applications Handbook*, Y. Sheng, (A. D., Poulariskas, Ed.) CRC and IEEE Press, Boca Raton, FL, 2000.

5. *Wavelets: Mathematics and Applications* : J. J. Benedetto and M. W. Frazier, CRC Press, Boca Raton, FL, 1994.

6. *The Fractional Fourier Transform with Applications in Optics and Signal Processing*: H. M. Ozaktas, Z. Zalevsky, and M. A. Kutay, John Willy, 2001.

7. *Fundamentals of Wavelets - Theory, Algorithms, and Applications*: J. C. Goswami and A. K. Chan, John Wiley, 2011.

8. *Wavelets—Theory and Applications*: A. K. Louis and P. Maab, John Wiley, 1997.

9. *Multiresolution Techniques in Computer Vision*: E. A. Rosenfeld, Springer-Verlag, 1984.

10. *Multiresolution Signal Decomposition*: A. N. Akansu and R. A. Haddad, Academic Press, 1992.

11. *Optical Signal Processing*: A. B. VanderLugt, John Wiley, 1992.

12. *Shape Detection in Computer Vision Using the Hough Transform*: V.F. Leavers, Springer Verlag, 1992.

13. *The Radon Transform and Some of Its Applications*: S. R. Deans, John Wiley, 1983.

Bibliography

[1] J. W. Goodman. *Introduction to Fourier Optics*. Roberts and Co., 2007.

[2] D. V. Vranic and D. Saupe. A feature vector approach for retrieval of 3D objects in the context of MPEG-7. In *Proc. Conf. ICAV3D-2001*. Greece, 2001.

[3] V. Namias. The fractional order Fourier transform and its application to quantum mechanics. *J. Inst. Maths. Applics.*, 25:241, 1980.

[4] H. M. Ozaktas, Z. Zalevsky, and M.A. Kutay. *The Fractional Fourier Transform: with Applications in Optics and Signal Processing*. John Wiley, 2001.

[5] V. A. Narayanana and K. M. M. Prabhu. The fractional Fourier transform: theory, implementation and error analysis. *Microproc. and Microsyst.*, 27:511–521, 2003.

[6] L. B. Almeida. The fractional Fourier transform and time-frequency representations. *IEEE Trans. Signal Process.*, 42:3084–3091, 1994.

[7] S. C. Pei and M. H. Yeh. Improved discrete fractional Fourier transform. *Opt. Let.*, 22, 1997.

[8] H. M. Ozaktas, O. Arkan, M. A. Kutay, and G. Bozdag. Digital computation of the fractional Fourier transform. *IEEE Trans. Signal Proces.*, 44:2141–2150, 1996.

[9] S. C. Pei, C. C. Tseng, M. H. Yeh, and J. J. Shyu. Discrete fractional Hartley and Fourier transforms. *IEEE Trans. Circuits and Syst.*, 45:665–675, 1998.

[10] S. G. Mallat. A theory for multi-resolution signal decomposition: the wavelet representation. *IEEE Trans. Pat. Ana. and Machine Intell.*, 11:674–693, 1989.

[11] I. Daubechies. The wavelet transform: time-frequency localization and signal analysis. *IEEE Trans. Inf. Th.*, 36(5):961–1005, 1990.

[12] G. Beylkin. Discrete Radon transform. *IEEE Trans. Acoust. Speech and Signal Proces.*, 35(2), 1987.

[13] A. W. Lohmann and B. H. Soffer. Relationships between the Radon-Wigner and fractional Fourier transforms. *J. Opt. Soc. Am. (A)*, 11:1798–1811, 1994.

[14] P. Milanfar. A model of the effect of image motion in the Radon transform domain. *IEEE Trans. Image Process.*, 8, 1999.

[15] M. Barva, J. Kybic, J. Mari, and C. Cachard. Radial Radon transform dedicated to micro-object localization from radio frequency ultrasound signal. In *Proc. IEEE Int. Conf. on Ultrasonics, Ferroelectrics*. IEEE, 2004.

[16] P. G. Challenor, P. Cipollini, and D. Cromwell. Use of the 3D Radon transform to examine the properties of oceanic Rossby waves. *J. Atmos. and Oceanic Tech.*, 18, 2001.

[17] P. V. C. Hough. Method and means for recognizing complex patterns. *Patent 3069654*, Patent 3069654, 1962.

[18] R. O. Duda and P. E. Hart. Use of the Hough transform to detect lines and curves in pictures. *ACM Comm.*, 55:11–15, 1972.

[19] J. Illingworth and J. Kittler. The adaptive Hough transform. *IEEE Trans. Pat. Ana. and Machine Intell.*, 9(5):690–698, 1987.

[20] M. Atiquzzaman. Multiresolution hough transforman efficient method of detecting patterns in images. *IEEE Trans. Pat. Ana. and Machine Intell.*, 14(11):1090–1095, 1988.

[21] Y. Li, H. H. Szu, Y. Sheng, and H. J. Caulfield. Wavelet processing in optics. *Proc. IEEE*, 84:720–732, 1996.

[22] V. F. Leavers. The dynamic generalized Hough transform. *Comp. Vis. Graphics and Image Process.*, 56(3):381–398, 1992.

[23] M. A. Lavin, H. Li, and R. J. Le Master. Fast Hough transform: A hierarchical approach. *Comp. Vis. Graphics and Image Process.*, 36:139–161, 1986.

[24] C. S. Weaver and J. W. Goodman. Technique for optically convolving two functions. *App. Opt.*, 5:1248, 1966.

[25] D. L. Flannery and J. L. Horner. Fourier optical signal processor. *Proc. IEEE*, 77:1511, 1989.

[26] D. Mendlovic and N. Konforti. Optical realization of the wavelet transform for two-dimensional objects. *App. Opt.*, 32:6542, 1993.

[27] Y. Sheng, D. Roberge, and H. Szu. Optical wavelet transform. *Opt. Eng.*, 31:1840–1845, 1992.

Chapter 7

Low-Level Photonic Information Processing

7.1 Introduction

Low-level processing, also known as early processing, is a subset of computer vision systems, and processing techniques and is in contrast to high-level processing which is concerned with cognitive level processing. Low-level processing is usually achieved with images captured by camera and processed by many image processing algorithms. The hardware implementation may be carried out in the electronic or in the photonic domain. In the photonic domain, the information about the image is usually fed into photonic devices and processed. The photonic domain processing may also be carried out in the electronic domain after photonic to electronic conversion.

During low-level processing, the raw data is processed by signal processing algorithms, usually generating feature descriptions or sets of symbolic descriptors which summarise characteristics of data in a quantitative way. Low-level processing reveals physical properties of an image, such as orientation, range, reflectance, and the like. Geometric representations of such images are used to capture all necessary information of two-dimensional (2D) and three-dimensional (3D) shape. On the other hand, high-level processing is related to the interpretation and reasoning from visual data. It is usually built on top of the low-level processing algorithms, taking feature-descriptors as input and generating abstract, qualitative descriptions about the content of the visual data. Ideally, the high-level processing can also act on the low-level processing, adjusting its parameters to improve the performance, creating a kind of feedback loop between the two levels (usually called bottom-up and top-down reasoning) [1]. High-level processing usually employs some sort of *a priori* knowledge about the visual interpretation to be made.

In general, signal features are low-level entities generated by signal (image or video) processing algorithms, like pixels, regions, and their properties. There is a set of common constructs for representing signal features, which is pervasive for almost all approaches [2]. An arbitrary set of signal data usually is the result of segmentation algorithms and is further characterised by attributes [3], [4]. The image data is further divided into image entities or feature-concepts for representing image data structures (e.g., regions, edges, and region graphs). Image concepts are then obtained for representing calculated image features (e.g., measures of size, position, shape, colour, and texture). The second group is composed of image processing functionality concepts, which aim to represent the kind of processing algorithms according to their intentions (e.g., segmentation, object extraction, etc.,). The block diagram shown in Figure 7.1 gives basic interconnection between low-level processing and high-level processing.

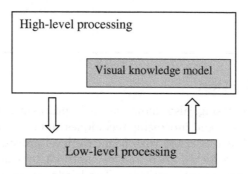

FIGURE 7.1: Mapping of one pixel with a certain gray value to a different value in a point operation

7.2 Low-level image processing

7.2.1 Point operators

If $g(x, y)$ gives the intensity at position (x, y), point operators alter the gray value of a pixel $g[x, y]$ in the input image to $g'[x, y]$, so that every pixel of input image maps to a different gray value in the output image, as shown in Figure 7.2.

Single-point processing is used in image enhancement or alteration [5]. This technique determines a pixel value in the enhanced or altered image that is dependent only on the value of the corresponding pixel in the input image. The process can be described with the mapping function M

$$s = M(r) \tag{7.1}$$

where r and s are the pixel values in the input and output images, respectively. The form of the mapping function M determines the effect of the operation. Point operators are also known as look-up table (LUT) transformations, because the mapping function, in the case of a discrete image, can be implemented by a look-up table.

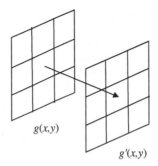

FIGURE 7.2: Mapping of one pixel with a certain gray value to a different value in a point operation

The various point operations are:

Thresholding: Thresholding of an image is the process of assigning those pixels a particular value, all of which are above a certain threshold level. For example, binary thresholding means assigning all values above a certain threshold as white, and others as black. The mapping function as defined can be decided in an ad-hoc manner. For example, a simple mapping function is defined by the thresholding operator as

$$s = 0 \ \ if \ \ r < T$$
$$= (L - 1) \ \ if \ \ r > T \tag{7.2}$$

Contrast stretching: Changing the contrast of an image means spreading out the gray level distribution. Visualised in the histogram, it is equivalent to expanding or compressing the histogram around the midpoint value. Mathematically, contrast stretching is expressed as

$$g'(x,y) = [g(x,y) - 0.5] \times contrast + 0.5 \tag{7.3}$$

The subtraction and addition of 0.5 is used for centring the expansion/compression of the range around 50% gray value.

Histogram equalisation: This is a general method of modifying intensity distribution of an image. The mapping function can also be defined in an ad-hoc manner, or it can be computed from the input image. For example, a simple mapping function is defined by a thresholding operator.

Brightness change: When changing the brightness of an image, a constant is added or subtracted from the intensity value of each pixel. This is equivalent to shifting the contents of the histogram left (subtraction) or right (addition).

Logarithm operator: This operator reduces the contrast of brighter regions.

Invert: Inverting the pixel values in the image produces the negative of the original image.

7.2.2 Group operations

Some of the group operations on pixels of an image include:

Scaling: The scale operator performs a geometric transformation which can be used to shrink or zoom the size of an image. Image size reduction, or sub-sampling, replaces a group of pixel values by an arbitrarily chosen value from within this group or by interpolation of pixel values in a local neighbourhood. Image zooming is achieved by pixel replication or by interpolation. Scaling changes the visual appearance in terms of the size of an image, and alters the quantity of information stored in a scene representation. Scaling is also a special case of affine transformation. For example, Figure 7.3 shows a 3×3 image reduced by taking the average of the pixels and zoomed by replicating each pixel by a 2×2 group having the same pixel value as that of the original pixel.

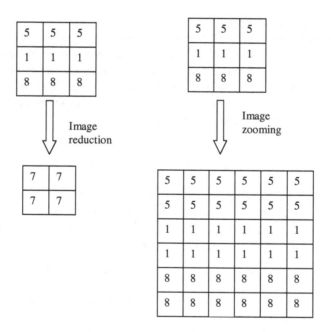

FIGURE 7.3: Scaling operation

Edge detection: Edge detection is low-level processing and is important for the success of many high-level processing. At low-level, edge detection reduces much of the redundant data by separating the target image from the scene. When pixels differ greatly from their neighbours, they are most likely the edge pixels of an image. Thus, simple edge detection is possible by subtracting neighbourhood pixels. Most of the edge detection schemes consists of three stages: filtering, differentiation, and detection. Filtering usually removes noise, while differentiation stage highlights the location of the image. In the detection stage, those points where intensity changes are significant are localised. The most simple filter used in edge detection is the mean filter, where the gray level at each pixel is replaced by the mean level in a neighbourhood round the pixel.

During differentiation, discrete approximation of a derivative is considered. Since an image is two dimensional, directional derivative in the direction of θ is used. Some edge detection algorithms use gradient operators, such as Prewit, Robert, Sobel, and Laplacian with an aim to reduce computational complexities.

7.2.3 Neighbourhood operations

Neighbourhood pixels include all those pixels surrounding a particular pixel, i.e., the neighbouring pixels are mostly a rectangle around a central pixel. Some of the neighbourhood operations include: (a) finding the pixel value at minimum or at maximum, (b) calculating median or average of a set of pixel values, and (c) detection of the edge of an image, or thresholding an image at a particular pixel value. Taking the average all of the pixels in a neighbourhood around a central value helps in smoothing the image. There are many image processing algorithms for edge detection and thresholding. For detains on all low-level processing discussed so far, the Readers may consult some books given in the list at the end of this chapter.

7.3 Morphological operation

Morphological operations are tools for extracting an image components that are useful in the representation and description of region shape [6]. As a discipline within imaging, mathematical morphology is concerned with the applications of its basic operators in all aspects of the image. The techniques are widely used for image analysis and have been a valuable tool in many computer vision applications, especially in the area of automated inspection [7]. Morphological processing techniques have relevance to conditioning, labelling, grouping, extracting, and matching operations on image. From low-level to high-level computer vision, morphological techniques are also important in operations such as noise removal and median filtering. Optical or photonic morphological processing are also implemented for feature extraction and shape description because of the inherent parallel processing capability of optical paths [8], [9], [10].

7.3.1 Binary morphological image processing

Conceptually, morphological image processing operations are rooted in planar geometric structure, which is altered by probing with a structuring element. Each operation uses the structuring element to determine the geometrical filtering process, satisfying four properties: translation invariance, anti-extensivity, increasing monotonically, and idempotence. The structuring element of a morphological operator is therefore a function defined in the domain of the spatial pattern. The value of each pixel of the domain is the weight or coefficient employed by the morphological operator at the pixel position. The selection of the size and shape of the structuring element is therefore an important step in morphological operations.

Fundamental morphological operations are erosion and dilation. These operations remove or add pixels from a binary image according to rules that depend on the pattern of the neighbouring pixels. The erosion operation reduces the size of an image, while the dilation operation enlarges the geometrical size of the image. Two other derived morphological operations are opening and closing. An opening is an erosion operation followed by dilation, and a closing operation is a dilation followed by an erosion operation.

The morphological operations are defined as set operations. Let x denote the set of pixel positions of the image in the whole Euclidean space E. A subset A is another symmetrical binary image, called the structuring element, in the two-dimensional Euclidean space. The classical Minkowsky set addition and subtraction of these two images X and A are defined in set notation as

$$X \oplus A = \{x + a : x \in X, a \in A\} = \cup X_a|_{a \in A} = \cup A_a|_{x \in X} \qquad (7.4)$$

$$X \ominus A = \cap X_{-a}|_{a \in A} \qquad (7.5)$$

where X_a is the translation of the image set along a and is defined as $X_a = \{X + a : x \in X\}$.

The transformations $X \to X \oplus A$ and $X \to X \ominus A$ are called a dilation and erosion operation by structuring element A. According to standard convention, the morphological dilation operation on the image X by the structuring element image A is given by

$$X \oplus \bar{A} = \{h \in E : (\bar{A})h \cap \frac{X}{\phi} \qquad (7.6)$$

The morphological erosion operation is given by

$$X \ominus \bar{A} = \{h \in E : A_h \subseteq X\} \qquad (7.7)$$

where h is an element in Euclidean space, and ϕ denotes the empty set. The reflected or the symmetric set $\bar{A}$ is related to A by $\bar{A} = \{-a : a \in A\}$.

Self complementation of the set yields a duality relation and the dilation operation is also expressed as

$$X \oplus \bar{A} = (X^c \ominus \bar{A})^c \qquad (7.8)$$

where X^c is the complement of the set constituting the image X.

By definition, morphological opening is erosion followed by dilation and the closing operation is viewed as the complimentary process of opening. Therefore the opening and closing operation are denoted as

$$Open(X, A) = (X \ominus \bar{A}) \oplus \bar{A} \qquad (7.9)$$

$$Close(X, A) = (X \oplus \bar{A}) \ominus \bar{A} \qquad (7.10)$$

The opening is the union of all translates of the structuring element that is included in the set X. Opening and closing are related to Boolean duality by

$$Close(X, \bar{A}) = Open(X^c, A)^c \qquad (7.11)$$

It may be noted that both opening and closing operations are translation invariant.

While applying the binary morphological image processing operations as defined, it is necessary to obtain a binary image of a test image, which in general is available as a gray image. Segmenting the gray-level pixel values of the test image into two gray-level values are all that is required to produce a binary image of the test image. Let $I(i, j)$ be the gray level pixel value at point (i, j) of the test image and $X(i, j)$ is the gray level of point (i, j) of the output image. If 1 represents white pixels, 0 represents black pixels, and T is the threshold value, then the binary image is obtained from the gray-level image by applying the rule which stipulates $X(i, j) = 1$, when $I(i, j) \geq T$, and 0 otherwise.

7.3.2 Gray-level morphological processing

Binary morphological operations, such as erosion, dilation, opening, and closing, are performed for the sets whose elements are vectors corresponding to pixel positions and therefore are set operations on set. A gray-scale image can be considered a three-dimensional (3D) set, where the first two elements are x and y coordinates of a pixel and the third element is gray-scale intensity value. The key issue is to use the infima/suprema (minima and maxima in discrete cases) to define gray-scale morphological operators. The structuring elements of the gray-scale morphological operations can have the same domains as those in binary morphology. However, a gray-scale structuring element is also possible having certain gray value instead of having only values 1 or 0. Gray-scale opening and closing are defined in a similar manner as in the binary case. The only difference when the operations are carried out is that the opening and closing use gray-scale dilation and erosion. As binary morphological operations do, gray scale opening is anti-extensive and gray scale closing is extensive.

The structuring elements for the gray-scale morphological operation can be of two types, namely, non-flat and flat. In the non-flat type structuring element the gray-scale intensity values are not uniform everywhere, and in the flat or binary type, the pixels are bi-valued 0 or 1.

7.3.3 Morphological processing by photonic technique

The morphological transformations are essentially parallel operations and thus the photonic systems which excel in parallel computations are well suited for morphological image processing [11], [12], [13]. An experimental arrangement, shown in Figure 7.4, is a hybrid photonic system, where pre-processing is done by photonic technique, and morphological operations are executed in the computer.

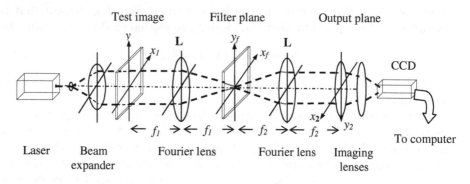

FIGURE 7.4: Fourier plane morphological operation

A He-Ne laser is used to form a parallel beam by using a collimator system. This collimated laser beam illuminates the test image (a fabric image in this case), placed at (x_i, y_i). A Fourier lens produces the diffraction pattern of the test image at the focal plane of the lens. A magnified and inverted image of the test image is reconstructed by using another Fourier lens of same focal length. The resultant image of the test fabric is captured by a CCD camera and is stored in a computer. Depending on a particular application, spatial filtering of the Fourier plane image may be necessary. With the help of a software programme, the gray-level image is converted into a binary image by thresholding. The threshold value for binary conversion is pre-selected according to the demands of cleanliness for that image. Morphological opening and erosion operations are then applied on the binary image of the test image with a properly selected structuring element.

The process has been applied for defect detection in fabric. In Figure 7.5, the test fabric image has a knot as a defect in the fabric. This defect is detected by morphological photonic operation in the set up shown in Figure 7.4. A 5×5 structuring element is used. Expansion of the size of the defect during the opening operation and shrinking of the size in the erosion operation have been observed. Nevertheless, the coarse grating structure of the warp and weft of the fabric is eliminated and the defect is detected.

Many alternative all-photonic arrangements are also possible, where the inherent parallel processing capability of the optical path is exploited in morphological operations to reduce processing time in real-time operations [14]. A lens-less shadow-casted projection system [15], [16] is one of the easiest

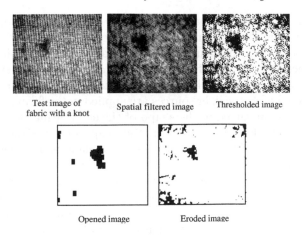

Test image of fabric with a knot   Spatial filtered image   Thresholded image

Opened image   Eroded image

FIGURE 7.5: Morphological operation on test fabric for defect detection

techniques to experiment with. The basic set up of the processor is shown in Figure 7.6. A test image in the input plane I is illuminated by point sources of light at the source plane L. The test image can be recorded in an electrically addressed transmittive type spatial light modulator (SLM). The source array is constructed by high intensity light-emitting diodes. Each point source casts a shadow of the object at the output plane O.

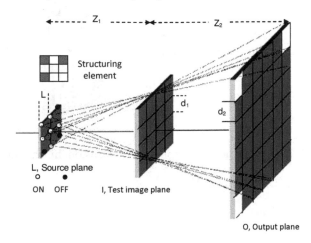

FIGURE 7.6: Photonic architecture for morphological operation

The shadow-casted system is designed in such a manner that the adjacent sources cast superposing shadows. By proper design, the shadows are offset by one cell in the corresponding direction at the output or the superposition plane, where a cell is defined as a unit area or one set element in the input object. To achieve the desired morphological operation, proper superposition of the sub-cells at the output plane has to be ensured.

The dimension of the source array depends on the dimension of the structuring element set A. The shape of A is determined by the switching states of the light sources. Due to the inherent nature of a shadow-casted projection system, instead of A, the light sources are applied in the form of (-A). Since the light sources are arranged in the shape of the structuring element, the casted shadows of each element are also in the shape of the structuring element. Therefore, dilation of the object at the input plane is inherently achieved at the output plane.

It is evident from the definition of the erosion and dilation operations that the erosion is a dual dilation operation and therefore erosion operation can be obtained from the dilation operation. Figure 7.7 shows a method of obtaining the erosion operation in terms of dilation. An object image is complemented and displayed in SLM. This image is then subjected to Ex-OR operation with the inverse of structuring element. The eroded image is obtained by complementing the result of Ex-OR operation.

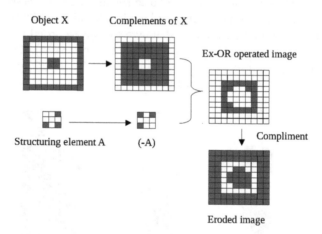

FIGURE 7.7: Schematic of morphological erosion in shadow-casted technique

7.4 Photonic techniques of transformation of images into graphics

The creation of 3D graphical models of complex objects or environments is becoming an increasingly important topic in the field of photonic information processing. The rapid development of imaging devices, processors, and light sources (mainly low cost lasers) fuelled many applications in computer vision systems. The first step in such applications is to convert the range and the profile information of the image into graphics for various types of processing. The application of image to graphic transformation has other utility in the accurate measurement of surface profile and, traditionally named surface or 3D profilometry [14], [17], [18]. In most of the cases, a structured laser light pattern is projected onto an object and then the image is captured and reconstructed for obtaining the three-dimensional range from the distortions in the reflected and captured image [19], [20]. The accurate registration of a set of such images is of major concern in the design and implementation of a 3D graphical model, where the model representation of the surface of an object is affected by the acquisition error and the registration error.

7.4.1 Range image acquisition

A range image offers a direct way to produce shape description of an object surface [21], [22]. Based on the measurement principle, range image acquisition techniques can be put into two broad categories: passive methods and active methods. Passive methods do not interact with the object, whereas active methods make contact with the object. A well-known passive approach is stereo-vision. Stereo-vision effects can be achieved by moving an optical sensor to a known relative position in the scene or by using two or more optical sensors which are previously fixed in known positions. In order to obtain 3D coordinates of a given point from a given projection point, correspondence geometry is studied. This point-correspondence problem imposes severe limitations on stereo-vision in practical applications. While active methods require more complicated structure designs, they offer dense and accurate range data (range images). Active 3D imaging methods using photonics technology have the advantages of non-contact operation mode and damage-free data acquisition.

Photonic approaches can be further divided into scanning techniques and full-field imaging techniques. Scanning techniques are represented by laser triangulation and laser radars. Laser triangulation uses either a laser point or a light-strip to illuminate the surface and obtain the 3D contour of the object based on the triangulation principle. On the other hand the laser radar technique is based on the measurement of the flight time (time required for travel) or phase of a pulsed or modulated laser. The scanning techniques required either one-dimensional or two-dimensional scanning to cover the entire surface of the object. This makes the systems more complicated and measurement more time-consuming.

7.4.2 Image registration and integration

The goal of image registration is to find the relative position and orientation of each view with respect to other views. Existing registration algorithms can mainly be categorized as either surface matching or feature matching. Surface matching algorithms start with an approximate registration, and then iteratively refine that registration by gradually reducing the error between overlapping areas in the range images. Among existing algorithms, the iterative closest point (ICP) algorithm has been proved to be one of the appropriate methods for accurately registering a set of range images. Advances in registration methods include the registration and transformation of parameters among multiple-range views by making use of the principal axes formed by three eigenvectors of the weighted covariance matrix of 3D coordinates of data points. Feature matching algorithms have also been developed, to map correspondences between a number of features in the range images by establishing a transformation function. Such algorithms, however, rely on the existence of predefined features.

Based on the accurate registration of range images, different structured or unstructured techniques have been developed for integrating or merging range images. Unstructured integration presumes that one has a procedure that creates a surface from an arbitrary collection of points in 3D space. These algorithms are well behaved in smooth regions of surfaces, but they are not always robust in regions of high curvature or in the presence of systematic range distortions. Structured integration methods make use of information about how each point in a range image is obtained. Thus, it is possible to gradually improve the surface description in regions of high noise. Several algorithms have been proposed for integrating structured data to generate implicit functions. Correct integration of overlapping surface measurements in the presence of noise is achieved using geometric constraints based on measurement uncertainty. Fusion of overlapping measurements is performed using operations in 3D space only, thus avoiding the local 2D projection.

7.4.2.1 Laser-based range acquisition by time-of-flight

There are two types of direct measurement strategies: time-of-flight methods and triangulation methods. The laser pulse time-of-flight (TOF) distance measuring technique refers to the measurement of time taken for a laser pulse to travel from its transmitter to an object and then back to the receiver. A TOF system therefore measures the round trip time between a light pulse emission and the return of the pulse echo resulting from its reflectance off an object. The range r is determined by

$$r = \frac{c\Delta t}{2} \tag{7.12}$$

where c is the velocity of light and Δt is the total time of flight. This method is particularly appropriate in applications involving distances longer than 1 m, and averaging enables millimetre or even sub-millimetre precision.

Light emission can be continuous or pulsed. Continuous laser range-finders operate on the principle of emitting a continuous beam and inferring range as a function of the phase shift between the outgoing and incoming beams [23]. A subset of continuous-wave laser ranging system is amplitude-modulated (AM) systems. These systems modulate the amplitude of a laser beam at a frequency $f_{AM} = \dfrac{c}{\lambda_{AM}}$, where λ_{AM} is the wavelength of laser. The phase difference $\Delta\phi$ between the incoming and outgoing signal is given by

$$\Delta\phi = 2\pi f_{AM}\Delta t = 4\pi f_{AM}\frac{r}{c} \tag{7.13}$$

The range is computed as

$$r = \frac{\lambda_{AM}}{4\pi}\Delta\phi \tag{7.14}$$

A problem involving AM laser range finders is evident when the range values exceed the ambiguity interval of $\lambda_{AM}/2$. This causes equivocal range measurements because it is impossible to determine the period in which the phase shift occurs.

Another type of continuous-wave laser range finder is frequency-modulated (FM). In an FM system, the optical frequency of a laser diode can be tuned thermally by modulating the laser diode drive current. If the transmitted optical frequency is repetitively swept linearly between $(v - \Delta v/2)$ and $(v + \Delta v/2)$, creating a total frequency deviation of Δv during the period $1/f_m$, the reflected return signal can be mixed coherently with a reference signal at the detector. The mixing of the return signal and the reference signal produces a beat signal. This coherent mixing is often a multiplication of the reference and return signals. The beat signal is demodulated to produce a beat frequency, f_b, which is a periodic waveform. The beat frequency is proportional to the number of zero crossings in one ramp period [24].

One of the major sources of noise in FM laser range-finders is speckle. Speckle effects originate from surface roughness at the microscopic level and are therefore related to properties of the material being scanned. When illuminated by a coherent wavefront of light, reflections tend to be out of phase, giving rise to random constructive and destructive interference, which results in a specular-looking reflection [25].

A pulsed TOF distance measuring device is also evolved which consists of a laser transmitter emitting pulses with a duration of 5 to 50 ns, a receiver channel including a p-i-n or an avalanche photodiode, amplifiers, an automatic gain control (AGC), and timing discriminators. The emitted light pulse initiates a start pulse which triggers the time interval measurement unit, and the reflected light pulse initiates a stop pulse to terminate the measurement. The detection of the stop pulse is used as a trigger for successive pulses which generates a pulse repetition rate at frequency f_p. The range is given by

$$r_m = \frac{c}{2f_p} \tag{7.15}$$

As in the case of FM system, pulsed laser range finders also suffer from the effects of speckle [26]. The same type of random constructive and destructive interference appears in the returned signals.

The selection of laser type depends on the intended measurement range and the required speed. For long distances (up to several kilometres), a Nd:YAG laser can be used, giving peak powers extending to the megawatt level. Low cost pulsed laser diodes, i.e., single heterostructure or double heterostructure (DH) type, capable of producing peak powers of tens of watts, enable measuring distances up to a few hundreds of metres or even longer using coherent summing. The repetition frequency of Nd: YAG lasers is low; whereas laser diodes can be used at rates of tens of kilohertz. The DH-type laser may even reach the megahertz level. The maximum range achievable with a laser range finder depends strongly on the visibility. Range performance is generally specified for clear air (20km visibility), while at lower visibility, the maximum range is reduced due to atmospheric attenuation. Under conditions of poor visibility, partial reflections may be received from a number of false targets before the true target range is reached [27].

7.4.3 Photonic profilometric techniques

Full-field optical or photonic methods do not require scanning for single view acquisition and use multiple stripes or patterns rather than a single laser spot or stripe [18]. One approach uses the sequential projection of a binary coded pattern [19]. Multiple frames of the coded pattern image are captured to encode a pixel on the CCD with its corresponding range. Fringe projection is another popular full-field imaging technique where a camera captures an image of a fringe pattern projected onto a target surface. If the surface is flat, the fringe lines will appear straight. If the surface is not flat, however, any surface features will alter the appearance of the fringe pattern and the deformation can be related to the object contour. Knowing the exact position, orientation, and angle of projection of the original fringe, each line on the surface is uniquely identifiable. The well-known triangulation method is then applied to the fringe pattern to calculate the profile of the surface [28].

Classical 3D acquisition devices use a single laser stripe scanned progressively over the surface of the target object, while the object is expected to be static so that all the stripe images are captured in one frame. For reducing the technological burdens of scanning and processing, multi-stripe and sinusoidal fringe patterns are used to illuminate the entire target surface at the same time [29]. But these multi-stripe patterns introduce ambiguities in the surface reconstruction around surface discontinuities, and can be sensitive to surface reflectance variations, such as albedo. The solution to the ambiguity and the albedo problem is to encode the surface repeatedly with multiple light-striped patterns with variable spatial frequencies. The individual patterns are spatially modulated along the orthogonal dimension, perpendicular to the phase dimension.

7.4.3.1 Laser triangulation technique

For determining range via triangulation, the baseline distance between source and sensor as well as sensor and source angles are used. Figure 7.8(a) shows the configuration for triangulation ranging. P_1 and P_2 represent two reference points (e.g., camera and laser source), while P_3 is a target point. The range B can be determined from the knowledge of the baseline separation A and the angles θ and ϕ. Using the law of reflections given by

$$B = A\frac{\sin\theta}{\sin(\theta + \phi)} \tag{7.16}$$

In practice, this simple equation for the range B is difficult to achieve because the baseline separation and angles are difficult to measure accurately. However, it is possible to obtain range information via laser triangulation without knowing A, θ, and ϕ. The schematic setup is shown in Figure 7.8(b). The camera is represented by the image plane, focal point, and optical axis. The laser is directly above the camera, represented by line CE, and is positioned in such a way that the path of the laser and the optical axis form a vertical plane.

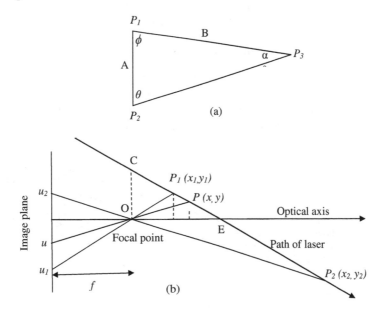

FIGURE 7.8: Schematic diagram of laser triangulation technique

Point P is the target of interest positioned at (x, y). The point x is the projection of point P on the optical axis, and u is the vertical point on the image plane. P_1 and P_2 are two points used in the calibration of the system. E is the point where the path of the laser intersects the optical axis. There is no need to know the baseline distance between camera and laser, nor the focal length f of the camera, however, x_1, x_2, u_1, and u_2 are known. From the geometry of similar triangles, the slope m of the laser path and the y-intercept (c, the height of point C) is given by

$$m = \frac{u_2 x_2 - u_1 x_1}{f(x_2 - x_1)} \tag{7.17}$$

and

$$c = \frac{u_2 x_2}{f} - m x_2 \tag{7.18}$$

Moreover, the line uP passing through O is represented by, $y = x\dfrac{u}{f}$, the laser path follows the equation of a straight line, and hence,

$$x = \frac{N}{(ud - k)} \tag{7.19}$$

where N, d, and k are obtained through calibration process. Further $d = (x_2 - x_1)$, $k = (u_2 x_2 - u_1 x_1)$ and $N = (u_1 - u_2)x_1 x_2$. During calibration, the target is placed at distances x_1 and x_2 from the camera, and the height u_1 and u_2 at which the laser spot strikes the target and appears on the image are recorded. Then d, k, and N are computed. The system is insensitive to the errors, if any, in determining the optical axis.

7.4.3.2 Fringe projection techniques

One of the features of the fringe projection techniques for non-contact type profilometric applications is their ability to provide high-resolution, whole-field 3D reconstruction of objects at video frame rates. A typical single line projection profilometric system is shown in Figure 7.9.

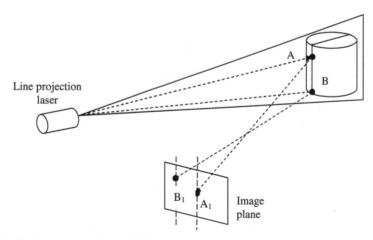

FIGURE 7.9: Schematic diagram of a fringe projection technique

The systems consist of a projection unit generating single strip or multiple strips of light, as the case may be, along with an image acquisition unit and a processing/analysis unit. The projection unit generates structured pattern (in this case a line AB), onto the object surface. In the image plane or camera plane the line AB is reflected as points A_1 and B_1. Recording the image of the light pattern is modulated by the object height distribution and is usually reflected in the phase of the acquired image. Calculation of the phase modulation is carried out by analysing the image with any of the existing fringe analysis techniques. Most of the methods generate wrapped phase distribution, and using a suitable phase unwrapping algorithm gives continuous phase distribution which is proportional to the variations in object height [30].

A fringe pattern can be generated by projecting a grating or laser interference or structured laser light. Some representative fringes used in the fringe projection technique are shown in Figure 7.10 where (a) and (b) are a sinusoidal fringe pattern and (c) is structured laser light. In order to improve

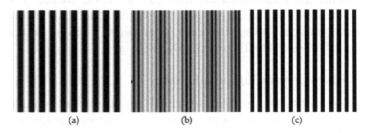

FIGURE 7.10: Some of the fringes used in profilometry

the measurement accuracy and to calibrate the system effectively, software-produced fringe patterns can be projected using spatial light modulators (SLMs) [31]. The other method of generating patterns is to use a digital micromirror device (DMD), which gives high fringe brightness and contrast, high image quality, and spatial repeatability. To acquire the coloured texture information corresponding to the range image, several approaches, such as a laser grating projection (LGP) system have also been investigated. The measurement of a 3D object can be obtained by the system, where a laser diode (LD) and a revolving polygon mirror (PM) are used to generate the grating patterns [32].

The processing in the fringe projection techniques has several steps, and in each step several variants are available. These techniques are classified mainly depending on the type of fringe analysis method that is employed. The often-used techniques are phase measuring profilometry (PMP) and Fourier transform profilometry (FTP). In FTP the phase recovered or estimated from the deformed fringe pattern by using most of the fringe analysis methods is limited to the interval $(+\pi, -\pi)$ corresponding to the principal value of arctan function. However, the true phase may range over an interval greater than 2π in which case the recovered phase contains artificial discontinuities which need to be removed by phase unwrapping [33]. Normal phase unwrapping is carried out by comparing the phase at neighbouring pixels and adding or subtracting 2π to bring the relative phase between the two pixels into the range of $\pm\pi$. In real measurements, the presence of shadows, low fringe modulations, non-uniform reflectivities of the object surface, fringe discontinuities, and noise makes the phase unwrapping process difficult and path dependent. Several advanced unwrapping algorithms have been developed to overcome most of the difficulties [34], [35].

An important step in the process of measuring the 3D height distribution using the fringe projection technique is system calibration [36], which facilitates the conversion of image coordinates (pixels) to the real-world coordinates and the mapping of unwrapped phase distribution to the absolute height distribution. The former task is often accomplished by adopting standard camera calibration techniques which normally involves the determination of the intrinsic parameters of the camera. The extrinsic parameters describe the transformation relationship between the 3D world coordinates and 2D image coordinates. For the latter task, the simple triangulation principle establishes the relation between the unwrapped phase distribution and the geometric coordinates of the 3D object [37], [38].

7.4.3.3 Phase-measuring profilometry

The phase-measuring profilometry (PMP)has several advantages, including its ease in pixel-wise calculation, resistance to ambient light, and reflection variation. Also, it can have as few as only three frames for whole-field depth reconstruction [17]. The system generally works with sinusoidal projection patterns which are projected and shifted by a factor of $\frac{2\pi}{N}$ for N times, as

$$I_{pn}(x_p, y_p) = A_p + B_p \cos\left(2\pi f_\phi y_p - \frac{2\pi n}{N}\right) \qquad (7.20)$$

where I_{pn} is the reflected intensity at projection coordinates, A_p and B_p are the projection constants, (x_p, y_p) are the projector coordinates, f_ϕ is the frequency of the sinusoidal wave, and the subscript $n = 1, 2, ...N$ represents the phase shift index where N is the total number of phase shifts.

The y_p dimension is in the direction of the depth distortion and is called the phase dimension. On the other hand, x_p dimension is perpendicular to the phase dimension, and is called the orthogonal dimension. The frequency f_ϕ of the sinusoidal wave is in the phase direction. The reflected intensity images from the object surface after successive projections are

$$I_n(x, y) = \alpha(x, y) \cdot \left[A + B \cos\left(2\pi f_\phi y_p + \phi(x, y) - \frac{2\pi n}{N}\right)\right] \qquad (7.21)$$

where (x, y) is the image coordinate and $\alpha(x, y)$ is the reflectance variance. The pixel-wise phase distortion $\phi(x, y)$ of the sinusoidal wave corresponds to the object surface depth. The value of $\phi(x, y)$ is determined from the captured patterns by

$$\phi(x,y) = \arctan\left[\frac{\sum_1^N I_n(x,y)\sin(2\pi n/N)}{\sum_1^N I_n(x,y)\cos(2\pi n/N)}\right] \qquad (7.22)$$

When calibrating the system, the phase map of the reference plane $\phi_r(x,y)$ is pre-calculated from the projections on the reference plane. The depth of the object surface with respect to the reference plane is easily obtained through simple geometric algorithms.

Figure 7.11 shows a schematic diagram of phase measuring profilometric system where the distance between the projector lens centre, O_p, to the camera lens centre, O_c, is d. Both the projector and the projector camera plane are at a distance l from the reference plane. The height h of the object at point A is calculated by

$$h = \frac{BC \cdot (\frac{l}{d})}{1 + \frac{BC}{d}} \qquad (7.23)$$

where BC is proportional to the difference between the phase at point B, ϕ_B, and the phase at point C, ϕ_c. Since $BC = \beta(\phi_C - \phi_B)$, where the constant β, as well as other geometric parameters, l and d are determined during the calibration procedure.

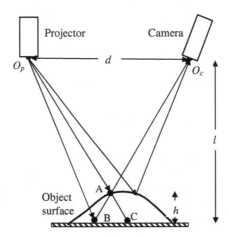

FIGURE 7.11: Geometry of the phase-measuring system

The phase value $\phi(x,y)$ is wrapped in the range value of $(-\pi, +\pi)$ independent of the frequencies in phase direction. The phase unwrapping procedure retrieves the non-ambiguous phase value out of the wrapped phase. With relatively higher frequencies in phase direction, the range data have higher signal-to-noise-ratio (SNR) after non-ambiguous phase unwrapping.

In photonic triangulation systems, the accuracy of the range data depend on proper interpretation of light reflections. Perturbations of the shape of the image illuminant occur whenever (a) the surface reflectance varies, (b) the surface geometry deviates from planar geometry, (c) the light paths to the sensor are partially occluded, or (d) the surface is sufficiently rough to cause laser speckle. One possible strategy for reducing these errors is to decrease the width of the beam and increase the resolution of the sensor [14].

7.4.3.4 Fourier transform profilometry

The Fourier transform profilometric technique has the merit of measuring rapidly changing scenes because only a single fringe image is required to recover a 3D shape [39], [40], [41]. In the FTP method, the phase is obtained by applying the Fourier transform to the fringe image, followed by applying a band-pass filter to preserve the carrier frequency component for phase calculation. Mathematically, a typical fringe pattern can be described as

$$
\begin{aligned}
I &= a(x,y) + b(x,y)\cos[\phi(x,y)] \\
&= a(x,y) + \frac{b(x,y)}{2}[e^{j\phi(x,y)} + e^{-j\phi(x,y)}]
\end{aligned}
\tag{7.24}
$$

where $a(x,y)$ is the average intensity, $b(x,y)$ is the intensity modulation or the amplitude of the carrier fringes, and $\phi(x,y)$ is the phase to solve for.

When a band-pass filter is applied in the Fourier domain only one of the complex frequency components is preserved, then

$$
I_f(x,y) = \frac{b(x,y)}{2}[e^{j\phi(x,y)}]
\tag{7.25}
$$

From this equation, the phase can be calculated as

$$
\phi(x,y) = \arctan\frac{Im[I_f(x,y)]}{Re[I_f(x,y)]}
\tag{7.26}
$$

where $Im[.]$ and $Re[.]$ represent the imaginary part and real part of the complex number. The phase obtained from the above equation ranges from $-\pi$ to $+\pi$. The continuous phase map can be obtained by applying any spatial phase unwrapping algorithm. Finally, 3D coordinates is obtained from the phase, if the system is properly calibrated.

7.4.4 Flow measurement by photonic technique

7.4.4.1 Laser Doppler anemometry

Laser Doppler anemometry (LDA) is a photonic technique used for measuring the velocity and turbulence in gas or liquid flow. The basic idea underlying the system is to measure the velocity of tiny particles transported by the flow. If these particles are small enough, their velocity is assumed to be that of the stream then LDA provides a measure of the local instantaneous velocity, the mean velocity as well as the turbulent quantities. Laser anemometry has the advantages of (a) non-contact measurement, (b) no need of calibration, (c) well-defined directional response, and (d) high spatial and directional response.

The Doppler shift is observed when light is reflected from a moving object, the frequency of the scattered light is shifted to higher frequency by an amount proportional to the speed of the object. So, one could estimate the speed by observing the frequency shift. The flow is seeded with small, neutrally buoyant particles that scatter light. The particles are illuminated by a known wavelength of laser light and the scattered light is detected by a sensitive photodetector. The photodetector generates a current in proportion to the absorbed photon energy. The difference between the incident and scattered light frequencies is called the Doppler shift. The fractional shift in frequency of scattered light is approximately V/c, where V is the group velocity of the particles and thus the flow, and c is the velocity of light. This shift can reach the amount of some hundred MHz, if the velocity is very high. The frequency shifted light is analysed directly by spectrometric methods, or by one of the LDA techniques known as differential method, reference beam method, or interference fringes method.

If the vector $\mathbf{V}$ represents the particle velocity, the unit vectors $\mathbf{e}_i$ and $\mathbf{e}_s$ describe the direction of incoming and scattered light, respectively. According to the Lorenz-Mie theory, the light is scattered in all directions at once, but it considers only the light reflected in the direction of the detector. The seeding particles act as moving transmitters, and the movement introduces additional Doppler-shift in the frequency of the light reaching the detector. Using the Doppler theory, the frequency of the light reaching the receiver is calculated as

$$f_s = f_i \frac{1 - e_i \dfrac{V}{c}}{1 - e_s \dfrac{V}{c}} \qquad (7.27)$$

where f_i is the frequency of incoming light.

Even for supersonic flows, the particle velocity V is much lower than the speed of light, meaning that $V/c << 1$. The above expression can be approximated to

$$f_s = f_i \left[1 + \frac{V}{c}(e_s - e_i) \right] = f_i + \Delta f \qquad (7.28)$$

The particle velocity V is then determined from the measurements of the Doppler shift Δf. In practice this frequency change can only be measured directly for very high particle velocities.

Since the Doppler shift is a very small fraction of the incident frequency, estimating a small value from the difference of two large values resulted in a high degree of uncertainty. To improve the estimate of Doppler shift, a method using two incident beams has been developed. In this configuration the incident beam is split into two beams of equal intensity. The beams are directed to intersect, and the point of intersection is the measurement volume. Particles that pass through the measurement volume scatter light from both beams.

The frequency shift of the light scattered from each beam will be different, because the orientation of the two beams relative to the photodetector and relative to the particle's velocity vector are different. The scattered beams are

$$f_{s1} = f_1 \left[1 + \frac{V}{c}(e_s - e_i) \right] \tag{7.29}$$

$$f_{s2} = f_2 \left[1 + \frac{V}{c}(e_s - e_i) \right] \tag{7.30}$$

When two wave trains of slightly different frequency are superimposed, it results in the phenomenon of a beat frequency which corresponds to the difference between the two-wave frequencies. Since the two incoming waves originate from the same laser, they also have the same frequency, $f_1 = f_2 = f_i$. This gives the general equation expressing the Doppler shift f_d in the frequency of the scattered light as a function of particle velocity $\mathbf{V}(v_x, v_y, v_z)$, as

$$f_d = f_{s1} - f_{s2} = \frac{f_i}{c}[|e_1 - e_2|.|V|\cos\phi] = \frac{2\sin(\theta/2)}{\lambda}v_x \tag{7.31}$$

where λ is the wavelength of laser light, θ is the angle between the incoming laser beams, and ϕ is the angle between the velocity vector $\mathbf{V}$ and the direction of measurement. The unit vector e_s is dropped out of the calculation, meaning that the position of the receiver has no direct influence on the frequency measured. However, according to the Lorenz-Mie theory, the position of the receiver has considerable influence on the signal strength.

The Doppler-frequency f_d is much lower than the frequency of the light itself, and it can be measured as fluctuations in the intensity of the light reflected from the seeding particle. Thus the particle velocity is

$$v_x = \frac{\lambda}{2\sin(\theta/2)}f_d \tag{7.32}$$

Many variations of the LDA setups are tried for different applications. In the reference beam mode, the light scattered from one laser beam is mixed (heterodynes) in the photodetector with the light from a reference beam (with original laser light). This is done by a coaxial superposition of the scattered beam and the reference beam at the photodetector. In the differential Doppler mode, two laser beams of equal intensity intersect in the measurement point. The light scattered from any direction is picked up by the photodetector.

The most used method is a fringe model laser Doppler anemometer. A laser beam is divided into two parallel beams of equal intensity by a beam splitter. By means of transmitting optics with the focal length f, the two beams intersect each other in the focal point of the optics forming the angle $\frac{\theta}{2}$. Two laser beams are coherent and monochromatic and the interference fringes are formed at the beam crossing. The distance between fringes d_f is given by

$$d_f = \frac{\lambda}{2\sin(\frac{\theta}{2}} \tag{7.33}$$

As a particle crosses the fringe pattern, the intensity of the scattered light varies with the intensity of the fringes. Thus, the amplitude of the signal burst varies with time scale $\frac{d_f}{V}$, where V is the velocity component perpendicular to the fringe pattern. The frequency of the amplitude modulation (which is nothing but the Doppler shift, is thus

$$\frac{V}{d_f} = \frac{2V}{\lambda} \sin\left(\frac{\theta}{2}\right) \qquad (7.34)$$

The Doppler frequency is not dependent on the position of the photodetector and depends only on the magnitude of V and not the direction. To remove this directional ambiguity, the frequency of one of the incoming beams is shifted by a known value. This causes the fringe pattern to move toward the incoming unshifted beam. The frequency recorded by the photodector is then dependent on the sign of V, that is up for positive V and down for negative V. An acoustic-optic modulator (Bragg cell) generates the required frequency shift.

Gas lasers operating in the fundamental optical mode are ideal sources for Doppler shift measurements. Laser light can be focused to the small spot and all laser energy can be concentrated in it. When particles go through the beam crossing, they scatter light in all directions. This scattered light can be collected from any direction by a detector.

The most important parameters of LDA systems relate to the calibration constant, the dimensions of the measuring system and the parameters such as laser wavelength, beam waist, LDA beam separation, beam expansion, and measuring distance. The primary result of a laser anemometer measurement is a current pulse from the photo detector. This current contains the frequency information relating to the velocity scattering centres. The photocurrent also contains different noises (shot, secondary electron noise, and pre-amplifier thermal noise).

Particles in the flow are the scattering centres for LDA, and they have an important role in measurements. The ordinary particles may not be sufficient to satisfy the requirements of LDA measurements. The natural concentration of particles is dominant factor and most often causes an undesirable shot noise. It can be avoided, or reduced by controlling the sizes and concentration of the seeding particles through filtering.

Books for further reading

1. *Handbook of Machine Vision*: S. Zhang (ed.) CRC Press, NY, 2013.

2. *LDA Application Methods*: Z. Zhang (ed.), Springer-Verlag, Berlin, 2010.

3. *Laser Doppler and Phase Doppler Measurement Technique*: H. E. Albrecht, M. Borys, N. Damaschke and C. Tropea, Springer-Verlag, Berlin, 2003.

4. *Principles and Practices of Laser-Doppler Anemometry*: F. Durst, F. Melling and J. Whitelaw, Academic Press, New York, 1981.

5. *Computer Vision-Algorithms and Applications*: R. Szeliski, Springer-Verlag, London, 2011.

6. *Concise Computer Vision*: R. Klette, Springer-Verlag, London, 2011.

7. *Emerging Topics in Computer Vision*: C. H. Chen (ed.), World Scientific Publishing, Singapore, 2012.

8. *Image Analysis and Mathematical Morphology*: J. Serra (ed.), Academic Press, London, 1988.

9. *Mathematical Morphology in Image Processing*: E. D. Dougherty (ed.), Marcel Dekker Inc., NY, 1992.

10. *Image Processing and Mathematical Morphology*: F. Y. Shih; CRC Press, NY, 2009.

11. *Digital Image Processing*: R. C. Gonzalez and R. E. Woods; Prentice-Hall, 2008.

12. *Digital Image Processing and Analysis*: S. E. Umbaugh, CRC Press, New York, 2011.

Bibliography

[1] B. Draper, A. Hanson, and E. Riseman. Knowledge-directed vision: control, learning, and integration. *Proc. IEEE*, 84(11):1625–1637, 1996.

[2] G. Papadopoulos, V. Mezaris, I. Kompatsiaris, and M. Strintzis. Combining global and local information for knowledge-assisted image analysis and classification. *Eurasip J. Adv. in Signal Process.*, 45842, 2007.

[3] A. Chella, M. Frixione, and S Gaglio. Conceptual spaces for computer vision representation. *Artificial Intelli. Rev.*, 16(2):137–152, 2001.

[4] J. Gonzalez, D. Rowe, J. Varona, and F. Xavier Roca. Understanding dynamic scenes based on human sequence evaluation. *Image and Vis. Comput.*, 27(10):1433–1444, 2009.

[5] J. Sanchez and M. P. Caton. *Space image processing*. CRC Press, 1999.

[6] P. Maragos. Tutorial on advances in morphological image processing. *Opt. Eng.*, 26(3):623–630, 1987.

[7] R. Haralick, S. Sternberg, and X. Zhuang. Image analysis using mathematical morphology. *IEEE Trans. PAMI*, 9(4):532–538, 1987.

[8] D. Casasent and E. Botha. Optical symbolic substitutions for morphological transformations. *App. Opt.*, 27(9):3806–3812, 1988.

[9] E. L. O'Neill. *Introduction to Statistical Optics*. Addison-Wesley, Reading, MA, 1963.

[10] J. F. Liu, X. Sun, R. Camacho-Aguilera, L. C. Kimerling, and J. Michel. Ge-on-Si laser operating at room temperature. *Opt. Let.*, 35:679–681, 2010.

[11] Y. Li, A. Kostrzewski, D. H. Kim, and G. Eichman. Compact parallel real-time programmable morphological image processor. *Opt. Let.*, 14(10):981–985, 1989.

[12] D. Casasent. Optical morphological processors. *Proc. SPIE*, 1350:380–384, 1990.

[13] R. Schaefer and D. Casasent. Optical implementation of gray scale morphology. *Proc. SPIE*, 1658:287–292, 1992.

Bibliography

[14] P. S. Huang, Q. Hu, and F. Chiang. Error compensation for a three dimensional shape measurement system. *Opt. Exp.*, 42 (2):341–353, 2003.

[15] J. Tanida and Y. Ichioka. Optical logic array processor using shadowgram: optical parallel digital image processing. *J. Opt. Soc. Am.*, 82:1275–1285, 1985.

[16] A. K. Datta and M. Seth. Multi-input optical parallel logic processing using shadow-casting technique. *App. Opt.*, 34(35):8164–8170, 1994.

[17] C. J. Chang-Hasnain. Tunable VCSEL. *IEEE J. Sel. Top. Quant. Elect.*, 6(6):978–987, 2000.

[18] Z. Zhang and S. Satpathy. Electromagnetic wave propagation in periodic structures: Bloch wave solution of Maxwell's equation. *Phy. Rev. Let.*, 60:2650–2653, 1990.

[19] J. Salvi, J. Pages, and J. Batlle. Pattern codification strategies in structured light systems. *Pattern Recognition*, 37 (4):827–849, 2004.

[20] Q. Fang and S. Zheng. Linearly coded profilometry. *Applied Optics*, 36 (11):2401–2407, 1997.

[21] M. C. Amann, T. Bosch, M. Lescure, R. Myllyla, and M. Rioux. Laser ranging: a critical review of techniques for distance measurement. *Opt. Eng.*, 40(1):10–19, 2001.

[22] T. Bosch and M. Lescure. Selected papers on laser distance measurement. *SPIE Milestone series*, MS115, 1995.

[23] W. L. Green B. L. Chase and M. Abidi. Range image acquisition, segmentation, and interpretation: A survey. *Techical Report, Department of Electrical Engineering, University of Tennessee at Knoxville*, 1995.

[24] K. Nakamura, T. Hara, M. Yoshida, T. Miyahara, and H. Ito. Optical frequency domain ranging by a frequency-shifted feedback laser. *IEEE J. Quant. Elect.*, 36:305–316, 2000.

[25] M. Koskinen, J. Kostamovaara, and R. Myllyla. Comparison of the continuous wave and pulsed time-of-fight laser range finding techniques. *Proc. SPIE: Optics, Illumination, and Image Sensing for Machine Vision VI*, 1614:296–305, 1992.

[26] R. Baribeau and M. Rioux. Influence of speckle on laser range finders. *App. Opt.*, 30:2873–2878, 1991.

[27] R. R. Clark. A laser distance measurement sensor for industry and robotics. *Sensors*, 11:43–50, 1994.

[28] F. Chen, G. M. Brown, and M. Song. Overview of three-dimensional shape measurement using optical methods. *Opt. Eng.*, 39(1):10–22, 2000.

[29] J. Harizanova and V. Sainov. Three-dimensional profilometry by symmetrical fringes projection technique. *Opt. and Laser Eng.*, 44(12):1270–1282, 2006.

[30] H. Du and Z. Wang. Three-dimensional shape measurement with an arbitrarily arranged fringe projection profilometry system. *Opt. Let.*, 32(16):2438–2440, 2007.

[31] C. R. Coggrave and J. M. Huntley. High-speed surface profilometer based on a spatial light modulator and pipeline image processor. *Opt. Eng.*, 38(9):1573–1581, 1999.

[32] K. Iwata, F. Kusunoki, K. Moriwaki, H. Fukuda, and T. Tomii. Three dimensional profiling using the Fourier transform method with a hexagonal grating projection. *App. Opt.*, 47(12):2103–2108, 2008.

[33] T. R. Judge and P. J. Bryanston-Cross. A review of phase unwrapping techniques in fringe analysis. *Opt. and Laser Eng.*, 21(4):199–239, 1994.

[34] H. O. Saldner and J. M. Huntley. Temporal phase unwrapping: Application to surface profiling of discontinuous objects. *App. Opt.*, 36(13):2770–2775, 1997.

[35] S. Su and X. Lian. Phase unwrapping algorithm based on fringe frequency analysis in Fourier-transform profilometry. *Opt. Eng.*, 40 (4):637–643, 2001.

[36] J. F. Liu, R. Camacho-Aguilera, and X. Sun. Ge-on-Si optoelectronics. *Thin Solid Films*, 520:3354–3360, 2011.

[37] X. Zhang, Y. Lin, M. Zhao, X. Niu, and Y. Huang. Calibration of a fringe projection profilometry system using virtual phase calibrating model planes. *J. of Opt. (A): Pure and App. Opt.*, 7 (4):192–197, 2005.

[38] A. Baldi. Phase unwrapping by region growing. *App. Opt.*, 42 (14):2498–2505, 2003.

[39] Q. Kemao. Windowed fourier transform for fringe pattern analysis. *App. Opt.*, 43 (13):2004, 2695-2702.

[40] J. F. Lin and X. Y. Su. Two-dimensional Fourier transform profilometry for the automatic measurement of three-dimensional object shapes. *Opt. Eng.*, 34 (11):3297–3302, 1995.

[41] X. Su and W. Chen. Fourier transform profilometry: A review. *Opt. Laser Eng.*, 35 (5):263–284, 2001.

Chapter 8

Photonics in Networking and Communication

8.1 Light propagation in optical fibre

The modern impetus in information communication with carrier waves at optical frequencies owes its origin to the invention and practical realisation of the lasers and optical fibres. As a result, today photonic communication is the most useful and advanced mode of information communication technology, fuelled by the ever-increasing demands of higher bandwidth [1]. This increasing trend, of the needs of large bandwidths, will continue in the future. In this chapter we shall use the term *photonic communication* instead of the commonly used optical communication (see Preface).

Like any other communication system, photonic communication systems operate on the same concept as other wired communication systems, as shown in Figure 8.1.

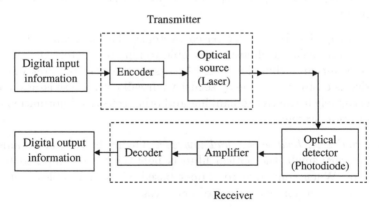

FIGURE 8.1: Fibre optic communication system

This too has the three basic modules: transmitter, receiver, and the channel or medium of communication. The input information is encoded and accordingly transmitted as a modulated optical signal through the optical fibre by an optical source at one end, and detected by a detector at the other end of the fibre. Encoding, multiplexing, and other electronic circuits necessary for converting information into the optical signal form the transmitter. The detected optical signal is amplified, and then decoded to obtain the output information. All the electronic circuits necessary to decode and demultiplex along with the detector and associated amplifier form the receiver. The optical source serves as the electrical-to-optical signal converter, which may be either a light emitting diode (LED) or preferably a semiconductor laser. The detector is a photodiode (e.g., p-i-n diode or avalanche photodiode) which converts the optical signal back to an electrical signal. Fibre optic communication systems operate in the frequency spectrum (frequencies between 10^{13} Hz and 10^{15} Hz).

The advantages of a photonic communication system are:

1. *Higher bandwidth of transmission*: Whereas the allowable bandwidth in copper cables is nearly 500 MHz and 700 MHz in radio systems, optical fibre-based communication systems provide a potential bandwidth in the order of GHz, since the optical carrier frequency ranges between 10^{13} and 10^{16} Hz. Thus, with a higher bandwidth and low-loss characteristic, fibres have a higher information-carrying capacity.

2. *No cross-talk or interference*: Since optical fibres are dielectric waveguides, they are free from electromagnetic interference (EMI) and require no EMI shielding. Hence data communication in an optical fibre is unaffected in an electrically noisy environment.

3. *Low loss*: In comparison to copper cables, fibres exhibit very low attenuation or transmission loss. Hence, fibre communication systems have wide repeater amplifier spacings, thus reducing system cost and complexity.

4. *Security*: Unlike metallic-based systems, the dielectric nature of optical fibre makes it almost impossible to remotely detect the signal being transmitted within the cable. Accessing the fibre requires intervention that is easily detectable by security surveillance. These issues make fibre extremely attractive in banking and other sensitive information communication systems.

5. *Small size, lightweight, and low cost*: Owing to the core diameter being very small, the optical fibre is lightweight as compared to copper wire. When compared to copper-based wires and cables, optical fibre and optical fibre cables are of lower cost.

6. *Ruggedness and system reliability*: The optical fibre is rugged, can withstand high temperature, and is of high tensile strength.

8.1.1 Optical fibre

Optical fibres are solid but concentric cylindrical waveguides made of an internal core and a surrounding clad layer. The core with refractive index n_1

serves as the medium for light propagation, while the cladding has a lower refractive index n_2. The basic phenomenon of guidance can be explained by ray model of optical fibre. When a light ray encounters the interface of a fibre, the ray is refracted and its direction of propagation changes according to Snell's law of refraction, such that rays are reflected back to the core. Light is guided through the total internal reflection, i.e., only rays with incident angles greater than the critical angle, at the clad-core interface, are transmitted, as shown in Figure 8.2.

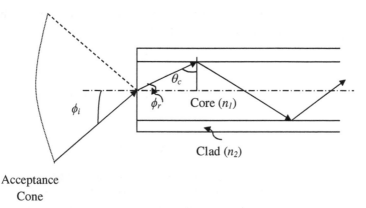

FIGURE 8.2: Launching of light and its propagation inside the fibre

The maximum angle of incidence at which the light is launched into the fibre and gets propagated inside the fibre is thus given by $\phi_i|_{max}$ and is called the acceptance angle. The cone formed by the acceptance angle is called the acceptance cone. Rays greater than the acceptance angle may enter the core but are refracted to the cladding and eventually lost by radiation [2].

Considering the refractive index of air to be unity, Snell's law for the air and core interface, as seen from Figure 8.2, is

$$\sin \phi_i = n_1 \sin \phi_r \tag{8.1}$$

where ϕ_i is the angle of incidence, ϕ_r is the angle of refraction at the air-core interface, and $\phi_r = (\pi/2) - \theta$.

Again, for angles larger than the critical angle θ at the core-clad interface, the following equation holds well.

$$n_1 \sin \theta = n_2 \tag{8.2}$$

For confining rays within the fibre core, this necessary condition has to be obeyed. Thus,

$$\sin \phi_i = n_1 \sin((\pi/2) - \theta)$$
$$= n_1 \cos \theta$$
$$= n_1 \sqrt{(1 - \sin^2 \theta)} \tag{8.3}$$
$$= n_1 \sqrt{\left(1 - \frac{n_2^2}{n_1^2}\right)}$$

where n_1 and n_2 are the refractive index of the core and clad, respectively.

The acceptance angle or the maximum angle of incidence ϕ_i is related to the refractive indices by

$$\phi_i|_{max} = \sqrt{n_1^2 - n_2^2} \tag{8.4}$$

An important fibre parameter is the numerical aperture (NA), which is simply equal to $\sin \phi_i|_{max}$. Thus

$$NA = (\phi_i)_{max} = \sqrt{n_1^2 - n_2^2} \tag{8.5}$$

NA represents the light gathering capacity of the optical fibre.

For propagation of light inside the fibre core, there are possibilities that the light may get transmitted as *meridional rays* or as *skew rays*. When a light ray is launched in a plane containing the axis of the fibre, the light ray after total internal reflection travels in the same plane in which it was launched. Thus the ray will always cross the axis of the fibre while propagating, and hence called the *meridional ray*. When the ray is not launched in a plane containing the axis of the fibre, i.e., launched at some angle not intersecting the axis of the fibre, then after total internal reflection the ray will never intersect the axis of the fibre. The ray essentially will spiral around the axis of fibre and thus called the *skew rays*.

As mentioned earlier, when the angle of incidence at the core and clad interface $\theta_{interface}$ is greater than the critical angle θ, a ray propagates through the optical fibre. So, rays having $\theta \geq \theta_{interface} \leq \pi/2$ will propagate along the fibre with some discrete velocities, known as modal velocities. The interference of the travelling waves inside the fibre gives rise to different modes within the fibre. A specific mode is obtained only when the angle between the rays and the core-clad interface has a particular value. Modes can therefore be termed the stable electric field distributions within the fibre core.

Modes in an optical fibre can be explained by the wave theory and principles of waveguide. Treating light as a transverse electromagnetic wave, during the propagation of meridional rays along the fibre, the electric and magnetic fields of all the rays superimpose to result in electric and magnetic field distribution which may be either transverse electric, TE_x, or transverse magnetic TM_x, in nature. The subscript x denotes the definite number of maxima and minima in the resultant light intensity pattern. The propagation of skew rays, on the other hand, results in a particularly special form of modes which are neither TE nor TM in nature and are called hybrid modes. Figure 8.3 shows different intensity patterns created by superposition of the wavefronts of all the light rays for transverse electric modes that propagate in an optical fibre. The fields that are shown in the cladding region are actually the evanescent fields that exist in the cladding owing to the boundary condition requirement at the core-cladding interface. For very low launching angles with respect to the axis of the fibre, the intensity pattern created is the one which is shown by TE_0 in Figure 8.3. There exists a maximum intensity region around the

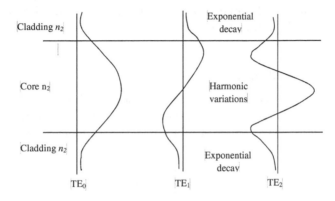

FIGURE 8.3: Different TE modes in an optical fibre

axis of the core and the fields start to decay towards the periphery of the core. These fields eventually decay down to negligibly low value in the cladding. If the launching angle is increased further, the intensity patterns are shown for TE_1 and TE_2 in Figure 8.3. The subscript of TE, in fact, indicates the number of destructive interferences in the pattern where the field intensity crosses the zero level, or in other words, creates an optically dark area. So, no dark area exists for TE_0. For TE_1 and TE_2, one and two dark areas are possible and for higher TE modes, more dark areas are possible.

The ray-model of light showed that the launching angle of the light ray must be smaller than the acceptance angle of the optical fibre core. But the consideration of the wavefronts establishes that this condition of the launching angle is not enough to ensure a successful propagation of light in an optical fibre. The launching angle must be such that the angle of refraction of the launched ray into the fibre must satisfy the phase condition given below for sustained propagation inside the optical fibre core

$$\frac{2\pi n_1 d \sin \theta}{\lambda} + \delta = \pi m \tag{8.6}$$

where n_1 is refractive index of core, λ is the wavelength of the light in the core, d is the diameter of the core, and δ is the phase change undergone in each total internal reflection of a ray and $m = 1, 2, 3....,$.

The different discrete values of the angle θ indirectly signify the different allowable launching angles of the light rays into the optical fibre and for $m = 0$, $\theta = 0$. This refers to the ray that propagates along the axis and it does not require any phase condition to be satisfied. Taking the the value of $m = 1$, the value of θ for first-order mode θ_1 and is given by

$$\theta_1 = \sin^{-1}\left[\frac{\lambda(\pi - \delta)}{2\pi dn_1}\right] \tag{8.7}$$

The value of θ_1 signifies the first annular ring of rays that propagates inside the fibre. Similarly , the other modes that propagate in the fibre by subsequent substitution of the corresponding values of m are obtained, until the condition $\theta \leq \alpha$ is reached.

8.1.1.1 Types of optical fibre

Fibres are classified as *step-index* (SI) or *graded index* (GI) fibres [3] depending on the refractive index profile of the core, as shown in Figure 8.4. Step-index fibres have a core of uniform refractive index profile and the profile is expressed as

$$n(r) = \begin{cases} n_1 & \text{if } r < a \\ n_2 & \text{if } r \geq a \end{cases} \tag{8.8}$$

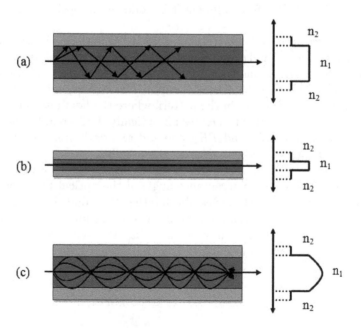

FIGURE 8.4: Types of fibre with their profiles and ray-paths within the fibre: (a) multimode step-index, (b) monomode step-index, and (c) graded index

However, depending on the modes of propagation, the step-index fibres may be monomode or multimode fibres [4]. The term multimode refers to the fact that multiple modes or light paths through the fibre are possible. The number of modes V supported in a fibre is determined by the refractive index, operating wavelength, and the diameter of the core. Monomode step-index fibre allows only one path, or mode, for light to travel within the fibre. In a multimode step-index fibre, the number of modes M_n propagating can be approximately expressed as

$$M_n = \frac{V^2}{2} \tag{8.9}$$

where V is known as the normalised frequency, or the V-number, and is given by

$$\begin{aligned} V &= 2\frac{\pi a}{\lambda}\sqrt{n_1^2 - n_2^2} \\ &= 2\frac{\pi a}{\lambda}(NA) \end{aligned} \tag{8.10}$$

where a is the core radius. From the above equation, it is seen that reducing the core-radius reduces the possible number of modes V passing through the fibre. The total number of guided modes M for multimode fibres is $M = 0.5V^2$ for multimode silicon fibre and $M \approx 0.25V^2$ for multimode germanium doped fibre. The analysis of how the V-number is derived is beyond the scope of this book, but it can be shown that by reducing the diameter of the fibre to a point at which the V-number is less than 2.405, higher-order modes can be effectively extinguished and monomode operation is possible.

The graded-index fibre or gradient-index fibre has a core with a refractive index that decreases with increasing radial distance from the optical axis of the fibre. Hence, light rays follow sinusoidal paths along the length. The refractive index profile for a graded-index fibre is nearly parabolic and this parabolic profile results in continual refocusing of the rays in the core, thus minimising modal dispersion. Graded-index fibre is characterised by its ease of use i.e., large core diameter and numerical aperture, similar to a step-index multimode fibre, and has greater information carrying capacity. Light travelling through the centre of the fibre experiences a higher index of refraction than does light travelling in higher modes. This means that even though the higher-order modes must travel farther than the lower-order modes, they travel faster, thus decreasing the amount of modal dispersion and increasing the bandwidth of the fibre.

Fibres are represented by their core-cladding diameter as 8/125 μm for monomode step-index fibres and 62.5/125 μm for multimode step-index fibres, where a and b are the radii of core and cladding, respectively. The base material for optical fibres is pure silica (SiO_2) in the form of fused quartz (amorphous). The refractive index of the base material can be modified by the addition of impurities. For lowering refractive index, B_2O_3 can be used, whereas for increasing the refractive index, P_2O_5 or GeO_2 can be used.

8.1.1.2   Signal distortion in optical fibre

When an optical signal is transmitted into an optical fibre, the signal is distorted owing to mainly two phenomena: dispersion and attenuation. The diagram in Figure 8.5 shows the decrease in input pulse height due to attenuation and the broadening of the pulse due to dispersion in the optical fibre.

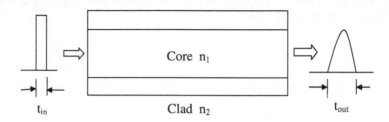

FIGURE 8.5: Pulse broadening due to dispersion in a fibre

(a) Dispersion: In digital photonic communication systems, information to be sent is first coded in the form of light pulses and then transmitted from the transmitter to the receiver, where the information is decoded. The larger the number of pulses that can be sent per unit time and still be resolvable at the receiver end, the larger will be the transmission capacity, or bandwidth of the system. A pulse of light sent into a fibre broadens in time as it propagates through the fibre. This phenomenon is known as dispersion. Dispersion, expressed in terms of the symbol Δt, is defined as pulse spreading in an optical fibre. As a pulse of light propagates through a fibre, the numerical aperture, or core diameter, or refractive index profile, or wavelength causes the input pulse to broaden, which in turn poses a limitation on the overall bandwidth of the fibre. Dispersion Δt can be written as

$$\Delta t = \sqrt{\Delta t_{out} - \Delta t_{in}} \tag{8.11}$$

where Δt_{out} and Δt_{in} are the output and input pulse widths through the fibre. The dispersion is also a function of fibre length.

Overall effect of dispersion on the performance of a fibre optic system is known as inter symbol interference, which occurs when the pulse spreading due to dispersion causes the output pulses of a system to overlap, rendering them undetectable. If an input pulse is caused to spread such that the rate of change of the input exceeds the dispersion limit of the fibre, the output data will become indiscernible. Different subcomponents responsible for dispersion are added together to form the total dispersion of an optical fibre of certain length. These are:

Intermodal dispersion: Intermodal dispersion is the pulse spreading caused by the time delay between lower-order modes (modes or rays propagating straight through the fibre close to the optical axis) and higher-order modes (modes propagating at steeper angles). Modal dispersion is the primary cause for bandwidth limitation in multimode fibre and it is not detrimental in monomode fibre, where only one mode is allowed to propagate.

Chromatic dispersion: Chromatic dispersion is pulse spreading due to the fact that different wavelengths of light propagate at slightly different speeds through the fibre. All light sources have finite line-widths and therefore, they emit more than one wavelength. Because the index of refraction of glass fibre is a function of wavelength, different wavelengths propagate at different speeds. This contributes to the effect of chromatic dispersion, which is typically expressed in units of nanoseconds or picoseconds per km. Chromatic dispersion again consists of two parts: material dispersion and waveguide dispersion. Material dispersion is due to the wavelength dependency on the refractive index of glass. Waveguide dispersion is due to the physical structure of the waveguide. Material dispersion and waveguide dispersion can have opposite signs (or slopes) depending on the transmission wavelength.

In the case of a step-index monomode fibre, material and waveguide dispersions may effectively cancel each other at 1310 nm, yielding zero-dispersion, which makes high-bandwidth communication possible at this wavelength. Because of mismatch between minimum attenuation at 1550 nm and zero dispersion at 1310 nm, special zero-dispersion-shifted fibres are evolved. Zero-dispersion-shifted fibre shifts the zero dispersion wavelength of 1310 nm to coincide with the 1550 nm transmission window of glass fibre by modifying the waveguide dispersion slope. The modification is accomplished by modifying the refractive index profile of the fibre in a way that yields a more negative waveguide-dispersion slope. When combined with a positive material dispersion slope, the point at which the sum of two slopes cancel each other can be shifted to a higher wavelength such as 1550 nm or beyond [5].

The total dispersion of an optical fibre, Δt_{tot}, can be approximated using

$$\Delta t_{tot} = [\Delta t_{modal}^2 + \Delta t_{crom}^2]^{1/2} \tag{8.12}$$

The transmission capacity of fibre is typically expressed in terms of bandwidth × distance. For example, the (bandwidth × distance) product for a typical 62.5/125 μm (core/cladding diameter) multimode fibre operating at 1310 nm might be expressed as 600 MHz·km. The approximate bandwidth BW of a fibre can be related to the total dispersion by the relationship $BW(Hz) = 0.35/\Delta t_{tot}$.

(b) Attenuation: The phenomena of attenuation in optical fibre is another phenomenon which causes signal distortion in optical fibre. Moreover, fibre attenuation dictates the selection of transmission wavelength of an optical fibre communication system. The attenuation is caused by various mechanisms such as Rayleigh scattering, absorption due to metallic impurities, OH ions in the fibre, and intrinsic absorption by the silica molecule itself. The Rayleigh scattering loss varies as λ^{-4}, i.e., longer wavelengths scatter less than shorter wavelengths.

Rayleigh scatter causes the dB loss/km to decrease gradually as the wavelength increases from 800 nm to 1550 nm. The two absorption peaks around

1240 nm and 1380 nm are primarily due to traces of OH ions and metallic ions in the fibre which may be induced during the manufacturing process. The level of purity required to achieve low-loss optical fibres is therefore very stringent. If these impurities are completely removed, the two absorption peaks may disappear. For $\lambda > 1600$ nm, the increase in the dB/km loss is due to the absorption of infrared light by silica molecules. This is an intrinsic property of silica, so no amount of purification can remove this infrared absorption tail.

Typical optical transmission wavelengths are 850 nm, 1310 nm, and 1550 nm. Figure 8.6 shows the spectral dependence of fibre attenuation (i.e., dB loss per unit length) as a function of wavelength of a typical silica optical fibre. There are two windows at which the dB/km loss attains its minimum value. The first window is around 1300 nm (with a typical loss coefficient of less than 1 dB/km) where the material dispersion is negligible. However, the loss coefficient is at its absolute minimum value of about 0.2 dB/km around 1550 nm. The latter window has become extremely important in view of the availability of erbium-doped fibre amplifiers.

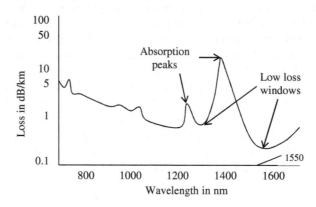

FIGURE 8.6: Attenuation in optical fibre at different wavelengths

For an optical fibre of length L (in kilometres), if P_{in} and P_{out} are the input and output optical power, respectively, then the attenuation constant α of the fibre in dB/Km is

$$\alpha = \frac{10}{L} \log_{10} \frac{P_{in}}{P_{out}} \tag{8.13}$$

8.1.2 Point-to-point fibre link

The so-called point-to-point fibre link is the basic building block of any photonic communication system. Most generic point-to-point communication links consist of two endpoints connected by a link. In a generic configuration, an endpoint system could be a computer or terminal, either in an isolated location or physically connected to a network. The term communications link refers to the hardware and software connecting the nodes of a network. The role of the link is to transport information, available in the form of a digital bit stream, very accurately from one point of the network to another point. For short-haul or short-distance communication, the link length may be less than a *km*, whereas for long-haul or long-distance links, the link length may be thousands of kms. The loss of signal becomes a matter of concern, when the link length exceeds a certain value depending on the operating wavelength. Hence it becomes necessary to compensate for the fibre losses by using signal amplifiers or repeaters/regenerators at regular intervals along the length of the links. A regenerator consists of a receiver followed by a transmitter.

The primary design criteria signify the most basic and fundamental information parameters to be made available by the user to the designer for designing a reliable fibre optic link. The first important information to be specified by the user is the desired bit rate of data transmission. However, the dispersion in the optical fibre exerts a limitation on the maximum achievable and realisable data rate of transmission. The next intricate information required for the design is the length of the optical link so as to enable the designer to ascertain the position of the optical repeaters along the link for a satisfactory data link. Along with the primary design criteria, there are some additional parameters which facilitate better design and quality analysis of the link. These factors consist of the scheme of modulation, the system fidelity, cost, commercial availability, etc.

In optical link design problems, the power of the optical source and the optical receiver are generally expressed in dBm. The dB-equivalent power of 1 mW power is taken as the reference, i.e., 0 dBm and the increasing powers are expressed in their equivalent dBm values by normalising them with respect to 1 mW. Every tenfold increase in the actual power, increases the dBm equivalent power by 10. For a typical optical transmitter with a laser source, the output power normally ranges between 3 to 5 dBm. A typical optical receiver requires about -30 to -40 dBm of detectable power for a bit error rate (BER) of 10^{-9}.

8.1.2.1   Loss budget

Having fixed the transmitter output power, P_s, and the sensitivity of the receiver, P_r, the maximum allowable loss that can occur is the difference between the transmitter and the received powers. The loss occurs in the different components connected in the system such as the connectors, splices, the optical fibre, and also losses expected in the system itself, which is known as the system margin. Generally, a system margin of about 6 dB is pre-set in practical systems. The total loss is the sum total of all the losses occurring in each of these components (calculated per unit length). Therefore, maximum allowable loss, α_{max} in dB, is $(P_s - P_r)$ and α_{max} is given by

$$\alpha_{max} = \alpha_{fibre} + \alpha_{connectors} + \alpha_{splices} + \alpha_{system} \qquad (8.14)$$

The maximum possible length L_{max} of the optical fibre that can be used in the design, without affecting the system BER, can then be determined as

$$L_{max} = \frac{\alpha_{fibre}}{loss/km} \qquad (8.15)$$

8.1.2.2 Rise-time budget

The next step in the design of an optical link is the rise-time budget calculations. Rise time of a system is the time taken by the system to attain 90% of the steady state response of the system (from the initial state) to particular input. The rise time of a system varies inversely with its bandwidth, thus rise-time analysis gives the effective bandwidth of the optical link. The system rise time t_{sys} is calculated as a root mean square value of the transmitter rise time t_{tx}, the receiver rise time t_{rx}, and the rise time associated with the optical channel dispersion $D\beta_\lambda$, where D is the dispersion parameter, β_λ is the spectral width of the optical source, and L is the length of the optical link. Therefore the rise-time budget is given by

$$t_{sys} = [t_{tx}^2 + D^2\beta_\lambda^2 + t_{rx}^2]^{1/2} \qquad (8.16)$$

For a satisfactory operation of the optical link, the system rise time should be less than or equal to 70% of the bit duration at the specified data rate.

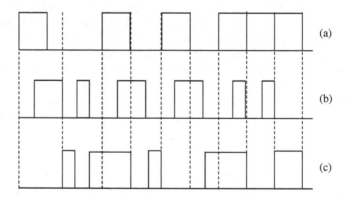

FIGURE 8.7: Some of the modulation codes used in optical communication

8.1.2.3 Modulation format

Some common code-formats are shown in Figure 8.7. The modulation format used in photonic communication is mainly known as the return-to-zero (RZ) or non-return-to-zero (NRZ) format. A simple variation of two-level block code of $nBmB$ type is 1B2B. This code is commonly known as Manchester or biphase code and is s preferred modulation format. In the RZ format, each optical pulse representing bit 1 is shorter than the bit slot, and its amplitude returns to zero before the bit duration is over. In the NRZ format, the optical pulse remains on throughout the bit slot and its amplitude does not drop to zero between two or more successive 1 bits. As a result, pulse width varies depending on the bit pattern, whereas it remains the same in the case of RZ format. An advantage of the NRZ format is that the bandwidth associated with the bit stream is smaller than that of the RZ format by about a factor of two, simply because on-off transitions occur fewer times. In 1B2B code, there are never more than two consecutive identical symbols, and those two symbols are transmitted for one information bit, giving 50% redundancy. Thus this code requires twice the bandwidth [6]. Many other improved codes are now available [7].

In order to decrease the optical spectral band occupied by a channel without decreasing the amount of information or data carried requires the use of an even more efficient modulation format such as alternate mark inversion (AMI) or duobinary (DB). In some cases the three states of the ternary scheme (i.e. binary 0, 1 and 2) may be identified by assigning three distinct electrical voltages values of 0, +V and -V.

8.1.2.4 System performance

If there is a random reception of unbiased digital data in the optical form, the quantity of interest is the average power in the optical signal and there has to be some minimum average power required in the signal for its reliable detection. Let, P_1 be the average incident power onto the photodetector when logic 1 is received, and P_0 be the average power incident onto the detector when logic 0 is received. Also, let I_1 and I_0 be the generated photocurrents corresponding to the two levels 1 and 0, respectively. Without compromising any generality, it can be assumed that $P_0 = 0$ such that $I_0 = 0$. If R is the responsivity of the photodetector then, $I_1 = \text{R}P_1$ and $I_0 = \text{R}P_0 = 0$. The average power received is given by $P_1/2$. The measure of the noise in the reception of the two logic levels 0 and 1 can be ascertained from the variances in the generated photocurrent in the two levels. For the logic 0 reception, there is no incident power and the thermal noise component dominates the noise component in the output signal of the photodetector. So, the total variance in the 0 level is $\sigma_0^2 = \sigma_T^2$, where σ_T^2 is the variance that characterises the thermal noise component in the reception. For the logic 1 level, the noise component in the output signal from the photodetector is composed of two types of noise: shot noise and thermal noise. Therefore, the total noise variance σ_1^2 is the sum of the variances of shot noise and thermal noise. If the average power of the received signal is large, however, the thermal noise becomes almost negligible in comparison to the shot noise. At low optical-power level, the average power required to achieve the desired BER is directly proportional to the square root of the bandwidth of operation. Since bandwidth is directly proportional to the data rate of transmission, as the data rate of information transmission increases, the minimum power required to achieve the desired

BER also increases. However, this increase is not rapid because the power varies as square root of the bandwidth.

8.1.3 Long-haul photonic system

The telecommunication applications can be broadly classified into two categories: long-haul and short-haul, depending on whether the optical signal is transmitted over relatively long or short distance compared with typical intercity distance. Short-haul telecommunication applications operate at low bit rates and cover intracity distances of less than 10 km, and long-haul telecommunication systems cover several thousand kilometre of distances as in underwater communication. Long-haul systems require high-capacity trunk lines and the optical signal is regenerated electronically using repeaters at regular intervals of 100-200 km, thus enabling the transmission at higher bit rates. Moreover, transmission distances of thousands of km can be enhanced by using optical amplifiers in fibre-optic communication systems. Long-haul communication systems are characterised by longer trunk distances between users (includes interstate, nationwide, or worldwide distances), higher traffic volumes and densities, large-size switching centres and central offices, as well as automatic switching for handling calls and messages without operator assistance [8]. Optical fibre-based communication technology has emerged as the preferred technological choice in terms of reliability and cost effectiveness for long-haul high-capacity networks. Due to attenuation and dispersion, the transmitted signal amplitude may fade with distance traversed. So, repeaters which basically consists of amplifiers are spaced equally along the length of a long-haul transmission line to amplify and boost the basic signal strength.

8.2 Photonic free-space communication

Free-space photonic communication has several advantages over that using optical fibre. Firstly, the atmosphere is an almost nonbirefringent medium which guarantees the preservation of photon polarisation. Thus the channel may be considered as a noiseless channel. Secondly, there is a relatively low absorption loss in the atmosphere for certain wavelengths. However, the varying and sometimes unpredictable atmospheric behaviour is an important issue that has to be considered especially in applications where the communication relies on correct detection of photons.

The atmosphere consists of various layers with different specifications. Generally, the propagation in each layer depends on the temperature, height, and turbulence. If the light trajectory is vertical, which is the case for astronomic observation, deep space, or satellite communication, all the effects such as turbulence inhomogeneities and temperature variations may be combined in the channel model. Therefore, the models end up in more complex calculations.

8.2.1 Near-ground free-space photonic link

The lowest layer of the atmosphere is called the troposphere, which is also the densest layer. In fact, if the density of the atmosphere was evenly distributed in height with a density equal to that of the troposphere, it would have ended at an elevation of 8 km. Since higher density results in more attenuation, a successful near-ground horizontal free space communication with a range of 8 km guarantees the possibility of establishing ground-satellite communication. Generally, the air impacts vary during the propagation, but in a temporally and locally correlated manner. A number of phenomena which influence the propagation and are described below.

8.2.1.1 Diffraction

The photonic source (in this case a laser) is considered as an electromagnetic Gaussian beam. This means that the optical intensity and transversal electric field distribution along the beam cross section radius is Gaussian. The beam radius can be considered at the beam waist by w_0 and the local beam radius and $x(z)$, respectively. The local beam radius is defined as the distance where the field amplitude drops to $\frac{1}{e}$ of the field value at the beam centre, or more precisely, propagation axis. The intensity distribution which is dependent on the radial distance from the centre axis of the beam r and the axial distance from the beam waist x can be written as

$$I(r, z) = I_0 \left[\frac{w_0}{w_z}\right]^2 \exp\left[\frac{-2r^2}{w(z)^2}\right] \tag{8.17}$$

where I_0 is a constant corresponding to the intensity distribution exactly at the source that is at, $w = r = 0$). The intensity is averaged over time, and $w(z)$ is expressed as

$$x(z) = w_0\sqrt{1 + \frac{z}{z_r^2}} \tag{8.18}$$

where $z_R = \frac{\pi}{w_0}^2\lambda$ is called the Rayleigh wavefront and this is maximal at this distance from the beam waist.

The diffraction results in spatial spreading of the the final beam. This phenomenon depends on the light wavelength but not the atmosphere, and is independent of atmospheric conditions. The beam spread can be calculated as a function of the communication range for a particular wavelength. The smaller the beam radius, the better the coupling and receiver detection efficiency. The analytical derivation gives the optimum value of $w_0^{opt} = \sqrt{\frac{\lambda D}{\pi}}$ for distance D and wavelength λ.

8.2.1.2   Absorption

The attenuation is a severe problem during beam propagation. It is highly correlated with the weather conditions and varies for different frequencies of light. For example, foggy weather brings about huge attenuation at optical frequencies but small attenuation at RF frequencies. However, light rain has low attenuation in the optical range but high loss for RF communication. The physical layer is said to have a loss exponent or extinction coefficient with the unit of dB/km. The loss exponent can be a fraction of one in clear weather with high visibility, and goes up to around one-hundred on some rainy or humid days or in polluted air. The intensity at a distance can be stated as

$$I(\lambda, z) = I_0(\lambda)e^{-z\alpha_{abs}(\lambda)} \tag{8.19}$$

where $I(\lambda, z)$ is the intensity at a distance z, I_0 is the intensity near the source, α_{abs} is the attenuation coefficient, and λ is the wavelength of light.

As can be seen, the atmosphere has quite good transparency for the optical window and radio window. This is the reason that they are appropriate frequency bands to be exploited in long-distance distance photonic communication in the atmosphere.

8.2.1.3   Scattering

The scattering phenomenon degrades the received optical energy like absorption but by two different mechanisms. In the first case, the elastic scattering is caused by the perturbation of molecules which have a small size compared with the wavelength. It is also called Rayleigh scattering, where the scattering is inversely proportional to the fourth power of the frequency. Consequently, by increasing the wavelength, this effect rapidly decreases. However, it cannot be neglected for the optical range which corresponds to relatively high frequencies. The second type of scattering, known as Mie scattering, is associated with aerosols, or the particles which have comparable sizes with the wavelength. The amount of scattering depends on the light frequency, the particle size, and the distribution and composition of aerosols.

8.2.1.4   Scintillation

In general the final size of the laser beam passing through the atmosphere is always larger than the expected size when only the diffraction is taken into account. This is due to the consecutive refractions of light, called scintillation that makes the propagation incoherent and diverging. The impact of scintillation on the laser beam is observed as two modifications which change both the amplitude and phase of the light. The first one is broadening, which is mainly because of small-scale eddies in the air. The second significant change is beam wandering, which is due to relatively large eddies. Both of them are big challenges for the alignment and beam tracking systems at the receiver.

8.3 Photonic networks

The term optical networks, or preferably photonic networks, denote high-capacity information communications networks based on photonics technologies and optical components that can provide capacity, provisioning, routing, grooming, or restoration at the wavelength level. The use and application of such networking reduces cost per transmitted bit per kilometre compared to other long-distance networking technologies, like wireless networking, or coaxial cable/power-lines-based wire-line networking [9].

Photonic networking solutions span the full spectrum of transport networks, from core backbone and metro down to the access network domains. High-bitrate data traffic covering long distances is the characteristics of backbone and metro networks, where photonics networking is dominant. The access network carries different kinds of data streams to and from small units, which are multiplexed/demultiplexed in nodes having fixed back-hauling connections to the transport core network. The use of photonics networking has started its expansion to the access network segment as well.

Two innovations in photonic transmission of information led to the use of multiple channels of light, or wavelengths which could be multiplexed together onto a single fibre. These are wavelength division multiplexing (WDM) techniques and the use of erbium-doped fibre amplifier (EDFA). Instead of using one amplifier per wavelength, EDFA allows amplifying all wavelengths carried on a fibre.

Photonic networks follow general network topologies used in electronics and computer communication networks. These networks consist of sub-networks having different topologies:

1. *Point-to-point link topology*: It refers to a communications connection between two nodes/stations or endpoints.

2. *Star topology*: It consists of a star coupler which splits the incoming power into a number of outputs in a star network. The coupler may be of active or passive type depending on whether the data is transmitted to only a specified node/station or to all nodes of the network. Hence, information passed by one node is passed to another node/station through the coupler.

3. *Bus topology*: All nodes/stations are connected to the bus. If a signal is transmitted by one node/station, it is transmitted to all other nodes/stations connected to the bus. The transmission may appear both ways along the bus.

4. *Ring topology*: All nodes/stations are connected point-to-point in a ring-like manner through optical couplers.

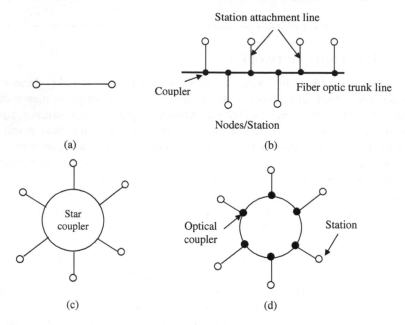

FIGURE 8.8: Various network topologies: (a) point-to-point, (b) bus, (c) star, and (d) ring type

Figure 8.8 shows the various network topologies. However, the concept of an all-optical node has been promoted, where the key component is the all-optical switch. Such a device has all-optical interfaces, and switches various wavelengths in the optical domain. Thus, those wavelengths carrying bypass traffic can remain in the optical domain through the node. All-optical nodes clearly enable the removal of a tremendous amount of electronics from the network [10].

Depending on the advancement of technologies, photonic networks [8] are divided into two generations:

1. The first-generation network operates on a single wavelength per fibre and is opaque, that is, the signal in a path is regenerated by repeaters. All signal switching and processing functions are implemented in the electronic layer. The technology uses a fibre distributed data interface (FDDI) and synchronous optical network and synchronous digital hierarchy (SONET/SDH). This framework has standardised line rates, coding schemes, bit-rate hierarchies, and functionality. Types of network elements and network architectures are also standardised.

2. The second-generation network utilises wavelength division multiplexing (WDM) technologies. Backbone core and metro optical networks rely heavily on high-capacity optical transmission links utilizing WDM. Besides offering large transmission capacities per optical fibre, second

generation systems have enabled the realisation of wavelength, routed networks with the use of either electronic or photonic switching nodes. Multi-wavelength optically routed networks are also characteristic of second generation systems [11].

The FDDI scheme uses two fibre pairs, each operating at 100 Mbits/sec, where the data bits are coded as bit-patterns, and is relatively expensive. FDDI is commonly used at the Internet service provider (ISP) peering points that provide interconnections between ISPs. FDDI can support multimode fibre for a maximum transmission distance of 2 km. FDDI also supports the use of monomode fibre having a transmission distance up to 40 km. Photonic transmitters in the wavelength range 850 nm, 1300 nm, and 1550 nm are used [12]. In general, FDDI uses the ring network topology, as seen in Figure 8.9. This has a two-fibre ring pair between nodes or workstations (WS) in a node-to-node ring network topology. One of the ring pairs serves as the primary ring and is active, while the other is a secondary ring that is kept on hold. In the event of a fibre break, both the rings are put to use, as seen in Figure 8.9(b).

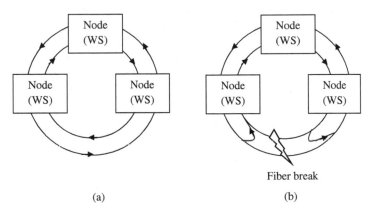

FIGURE 8.9: (a) A two-fibre ring pair in fibre distributed data interface (FDDI) network and (b) use of primary and secondary fibres in the case of a fibre break in an FDDI ring

SONET is used in North America, whereas the SDH network is used in Europe, Japan, and the rest of the world. The transmission bit rate of the basic SONET signal is 51.84 Mb/sec, whereas the basic SDH rate is 155.52 Mb/sec and is called the synchronous transport module level 1 (STM-1). The SONET/SDH network topologies or configurations may be a

1. *Point-to-point topology*: In this topology, the nodes may be terminal multiplexers (TM) or line terminating equipment (LTE).

2. *Linear topology*: Here the add/drop multiplexers (ADM) are inserted between TM in point-to point-links.

3. *Ring topology*: The nodes may be accessed through unidirectional path-switched rings (UPSRs) or bidirectional line-switched rings (BLSRs).

A SONET/SDH ring can be unidirectional or bidirectional. SONET/SDH rings are self-healing rings because the traffic flowing along a certain path can be switched automatically following failure or degradation of the link segment. In self-healing SONET/SDH rings, when a failure occurs in the link due to failure of the transmission or receiver equipment or the fibre breaks, the service is automatically restored using the automatic protection switching (APS) protocol. The restoration time is less than 60 ms. The protection or restoration is possible since the two devices at the nodes are connected with two different fibres (termed the working and protection fibres) such that when a link failure occurs, the SONET/SDH signal is split and simultaneously transmitted over both fibres. The destination selects the best of the two signals. Protection on a SONET/SDH ring can be at the level of a line or a path.

8.3.1 Wavelength division multiplexing (WDM)

The main function of a photonic multiplexer is to couple two or more wavelengths into the same fibre. Optical fibres offer large bandwidth (∼100 THz). Wavelength-division multiplexing (WDM) is a technology which multiplexes a number of optical carrier signals onto a single optical fibre by using different wavelengths of laser light [13]. The source wavelength may vary between 1530 to 1610 nm [14]. Thus, a number of different peak wavelength optical signals can be transmitted through a single fibre, as shown in Figure 8.10. WDM allows bidirectional communications over the fibre, as well as enhances the transmission capacity.

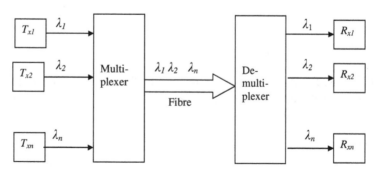

FIGURE 8.10: WDM in an optical fibre communication system

The transmitter (Ts) blocks in the figure are the individual optical transmitters. Ts blocks, in themselves, may be multiplexed output of several other transmitter networks, thereby forming some sort of a hierarchical multiplexed architectural system. Also, the transmitter block may consist of a time-division multiplexed type of system where the data signals to be transmitted are multiplexed in the time domain. The output signals from these transmitters at their corresponding wavelengths are then multiplexed in the wavelength domain (in accordance with the ITU G.692 standard) by the wavelength multiplexer. The wavelength multiplexer combines all the output signals and combines them to be transmitted along the optical fibre to reach the receiving end. The architecture consists of multiple transmitters and receivers, as far as the type of communication is concerned, although it is still a point-to-point link. Periodic amplifiers and repeaters may be installed along the optical link to establish a secure and satisfactory photonic communication link with the minimum possible BER. At the receiver side, the multiplexed transmitted signal is received and then de-multiplexed by a wavelength de-multiplexer. The respective receivers collect their intended data signals, and further processing takes place on these receivers before the signal is actually delivered to the end-user [15].

In a WDM system, therefore, bit streams from several transmitters are multiplexed in the wavelength domain together, and a wavelength demultiplexer separates channels and feeds them into individual receivers, with channel spacing varying in the range 25 and 100 GHz. Modern WDM systems can handle almost 160 signals and can thus expand a basic 10 Gbit/s system over a single fibre pair to over 1.6 Tbit/s. Usually WDM systems operate on monomode fibre cables but WDM can also be used in multimode fibre cables which have core diameters of 50 or 62.5 μm [16].

WDM systems may be either the coarse (CWDM) or the dense (DWDM) type. CWDM systems provide up to 8 channels in the C-band at around 1.550 μm of silica fibres, and DWDM systems provide denser channel spacing in the same transmission window. Therefore, CWDM in contrast to DWDM uses increased channel spacing to allow less stringent and thus cheaper transceiver designs. In CWDM wideband, optical amplification does not occur, limiting the optical spans to several tens of kilometres. DWDM may have 40 channels at 100 GHz spacing or 80 channels with 50 GHz spacing or even higher number of channels with 12.5 GHz spacing. Some of the advantages of DWDM systems are (a) capacity for upgradation with minor changes; (b) transparency as the actual timing information of the data is irrelevant to any up gradation in the channel capacity; (c) wavelength itself can be used as destination address for routing; and (d) wavelength switching can occur, whenever the required wavelength is non-available due to congestion or other network factors.

8.3.1.1 Routing topologies

Routing is the process of selecting the best paths in a network. Wavelength routers are an important component in WDM and are often used in most long-haul networks. Wavelength routing requires photonic switches, cross connects, etc. The early development of all-optical networks was typically based on core optical devices in the network nodes, which are passive splitters, combiners, and broadcasting stars. Such passive networks support a relatively small number of connections for a given number of wavelengths.

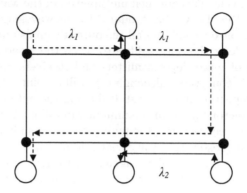

FIGURE 8.11: Wavelength routing network

Wavelength routing networks as shown in Figure 8.11 are circuit-switched networks. Based on the routing of information delivery at the nodes/stations, the routing schemes as shown in Figure 8.12 can be: (a) unicast, (b) broadcast, (c) anycast, (d) multicast, and (e) geocast. The unicast router delivers information to a single specific node/station, while a broadcast router delivers information to all nodes in the network. Similarly the anycast system delivers information to anyone out of a group of nodes, typically to the one nearest to the source node in the network. A multicast router provides information to a group of nodes that are interested in receiving the information. A geocast router delivers information to nodes in a large geographical area.

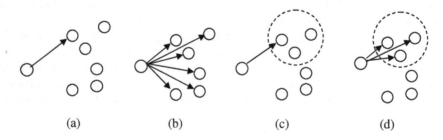

(a)   (b)   (c)   (d)

FIGURE 8.12: Routing schemes: (a) unicast, (b) broadcast, (c) anycast, (d) multicast, and (e) geocast

Routing topologies can also be static/non-adaptive or adaptive type depending on whether the routing tables are constructed manually or automatically. For larger and complex networks, such manually pre-computed tables are not feasible, hence the need for a dynamic or adaptive routing table arises, and these are automatically created as per the need. Adapting routing is carried out by computation of the optimised path length between one node and its neighbouring nodes using algorithms. Some of the methods are given below:

Distance vector algorithms: This approach assigns a cost factor to each of the links between each node in the network. The cost is optimised such that a node will send information to another node via the path that results in the lowest total cost. To do this, a node computes the cost factor (or the distance) between all its neighbouring nodes. This distance information and the total cost to each node, and the next hop to send data to get there, make up the routing table, or distance table. Over time, all the nodes in the network will discover the best next hop for all destinations, and the best total cost. When one network node fails, the next hop to the failed node is discarded and new routing-table is created. These nodes convey the updated routing information to all adjacent nodes, which in turn repeat the process.

Link-state algorithms: For these algorithms, each node needs a graphical map of the network. To produce the map, each node floods the entire network with information about the other nodes it can connect to and then assembles this information into a map. Using this map, each router independently determines the least-cost path or the shortest path from itself to every other node. This serves to construct the routing table, which specifies the best next hop to get from the current node to any other node.

Optimised link state routing algorithm: A link-state routing algorithm is optimised for mobile adhoc networks. Each node discovers 2-hop neighbour information and elects a set of multipoint relays (MPRs) for routing.

Path vector protocol: Path vector routing is used for inter-domain routing, unlike the other three algorithms. Similar to distance vector routing, one node called the speaker node in each system acts on behalf of the entire autonomous system. The speaker node creates a routing table and advertises it to neighbouring speaker nodes. The idea is the same as with distance vector routing except that only speaker nodes in each autonomous system can communicate with each other. The algorithm derives its name from the fact that a route is defined as a pairing between a destination and the attributes of the path to that destination.

8.3.1.2 Photonic cross-connects

A photonic cross-connect is an all-optical switch. It isolates a wavelength on its input port, converts it, and multiplexes it onto the output port. These devices can be interconnected in n numbers of arbitrary mesh. Wide area WDM require dynamic routing of the wavelength channels so as to reconfigure the network, while maintaining its non-blocking nature. To support the transmission line rate of several Gbits/second, the fibre transmission capacity has been increased by optical amplification and dense wavelength division multiplexing. But to meet the ever-increasing challenge of accommodating the higher bandwidth, photonic cross-connects capable of transmitting signals on a per-wavelength basis are being developed. Photonic cross-connects are classified as per the transmission medium as well as the switching mechanism, and thus they are divided into two classes, namely free-space and waveguides cross-connects. These photonic cross-connects are advantageous over the digital interconnects used at the nodes, because of negligible cross-talk and high bandwidth.

Photonic cross-connect devices consist of N input ports and N output ports, with a matrix that defines the connectivity between input and one or more outputs. Mathematically, this model may be represented by a matrix of a cross-connecting device. If I_K is the amplitude of light at input port K, O_L is the amplitude of light at output port L, then T_{KL} is the transmittance matrix governed by the absorption and dispersion characteristics of the connecting medium, as shown in Figure 8.13 and expressed as

$$O_L = \sum T_{KL}I_L \qquad (8.20)$$

Ideally, the T_{KL} terms may be either 1 or 0, signifying connect or no connect resulting in zero connectivity loss and zero dispersion.

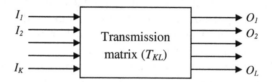

FIGURE 8.13: Modeling a photonic cross-connect

Although current data, voice, or video is transmitted using digital cross-connects, multiplexers, routers, and high-performance switches, in the future, photonic cross-connects promise to cope with the increasing bandwidth capacity and ultra-fast switching.

Channel cross-connection is a key function in most communications systems. In electronic systems, massive interconnection is made possible by integrated circuitry [17]. Similarly massive interconnection is needed in optical communication systems. Optical channel cross-connection may be accomplished in two ways:

1. Convert optical data streams into electronic data, then use electronic cross-connection technology, and then re-convert electronic data streams back into optical data. This is known as the *hybrid approach*.

2. Cross-connect optical channels directly in the photonic domain. This is known as *all-optical switching*.

The hybrid approach is currently popular because of existing expertise in designing high-bandwidth multichannel (NXN) non-blocking electronic cross-connects, where N may be on the order of thousands. All-optical switching is used in high-bandwidth, few-channel cross-connecting fabrics (such as routers), where N may be a maximum of 32. However photonic cross-connects with N up to 1000 seems quite a promising challenge.

8.3.1.3   Add-drop multiplexer

When one or more channels need to be added or dropped in a WDM architecture, preserving the integrity of other channels, then an add/drop multiplexer is used [18]. Such add/drop multiplexers can be used at different points on the communication link to add or remove selected channels, thus increasing the network flexibility. The basic operation of an add-drop multiplexer is illustrated in the block diagram shown in Figure 8.14. It generally involves demultiplexing of the input WDM signal and then, after selection of the channel to be dropped or added by means of optical switches, as shown in Figure 8.15, the entire signal is finally multiplexed again.

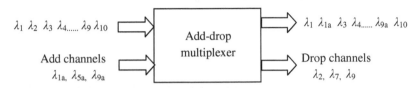

FIGURE 8.14: Block diagram of an add/drop multiplexer

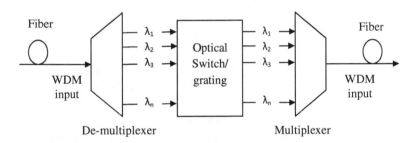

FIGURE 8.15: Add/drop multiplexer using optical switch or grating

Photonic add/drop multiplexers can be of fixed or reconfigurable types, depending on whether the communication link is dedicated or not. If the communication network is a dedicated one, then there exist fixed data channels that need to be added or removed. Reconfigurable type photonic add/drop multiplexers can electronically add or remove a selected channel as desired by the network system, thereby helping in bypassing faulty links, and upgrading the system from time to time.

8.3.1.4   Add-drop filter

The add/drop filter is so named because it can filter out a specific wavelength channel without affecting the WDM signal. Instead of demultiplexing all the wavelength channels with an add/drop multiplexer, if a single channel only needs to be demultiplexed, then an add/drop filter can be used in WDM system. It basically consists of a multi-port device, which can separate a single channel through one port, while the rest of the channels are transferred to another port. The wavelength selectivity property of fibre gratings can be utilised for an add/drop filter, as depicted in Figure 8.16. The isolators are used to separate out the channel to be dropped/added.

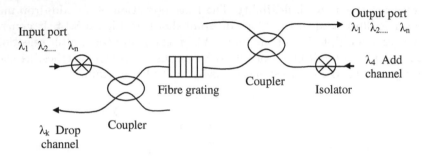

FIGURE 8.16: Add/drop filter using fibre grating, couplers, and isolators

8.3.1.5 Optical clock and bus

The point-to-point high-speed serial bus seems to offer the best solution. Multiple serial buses connecting functional units would distribute the load in a balanced manner. The number of serial buses connected to any given functional units has to be such that the aggregate bandwidth is maintained. The latency time would be increased, but tolerable by modern processors.

Alternate approaches have been studied. In a multiprocessor system with serial links to multiport memory modules, the interconnection network consists of a set of high-speed serial links between M memory modules and N processor modules. On the other hand, the memory module is serially multi-ported in order to feed the network bandwidth requirements [19]. The distribution of a high-quality clock signal from a central point to all modules irrespective of the connection links seems more advantageous. Optical clock distribution allows the provision of such a signal with simple phase-recovery circuits.

The benefits of a serial link in the multi-GHz region are appealing, since clock distribution and data transfer need to be performed with a perfect and stable clock at such a frequency of operation. Optical clock distribution produces low jitter over long distances and stable regions on a chip and multi-chip level. High-speed serial links offer the best usage of memory bus and the reduction of the bus width seems cost-effective.

8.3.1.6 Access network

An access network is the part of a telecommunications network which connects subscribers to their immediate local exchange of the service provider or the backbone of the network. The local exchange contains numerous pieces of automated switching equipment to connect the caller to the receiver through the exchange. It is in contrast to the core network which connects local providers directly to each other. An access network encompasses connections that extend from a centralized switching facility to individual homes. Tree networks are widely deployed in the access front.

The process of communication in an access network begins when one or more users interact with the local exchange to initiate user information transfer. The attempt to initiate interaction results either in a successful access or in a failed access. A successful access enables user information transfer between the intended caller and the receiver destination, whereas an unsuccessful access results in termination of the initiated attempt within the specified maximum access time.

The access network thus connects to the backbone, which consists of a high-speed core switching point. The end points/nodes of the access network could be buildings on a campus or cities in a country, connected to the central core switch or backbone through routers, as shown in Figure 8.17.

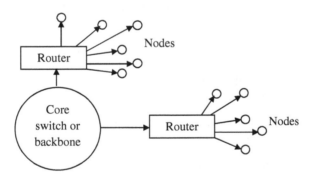

FIGURE 8.17: An access network

8.3.1.7 Transport network

Waveguide communication is preferred over free-space communication, where losses are quite high. So, the transmission medium plays an important role in deciding the speed and reducing the losses during communication. The role of photonic technologies in enhancing network performance is a major research area today. The intimate integration of optics with electronics (hybrid approach) on the same chip enables massive bandwidth up-scaling. In most existing networks, optics is used only on links to transport signals, whereas electrical processing is done at the nodes of the network, which demands many optical-to-electronic and electronic-to-optical conversions. A simple example is the synchronous digital hierarchy ring network as shown in Figure 8.18.

A photonic transport network consists of three layers: (a) photonic channel layer, which is a link between two users via an entire lightpath, (b) pho-

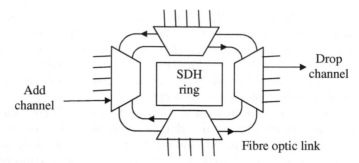

FIGURE 8.18: Synchronous digital hierarchy (SDH) ring-network

tonic multiplex section, where several wavelength channels are multiplexed and transmitted as a single signal over a fibre. It is the section between a multiplexer and a demultiplexer that carries the combined signal, and (c) the photonic transmission section, which transports the signal between two access points over which the multiplexed signal is transmitted. The transport network challenges [20] are:

1. To minimise the transport cost per bit by maximising the range of unregenerated signal through better optical amplification, and increasing the capacity of fibre through inclusion of more wavelengths (through WDM) or increasing the bit-rate per wavelength.

2. To use coherent optical systems to optimise cost and potentially photonic integrated circuits (PICs).

3. Using an all-photonic network architecture, where transport and processing is completely optical, instead of the point-to-point architecture. In the all-photonic architecture, optics technology is used to transport and process signals, thus reducing the OEO conversions, but control of processing remains in electronics domain.

4. To reduce cost, size, and power dissipation at the interfaces.

8.3.2 Erbium-doped fibre amplifier (EDFA)

In a generic sense, the optical amplifier unit amplifies an optical signal in the same way an electronic amplifier amplifies an electronic signal. An optical amplifier generally consists of an active medium which is excited by a pump-source, as shown in Figure 8.19. The incident optical input signal is coupled to the active medium via fibre-to-amplifier couplers, and the optical signal gets amplified inside this medium by virtue of stimulated emissions. The amplified optical output is then re-coupled to the optical fibre via another fibre-to-amplifier coupler. When the input optical signal travels through the active

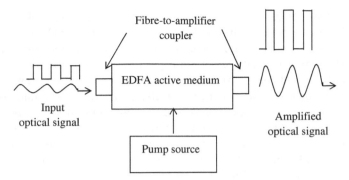

FIGURE 8.19: Generic optical amplifier

medium, the power from the pump gets transferred to the signal power and thus an amplified optical output is obtained.

Erbium-doped fibre amplifier (EDFA) makes use of rare-earth elements as a gain medium by doping the fibre core during the manufacturing process. Amplifier properties such as the operating wavelength and the gain bandwidth are determined by the dopant rather than by the silica fibre, which plays the role of a host medium. Though many dopants have been tried, Erbium doping is found to be suitable as they operate in the wavelength region near 1550 nm. Pumping at a suitable wavelength provides gain through population inversion. The gain spectrum depends on the pumping scheme as well as on the presence of other dopants, such as Ge and Al, within the fibre core. The amorphous nature of silica broadens the energy levels of Er^{3+} into bands. Figure 8.20 shows a few energy levels of Er^{3+} in silica glasses.

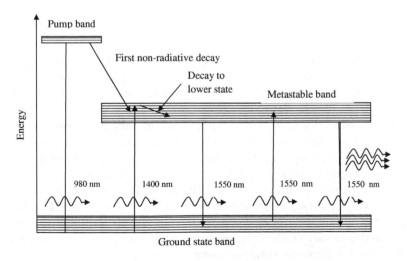

FIGURE 8.20: Energy levels in erbium

Many transitions can be used to pump an EDFA and efficient pumping is possible using semiconductor lasers operating near 980 nm and 1480 nm wavelengths. Once the amplifier material is excited by an appropriate pump source, there are primarily two main types of possible radiations due to downward transitions of the excited electrons: (a) the spontaneous emission, which is highly incoherent and occurs naturally, and (b) the stimulated emission, which is coherent and takes place only when stimulated by incident signal photons. The stimulated emission is important as far as the optical amplification is concerned. However, spontaneous emissions, though can be reduced, cannot be totally eliminated, and it is these emissions which lead to a noise component in the optical amplifier output.

The pump-source creates and maintains a population inversion condition in the optical amplifier material by pumping it with energy corresponding to wavelength λ_p. The input weak optical signal is combined with λ_p and the two wavelengths travel through the Erbium core of an optical fibre which acts as the amplifying material. The signal wavelength then gets amplified inside the optical fibre in a similar way as in the case of a laser and the amplified optical signal is then obtained at the other end. However, the output at the EDFA consists of some portion of the pump wavelength too, which is filtered out using a wavelength de-combiner at the output of the EDFA. The interaction of the pump-photon and the signal photon in the EDFA is independent of their relative direction of travel. In other words, if the pumping photons are input to the EDFA from the output end of the device, i.e., opposite to the direction of flow of the signal photons, the EDFA would, still, have the same performance characteristics.

In the case of low optical input signal power, $P_{s(input)} \ll P_{pump}$ and the gain of the amplifier is given by

$$G \approx \frac{\lambda_p}{\lambda_s} \frac{P_{pump}}{P_{s(input)}} \tag{8.21}$$

This equation explains the initial high gain of the EDFA at low input signal power levels. The gain, however, decreases as the input signal power level rises. The gain of the EDFA is also a function of the length of the Erbium doped fibre.

The possibilities of photonics integration in enhancing the transport network can be accomplished using the WDM photonic integrated circuit module, monolithic tunable optical router, high capacity optical switching core, highly scalable integrated optical switch matrix, and client interfaces using so-called electro-photonics integrated circuits (EPIC) [21]. Further, the entire coherent receiver consisting of several photodiodes can be integrated in a photonic integrated circuit (PIC); the numerous components are all lithographically connected and can much more easily be path-length matched and balanced. Enhancing the photonic transport networks using some cutting-edge technologies, such as super-continuum light sources and a coherent optical amplifier, have also been demonstrated by researchers.

8.4 Photonic secure communication

Data security is an extremely important issue that needs to be addressed for long-haul communication. Usually the data is encrypted or coded before transmission and finally decrypted at the receiving end. Information can be communicated over long distances through optical fibres at the speed of light, yet data encryption is bottlenecked by electronics. Processing delays in encrypting a packet of data makes the provision of data security costly for high-speed transactions. So, research has been directed towards optical encryption in which data is encrypted at the speed of light. Technology development in this area could potentially lead to ultrafast, secure data transmission, useful in applications ranging from high-security communications to high-frequency trading.

An encryption scheme is an algorithm that is undertaken by the sender using the input and a secret key and the encrypted data is transmitted. This is followed by a decryption algorithm run by the receiver, which returns the originally sent data by the use of the secret key and encrypted data. The basic idea behind encryption is that if the secret keys used by sender and receiver are the same, the data recovered by the receiver is precisely what was sent by the sender.

8.4.1 Phase and polarisation encryption

In all-optical encryption, photon polarisation states are used to transmit the information [22]. Many optical fibre-based secured communication schemes use the BB84 protocol as phase encoded states. The sender (traditionally referred to as Alice) and the receiver (referred to as Bob) are connected by a communication channel which allows polarisation states to be transmitted. This secure-key-based transmission can further communicate via a public classical channel using broadcast radio or the Internet.

The security of the protocol comes from encoding the information in non-orthogonal states, which in general cannot be measured without disturbing the original state. BB84 uses two pairs of states, with each pair conjugate to the other pair, and the two states within a pair orthogonal to each other. Pairs of orthogonal states are referred to as a basis. The usual polarisation state pairs used are either the rectilinear basis of vertical 0° and horizontal 90°, or the diagonal basis of 45° and 135°. Any two of these bases are conjugate to each other, and so any two can be used in the protocol. Figure 8.21 shows the rectilinear and diagonal bases are used.

The BB84 protocol for transmission is described as follows: Alice creates a random bit (0 or 1) and then randomly selects one of her two bases with a photon polarisation state depending both on the bit value and the basis. For example a 0 is encoded in the rectilinear basis (+) as a vertical polarisation

Basis	0	1
+	↑	→
X	↗	↖

FIGURE 8.21: Orthogonal states or basis of BB84 protocol

state, and a 1 is encoded in the diagonal basis as a 135° state, as shown in Figure 8.21. This process is then repeated with Alice recording the state, basis, and time of each photon sent.

8.4.2 Decryption techniques and error control

In the BB84 secure-key protocol, as Bob does not know the basis the photons were encoded in, all he can do is to select a basis at random to measure in, either rectilinear or diagonal. He does this for each photon he receives, recording the time, measurement basis used, and measurement result. After Bob has measured all the photons, he communicates with Alice over the public classical channel. Alice broadcasts the basis each photon was sent in, and Bob the basis each was measured in. Both discard photon measurements (bits) where Bob and Alice used different basis, which is half on average, leaving half the bits as a shared key.

To check for any intruder, Alice and Bob now compare a certain subset of their remaining bit strings. If a third party has gained any information about the photons' polarisation, this introduces errors in the receiver's (Bob's) measurements. If more than a specified bit differs they abort the key and try again.

8.4.3 Photonic image encryption techniques

For secured image transmission, multiple images that need to be transmitted at an acceptable transmission rate require the use of encryption and compression techniques. Many image encryption algorithms have been proposed in recent years [23], [24], [25]. Some of these algorithms can be implemented using photonic techniques taking advantage of both the two-dimensional imaging capabilities and inherent parallelism of optics. Among these methods, the coherent optical one is very promising [26].

One of the most used optical encryption algorithms is the Fourier plane encoding algorithm [23], [27]. This algorithm encodes an image as a stationary white noise by the use of two independent random phase codes in the input plane and Fourier plane. The image is multiplied by one of the random phase codes. After Fourier transform of this product, it is multiplied by the second random phase code, as shown in Figure 8.22. A second Fourier transform gives the encrypted image. The implementation is shown in a 4f setup, where a parallel beam of light illuminates the input plane, which contains the product of the input image and the random security code. An encryption is said to be strong only when the attacker is unable to find the key used for a given encryption. So this technique gives a robust method of encrypting data or images to be transmitted. The image is decrypted using the same 4f setup and the conjugate of the same random phase codes.

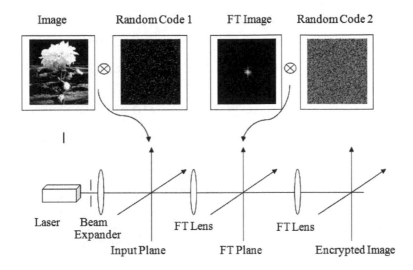

FIGURE 8.22: Encoding by double random phase code in a 4f setup

A photonic double random-phase encryption method using a joint transform correlator (JTC) architecture (to be described in Chapter 10), was proposed [28], wherein the joint power spectrum of the image to be encrypted and the key codes are recorded as the encrypted data. As shown in Figure 8.23, if $f(x)$ represents the input image and $h(x)$ is the key used for encryption, then the joint power spectrum of $f(x)$ and $h(x)$ gives the encrypted data. A parallel beam of light is incident on the joint image to produce the joint power spectrum. Unlike the case with classical double random-phase encryption, the same key code $h(x)$ is used to both encrypt and decrypt the data, and the conjugate key is not required. For decryption, the key is placed at the input plane, which is illuminated by a parallel beam of light from a beam expander. It has been shown [29] that it is possible to significantly improve the quality of the decrypted image by introducing a simple nonlinear operation in the joint power spectrum containing the encrypted function. This nonlinearity also makes the system more resistant to plain text attacks. The recording of joint power spectrum is done in a photo-refractive crystal.

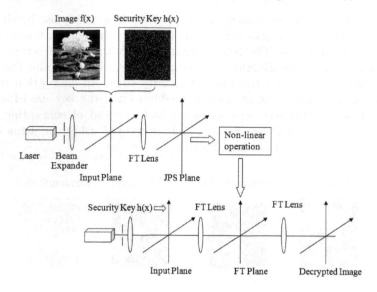

FIGURE 8.23: Encoding by double random phase code in a JTC architecture

Another encryption method employing JTC architecture is cited in [30]. This utilises a protocol for image encryption and decryption without the need for an external reference wave. It has been found that the proposed method retrieves images with better quality compared to conventional methods.

Books for further reading

1. *Optical Fibre Communication*: G. Keiser, McGraw Hill, 3rd edition, 2000.

2. *Fibre Optic Communication Systems*: G.P. Agrawal, Wiley, New York, 1997.

3. *Optical fibre Communications*: J. M. Senior, Prentice Hall, Englewood Cliffs, NJ, 1992.

4. *Optical Fibre Telecommunications, Volume-II*: S. E. Miller and I. P. Kainow (ed.), Academic, New York, 1988.

5. *Optical Fibre Telecommunications, Volume-III, A and B*: I. P. Kainow and T. L. Koch (eds.): Academic, New York, 1997.

6. *Fibre Optic Communications*: J. C. Palais, Prentice Hall, New York, 1998.

7. *An Introduction to Fibre Optic Systems*: J. Powers, S. Irwin, Chicago, 1997.

8. *Optical Networks*: R. Ramaswami and K. Sivarajan, Morgan Kaufmann, San Francisco, 1998.

9. *Optical Communication Networks*: B. Mukherjee, McGraw Hill, New York, 1997.

10. *Passive Optical Components for Optical Fibre Transmission*: N. Kashima, Artec House, Norwood, MA, 1995.

11. *Optical Networks and Their Applications*: R. A. Barry, Optical Society of America, Washington, DC, 1998.

12. *Design of Optical WDM Networks*: B. Ramamurthy, Kluwer Academic, Norwell, MA, 2000.

13. *Understanding SONET/SDH and ATM*: S. V. Kartalopoulos, IEEE Press, Piscataway, NJ, 1999.

14. *Optical Networking with WDM*: M. T. Fatehi and M. Wilson, McGraw-Hill, New York, 2002.

Bibliography

[1] X. Liu, D. M. Gill, and S. Chandrasekhar. Optical technologies and techniques for high bit rate fibre transmission. *Bell Labs Tech. J.*, 11(2):83–104, 2006.

[2] C. Pask and R. A. Sammut. Developments in the theory of fibre optics. *Proc. IEEE*, 40(3):89–101, 1979.

[3] A. Ghatak and K. Thyagarajan. Graded index optical waveguides: a review. In E. Wolf (ed.), *Progress in Optics*, North-Holland, 1980.

[4] A. W. Snyder. Understanding monomode optical fibers. *Proc. IEEE*, 69(1):613, 1981.

[5] M. Nishimura. Optical fibers and fiber dispersion compensators for high-speed optical communication. *J. Opt. Fiber Comm.*, 2(2):115139, 2005.

[6] C. C. Chien and I. Lyubomirsky. Comparison of RZ versus NRZ pulse shapes for optical duobinary transmission. *J. Lightwave Tech.*, 25(10):29–53, 2007.

[7] J. P. Fonseka, J. Liu, and N. Goel. Multi-interval line coding technique for high speed transmissions. *IEE Proc. Commun.*, 153(5):619–625, 2006.

[8] M. P. Clark. *Networks and Telecommunications: Design and Operation.* John Wiley, New York, 1997.

[9] A. Rodriguez-Moral, P. Bonenfant, S. Baroni, and R. Wu. Optical data networking: protocols, technologies, and architectures for next generation optical transport networks and optical internetworks. *J. Lightwave Tech.*, 18(12):1855–1870, 2000.

[10] D. F. Welch et al. Large-scale InP photonic integrated circuits: enabling efficient scaling of optical transport networks. *IEEE J. Selected Topics in Quant. Elect.*, 13(1):22–31, 2007.

[11] M. P. McGarry, M. Reisslein, and M. Maier. WDM ethernet passive optical networks. *IEEE Commun. Mag.*, 44(2):15–22, 2006.

[12] A. VanNevel and A. Mahalanobis. Comparative study of maximum average correlation height filter variants using LADAR imagery. *Opt. Eng.*, 42(2):541–550, 2003.

[13] J.M. Senior. *Optical Fiber Communications: Principle and Practice.* Prentice Hall of India. New Delhi, 2002.

[14] F. Forghieri, R. W. Tkach, and A. R. Chraplyvy. WDM systems with unequally spaced channels. *J. Lightwave Tech.*, 13(5):889–897, 1995.

[15] R. Davey, J. Kani, F. Bourgart, and K. McCammon. Options for future optical access networks. *IEEE Commun. Mag.*, 44(10):50–56, 2006.

[16] G. J. Pendock and D. D. Sampson. Transmission performance of high bit rate spectrum- sliced WDM systems. *J. Lightwave Tech.*, 14(10:21412148, 1996.

[17] W. Wei, Z. Qingji, O. Yong, and L. David. High-performance hybrid-switching optical router for IP over WDM integration. *Photonics Network Comm.*, 9(2):139–155, 2005.

[18] C. R. Giles and M. Spector. The wavelength add/drop multiplexer for lightwave communication networks. *Optical Networking*, 4:207, 1999.

[19] D. Litaize, A. Mzoughi, P. Sainrat, and C. Rochange. The design of the m3s project: a multiported shared memory multiprocessor. In *Supercomputing 92*, Minneapolis, USA, pages 326–335, 1992.

[20] K. I. Sato. Photonic transport networks based on wavelength division multiplexing technologies. *Phil. Trans. R. Soc. Lond. A*, 358:2265–2281, 2000.

[21] D. F. Welch et al. Large-scale inp photonic integrated circuits: enabling efficient scaling of optical transport networks. *EEE J. Sel. Top. Quantum Electron*, 13(1):2231, 2007.

[22] C. H. Bennett and G. Brassard. Quantum cryptography: Public key distribution and coin tossing, *Proc. IEEE Int. Conf. on Computers, Systems and Signal Processing*, 175: 8, 1984.

[23] P. Refregier and B. Javidi. Optical image encryption based on input plane and Fourier plane random encoding. *Opt. Let.*, 20:767–769, 1995.

[24] G. Unnikrishnan, J. Joseph, and K. Singh. Optical encryption by double-random phase encoding in the fractional Fourier domain. *Opt. Let.*, 25:887–889, 2000.

[25] B. M. Hennelly and J. T. Sheridan. Optical image encryption by random shifting in fractional Fourier domains. *Opt. Let.*, 28:269–271, 2003.

[26] T.J. Naughton, B.M. Hennelly, and T. Dowling. Introducing secure modes of operation for optical encryption, *J. Opt. Soc. Am. (A)*, 25: 2608-2617, 2008.

[27] P. Kumar, J. Joseph, and K. Singh. Known-plaintext attack-free double random phase-amplitude optical encryption: vulnerability to impulse function attack. *J. of Opt.*, 14:045401, 2012.

[28] T. Nomura and B. Javidi. Optical encryption using a joint transform correlator architecture. *Opt. Eng.*, 39:2031–2035, 2000.

[29] J. M. Vilardy, M. S. Millan, and E. P. Cabre. Improved decryption quality and security of a joint transform correlator-based encryption system. *J. of Opt.*, 15:025401, 2013.

[30] Y. Qin and Q. Gong. Optical encryption in a JTC encrypting architecture without the use of an external reference wave. *Opt. and Laser Tech.*, 51:5–10, 2013.

Chapter 9

Photonic Computing

9.1 Introduction

Ever-increasing needs for information processing with higher computational speed have resulted in the improvement in very large scale integration (VLSI) technology where devices are becoming smaller in dimensions with greater complexity. However, VLSI technology is fast approaching its fundamental limit in sub-micron miniaturisation, and further reduction in dimensions may invite several problems of lithography, such as dielectric breakdown, hot carriers, and short channel effects. Even if most of the problems are addressed, the demand for further miniaturisation will continue as long as increasing demands for higher bandwidth prevails in information processing. Therefore, an alternative solution to the problem is needed. The photonics platform for computation and communication offers an alternative approach to the microelectronics-based system. The high speeds, high bandwidth, and low cross-talk achievable by photonics technology are well suited for an ultra-fast computing scheme with high interconnection densities. In addition, the high efficiencies of photonic devices allow such implementations to match or even offer better performance than equivalent electronic systems.

9.2 Requirements of high-speed computing and parallel processing

It is necessary to consider the abstract concepts of parallel computation and corresponding complexities to highlight the relevant constraints which the photonic parallel processing and computing system must cope with. In a given architecture, the complexity function f is the measure of time and space (memory) required for the execution of a given algorithm of size n. This helps in deciding the performances of a given architecture and technology. The asymptotic behaviour of the complexity function is of much interest at large values of n. The major concern is, however, with the performance of an algorithm which is defined as the order of f, i.e., $O(f)$. For example, the time complexity function $t(n) \geq n$, for the most primitive sequential machine, i.e., the Turing machine, whereas for a less primitive one, i.e., the random access machine, $t(n) \geq O(T(n)logn)$.

The hardware classifications seem to be the natural choice when an architectural model of computer is selected or evolved for a particular application. The representations are considered in terms of unidirected acyclic graphs, called combinational Boolean circuits. Each circuit consists of, in general, several gates of which NOT, OR, and AND are termed the basic gates. A circuit with n input variables and m output variables computes a function $f : (0,1)^n \rightarrow (0,1)^m$ for a particular problem. The size of a circuit is defined as the number of gates in the circuit, whereas the depth of a circuit is the number of gates in the longest path from an input node to an output node. Further, $c = (c_0, c_1, c_2, ...)$ is said to be a family of circuits, if c_n is a circuit with n input nodes for each $n \geq 0$. A function f is said to be of size complexity $Z(n)$, if it is computable by a family of circuits of size $Z(n)$. The family c of circuits of size complexity $Z(n)$ is said to be a uniform family of circuits, if an $O(logZ(n))$ space-bounded deterministic Turing machine can compute $(1^n, c_n)|n \geq 0$. The *size* of circuits is a major resource for parallel computation as *time* in sequential computation. Proposals and demonstrations exist at many level of parallelism. For less than $O(10)$ parallel connections the parallelism is low, for $O(102)$ parallel connections, the level is medium, and for greater than $O(103)$ parallel connections, the level is said to be high.

Unlike the case for abstract or actual architectures of sequential machines, however, the way to relate the different architectures of parallel machines is not obvious. An abstract but useful classification cited by Flynn [1] depends on the differences in data and instruction streams and the degree of parallelism. For information processing, four classifications are used. They are SISD (single instruction single data), SIMD (single instruction multiple data), MISD (multiple instruction single data), and MIMD (multiple instruction multiple data) operations. A conventional computer of Von Neumann architecture is based on the SISD method, where the instruction stream is controlled by a single programme counter and the data is accessed with the address in the instructions.

Though the scheme is very successful, constant efforts are still being made by using various architectural concepts to improve upon the performances by eliminating the so-called Von Neumann bottleneck. Some cellular logic processing architectures [2] adopt SIMD operation, in which all processor elements start the same operation by a single instruction from a control computer [3]. This implies that all processors address the same position in their local memory; this address is also broadcast by the central controller. A computer with pipeline architecture is classified as an MISD machine.

The terminology concerning *single data* in the case of the MISD is somewhat misleading, since several different streams of data can actually exist concurrently. The major constraint in the MISD is the necessity of raster processing. In MIMD machines, independent instructions are executed on data streams in parallel. Therefore the MIMD architecture operates on the real multiprocessor techniques. Each processor has its own programme memory and the instruction flow control. The SIMD architecture has the lowest degree of parallelism, while the MIMD is the highest. The SIMD and MIMD architectures are suitable for parallel processing of regular data, such as arrays and image data.

9.2.1 Single-instruction multiple data architecture

An SIMD architecture-based computer comprises many processors usually called processing elements (PE), which all obey a sequence of instructions broadcast from a single central control unit as shown in Figure 9.1. Flow of information is bidirectional, i.e., instruction sequences fan-out to each processing element and control information. Usually a single-status bit fans-in to the controller, so as to make it able to control decisions on global data. The class of parallel architectures operated on SIMD architecture is useful

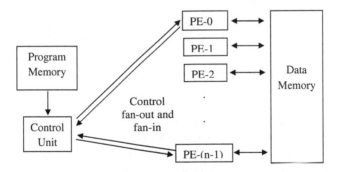

FIGURE 9.1: Schematic of SIMD architecture

for local processing. There are no interactions during a given cycle between the participating local neighbourhood data arrays. Cellular logic array (CLA) constitutes one of the basic SIMD architectures. Although the initial thrust of the architecture is on fast arithmetic computation, a cellular logic array processing has been extended in the parallel hierarchical and multifunctional systems of the pyramidal type [4].

The other example of SIMD-like, interconnection-intensive architecture is the connection machine (CM). The rationale for the CM comes partly from semantic network and knowledge representation language. The knowledge available is implicitly stored and one needs to use inference to make it explicit. Fast performance, however, requires a large number of processing elements and programmable interconnections, and therefore, the topology of the architecture has to be reconfigurable. The methodology has direct application in the photonic computing technique for artificial intelligence systems [5].

9.2.2 Multiple-instruction multiple-data architecture

A generic MIMD computer, shown in Figure 9.2, has a conventional microprocessor core architecture, which includes a local memory system. The communication is synchronized with other processors. This type of architecture provides massive parallelism, requiring individual control units for each processor, where individual processors are allocated individual tasks that can be performed simultaneously. The MIMD architecture is appropriate for high-

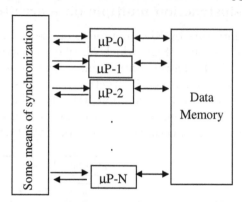

FIGURE 9.2: Schematic of MIMD architecture

level visual processing tasks, which require different algorithms to be executed on disjoint data sets. MIMD is concerned with interactive process that shares resources and is thus characterized by asynchronous parallelism. Message passing multi-computing architecture (i.e. transputer), and the shared memory multiprocessors are two important examples of the MIMD machine. SIMD and MIMD can, however, simulate each other. An SIMD machine could interpret the data as different instructions, while MIMD could execute only one instruction rather than many. There are no strict boundaries between the architectures, and the basic requirement for solving a specific problem is guided by the questions of efficiency and cost.

9.2.3 Pipelined and parallel architecture

One architecture of our interest is the pipelined processor, which provides overlapped parallelism in space and time. The task is properly divided into subtasks and the stages of the pipeline then execute the original task in a concurrent fashion. There are three major classes of pipeline architecture. The first one, vector super computer processing technique, is extensively used in arithmetic processing and is useful for its completeness. The second one, known as multifunctional pipeline, is generally used for low-level image processing, where operations are performed in real time. The third variety of pipeline is known as systolic architecture, which expands on pipelines by allowing the inputs and the partial results to flow from stage to stage. A systolic architecture could implement basic convolution operations.

One of the parallel configurations that has become important is the data flow architecture. Such architecture departs from the control-driven characteristics of Von Neumann computation and is also known as the data driven computing machine. A data flow scheme proceeds by instructions or data flow graphs (like the PETRI nets), activated by the availability of data operands, and is thus not too different from the fine-grained multi-computers [6]. The design issues related to data-flow computation are concerned with the decomposition (or compilation) of a given programme into some data-flow graphs exhibiting true parallelism and/or concurrency via an appropriate programming language [7]. Parallel distributed processing (PDP) is another major paradigm of computation. Studies have indicated that parallel distributed processing may provide speed at the expense of growth of parallel network [8]. The PDP is usually reconfigurable and the hardware organization is flexible enough to accommodate different computational tasks.

It seems appropriate to conclude by suggesting that one way of coping with constraints of computations is to employ truly distributed and hierarchical representations and related processing algorithms. A good match between representation and algorithm is also desirable. It is also evident that distributed parallel processing trades space for time, and that a major advantage in using it would rest in its inherently large interconnection capacity. The concept of computation and processing can then go beyond the traditional Von Neumann architecture. One can then contemplate photonic architectural concepts of computing because of its inherent capability of parallel processing. On a system level, these computational techniques are distributed in nature and will have the capability of making logical inferences.

9.3   Limitations of electronic computation

Performance of an electronic computer depends very strongly not only on the algorithm but also on the hardware and interconnection capabilities. With the advent of sub-micron technology, the switching delays of a logic gate is likely to fall below 100 ps and clock speed close to a few GHz. However, it seems that the interconnection capabilities and the system bus technology appear to have reached a final limit. The reasons are due to the physical limitation of the time-bandwidth product of the electronic wires, the topological limitations of the physical layout, and the architectural limit known as Von Neumann bottleneck. Present-day computers have been able to attain high processing rate to some extent (i.e., in the tera flops range), by decentralizing the workload amongst several processors, working simultaneously in parallel mode. That is not sufficient for future demand of information processing on large volume of data, involving arithmetic and logical operations or image processing operations, where a high degree of parallelism is required. Future computational operations are expected to require data processing rates in the range of several giga flops.

Parallel processing reduces the time required for complete execution of a particular operation, yet it has been observed that speed of the electronic arithmetic logic unit (ALU) is still slowed down due to several reasons: delay in the device response, the circuit delay, and the delay in the interconnections. Very large scale integration (VLSI) technology has provided a reduction in the topological dimensions, by reducing the interconnection length. This in turn increases the gate count in the integrated circuit, resulting in the accommodation of more gates in a single processor.

With the improvement of technology, the speed is likely to increase further than what is available today. However, when the operation of a computer is taken into account in totality, the performance is limited by (a) the number and bandwidth of the inter- and intra-processor communication, and (b) data storage and retrieval rates. Several communication problems mainly related to the physical topology and architectural limitations are still being addressed in electronic computers [8]. Some of these are inherent limitations, and therefore more elaborations are needed on these issues.

9.3.1   Physical limitations

The time-bandwidth product is inherently limited due to capacitance and inductance of interconnections within a VLSI processor chip. It has been established that the local and global distribution length follows the relations $0.1\sqrt{A}$ and $0.5\sqrt{A}$, where A is the chip area. The length of wire, whether it is local or global, offers a time delay to the signal for its travel due to capacitance and inductance. This time delay, however, can be minimised by designing several layers of hierarchy in the selection of length and width of the interconnections. A processor based on submicron technology uses six or more such layers to reduce the time delay, though the delay can never be reduced to zero. For chip-to-chip connections, the time delay is more pronounced because of the long physical length of interconnection. The problem is severe when bus interconnections are considered and padding is necessary to balance the different time delays associated with different circuit boards connected to the bus.

9.3.2   Topological limitations

Topological limitation mainly arises due to the number of active devices or logic gates per chip. When sub micron technology is used, it is not impossible to put 10^9 transistors or logic gates within a chip of 10 mm × 10 mm area. The density of logic gates within a chip imposes another problem of taking out connections from the chip for full exploitation of functional capacities.

However, in practice less than 1000 connections can be realised. Therefore, further improvement of packaging density will not help in solving the geometrical problem of pin connections. As the area of chip increases, the packaging density or number of devices in the chip increases proportionally. Moreover, the time delay increases with the square root of the chip area.

9.3.3 Architectural bottleneck

The classical Von Neumann architectural bottleneck is somewhat eased in a computer by the use of a multiprocessor, or other architectures already discussed. However, the cycle time mismatch between the memory and processor in a sequential or parallel addressing machine is still a problem. The use of cache memory offered some advantages, yet the updating of cache cannot match the processor cycle. Further increase will require large bus width of 1024 bits which seem difficult at the level of present-day technology. The use of electronic crossbar switching has been considered as an alternative to the common bus structure.

The problems in designing interconnections for electronic systems at high data rates, interconnection length, and input/output densities can be summarized as:

1. Clock and signal skew

2. Power dissipation and thermal management

3. Signal distortion and signal latency

4. Sensitivity to electromagnetic interferences (EMI)

5. Interconnect reliability

9.3.4 Fundamental limits

At this stage, it might be of interest to explore the possibilities of electronics and photonics on the basis of fundamental physical limits. The techniques of presentation of data and the complexities in the algorithm are the two basic guidelines when the costs of computation are compared. The theoretical minimum possible energy cost for representing a variable of dynamic range of 1000 has been estimated for both the electronics and optics/photonics technology. In a digital electronic computer, at least 10 electron per bit are required to represent a data bit in order to guarantee a bit error rate (BER) of 10^{-9}. A minimum of approximately $2 \times 10^3 \ k_B T$ of energy, where T is the temperature and k_B is the Boltzmann constant, would be required to represent a 10-bit number in a digital electronic computer. In a digital photonic computing system, each electron must generate at least one photon, and vice versa. Thus, at least 100 photons are required to store a 10-bit number, if the ideal detector is used. If a laser diode operating at a 1.5 eV band gap is used, the digital representation would require 10^4 $k_B T$ of energy at the same BER. Therefore very little would be gained unless the benefits offered by the parallel processing algorithms are carefully exploited in the case of digital photonic computing. It must be mentioned, however, that the available and foreseeable devices in both the digital electronics and digital optic domain operate too far from the physical limits. Therefore, it is of little use to compare the performances on the basis of present-day device technologies. The criteria for comparison must be dominated by the practical systems now available, or would be available in the near future under the constraints of the device technologies.

9.4 Photonics computing

Since the 1970s, many years of research have been devoted to photonic implementation of traditional models of computation, but due to power consumption, volume, and scaling issues, interest decayed in the 1980s. Neither analog nor digital approaches have proven scalable due primarily to challenges of cascadability and fabrication reliability. Analog photonic processing accumulates phase noise, in addition to amplitude noise. This makes cascaded operations particularly difficult. Digital logic gates that suppress noise accumulation have also been realised in photonics, but photonic devices have not yet met the extremely high fabrication yield required for complex digital operations. Most of the components that are currently very much in demand are all-optical switching components, the development of which was based on nonlinear optical materials has been relatively slow. Moreover, most of the photonic components require a high level of laser power to function.

The desire for faster computation demands the increase of computing power at a rapid rate. Parallel processing techniques can reduce the time required for complete execution of a particular operation, yet it has been observed that the speed of the arithmetic logic unit (ALU) is still slow due to several reasons cited earlier. Moreover space and power penalties for large values of fan-in and fan-out requirements for electronic devices [9], [10] have greatly restricted the use of electronics in computing. As improvements in device performance and inter-processor communication are likely to be slow, large improvements in overall performance must increasingly come from architectural innovations. Increased parallelism is likely to dominate the area of computation in the near future.

An altogether different concept of computing has therefore been attempted, where photons are used as the information carrier instead of electrons. The concept that optics can be a possible replacement for electronics in the field of digital image processing and computing has been probed since the invention of laser. The notion that optics and photonics could surpass electronic data handling capability and performance gained serious attention in the mid 1970s [11], [12]. These attempts have resulted in the evolution of technology demonstration systems and have sparked a flurry of research activities and digital optical/photonic computing for information processing.

At this stage it is necessary to standardize the nomenclature that is going to be used. It has been seen that digital optical/photonic computing systems are mainly based on the principles of optical interconnects. However, communication and interfacing with the outside world still have to remain in the electronic domain. Therefore, when a computing system is considered optical domain, the term photonic computing is supposed to be a better option. However, when the question of interconnection appears in optical domain, the term optical interconnection shall be maintained.

Photonic computing actually means the photonic manipulation of discrete numerical data. This has significant potential for multiple-order-of-magnitude performance improvements in areas such as speed, non-interfering interconnections, massive parallelism, reliability, and two-dimensional data representation, as well as in fault tolerance, compared to all-electronic computing. The major attraction of photonic computing lies in its ability to represent two-dimensional arrays of discrete binary data by on/off states of light sources, indicating one or zero of binary data, respectively. The speed and parallelism inherent in photonic systems have proved beneficial for a variety of computationally intensive problems, which include signal and image processing, equation solving, numerical processing and for the implementation of digital logics, among other fields.

The basic requirements of a photonic computing architecture (shown in Figure 9.3) for information processing are: (i) optical sources and a method of modulating the sources, (ii) techniques of input information processing, (iii) optical interconnections, and, finally, (iv) the information processing or decision-making stage. In short, processing of information in the photonic domain involves generation, propagation, modulation, and detection of light. The optical storage is desirable but not essential for the system.

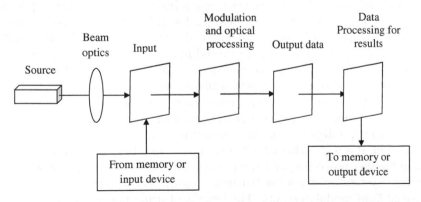

FIGURE 9.3: Schematic of a photonic computing architecture

Photonic digital computing and processing advocates the new and radically different computer architecture predicted upon the two-dimensionality of photonic systems, speed of photonic devices, and interconnects. The architectural concepts include everything from implemented programmable photonic logic arrays using smart pixels, and special purpose cellular image processing, to digital photonic computers based on non-linear logic with new types of optical logic gates.

Some basic building blocks that may be identified in order to utilise photons for computing are: (a) a way in or a trace for light to enter, (b) a module that would be guiding the light, (c) a module that would be encoding data into optical bits, (d) an interpreter of optic bits, and (e) basic computational algorithms. Some of the advantages of using the photonics paradigm in computing are that it can provide very high space-bandwidth and time-bandwidth, thus offering potential for high throughput. Photonic signals can propagate through each other with essentially no interaction and crosstalk, a feat that electronics cannot match. The photonic processors are inherently two-dimensional and work in parallel, and therefore can perform pattern matching correlations, and other image processing operations very well, and are used frequently in unification processes in logic programming.

Three properties of optics make it an attractive candidate for computing. First is the large bandwidth of optical sources, which may approach to giga hertz. Second is the large space bandwidth product. A two-dimensional (2D) optical system has an extremely high number of resolvable elements, each of which may be considered an individual parallel communication channel having a high degree of connectivity. The third characteristic is related to the non-interfering connectivity of optical paths. Modern-day lasers serve as powerful, highly intense and coherent photonic sources, replacing incoherent and less intense sources such as LEDs. Images at the input in 2D plane format can be loaded from the memory or from the external world. Switching devices like non-linear electro-optic devices or spatial light modulators enable the recording of input information as well as modulation of light passing through them. The information emerging from the modulating device is stored in photodetectors or CCD arrays. The information-processing and decision making process is then easily handled by the computer directly interfaced to the system. If required, the processed information is again converted to optical information by photonic converters and fed to the system for further processing. The non-interacting parallelism of optics and information modulation by two-dimensional light modulators can compete with electronics in areas of arithmetic and logical processing operating in SIMD mode.

Basically, a digital photonic computing architecture is an embodiment of reconfigurable and addressable optical interconnects in 2D or 3D for large volume parallel processing and computing. Interconnections in a photonic computing system are implemented using lenses, lens-let arrays, beam-splitters, spatial light modulators, etc. The benefit of using free-space optical imaging systems is mainly due to constant path lengths over the entire field, where the transit time differences are extremely small, on the order of femtoseconds. Interconnections based on imaging not only enjoys the advantage of parallelism of optical paths but also brings in a certain degree of regularity. Tolerance caused by aberrations is almost negligible, i.e., on the order of less than a wavelength. If data switching is necessary, the pipelined architecture can be efficiently utilised to break up the computations into data blocks where registering and storing of data seldom occurs.

Another area of recent interest in digital photonic computing is intelligent processing, which involves such concepts as knowledge, reasoning, learning, and adopting, that one should follow. Knowledge-based expert systems and their photonic implementations are ideally suited for carrying out complex computations of the kind encountered in mapping, associations, cognition, and control. Advances made in the study of artificial neural networks have also spurred photonic computing research for intelligent pattern identification.

9.4.1 Throughput issues

Parallel digital optical computers are often characterized by their asymptotic maximum throughput $r^\star$ and a parallelism parameter $n^\star$ to indicate the efficiency of the machine for small problems. It might be of interest to estimate $r^\star$ for an SIMD optical array. We hypothesize that an arithmetic unit comprising several SLMs, each having $p \times q$ processing elements or cells, is capable of performing full additions or logic operations at a frame rate S_f. The memory unit will have a matching cycle time. The raw processing power therefore is $P_p = pqS_f$ elementary operations (full additions or logic) per second. Assuming that the floating point operations require P_e elementary operations, we arrive at an estimate of maximum throughput $r^\star = P_p/P_e$ floating point operations per second (FLOPS). For an SLM having 512×512 picture elements operating at a frame rate of 1 MHz, the raw processing power would be about 26×10^{10} operations per second. Assuming that the floating point operations require 10^3 elementary operations, the maximum throughput would be only 260 MFLOPS per second. The potential throughput of the arithmetic unit can then be entirely realised because there is no bottleneck. This is due to the fact that the parallelism of the optical data movement and storage is as great as that of the arithmetic unit.

The second parameter of interest is $n^\star$, the vector length for which the machine throughput is $r^\star/2$. For a non-pipelined SIMD array, $n^\star$ is just half the array size. This means the processor will easily execute at least a million operations in parallel. Unfortunately, $r^\star$ and $n^\star$ do not completely characterize the performance of the parallel optical processor, because they are insensitive to the degree of communication available among the processing elements. Some of the communication-oriented classifications are possible for parallel architectures. The major issues in such architectures are diameter (the worst-case number of communication cycles required to send data), bandwidth (total number of messages that can be sent/received in one communication cycle), broadcast time, and maximum finding time.

9.5 Digital optics

Light can be represented as a signal having continuously variable intensity levels like the analogue signals in electronics. As digital representation of a signal is discrete in nature, digital electronics, or digital (electronic) circuits, represent signals by discrete bands of analogue levels, rather than by continuously varying signal levels. All levels within a band represent the same signal state. In most cases the number of these states is two, and they are represented by two voltage levels: one is the ground or 0 volts or the reference voltage value and the other is the supply voltage. These signals correspond to the false ($= 0$) and true ($= 1$) values of the Boolean domain, respectively.

Data handling in a computer is generally taking place at three levels. The lowest is at the gate level, wherein binary switches perform Boolean logic operations. The highest is at the processor level where the entire algorithm is executed as a single piece. A traditional digital optical processor operates at this level. At the middle level or at the register level, numbers or blocks of numbers are manipulated with synergism that is simple enough, yet they can be combined to form a large number of higher-level operations. It has been indicated that the complexity of each processing element tends to decrease as the number of interconnections increases [13]. Eventually, the processing elements are reduced to simple gates, and the processing structure approaches the realm of combinational logic. Here the system has no memory, except in the sense of a make-or-break connection pattern. Several methods of implementing combinational and sequential digital logic gates by optical techniques have been investigated [14], [15].

The development of digital optical logic processing at a reasonable level of sophistication has proved relatively elusive, in part because of fundamental difficulties associated with the power dissipation in optical switching elements. At the outset it must be mentioned that except at a very high switching speed, optical logic offers no particular advantage over electronic logic, when energy, power, and bandwidths are considered.

9.5.1 Array logic, multivalued logic, and programmable logic

Apart from the basic division between combinational and sequential logic based on their operational characteristics, other types of logic circuits have become important as processing techniques grow. Special interest would lie with the logical processing of data arranged in two-dimensional formats. The other important logic networks of interest for applications in the photonic domain are array logic, multivalued logic and threshold logic. Basically, array logic is a technique to construct any logic circuit using circuit elements in an array structure. The circuit element of this architecture is sometimes called a logic array. In a logic array, a set of input bit signals corresponds to an address signal, and a set of output signals corresponds to a data signal. An address is selected by an address decoder and the data designated by the address is accessed. To obtain output signals with value 1 for a specific set of input signals is simply to define a logical function describing the relation between the input and output signals. Implementation of a two-level network with AND and OR gates is important for minimal and reliable realisation. Procedures for implementing any logic by using AND and OR arrays is in general dependent on the use of a decoder or crossbar. Two- or three-level combinational logic arrays are capable of forming logically complete sets and can implement general logic operations.

Programmable logic devices (PLDs) are standard, off-the-shelf parts that offer customers a wide range of logic capacity, features, speed, and voltage characteristics. The performance of the devices can be changed at any time to achieve any number of functions, to quickly develop, simulate, and test their designs. Then, a design can be quickly programmed into a device, and immediately tested in a live circuit. PLAs are based on re-writable memory technology. PLAs are subsets of logic array and are used to implement combinational logic circuits. A PLA has a set of programmable AND gate planes, linked to a set of programmable OR gate planes, which are conditionally complemented to result in an output. This layout allows for a large number of logic functions to be synthesized in the sum of products (and sometimes product of sums) canonical forms.

9.5.1.1 Optical programmable array logic (OPAL)

Optical array logic is a technique for execution of parallel logic operations on 2D image data and is the optical version of electronic array logic. One of the most primitive architectures of digital optical parallel processing systems, is composed of many 2D parallel logical gates, and is called a parallel random logic system [16]. A major feature of this system is that the theories and techniques which are applicable in the field of electronics can be directly utilised into it. The logical operations are performed over two-dimensional input patterns and yield the processed output patterns. Unfortunately, in this architecture, a large number of parallel logic gates are required, therefore light attenuation and the time delay encountered result in poor performance.

The mechanism of constructing array logic circuits in analogy to programmable logic array (PLA) in electronics is termed optical programmable array logic (OPAL). The general optical architecture was realised by Tanida in 1983 [17]. Just as array logic with 2-bit decoders in electronics can implement any logical operation for the input signals by using the AND and the OR arrays, similarly the operation for input data coded into patterns corresponds to a logic array operation with a 2-bit decoder pattern. The architecture as shown in Figure 9.4 is capable of executing a shift operation in parallel to transmit data between a pixel and neighbouring ones. Using this property of logical neighbourhood operations and introducing the concept of array logic, data of a specific pixel and those of the neighbours are used as input signals to a logic array for a pixel by the system.

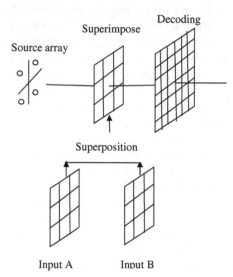

FIGURE 9.4: Schematic of an OPAL architecture

The advantage of this arrangement lies in the fact that maintaining the same structure of the input data arrangement, the system operates in an SIMD architecture on all the input pixels in parallel. For an output image O, the value O_{ij} of the ij-th pixel in it is expressed by

$$O_{ij} = L(A_{ij}, B_{ij}), (i, j = 1, 2....N) \tag{9.1}$$

where L is the logical operation, A_{ij} and B_{ij} are values of the ij-th pixel, and their neighbours in inputs A and B, respectively, and N is the pixel size of both inputs. In the system, N^2 logical neighbourhood operations are executed simultaneously in parallel. Thus, sixteen basic combinational logic operations of two input variables in parallel windows of a large block of data are carried out.

Optical technique is used for 2D correlation, whereas electronics is used for encoding-decoding, sampling, and inversion operations [18]. Applications of image logic algebra for OPAL processing have led to wire routing and numerical data processing [19] in shift operations, image casting, multiple imaging, and morphological processing. In addition to parallel logic operation, the processor has been applied in binary and gray image processing, arithmetic operations, and realisation of the Turing machine.

9.6 Photonic multistage interconnection

The prime mover in digital photonic computing is the interconnection technology. The subject of free space optical interconnects is closely related to space switching networks. A switching consists of an array of inputs and output devices interconnected by switching nodes. The input and output arrays are spatially connected optically by waveguides or through free space. For computing and switching purposes, implementation of both the space variant and space invariant interconnections are required. A system is said to be space variant if the array of lines emerging from an input element varies from input to input, whereas in a space invariant system, every input position generates the same output pattern, as is evident from Figure 9.5 It has been observed that though space variant operations limit the achievable space-bandwidth product, space variant interconnections are preferred over space invariant interconnections.

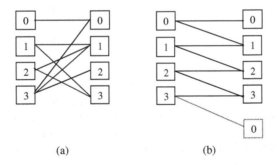

(a)   (b)

FIGURE 9.5: Schematic diagram of (a) space variant and (b) space invariant networks

Two very important switching networks are the crossbar network and the perfect shuffle network. Every input is connected to every output node in a crossbar and the point of intersection of every input and output forms a switching node. Likewise for a network containing N inputs and M outputs, a total of NM switching nodes are obtained. The two parameters that can describe a switching network are the fan-out of the switching nodes and the diameter of the network. The fan-out F is the number of input or output nodes, i.e., $F = N$ if it is assumed that $M = N$, as is the case in a crossbar network. The diameter D, on the other hand, is the number of switching stages required to link any input to any output. The crossbar therefore allows routing each of the inputs to each of the outputs. The most significant disadvantage of using a crossbar as a switching network is the huge number of switching nodes. The crossbar network for a 2D input and output provides the best example of a space variant network for routing over one another without interferences. A space invariant crossbar network can be efficiently implemented by an optical imaging setup as shown in Figure 9.6.

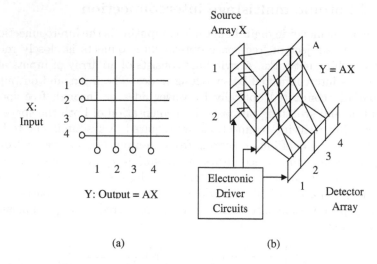

(a)   (b)

FIGURE 9.6: Schematic diagram of (a) space variant and (b) space invariant networks

The multistage interconnection networks (MINs) are also another interesting form of switching networks. The perfect shuffle network is a good example of a space invariant switching network, which provides a technique of shuffling an array of N input nodes by interleaving the upper and lower halves with one another. The initial order of the input nodes are re-established after the operation is repeated many times, as required. Hence for a 1D input with $N = 8$ as seen from Figure 9.7, the original order is restored after three stages of operation.

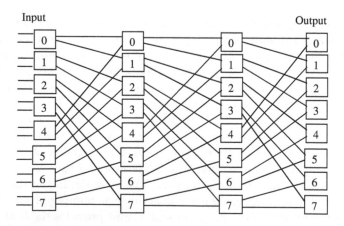

FIGURE 9.7: Schematic diagram of the perfect shuffle network with $N = 8$

Other multistage interconnection networks equivalent to the perfect shuffle are the banyan network and the crossover network. Figure 9.8 shows a single stage banyan and crossover network, respectively. The crossover network implements a spatial inversion of a particular section of the input array. Fortunately, most interconnection networks utilised for optical computing and switching show some amount of regularity, which enhances efficient implementation of the networks in terms of the space-bandwidth product.

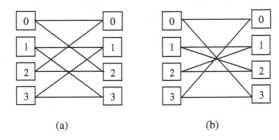

(a)   (b)

FIGURE 9.8: Schematic of (a) banyan network and (b) crossover network

9.7 Number systems for parallel photonic computing

To exploit massive parallelism of optical paths for computations, it is desired that the arithmetic operations be simultaneously performed on all operands. Hence photonic arithmetic units are usually classified as single-instruction multiple-data (SIMD) parallel processing systems. Compared to analogue photonic processing, which uses the intensity and/or locations of bright points as information to be processed in a two-dimensional plane, resulting in quantization errors and limitations in space-bandwidth product, the overall dynamic range is improved if the input information is represented as a sequence of digits, which can preferably be represented by the presence or absence of light or some of its distinguishable properties. Simple binary representation of data increases the dynamic range, yet binary arithmetic operations result in an undesirable serial carry or ripple carry propagation which has to be efficiently accommodated or eliminated for photonic computation, thus requiring suitable number representation.

In digital photonic arithmetic operations, data are presented either in two-dimensional or three-dimensional form. The need for carry or borrow propagation for addition and subtraction, respectively, results in significant time delay and does not fully utilise the parallelism offered by optics technology. To reduce the delay produced by serial carry propagation, the carry is to be either restricted to a few processing steps, or the arithmetic operations have to be carry-less.

Therefore, two approaches are attempted for efficient evolution of a photonic digital arithmetic processor. One approach is related to the development of a suitable number system where carry-free parallel algorithms for arithmetic operations can be performed. The other approach is to use architectural novelty, again for carry-free parallel operations. An efficient photonic arithmetic processor is expected to accommodate a balanced combination of both these approaches, since neither can individually yield optimally configured techniques, mainly because of limitations in present-day device technology.

Machine arithmetic differs from so-called real arithmetic primarily in precision. Because of fixed word length, arithmetic processors can produce results of finite precision, whereas real arithmetic can produce results to any degree of precision. Therefore, rounding-off of results is required in all computing machines. The existing schemes for number representation for digital arithmetic operations in computing are all positional number systems [20], [21], where a number is encoded as a vector of a few bits (or digits). Each bit is weighted according to its position in the vector. Associated with each number system is a radix (or base) r. Basically, digital number systems used for computation can be divided into two types, as binary and non-binary. For representing negative numbers in the binary category ($r = 2$), number systems with positive radices such as 2's-complements and redundant binary are used. To represent both positive and negative numbers, a system with a negative radix is utilised.

The emphasis on parallel processing capabilities has opened up renewed interest in suitable digital arithmetic (signed digit number). A major problem of binary computation lies with the limitation in the speed of operations because of long carry/borrow propagations. Therefore, alternative proposals for number systems termed non-binary number systems have been made. Another earlier approach was to use the residue number arithmetic system, which originated in China for parallel arithmetic operations [22].

Important number systems used for arithmetic operation in photonic computations include:

1. Residue number system

2. Binary number system

3. Redundant binary number system

4. Generalized signed-digit number system

    a. Modified signed-digit number system
    b. Mixed modified signed-digit number system
    c. Trinary signed-digit number system
    d. Quaternary signed-digit number system

5. Negabinary number system

It may be noted that except the residue number system, all number representations can be generalized as positional number representation.

9.7.1 Residue number system

A residue number system is a prominent candidate for number representation in non-binary form. According to the definition, residue numbers are inherently integer systems. In a residue number system, a set of relatively prime numbers called moduli is chosen to represent integers. The residue representation of an arbitrary number x is given by $(R_n, R_{n-1},...R_0)$ corresponding to the set of moduli $m_n, m_{n-1},...m_0$, where R_i is the residue of a modulo m_i. Each residue R_i is defined as the smallest possible integer remainder of (x/m_i), so that

$$x_i = x - m_i[x/m_i] \tag{9.2}$$

The system unambiguously represents the integers over the range M given by

$$0 \leq M \leq (\prod_{i=0}^{n} m_i) - 1 \tag{9.3}$$

The product of the moduli considered limits the largest decimal number that can be represented in a particular domain. Any arithmetic operation between two decimal numbers is possible as long as the result lies within the number domain. There has been an interesting proposition of representing a decimal digit at any position of a number with only two moduli 2 and 5 [23]. Since the product of 2 and 5 is 10, decimal numbers from 0 to 9 can be represented uniquely in this domain. Following this suggestion, a positional residue system was developed where both the residues and quotients of each digit of the decimal number with respect to 2 and 5 are recorded [24].

The residue number system and its modified varieties when used for photonic computing can confine carry propagation to fewer steps [25], [26]. The residues and quotients of each number are added digit-wise and then reconverted to decimals to obtain the carry-less result of addition through a four-step process. Truth-table conversions [27], [28] are widely used in photonic implementation of computer algorithms based on residue numbers. These techniques enable carry-free operation, yet suffer from the fact that the size of the truth-table increases greatly as the numerical range involved becomes large. Large primo-modulo elements must then be used for large operands. Operations on both residues and quotients can be implemented in parallel in photonic architecture; however, these systems are less advantageous when multiple number crunching is necessary. Another major drawback lies with the conversion of decimal to residue and vice versa. Moreover, the systems based on residue arithmetic are not suitable for single-instruction multiple-data operations.

9.7.2   Binary number system

A binary number system is a special case of a generalized positional number system where radix $r = 2$, $x_i \in 0, 1$. Therefore, any positive decimal number x is represented in the binary form as

$$x = \sum_{i=0}^{N-1} x_i 2^i \tag{9.4}$$

To represent a bipolar (both positive and negative) number in the binary system, sometimes an extra bit termed the sign-bit is added to the binary form. Another alternative is to convert the number into either 1's complement or to 2's complement form. In the sign-bit representation, the most significant bit indicates the sign of the integer number (0 for positive numbers and 1 for negative numbers). The magnitudes of numbers of opposite polarity are the same. For an n-bit word, a positive number has a range of 0 to $(2^{n-1} - 1)$ and a negative number has a range of 0 to $-(2^{n-1} - 1)$. The addition of sign-bit number requires higher storage capacity. Another problem of sign-bit representation is related to the determination of the sign of the result when two numbers of opposite sign are added.

Negative numbers are also represented by their 1's complement or 2's complement, the latter being mostly preferred. In 1's complement representation, the number in binary form (excluding the sign bit), is complemented bit-wise, i.e., 1's are changed to 0's and vice versa. This technique sometimes produces two possible undesirable repetitions of 0. In 2's complement notation, the radix $r = 2, x_i \in 0, 1$ and $\lambda_i = (-1, 1, ...1)$. Therefore, addition of 1 to the 1's complement number can represent a negative number. A negative number has to undergo a two-step process for representation in the 2's complement notation. The 2's complement notation is given by

$$x = -x_{N-1} 2^{N-1} + \sum_{i=0}^{N-2} x_i 2^i \tag{9.5}$$

$x_i \in 0, 1$, x_{N-1} where x_{N-1} is the sign bit.

Photonic implementation using architectures operating on the binary number have been attempted [29], [30]. However, their usefulness is mainly limited to carry propagation. Several approaches for efficient carry propagations have been proposed, which include carry-look-ahead techniques [31], [32]. Higher-order modular approach using symbolic substitution was attempted, where several bit pairs are operated in parallel [33].

9.7.3   Signed-digit (SD) numbers

Signed-digit number systems [34] permit parallel arithmetic operations where the redundancy of the basic representation is utilised, and therefore signed-digit number systems are extensively used in photonic computation.

The carry propagation can be limited to a few adjacent digit positions and thus some forms of carry-free arithmetic operations are also possible using signed-digit representations. Several varieties of signed-digit number representation used in photonic computation include the modified signed-digit (MSD), mixed modified signed-digit (MMSD), trinary signed-digit (TSD), and quaternary signed-digit (QSD). A generalized signed-digit (GSD) representation is proposed [35], from which all the above varieties of signed-digits can be derived.

In generalized signed-digit representation, the radix r can either be positive or negative. A decimal number x is represented in terms of an N-digit GSD representation as

$$x = \sum_{i=-M}^{N-1} x_i r^i \tag{9.6}$$

$x_i \in (-a, ., -1, 0, 1.a)$, where a is a positive integer and $r \geq 2$ is the radix of the SD, the digit x_i takes up values from the set $(-a, ..., -1, 0, 1, ...a)$.

No unique representation of a number is possible in signed-digit representation, and the redundancy depends on the selection of a. Minimum and maximum values of a can also be given. Since the digit set is quite large, many substitution rules are required for arithmetic operations. Hence, signed-digit arithmetic operations are often implemented through photonic symbolic substitution [36].

During implementation of photonic arithmetic operations, the higher radix and multi-input signed-digit [37], [38] have been able to confine carry to within one or two positions. The carry-propagation chains are eliminated by the use of redundant representation of the operands. Moreover, these higher-radix systems allow higher information storage density, less complexity, fewer system components, fewer gates, and fewer operations.

9.7.3.1 Modified signed-digit (MSD) representation

Any decimal number x, whether positive or negative, is represented by an N-digit modified signed-digit (MSD) number [39], [40] with a radix $r = 2, a = 1$ as

$$x = \sum_{i=0}^{N-1} x_i 2^i \tag{9.7}$$

$x_i \in \bar{1}, 0, 1$ where the MSD variable x_i is a member of the digit set $\bar{1}, 0, 1$.

It is to be noted that no unique representation is possible in this MSD system. For example, $(5)_{dec} = (1\bar{1}01)_{MSD} = (101)_{MSD} = (10\bar{1}\bar{1})_{MSD} = 1\bar{1}0\bar{1}\bar{1}_{MSD}$. Therefore, the MSD number representation can be considered as a redundant number (RN), since there is more than one way of representing a given number. This redundancy in representation of MSD is often utilised in performing arithmetic operations [41]. Evidently, the MSD system is a special case of the GSD system, where the base $r = 2$ and $a = 1$. An MSD negative number is designated as the MSD complement of the MSD positive number.

In a modified signed-digit number system, the digit set used is $\bar{1}, 0, 1$. Photonic symbolic substitution is often used in dealing with the digit-set for arithmetic operations [42], [43]. Since the sum of digits 1 and $\bar{1}$ results in the digit 0, MSD addition possesses the property of symmetric cancellation. MSD addition and subtraction generates primary transfer and weight digits in parallel, similar to the carry and sum, respectively, in binary addition. These transfer and weight digits are used to generate a secondary set of transfer and weight digits, which in turn generates bits that are summed up to produce the final summation result. The carry generated during the addition operation is propagated to two steps and the addition is completed in three stages within fixed time. The fundamental bottleneck of computation speed is thus eliminated. Because of these advantages, MSD number system has been very widely used in photonic computing using several architectures and techniques [44], [45]. Even techniques of arithmetic operations based on binary logic operations have been attempted using the MSD number system [46].

9.7.3.2 Mixed modified signed-digit (MSD) representation

Another representation of this MSD is mixed modified signed-digit (MMSD) representation. In this form, a decimal number x when represented in the N-digit binary MMSD scheme [47] takes the form

$$x = 2^N + \sum_{i=1}^{N} x_i 2^{i-1} \tag{9.8}$$

where the radix $r = 2$ and the coefficient x_i is a member of the digit set $\bar{1}, 0$.

However, the most significant bit (MSB) of the MMSD number is always 1. Therefore, the MMSD number contains digits $\bar{1}$, 0, and 1, and hence $a = 1$. The number gets increased by 2. The conversion of a decimal number to the MMSD form involves a three-step process: decimal to binary, binary to trinary, and finally trinary to MMSD. For addition using the MMSD system [48] in the photonic domain, one of the operands is converted to binary and the other to the MMSD system. Since the MMSD number contains a 1 as the most significant bit (MSB) with all other bits being either 0 or $\bar{1}$, the carry is completely eliminated in a single step. The main drawback in this technique lies in its conversion procedure, since conversion of a decimal number to the MMSD representation is quite complex and involves four steps.

9.7.3.3 Trinary signed-digit (TSD) representation

Any decimal number x is represented by an N-digit TSD number [49] as

$$x = \sum_{i=0}^{N-1} x_i 3^i \tag{9.9}$$

In the TSD system, the radix $r = 3$, $a = 2$ and the digit set used is $\bar{2}, \bar{1}, 0, 1, 2$. Since a number in this system can have more than one representation, the

TSD system is designated as a redundant number system [50]. For example, $(11)_{dec} = (2\bar{2}\bar{1})_{TSD} = (11\bar{1})_{TSD} = (102)_{TSD}$. The degree of redundancy usually increases with the increase of the radix. While applying the TSD representation for the addition of two single digits [51], [52], it is found that the digit combinations $(1, 2)$ or $(2, 2)$ or $(\bar{1}, \bar{2})$ or $(\bar{2}, \bar{2})$ cause a carry propagation to the next higher-order bit. In this two-step process, the generation of the carry is therefore avoided by mapping the two digits in question into an intermediate sum and intermediate carry such that the aforementioned combinations can never occur. Full adders are usually used for practically implementing TSD-based addition operation. The photonic symbolic substitution technique is often used in dealing with the increased digit set [53].

9.7.3.4 Quaternary signed-digit (QSD) representation

A decimal number x is represented by an N-digit QSD number [54] as

$$x = \sum_{i=0}^{N-1} x_i 4^i \qquad (9.10)$$

$x_i \in \bar{3}, \bar{2}, \bar{1}, 0, 1, 2, 3$.

As is evident, the radix $r = 4$, $a = 3$ and the digit set used is $\{\bar{3}, \bar{2}, \bar{1}, 0, 1, 2, 3\}$. The QSD system is also a redundant number system as seen from the representation of decimal eleven as $(11)_{dec}=(23)_{QSD}=(3\bar{1})_{QSD}=(1\bar{1}\bar{1})_{QSD}=(13\bar{1}\bar{1})_{QSD}$. The digit combinations $(1, 2)$, $(2, 2)$, $(1, 3)$, $(2, 3)$, $(3, 3)$, $(\bar{1}, \bar{2})$, $(\bar{2}, \bar{2})$, $(\bar{1}, \bar{3})$, $(\bar{2}, \bar{3})$, $(\bar{3}, \bar{3})$ cause a carry propagation to the higher-order bit, and hence these combinations are avoided during addition operation using the quarternary signed digit (QSD) representation. This is achieved by a method of symbolic substitution, which requires a number of substitution rules [55]. QSD arithmetic has been widely used for operation of optical content addressable memory, particularly in shared mode [56].

9.7.4 Negabinary number system and its variants

Signed-digit number systems are specially designed to represent bipolar numbers at the cost of redundancy. The use of a higher radix number only increases the digit set and does not contribute to the reduction of redundancy. Moreover, the increased digit set leads to the use of a large number of substitution rules when arithmetic operations are attempted. These difficulties have prompted the use of a number system that can provide unique representation and can also easily represent both positive and negative numbers as well as fractions without the use of substitutions rules. The number system that has proved to be a solution to the problems is designated as the negabinary number system.

A negabinary number system [57], [58] is defined as a number system having a radix (-2). The negabinary number system is a unique non-redundant way of representing both positive and negative numbers that can be expressed without using a sign bit or 2's complement. The negabinary number system has two forms, the unsigned and mixed. For the purpose of implementing carry-less arithmetic operations, a number system termed the signed negabinary number is utilised [59]. The negabinary number system representation can now be either (i) unsigned, (ii) signed, or (iii) mixed. The three forms of the number system have been classified on the basis of the digits used. For example, the unsigned number system uses the digits (0, 1), the signed uses the digits $(0, \bar{1})$, and the mixed form uses the digits $(0, \bar{1}, 1)$. The mixed form of negabinary number is sometimes called negabinary signed digit (NSD) representation, where radix r = -2. Again, considering $r = (-2)$ and $a = 1$ makes the mixed negabinary number system a special case of generalized signed digit representation.

9.7.4.1 Unsigned negabinary number system

In an unsigned negabinary number system, a decimal number x has unique representation $x_{N-1}......x_1 \ x_0 \ x_{-1} \ x_{-2}.....x_{-M}$ taking the form

$$x = \sum_{i=-M}^{N-1} x_i(-2)^i \qquad (9.11)$$

$x_i \in 0, 1$ where M and N are positive integers. Positive values of i result in integers, whereas the negative values result in fractions. As in a binary system, this form uses the digits 0, 1. The largest positive integer x is one having all its even position bits 1 and odd position bits 0. The smallest negative value of integer x is one having even position bits 0 and odd position bits 1. The range of number representation for even N is $[\frac{2}{3}(1 - 2^N), \frac{1}{3}(2^N - 1)]$, and for odd N is $[\frac{2}{3}(1 - 2^{N-1}), \frac{1}{3}2^{N+1} - 1)]$. Hence, for an 8-bit number, this scheme can represent numbers from (-170) to 85 and for 16-bit representation the range lies between (-43690) and 21845.

For representing decimal fractions for even M, the range is given by $[2^{-M}, \frac{1}{3}(1 - 2^{-M})]$ for positive fractions and $[-(2^{-M}), \frac{2}{3}(2^{-M} - 1)]$ for negative fractions. For odd values of M, the range is $[-(2^{-M}), \frac{1}{3}(1 - 2^{-(M-1)})]$ for positive fractions and $[2^{-M}, \frac{2}{3}(2^{-(M+1)} - 1)]$ for negative fractions. Therefore, 8-bit positive fractions ranging from (0.00390625) to (0.33203125) and negative fractions ranging from (-0.00390625) to (-0.6640625) can be represented by negabinary numbers.

Unsigned negabinary arithmetic is a two-step process of addition and subtraction in which the augend and the addend can have four combinations, namely (0, 0), (0, 1), (1, 0) and (1, 1). The result of addition of the operands generates an intermediate carry and an intermediate sum. The carry produced at one bit position is then subtracted from the sum of the next higher bit to yield a carry-less result in the second step and therefore the carry is termed a negative carry. This technique requires no substitution rules for its operation.

But restrictions are still imposed on the speed of operation due to additional steps involved, as well as on the coding techniques when implemented in photonic architectures.

9.7.4.2 Mixed negabinary number system

In the mixed negabinary number system, a decimal number x has a representation $x_{S-1}...x_1x_0$ taking the form

$$x = \sum_{i=-R}^{S-1} x_i(-2)^i \qquad (9.12)$$

$x_i \in 0, \bar{1}, 1$ where R and S are positive integers. Also there exists no unique representation of decimal fractions in this scheme.

The largest positive x has 1 in all its even position bits and $\bar{1}$ in all its odd position bits. The smallest negative value of x is one having $\bar{1}$ in even position bits and 1 in odd position bits. Therefore, the range of integers that this number system can represent is $[1-2^S, 2^S-1]$ for even S and $[1+2^S, -(1+2^S)]$ for odd S. For 8-bit form, the range is from (-255) to 255 and for 16 bit, the range is (-65535) to (65536).

In the case of fractions, the ranges that can be represented for positive and negative values are $(2^{-R}, 1 - 2^{-R})$ and $(-2^{-R}, 2^{-R} - 1)$, respectively, when R is even. For odd values of R, the ranges for positive and negative values are $[-2^{-R}, (1+2^{-R})]$ and $[2^{-R}, -(1+2^{-R})]$, respectively. For 8-bit fractions, the positive range becomes (0.00390625 to 0.99609375), while the negative range becomes (-0. 00390625 to -0.99609375).

In the mixed negabinary form, addition is carried out in each bit position in parallel to yield an intermediate sum and an intermediate carry. The $(1, 1)$ and $(\bar{1}, \bar{1})$ digit combinations generate a carry of $\bar{1}$ and 1, respectively, while $(1, \bar{1})$ and $(\bar{1}, 1)$ generate a carry of 0. The rest of the combinations generate redundant carries. By exploiting this redundancy, addition is performed in two steps. Rules are determined for the intermediate sum and carry in the first step so that the carry-less result is obtained in the second step.

9.7.4.3 Signed negabinary number system

In a signed negabinary number system, a decimal number x has unique representation $x_{Q-1}.....x_1x_0$, taking the form

$$x = \sum_{i=-P}^{Q-1} x_i(-2)^i \qquad (9.13)$$

$x_i \in 0, \bar{1}$ where P and Q are positive integers.

The largest positive integer x is one having all its odd position bits $\bar{1}$ and even position bits 0. The smallest negative integer value of x is one having odd position bits 0 and even position bits. The range of numbers that can be represented for even and odd Q are therefore $[\frac{1}{3}(1-2^Q), \frac{2}{3}(2^Q-1)]$ and $[\frac{1}{3}(1-2^{Q+1}), \frac{2}{3}(2^{Q-1}-1)]$, respectively. Hence, for an 8-bit number, this scheme can represent numbers from (-85) to 170. For a 16-bit number the range is between (-21845) and 43690.

For representing decimal fractions for even P, the range is given by $[2^{-P}, \frac{2}{3}(1-2^{-P})]$ for positive fractions and $[-(2^{-P}), \frac{1}{3}(2^{-P}-1)]$ for negative fractions. For odd values of P, the range is $[-(2^{-P}), \frac{2}{3}(1-2^{-(P+1)})]$ for positive fractions and $[2^{-P}, \frac{1}{3}(2^{-(P-1)}-1)]$ for negative fractions. Therefore 8-bit positive fraction numbers range from (0.00390625) to (0.6640625) and negative fraction numbers range from (-0.00390625) to (-0.33203125).

For converting a decimal number to its signed negabinary form, the number (or the quotient of division after the first iteration) is recursively divided by (-2). For converting a decimal fraction to its signed negabinary form, the decimal fraction is recursively multiplied by (-2). For example, $-(0.15625) \equiv 0.0\bar{1}\bar{1}\bar{1}\bar{1}$ and $(0.15625) \equiv 00\bar{1}0\bar{1}$.

9.7.4.4   Floating point negabinary notation

Floating-point notation is a very useful tool for representing large numbers in binary form [21]. A floating-point representation usually takes the form $\pm M$ $r^{\pm E}$, where M is the mantissa or significand, r denotes the radix or base, and E is the exponent. It has been observed that numbers expressed in the floating-point format are redundant numbers, i.e., they can be expressed in many ways. For example, the various floating point unsigned negabinary representations of decimal number 8 are $0.110\,(-2)^5 \equiv 0.0110\,(-2)^6 \equiv 110\,(-2)^2$. This is because $(0.110(-2)^5)_{unsignega} = (1 \times (-2)^{-1}+1(-2)^{-2}+ 0(-2)^{-3})(-2)^5 = 8_{decimal}$. Similarly, the other redundant representations also give the same value of 8. Since it can represent both positive as well as negative numbers, the floating-point notation of negabinary number systems increases the range of numbers that can actually be used in arithmetic operations.

9.8   Photonic computing architectures

The concept of photonic computing may not become a reality unless the technology can provide convincing evidence that basic arithmetic, logic operations, and other computations like matrix-vector multiplication can be handled efficiently in the photonic domain. Since addition is the most fundamental operation in any arithmetic computation, much interest is shown in implementing addition operation by photonic techniques. Other operations, such as subtraction, multiplication, and division, which can all be realised through addition, also have been tried.

It is expected that a photonic arithmetic system will be capable of performing high-speed computations utilizing the high parallelism and massive interconnection of optics, and will also take advantage of interfacing with the electronic devices. Therefore, two approaches have become popular: exploiting carry-free parallel algorithms and developing efficient photonic architectures. An efficient photonic arithmetic processor is expected to accommodate a balanced combination of both approaches, since neither can individually yield the desired result. Moreover, a photonic computing technique is required to perform basic logic operations. There is also a possibility of implementing arithmetic operations by using logic operations.

9.8.1 Optical shadow-casting architecture

The lens-less optical shadow-casting (OSC) technique completely exploits the advantages of optics in performing combinatorial logic operations [17]. This optical parallel processor that performs OAL processing is termed as the optical logic array processor (OLAP). It operates not only as a parallel logic gate, but also as a parallel-processing unit that is capable of dynamically controlling the operations. Since the operations in this scheme are linear, the computation speed is limited only by the time taken by light to travel from the input to the output plane.

The basic approach of OSC was later modified for programmed image processing applications. The encoding of two input 2D image patterns and processing of output image pattern are, however, performed by electronic techniques, which include delay and memory elements. Further, four kinds of wavelengths are used in the system, so that four independent product term operations can be executed simultaneously. In this way, 16384 product term operations of sixteen variables are achieved for two inputs consisting of (64 × 64) pixels. For correlation, a multiple imaging system consisting of a lens, an array of micro-prisms, and a reflective correlator with segmented mirror are used.

Most of the work carried out in the area of OSC techniques is related to the development of coding techniques, and many technologies are applied to code and decode the data in the form of a 2D image pattern. In the original coding methodology of two input image patterns proposed by Tanida [17], each input image pattern is divided into ($N \times N$) square areas or cells. Each cell in the coded pattern is divided into four sub-cells where only one sub-cell position can be bright, depending on the logical combinations of two inputs. Coding of the pattern is computer controlled and is generally intensity based. Instead of two inputs, multiple inputs can be accommodated by proper coding of inputs.

polarisation encoding using the birefringence property of crystals has been an attractive area of research and many research reports are available. Some of these are based on symbolic substitution, whereas others are based on the shadow-casting scheme. The OSC system has been extended to incorporate polarised codes for encoding the input pixels and output mask transparency. This has been widely used for designing adders, logic circuits, and also for multiprocessor design. Design minimisation is especially required for polarisation-encoded OSC-based circuits, since the size of the input overlap pattern, LED source pattern, and the output overlap pattern is directly dependent on the number of minterms and their proper spatial allocation. Some modified schemes have tried to increase the degrees of freedom for the purpose of coding, thereby making it possible to design complicated logic circuits. However, practical difficulties associated with polarisation-based system in terms of accurate detection of polarisation states were later removed.

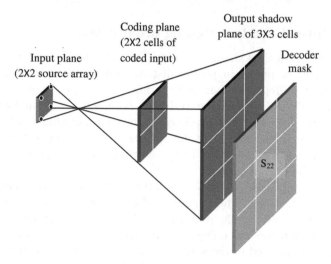

FIGURE 9.9: (a) Single cell operation of optical shadow-casting architecture and (b) coding of input A and B

Figure 9.9 is a schematic diagram of two input lens-less shadow casting system for one-cell operation. The operation kernel is applied at the source plane by the on/off conditions of the light source array consisting of elements l. For logic and arithmetic operation, a (2×2) source array was used and therefore $i = 1, 2$ and $j = 1, 2$. The input data are coded into two-dimensional images and two input images are superimposed to form a coded input image. The coding technique is used where each cell of the coded image consisting of four elements c_{ij} (where $i = 1, 2$ and $j = 1, 2$) can be expressed by a (2×2) positional matrix. The coded image at the input plane is correlated with an operation kernel that specifies the desired operation. When all four sources are at on state, on correlation of the (2×2) input matrix with the operational kernel, a (3×3) superposed pattern is formed, where each element is given by S_{ij} ($i = 1, 2, 3$ and $j = 1, 2, 3$). The element S_{22} of the positional matrix contains the information of all possible correlations between the coded sub-cell and corresponding on/off states of light sources, i.e., only the sub-cell S_{22} contains the conclusive result of input cells for a particular processing dictated by a particular operational kernel. Intensity at the sub-cell is recorded by a decoding mask (transparent at S_{22} position) and is given by

$$S_{22} = c_{11} \wedge l_{11} + c_{12} \wedge l_{12} + c_{21} \wedge l_{21} + c_{22} \wedge l_{22} \qquad (9.14)$$

where $'\wedge'$ is the symbol for 'operated on' and the symbol '+' denotes logical OR operation.

The situation is generalized for every other cell over the entire image pattern. Evidently, c_{ij} is the (i,j)-th cell of the coded pattern and l_{ij} is the (i,j)-th element of the operation kernel. The processed output is sampled and decoded by a decoder mask from all cells of the correlated pattern. Multi-cell operation two inputs A, and B forms a coded input pattern according to a specific coding technique. The coded input is correlated with the operational kernel in the form of on/off states of light sources. The correlated input is sampled to yield the processed image according to the operational kernel. The operation is shown in Figure 9.10.

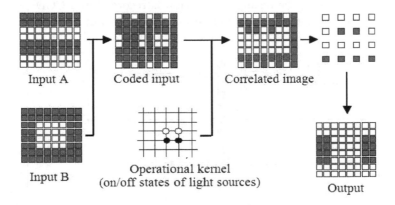

FIGURE 9.10: Multicell operation of optical shadow casting technique

9.8.2 Optical shadow-casting-based unified arithmetic and logic-processing architecture

A major requirement of photonic computing is carry-free arithmetic operations. Different ways of advancing the carry signals have been proposed for arithmetic operations, which include carry-look-ahead and carry-save addition approaches. However, the sequential nature of binary addition cannot be changed. Carry-limited or carry-free arithmetic operations using other number systems have long been investigated. Photonic processors developed for parallel addition using control operators in the form of a two-dimensional matrix require large memory and hence their widespread approval is limited, since code conversions are required to make such systems compatible with the conventional decimal number systems. In all techniques of carry-less arithmetic operations, complex multiple substitution rules involving multiple-step algorithms are required. Hence, these techniques are limited by higher processing time.

A photonic architecture which can perform arithmetic and logic operations in parallel and in parallel channels is shown in Figure 9.11(a). Based on the signed negabinary number system, the techniques of carry-less addition and subtraction of positive and/or negative numbers are carried out and achieved in a single step. The architecture is capable of simultaneously implementing logical operations. The logic operations are also carried out on a negabinary system. The sequential flip-flop operations can also be achieved using the same architecture, the only difference from the combinatorial logic operations being that it does not use the negabinary system for coding the inputs. The other advantage of the unified photonic architecture lies in the use of a single liquid crystal display (LCD) panel for representing spatially encoded data, operational kernels, and decoding mask in the form of two-dimensional images. The image representing the result of operation is recorded in a CCD camera and processed in a computer for presentation in decimal form.

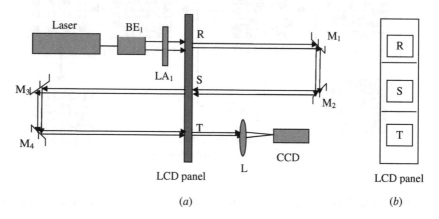

FIGURE 9.11: Schematic diagram of (a) architecture for implementing arithmetic and logical operations and (b) partition of SLM into three areas R, S, and T distances

The scheme for addition and subtraction basically follows the shadow-casting principle with some modifications in the spatial coding technique as well as in the structure of the decoding mask. For controlling the operational kernels and for presenting the input and decoding mask in the processing light path, at least three spatial light modulators are required in the conventional shadow-casting system. In the architecture, the use of a single liquid crystal display (LCD) panel and folding of the optical path has considerably reduced the complexity of the architecture. This is possible by dividing the active area of the LCD panel into three square areas designated as R, S, and T as shown in Figure 9.11(b). The R portion is used for introducing the operational kernel, the S portion for presenting coded input image, and the T portion is used for introducing the decoder mask.

In the architecture, light from a laser source is expanded by a beam-expander BE_1. The expanded parallel beam is incident on a (2×2) lenslet array LA_1. Each lens of the array acts as point source. The lenslet array in combination with the addressable opaque and transparent cells at the R region of the LCD panel introduces the operational kernels of the OSC system. The spatially coded cell pattern of each bit of the two operands is superimposed and the resulting superimposed pattern is recorded in the input plane at the location S of the LCD panel. A light beam from the effective source array, after being folded by mirrors M_1 and M_2, is incident on the input plane at S of the LCD panel. A shadow of the superimposed pattern is formed at the location T of the LCD panel by folding the beam using M_3 and M_4, where the decoding mask for all bits has been pre-recorded. The folding of light beam through mirrors helps in incorporating a single LCD panel instead of three spatial light modulators for the control of the operational kernels and for presenting the input and decoding mask. Light emerging from the decoding mask at T is condensed by the condenser lens L and then is incident on a CCD camera, connected to a computer via a frame grabber card. The image recorded by the camera is scanned and decoded to obtain the results of a particular operation. One of the advantages of the architecture is that it operates in single-instruction multiple-output parallel-processing mode. Several pairs of numbers can be added or subtracted in parallel in different channels. Further, two or more logical operations can be performed simultaneously in different channels.

The relations among the source separation l, input cell size d_1, the shadow cell size d is given by

$$d = \frac{ld_1}{l - d_1}z + z_1 = \frac{lz}{l - d_1}, d_1 < l \tag{9.15}$$

where the distance between input plane (S location of LCD panel) and source (or lenslet array, i.e., R location of LCD, effectively) plane is z (= a + b +c) of Figure 9.11. The distance between the shadow plane (T location of the LCD panel) and input plane (S of LCD panel) is $z_1 = z_2 + z_3 + z_4$, for a half-cell shift of the input in the shadow-casted plane at T.

9.8.2.1 Negabinary carry-less arithmetic processor

For arithmetic operations, the coding of the two inputs and their superimposed coded patterns follows a rule, i.e., one of the input is expressed in unsigned negabinary and the other in signed negabinary form. The decoding mask should be designed so as to interpret $\bar{1}$, 0, and 1 when a negabinary number system is used.

The source array mask and decoding mask are shown in Figure 9.12, where the cell structure generated effectively modulates the source array to perform signed Ex-OR operation on the operands. Further the (4 × 4) cell decoding mask for each bit of the projected superimposed input is also shown.

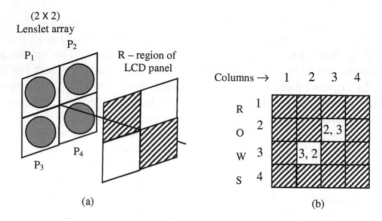

(a)

(b)

FIGURE 9.12: (a) Cell structure generated effectively modulates the source array to perform signed Ex-OR operation on the operands. (b) (4 × 4) cell decoding mask for each bit of the projected superimposed input.

The coding scheme for bit-wise superposition is shown in Figures 9.13 and the resultant input pattern is shown in 9.14. The presence of light in the cell (2, 3) of each bit of the decoding mask is interpreted as 1, and the presence of light in the cell (3, 2) of each bit of the decoding mask is interpreted as $\bar{1}$. The absence of light in any of these two cells, i.e., (2, 3) and (3, 2) of the decoding mask, is interpreted as 0. Each cell of the decoding mask has a size equal to the size of a single cell of the shadow formed.

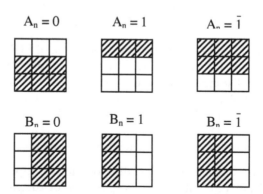

FIGURE 9.13: Bit-wise superimposition of the (3 × 3) patterns of the two input operands

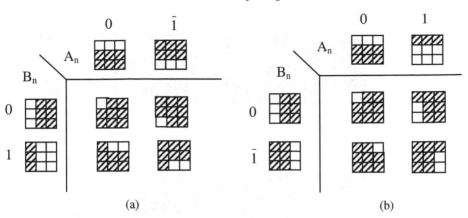

FIGURE 9.14: Resultant superimposed input combinations of each bit of A and B when (a) operands A and B are converted to signed and unsigned negabinary forms, respectively, and (b) operands A and B are converted to unsigned and signed negabinary forms, respectively

Several channels can add different pairs of numbers simultaneously by one single instruction. For addition of two numbers in C number of different channels by the proposed system, each number is assumed to be L bits in length, in negabinary notation. Therefore, the number of cells needed for each channel is $(3 \times 3L)$, since each bit of the numbers in negabinary notation is spatially coded as a (3×3) pattern. The required number of elements in the detector array is $(4 \times (3L + 1))$. For C number of channels, the required cell size for introducing the superimposed coded input is $(3C \times 3L)$ and the size of the detector array should be $((3C + 1) \times (3L + 1))$. Only one cycle is required to complete the whole addition operation and the process is independent of the number of bits in negabinary notation or the number of channels. Since it has been assumed that the bits pass through the system at f_c clock cycles of the interfaced computer, the processing time to compute the addition process in C number of channels is $1/f_c \ \mu s$. The number of channels that can be accommodated is limited only by the beam diameter and the camera aperture.

The results of addition of (170+85) and (100+(-50)) for the two channels are displayed in the result boxes which can be seen in Figure 9.15.

9.8.2.2   Negabinary-based logic processor

For logic operations, the input operands A or B can take on values of either 0 (= absence) or 1 (= presence) in the decimal number system. In the signed negabinary number system the decimal numbers 0 and 1 are represented as 00 and $\bar{1}\bar{1}$, respectively. Hence, the rules for logical operations in a signed negabinary number system is emulated as $0 + 0 = 0$; $0 + \bar{1} = \bar{1}$; $\bar{1} + \bar{1} = \bar{1}$;

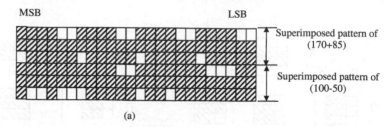

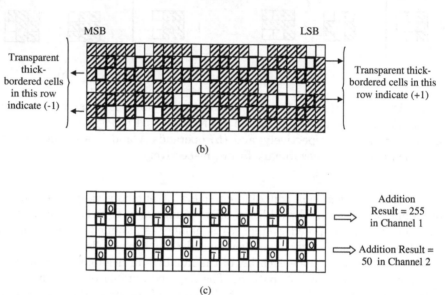

FIGURE 9.15: Resultant superimposed input combinations of each bit of A and B: (a) superimposed patterns of the addend and augend in two channels, (b) projected pattern of superimposed pattern for addition in the two channels (thick-bordered cells represent the cells to be scanned), and (c) scanned coefficients in the thick-bordered cells

(OR operation) $0 \bullet 0 = 0$; $0 \bullet \bar{1} = 0$; $\bar{1} \bullet \bar{1} = \bar{1}$; (AND operation) $0 \oplus 0 = 0$; $0 \oplus \bar{1} = \bar{1}$; $\bar{1} \oplus \bar{1} = 0$. (Ex-OR operation).

Photonic implementation of different logical operations requires proper selection of operational kernels. In the shadow-casting architecture, the operational kernels are in the form of a (2×2) array of sources in combination with the cell structure on an LCD panel. Any desired logical operation is implemented by controlling the transmission properties of the LCD panel, which in effect virtually controls the on/off states of the light sources. It has been observed that the number of 1s obtained at the output for the four input combinations $(0,0)$, $(0,1)$, $(1,0)$, and $(1,1)$ determines the number of effective sources that are to be switched on. Therefore, only one cell of the (2×2) cell structure designed on the LCD is kept transparent for implementing the

AND operation, whereas three cells are to be kept transparent for the logical OR operation. The design of cell structure for a few logic operations is shown in Figure 9.16.

Logical Function	Cell structure on LCD
A	
B	
$\overline{A}$	
$\overline{B}$	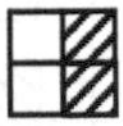
A + B	
A • B	
A ⊕ B	

FIGURE 9.16: Design of cell-structure in front of the lenslet array for some logic operations

The output when multiplied by corresponding negabinary radix gives the result in decimal number for desired operation.

9.8.2.3 Sequential photonic flip-flop processor

As described in [60], the same shadow-casting architecture has been utilised to achieve flip-flop sequential operation. In this operation, the previous state of the flip-flop is considered to be the third input along with the two inputs to the system, and the three inputs thus considered are spatially coded and superimposed as shown in Figure 9.17. The source, which is a (3 ×3) pattern with the central source always off, is modulated by the SLM as shown in Figure 9.18. The modulation of the source and decoding masks for different types of flip-flops are shown in Figure 9.19.

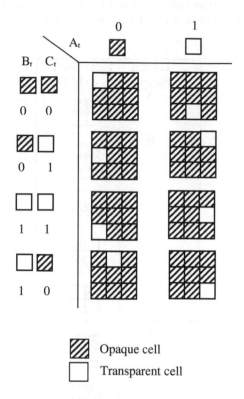

FIGURE 9.17: Coding of the three inputs to flip-flop

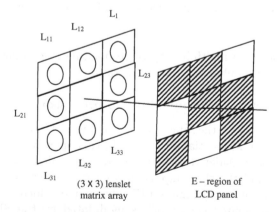

FIGURE 9.18: (3×3) source modulated by SLM

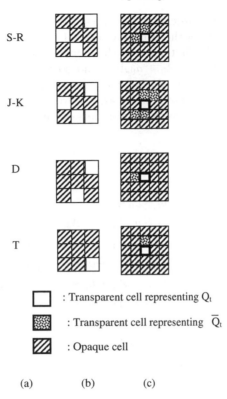

S-R

J-K

D

T

☐ : Transparent cell representing Q_t

▨ : Transparent cell representing $\overline{Q}_t$

▨ : Opaque cell

(a)   (b)   (c)

FIGURE 9.19: Source pattern and decoding masks for different flip-flops: (a) type of flip-flop, (b) source pattern, and (c) decoding mask

9.8.3 Convolution-based multiplier architecture

The product h of two L digit numbers a and A, expressed in base S, is represented by,

$$h = aA = \sum_{i=0}^{L-1} \sum_{j=0}^{L-1} a_i A_j S^{(i+j)} \tag{9.16}$$

where $a = \sum_{i=0}^{L-1} a_i S^i$ and $A = \sum_{j=0}^{L-1} A_j S^j$

It has been demonstrated that the multiplication of two numbers in binary form basically involves a convolution operation, and this process of multiplication by convolving digital data is termed the digital multiplication by analogue convolution (DMAC) technique. To avoid the occurrence of a carry, the convolution coefficients obtained by digital multiplication are expressed in mixed binary format. These coefficients are added up in decimal notation after multiplication with the respective base to yield the result of addition. To illustrate the DMAC algorithm, let us multiply two 8-bit binary numbers a and A represented as $a = a_n...a_1 a_0$ and $A = A_n...A_1 A_0$. The digits of the number a are written in the reverse order and slid past the digits of A. As shown below, the overlapping digits of the two numbers get multiplied and give the convolution coefficients. The first value of the convolution coefficient thus obtained is a_0A_0, the second is $(a_1A_0 + a_0A_1)$, and so on, with the final coefficients being a_7A_7. Evidently, convolution coefficients will result from multiplication of two 8-bit numbers. Therefore, if a and A are two binary numbers, then the convolution coefficients h for 8-bit numbers are expressed as

$$
\begin{aligned}
h_0 &= a_0A_0 \\
h_1 &= a_1A_0 + a_0A_1 \\
h_2 &= a_2A_0 + a_1A_1 + a_0A_2 \\
h_3 &= a_3A_0 + a_2A_1 + a_1A_2 + a_0A_3 \\
h_4 &= a_4A_0 + a_3A_1 + a_2A_2 + a_1A_3 + a_0A_4 \\
h_5 &= a_5A_0 + a_4A_1 + a_3A_2 + a_2A_3 + a_1A_4 + a_0A_5 \\
h_6 &= a_6A_0 + a_5A_1 + a_4A_2 + a_3A_3 + a_2A_4 + a_1A_5 + a_0A_0 \\
h_7 &= a_7A_0 + a_6A_1 + a_5A_2 + a_4A_3 + a_3A_4 + a_2A_5 + a_1A_6 + a_0A_7 \quad (9.17) \\
h_8 &= a_7A_1 + a_6A_2 + a_5A_3 + a_4A_4 + a_3A_5 + a_2A_6 + a_1A_7 \\
h_9 &= a_7A_2 + a_6A_3 + a_5A_4 + a_4A_5 + a_3A_6 + a_2A_7 \\
h_{10} &= a_7A_3 + a_6A_4 + a_5A_5 + a_4A_6 + a_3A_7 \\
h_{11} &= a_7A_4 + a_6A_5 + a_5A_6 + a_4A_7 \\
h_{12} &= a_7A_5 + a_6A_6 + a_5A_7 \\
h_{13} &= a_7A_6 + a_6A_7 \\
h_{14} &= a_7A_7
\end{aligned}
$$

The rules of binary multiplication with respect to $0 \times 0 = 0, 0 \times 1 = 0, 1 \times 0 = 0$, and $1 \times 1 = 1$ are applicable. Hence the convolution coefficients can only take on values as 0 and 1. The weighted convolution coefficient is therefore in decimal form, which is obtained by adding the convolution coefficients. The result of multiplication, however, is obtained by multiplying the weighted convolution coefficients with the corresponding binary base. In binary representation the factor takes the form of $(2)^n$. The final result of multiplication is in decimal form and is given as

$$
\begin{aligned}
h = {} & h_{14}2^{14} + h_{13}2^{13} + h_{12}2^{12} + h_{11}2^{11} + h_{10}2^{10} + h_9 2^9 + h_8 2^8 \\
& + h_7 2^7 + h_6 2^6 + h_5 2^5 + h_4 2^4 + h_3 2^3 + h_2 2^2 + h_1 2^1 + h_0 2^0
\end{aligned}
\quad (9.18)
$$

It may be noted that the addition in a binary system might produce carries. However, no carry is produced when the addition needed for multiplication is carried out by DMAC procedure. The procedure is a mixed representation (i.e., decimal weight factor h_n and binary radix 2^n), as indicated.

Photonic implementation of the DMAC algorithm can be performed either in the frequency domain or in the time domain. In the Fourier domain or frequency domain convolution, the Fourier transform of one of the function is stored on a hologram and the Fourier transform of the second function is made to incident on the hologram. Another lens collects the transmitted beams and routes them towards the detector. The product of the two functions thus obtained is re-transformed to yield the convolution operation. It has been demonstrated that the Fourier domain convolution worked.

In the time domain convolution technique, a time-reversed version of one function is held constant whereas the other function is slid past it. The product at each overlapping point gives the convolution coefficients. This technique has become particularly attractive owing to the availability of spatial light modulators and because of its compatibility with systolic processing. The time domain photonic convolvers can again operate either in the space-integrating or the time-integrating configuration. In the space integrating type, one of the operands is entered as a spatial variation $g(x)$ across an array of light sources, and is held constant. The other function $H(t)$ is presented across SLM or a CCD array which serves as a shift register, moving the function in discrete steps across $g(x)$. The light leaving the shift register is summed up spatially using a converging lens onto a single detector.

While the function $g(t)$ is fed as a time-varying signal on a single source, the other function $H(-t)$ is entered on an SLM, which serves as the shift register for the time-integrating configuration. The function $g(t)$ is uniformly spread over $H(-t)$ using a lens with $H(-t)$ shifted in discrete steps. The light leaving the register is summed up discretely in time onto discrete elements of a detector array.

9.8.3.1 Matrix-vector multiplication architecture

Matrix-vector (M-V) multiplication is an important linear algebraic operation where a matrix is multiplied with a vector. The operation is designated $y = Hg$, where g is the input vector composed of N number of elements, H is a two-dimensional matrix composed of $(M \times N)$ elements a_{ij}, and y is the output vector composed of M number of elements y_j. Regardless of the technique selected, M-V multiplication involves a series of multiplications followed by additions.

Many algorithms are available to perform multiplier operations in digital computers, however, the algorithms are guided by the electronic hardware that is used. Fixed-point multiplication, better known as paper-and-pencil method, is often implemented in computers by a binary version of the manual multiplication algorithm for decimal numbers based on repeated addition and shifting in registers. In fact, multiplication of k-bit integer with m-bit integer requires $(m \times k)$ full add and logical AND operations. Fixed-point multiplication therefore requires more hardware than addition, and as a result, is not included in the instruction set of yesterday's smaller computers.

Multiplication of a sign-magnitude number can be achieved by a straightforward extension of the unsigned paper-and-pencil method where the sign is computed separately. One can also use 2's complement technique, which forms the basis for many algorithms. Advances in VLSI technology have made it possible to built combinational circuits that can perform fairly large multiplications. For example, a multiplier chip can multiply two 16-bit numbers in 16 ns. They are composed of arrays of simple combinational elements, each of which implements an add-subtract-and-shift operation for small slices of the multiplication operands.

In the case of floating point multiplication, processor throughput is increased without large addition of hardware by the technique of pipeline processing. Multipliers often employ a technique called *carry-saved addition*, which is particularly well suited for pipeline processing. In implementing matrix-vector multiplication in photonic domain, it has been observed that one-to-one replacement of electronic techniques and algorithms does not always yield good results and in most cases fails to exploit the added advantages of optics technology. The major advantage of optics technology that can be exploited in the case of M-V multiplication is the representation of the data in two dimensions. Two-dimensional optical systems provide a large number of resolvable pixel points (or cells), each of which can be addressed and interconnected independently as an individual parallel communication channel. The large time bandwidth of the photonic sources provides added advantage of high-speed operations.

The basic approaches can be broadly divided into analogue and digital photonic techniques. Since the intensity of light is of interest in analogue technique, incoherent light can be used. In optical terms, the intensity levels or the transmission properties of the light modulator can represent the numbers to be operated. Therefore, summing two or many beams on a single detector can perform the addition. Many photonic matrix-vector multipliers are based on this principle.

In a typical generic photonic system needed for analogue M-V multiplication, coherent or incoherent light from the input source is modulated (switched on and off in the case of digital data) by input propagation structure to input the elements of the vector. The matrix elements are entered in a two-dimensional modulator, which controls the output propagation structure. The whole architecture is designed according to the technology and devices needed for input propagation and output propagation structures suitable for multiplication. Finally, the results of multiplication are summed up on an output detector element or a detector array. However, these intensity-dependent processors show a deteriorated performance with time, because the light intensity levels cannot be controlled and maintained throughout. Moreover, the limited dynamic range of analogue optical processing restricts the performance. Even though the light levels can be produced and detected in about five hundred discrete levels, the desired accuracy is somewhat restricted by the measurement of intensity by the detector. Therefore errors in the processing accumulate and result in lower numerical accuracy. These classes of analogue processors can take advantage of the high speed and massive parallelism of optics technology, yet their widespread applications are severely hindered by the achievable accuracy.

Several techniques have been proposed to increase the numerical accuracy by using digital representation of data instead of analogue representation. The so-called digital multiplication by analogue convolution (DMAC) techniques and their variants are useful methods for high accuracy multiplication. A modified version of DMAC technique is utilised for M-V multiplication. In this technique, on-off states of light sources are utilised as vector input to the system and a two-dimensional data plane is utilised instead of the detector array for displaying the result, which can be scanned for digital to decimal conversion. Matrix elements in digital form are entered on a liquid crystal panel (used as two-dimensional spatial light modulator) by controlling the transmission properties. The system gives the advantage of higher throughput at the expense of time required for electronic scanning. A (4 × 3) matrix and (3 × 1) vector multiplication of the form shown below is used to demonstrate negabinary M-V multiplication:

$$
\begin{pmatrix} a & b & c \\ d & e & f \\ g & h & i \\ j & k & l \end{pmatrix}
\begin{pmatrix} A \\ B \\ C \end{pmatrix}
=
\begin{pmatrix} aA + bB + cC \\ dA + eB + fC \\ gA + hB + iC \\ jA + kB + lC \end{pmatrix}
\tag{9.19}
$$

Here the elements of the matrix and the vector may be of either polarity. For simplicity, all elements are assumed to be 8-bit numbers. The first row of the matrix-vector product $(aA + bB + cC)$ can be expressed in terms of convolution coefficients h, h' and h'', given by

$$
\begin{aligned}
aA = h =\ & h_{14}(-2)^{14} + h_{13}(-2)^{13} + h_{12}(-2)^{12} + h_{11}(-2)^{11} + h_{10}(-2)^{10} \\
& + h_9(-2)^9 + h_8(-2)^8 + h_7(-2)^7 + h_6(-2)^6 + h_5(-2)^5 + h_4(-2)^4 \\
& + h_3(-2)^3 + h_2(-2)^2 + h_1(-2)^1 + h_0(-2)^0 \\
bB = h' =\ & h'_{14}(-2)^{14} + h'_{13}(-2)^{13} + h'_{12}(-2)^{12} + h'_{11}(-2)^{11} + h'_{10}(-2)^{10} \\
& + h'_9(-2)^9 + h'_8(-2)^8 + h'_7(-2)^7 + h'_6(-2)^6 + h'_5(-2)^5 + h'_4(-2)^4 \\
& + h'_3(-2)^3 + h'_2(-2)^2 + h'_1(-2)^1 + h'_0(-2)^0 \\
cC = h'' =\ & h''_{14}(-2)^{14} + h''_{13}(-2)^{13} + h''_{12}(-2)^{12} + h''_{11}(-2)^{11} \\
& + h''_{10}(-2)^{10} + h''_9(-2)^9 + h''_8(-2)^8 + h''_7(-2)^7 + h''_6(-2)^6 + h''_5(-2)^5 \\
& + h''_4(-2)^4 + h''_3(-2)^3 + h''_2(-2)^2 + h''_1(-2)^1 + h''_0(-2)^0
\end{aligned}
\tag{9.20}
$$

Bit-wise collection of convolution coefficients for the first row of the matrix-vector product would yield the same equation set.

The negabinary bits of the vector elements A, B, and C are multiplied with the negabinary equivalents of the first row of the matrix comprising a, b, and c. During the first cycle, the LSBs A_0, B_0, and C_0 of the vector elements are multiplied with the matrix elements. During the second cycle, the next higher order bits A_1, B_1, and C_1 of the vector elements are multiplied. This is continued till the MSB's of the vector elements are multiplied with the same matrix elements. Convolution coefficients for matrix multiplication can now be obtained by column-wise addition of coefficients. The operations comprising these eight cycles and finally the column-wise summation yield the first row of the matrix-vector product. Similar operations are required to yield the other three rows of the product. Similar to the binary system, the unsigned negabinary bits in this system take up values 1 and 0 only and hence can be represented by on/off states of light. The schematic diagram of the photonic architecture for (4×3) matrix-vector multiplication is shown in Figure 9.20.

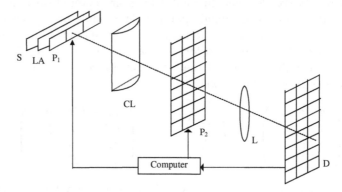

FIGURE 9.20: Schematic diagram of the matrix-vector multiplication process. S: laser source; LA: lenslet array; P1: source control SLM; CL: cylindrical lens; L: condensing lens; P2: SLM; D: detector array

9.8.4 Hybrid photonic multiprocessor

This multiprocessor architecture, better termed the connection machine, consists of an array of printed circuit boards (PCBs), each containing 512 processing elements divided equally among 32 electronic chips. The architecture is shown in Figure 9.21 with four chips on board, where each board contains a frequency-selective filter. The photonic chips contain a semiconductor laser, a photodiode, and reconfigurable diffraction gratings. The diffraction gratings perform the switching operation, enabling communication between the chips. This type of architecture employs wavelength division multiplexing (WDM) to direct the different wavelengths through the chips.

9.8.5 All-optical digital multiprocessor

An all-optical digital multiprocessor [61] is shown in Figure 9.22. The input to the system is either through an array of laser sources or by a two-dimensional spatial light modulator (SLM) placed in front of a laser source

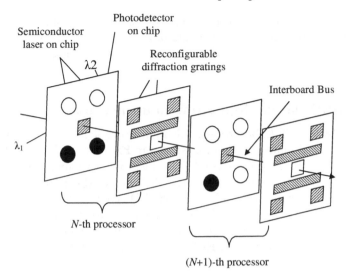

FIGURE 9.21: Schematic diagram of a hybrid photonic multiprocessor

array. The gate array placed in front of the array of sources may be another SLM or an array of bistable optical device (BOD).

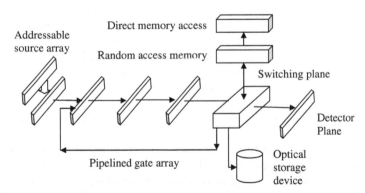

FIGURE 9.22: Schematic diagram of an all-optical digital multiprocessor

A BOD is comprised of a Fabry-Perot cavity with a material whose refractive index varies non-linearly with optical intensity. A BOD is preferable because of its high switching speed. The beam controller in front of the BOD employs reconfigurable diffraction gratings for switching and interconnection. Because of the large number of arrays required in this system, real-time hologram arrays are utilised. As seen in Figure 9.22, the beam controller directs the emerging beam to either the addressable detector array, or the random access memory, or the input of the gate array. Several logic elements can be interconnected through the beam controller to form a processing element or node, which represents individual elements in the gate array. Thus individual elements in the gate array can perform operations such as logic unit, clock, cache memory, etc.

9.8.6 S-SEED-based all-photonic multiprocessor

Developed by AT&T Laboratories, this multiprocessor [62] is a variant of the all-optical digital multiprocessor, as shown in Figure 9.23. This multiprocessor operates at (10^6) cycles/second. The bistable switching element is a GaAs/AlGaAs symmetric self-electro-optic effect device (S-SEED) which can provide a switching energy of 1 pJ. The S-SEEDs contain two mirrors with controllable reflectivity. The S-SEEDs are formed into 32 device arrays where each array contains two injection laser diodes emitting at $0.85\mu m$ wavelength. Each processor consists of four S-SEED arrays and each S-SEED drives two inputs.

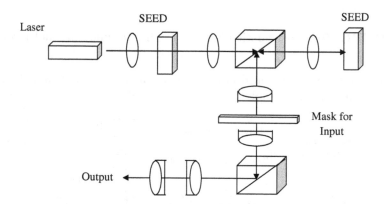

FIGURE 9.23: Block diagram of AT&T digital optical multiprocessor

Light from the injection lasers helps to communicate between the four arrays through lenses and masks. The masks are glass slides containing transparent or opaque spots to allow or stop light transmission. The output from one S-SEED array serves as the input to the next S-SEED array. The state of each S-SEED does not change until it is processed. The transmission of information occurs through optical fibres and free-space.

Books for further reading

1. *Optical Computing— a Survey for Computer Scientists*: D. G. Feitelson, MIT Press, New York, 1992.

2. *Optical Computing—Digital and Symbolic*: R. Arrathoon, Marcel Dekker, Inc., New York, 1989.

3. *The Connection Machine*: W. D. Hills, MIT Press, MA, 1985.

4. *Optical Computing Hardware*: J. Jahns and S. H. Lee, Academic Press, London, 1994.

5. *Computing with Optics, Introduction to Information Optics*: G. Li and M. S. Alam. Academic Press, CA, 2001.

6. *A Digital Design Methodology for Optical Computing*: M. Murdocca, McGraw-Hill, New York, 1990.

7. *Computer Arithmetic Systems—Algorithms, Architecture and Implementations*: A. Omandi, Prentice Hall, NJ, 1994.

Bibliography

[1] M. J. Flynn. Some computer organisation and their effectiveness. *IEEE Trans.*, C 21:948, 1972.

[2] R. C. Minnick. Cutpoint cellular logic. *IEEE Trans. Elec. Comp.*, EC 13:685, 1968.

[3] J. H. Siegel. A model of SIMD machine and a comparison of various interconnection networks. *IEEE Trans. Comp.*, 28:907, 1979.

[4] S. L. Tanimoto. A hierarchical cellular logic for pyramid computer. *J. Parallel and Dist. Comp.*, 1:105, 1984.

[5] G. Eichman, H. J. Caulfield, and I. Kadar. Optical artificial intelligence and symbolic computing. *App. Opt.*, 26:18–27, 1987.

[6] A. Veen. Data flow architecture. *CS*, 18:365, 1986.

[7] S. Kung, S. C. Lo, S. N. Jean, and R. Leonardi. Wavefront array processors-concepts to implementations. *Computers*, 20:18, 1987.

[8] S. Asia and Y. Wada. Technology challenges for integration near and below 0.1 micron. *Proc IEEE*, 85:505, 1997.

[9] F. E. Kiamilev. Performance comparison between opto electronics and vlsi multistage interconnection networks. *J. Light Wave Tech.*, 9:1742, 1991.

[10] J. Tanida, K. Nitta, T. Inoue, and Y. Ichioka. Comparison of electrical and optical interconnection for large fan-out communication. *J. Opt. A: Pure Appl. Opt.*, 1:262–266, 1999.

[11] D. Psaltis and R. A. Athale. High accuracy computation with linear analog optical systems: A critical study. *App. Opt.*, 25:3071, 1986.

[12] R. Arrathoon. *Optical Computing: Digital and Symbolic*. Marcel Dekker, Inc., New York, 1989.

[13] K Huang and F. A. Briggs. *Computer Architecture and Parallel Processing*. McGraw Hill, New York, 1984.

[14] Y. Ha, B. Li, and G. Hichman. Optical binary symmetric logic functions and their application. *Opt. Eng.*, 28:380, 1989.

[15] M. Murdocca. *A Digital Design Methodology for Optical Computing.* McGraw-Hill, NY, 1990.

[16] P. S. Guilfoyle and W. J. Wiley. Combinatorial logic based digital optical computing architectures. *App. Opt.*, 27:1661, 1988.

[17] J. Tanida and Y. Ichioka. Optical logic array processor using shadow-grams. *J. Opt. Soc. Am.*, 73:801–809, 1983.

[18] J. Tanida and Y. Ichioka. Modular components for an optical array logic system. *Appl. Opt.*, 26:3954, 1987.

[19] M. Fukui and K. Kitayama. Applications of image-logic algebra: wire routing and numerical data processing. *Appl. Opt.*, 31:4645–4656, 1992.

[20] J. J. H. Cavanagh. *Digital Computer Arithmetic: Design and Implementation.* McGraw-Hill, New York, 1985.

[21] A. Omandi. *Computer Arithmetic Systems: Algorithms, Architecture and Implementations.* Prentice Hall, New Jersy, 1994.

[22] N. S. Szabo and R. T. Tanaka. *Residue Arithmetic and Its Applications to Computer Technology.* McGraw Hill, New York, 1967.

[23] S. Mukhopadhyay, A. Basuray, and A. K. Datta. New technique of arithmetic operation using positional residue system. *App. Opt.*, 29:2981–2893, 1990.

[24] M. Seth, M. Ray (Shah), and A. Basuray. Optical implementation of arithmetic operations using the positional residue system. *Opt. Eng.*, 33:541–547, 1994.

[25] D. Psaltis and D. Casasent. Optical residue arithmetic: A correlator approach. *App. Opt.*, 18:163–171, 1979.

[26] C. D. Capps, R. A. Falk, and T. K. Houk. Optical arithmetic/logic unit based on residue arithmetic and symbolic substitution. *App. Opt.*, 27:1682–1686, 1988.

[27] A. P. Goutzoulis, E. C. Malarkey, D. K. Davis, J. C. Bradley, and P. R. Beaudet. Optical processing with residue LED/LD lookup tables. *App. Opt.*, 27:1674–1681, 1988.

[28] M. L. Heinrich, R. A. Athale, and M. W. Huang. Numerical optical computing in the residue number system with outer-product look-up tables. *Opt. Let.*, 14:847–849, 1989.

[29] S. Mukhopadhyay, A. Basuray, and A. K. Datta. A real-time optical parallel processor for binary addition with a carry. *Opt. Commun.*, 66:186–190, 1988.

[30] A. K. Datta and M. Seth. Parallel arithmetic operations in an optical architecture using a modified iterative technique. *Opt. Comm.*, 115:245–250, 1995.

[31] R. Golshan and J. S. Bedi. Implementation of carry-look ahead adder with spatial light modulator. *Opt. Commun.*, 68:175–178, 1988.

[32] A. Kostrzewski, D. H. Kim, Y. Li, and G. Eichman. Fast hybrid parallel carry-look-ahead adder. *Opt. Let.*, 15:915–917, 1990.

[33] S. P. Kozaitis. Higher order rules for symbolic substitution. *Opt. Comm.*, 65:339, 1988.

[34] B. Parhami. Carry-free addition of recoded binary signed-digit numbers. *IEEE Trans. on Computers*, 7:1470–1476, 1988.

[35] B. Parhami. On the implementation of arithmetic support functions for generalized signed-digit number systems. *IEEE Transactions on Computers*, 42:379–384, 1993.

[36] A. A. S. Awwal. Recoded signed digit binary addition subtraction using opto-electronic symbolic substitution. *App. Opt.*, 31:3205–3208, 1992.

[37] A. K. Cherri. Signed-digit arithmetic for optical computing: digit grouping and pixel assignment for spatial encoding. *Opt. Engg.*, 38:422–431, 1999.

[38] G. Li and F. Qian. Code conversion from signed-digit to complement representation based on look-ahead optical logic operations. *Opt. Eng.*, 40:2446–2451, 2001.

[39] Y. Li and G. Eichmann. Conditional symbolic modified signed-digit arithmetic using optical content-addressable memory logic elements. *App. Opt.*, 26:2328–2333, 1987.

[40] S. Zhou, S. Campbell, P. Yeh, and H. K. Liu. Two-stage modified signed-digit optical computing by spatial data encoding and polarization multiplexing. *App. Opt.*, 34:793–802, 1995.

[41] H. Huang, M. Itoh, and T. Yatagai. Modified signed-digit arithmetic based on redundant bit representation. *Appl. Opt.*, 33:6146–6156, 1994.

[42] R. P. Bocker, B. L. Drake, Lasher M. E., and T. B. Henderson. Modified signed-digit addition and subtraction using optical symbolic substitution. *App. Opt.*, 25:2456–2457, 1986.

[43] A. K. Cherri, M. S. Alam, and A. A. S. Awwal. Optoelectronic symbolic substitution based canonical modified signed-digit arithmetic. *Opt. and Laser Tech.*, 29:151–157, 1997.

[44] H. Huang, M. Itoh, and T. Yatagai. Optical scalable parallel modified signed-digit algorithms for large-scale array addition and multiplication using digit-decomposition-plane representation. *Opt. Eng.*, 38:432–440, 1999.

[45] F. Qian, G. Li, and M. Alam. Optoelectronic quotient-selected modified signed-digit division. *Opt. Engg.*, 40:275–282, 2001.

[46] F. Qian, G. Li, H. Ruan, and L. Liu. Modified signed-digit addition by using binary logic operations and its optoelectronic implementation. *Opt. Laser Tech.*, 31:403–410, 1999.

[47] A. K. Datta, A. Basuray, and S. Mukhopadhyay. Arithmetic operations in optical computations using a modified trinary number system. *Opt. Let*, 14:426–428, 1989.

[48] A. K. Datta, A. Basuray, and S. Mukhopadhyay. Carry-less arithmetic operation of decimal numbers by signed-digit substitution and its optical implementation. *Opt. Comm.*, 88:87–90, 1992.

[49] M. S. Alam. Parallel optical computing using recoded trinary signed-digit numbers. *App. Opt.*, 33:4392–4397, 1994.

[50] A. K. Cherri, N. I. Khachab, and E. H. Ismail. One-step optical trinary signed-digit arithmetic using redundant bit representations. *Opt. Laser Tech.*, 29:281–290, 1997.

[51] J. U. Ahmed, A. A. S. Awwal, and M. A. Karim. Two-bit trinary full adder design based on restricted signed-digit numbers. *Opt. Laser Tech.*, 26:225–228, 1994.

[52] M. S. Alam. Parallel optoelectronic trinary signed-digit division. *Opt. Eng.*, 38:441–448, 1999.

[53] M. S. Alam, M. A. Karim, A. A. S. Awwal, and J. J. Westerkamp. Optical processing based on conditional higher-order trinary modified signed-digit symbolic substitution. *Appl. Opt.*, 31:5614–5621, 1992.

[54] M. S. Alam, K. Jemili, and M. A. Karim. Optical higher-order quaternary signed-digit arithmetic. *Opt. Eng.*, 33:3419–3426, 1994.

[55] A. K. Cherri and N. I. Khachab. Canonical quaternary signed-digit arithmetic using optoelectronics symbolic substitution. *Opt. Laser Tech.*, 28:397–403, 1996.

[56] G. Li, L. Liu, H. Cheng, and X. Yan. Parallel optical quaternary signed-digit multiplication and its use for matrix-vector operation. *Optik*, 107:165–172, 1998.

[57] C. Perlee and D. Casasent. Negative base encoding in optical linear algebra processors. *App. Opt.*, 25:168–169, 1986.

[58] G. Li, L. Liu, L. Shao, and Z. Wang. Negabinary arithmetic algorithms for digital parallel optical computation. *Opt. Let.*, 19:1337–1339, 1994.

[59] A. K. Datta and S. Munshi. Signed-negabinary-arithmetic-based optical computing by use of a single liquid-crystal-display panel. *App. Opt.*, 49:1556–1564, 2002.

[60] A. K. Datta and S. Munshi. Optical implementation of Flip-Flops using single LCD panel. *Opt. and Laser Tech.*, 40:1–5, 2008.

[61] B. K. Jenkins, P. Chavel, R. Forchheimer, A. A. Sawchuk, and T.C. Strand. Architectural implications of a digital optical processor. *App. Opt.*, 23:3465–3474, 1984.

[62] OE Reports. Optical computer: Is concept becoming reality. *SPIE Int. Soc. Opt. Eng.*, 75:1–2, 1990.

Chapter 10

Photonic Pattern Recognition and Intelligent Processing

10.1 Introduction

The impetus of using photonics technologies and devices for pattern recognition started with a pattern correlator developed by Vander Lugt in 1963. With the progress of photonic devices for interfacing data and images, new techniques and algorithms slowly emerged. Photonic implementation for the recognition of pattern can basically be classified into two approaches. The first one is the correlation approach based on photonic implementation of Fourier-domain matched filters, or space-domain linear and nonlinear filters. The second approach is based on the implementation of various artificial neural network (ANN) models using photonic devices and architectures.

10.2 Correlation filters for pattern recognition

Correlation refers to the matching of an input target to the reference signal or image. Correlation filters have been widely used for pattern matching problems. A matched filter is used to detect the presence as well as position of known objects in a scene. The cross correlation $c(x, y)$ of a 2D target or test image $t(x, y)$ with a reference image $r(x, y)$ is expressed as

$$c(x, y) = t(x, y) \otimes r(x, y) \tag{10.1}$$

This equation can be rewritten in Fourier space as

$$C(u, v) = T(u, v)R^*(u, v) \tag{10.2}$$

where $T(u, v)$=FT$[t(x, y)]$, and $R(u, v)$=FT$[r(x, y)]$, and FT denotes the Fourier transform of space-domain signal. (x, y) are the coordinates in the space-domain, whereas (u, v) are the coordinates in the frequency domain. The result of correlation can then be calculated by inverse Fourier transforming $C(u, v)$.

The correlation technique can be extended if N multiple test images are present in the scene. The scene may also have noise $n(x, y)$. If $t(x, y) = \sum_{i=1}^{N} = r(x-x_i, y-y_i) + n(x, y)$ then the cross correlation of $r(x, y)$ and $t(x, y)$ results in the output $\sum_{i=1}^{N} c(x-x_i, y-y_i) + [n(x, y) \otimes t(x, y)]$. Therefore the correlation output is a noisy version of the sum of N correlations centred at coordinates of the target. Therefore correlation is viewed as a method to carry out instant decisions on the presence of targets form a scene.

Frequency domain correlation techniques are usually carried out using matched filters. Apart from a host of correlation matched filters, there are two well-known optical correlation architectures, namely the Vander Lugt correlator (VLC) and joint transform correlator (JTC). Both of them can be executed in the 4f optical setup and are based on the comparison of target or test image $t(x, y)$ and a reference image $r(x, y)$. The similarity between the two images is achieved by the detection of an auto-correlation peak. Whereas a Fourier-domain spatial matched filter is needed in VLC architecture, JTC avoids filter synthesis, but requires a spatial-domain impulse response for correlation.

10.2.1 Vander Lugt correlation

In Vander Lugt architecture, correlation can be thought of as the output from a matched filter, whose impulse response is the reflected version of the reference image. The correlator has an input signal which is multiplied by some pre-synthesised filter in the Fourier domain. The filter $H(u, v)$ maximises the signal-to-noise ratio and is obtained by using the Fourier transform of the test image $t(x, y)$. Thus $H(u, v) = T^*(u, v)$ where $T(u, v)$ is the Fourier transform of the test image $t(x, y)$.

During photonic implementation, the FT of the test or target image gets optically convolved with the filter in the Fourier domain. The convolved product is then optically inverse Fourier transformed to obtain the correlation peaks. Figure 10.1 gives a schematic diagram of VLC architecture where a filter image is placed in the FT plane.

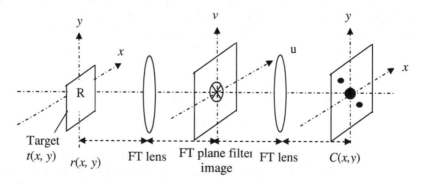

FIGURE 10.1: Schematic diagram of a VLC architecture

Detecting a target image requires critical fabrication of a complex-valued spatial filter which is matched in both amplitude and phase to the spectrum of the image. Thus alignment in VLC is extremely critical as it needs precise pixel-wise matching between the target FT and frequency plane filter. In photonic implementations the advantages of obtaining Fourier transform optically using a simple lens system is utilised. The photonic arrangement for Vander Lugt correlator is shown in Figure 10.2.

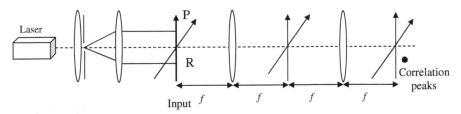

FIGURE 10.2: Optical arrangement of the VLC architecture

Light from a laser is expanded using a beam expander and then collimated using a lens and pinhole combination. The collimated beam is incident on the input image, which is presented through an electrically addressed SLM. The 4f system is then used to produce correlation peak at the output plane which is captured in a CCD camera. The Fourier-plane filter of the input image is first obtained. The filter so obtained is displayed on an SLM, illuminated by a collimated beam. The most important criterion for Vander Lugt correlator (VLC) is the stringent alignment so that the input FT exactly matches the filter in the frequency plane.

10.3   Frequency domain correlation filters

An ideal correlation filter when applied for recognition task would yield a sharp correlation peak for a perfect match of the correlation filter with the test image present in the database. Such a test is generally labelled as an authentic image. On the other hand, if no such peak is found in the correlation plane, the corresponding images are labelled as false. The correlation output is searched for peak and the relative height of this peak is analysed to determine whether the test image is recognized or not. The authentication of test face is generally guided by a metric, called the peak-to-sidelobe ratio (PSR), which is measured from the correlation plane. In an ideal case, a correlation peak with high value of PSR is obtained, when any Fourier-transformed test image is perfectly matched with the reference image with the FT image of the reference image. For practical purposes a PSR value of 10 is heuristically taken for perfect authentication.

Figure 10.3 describes pictorially how the frequency domain correlation technique is carried out for face recognition tasks using a correlation filter. As shown in Figure 10.3 the information of N number of training images from kth face class ($k \in C$), out of the total C number of face classes for a given database, is Fourier transformed to form the design input for kth correlation filter. The right-hand side of Figure 10.3 shows the nature of a typical correlation plane in response to authentic (true) and false images.

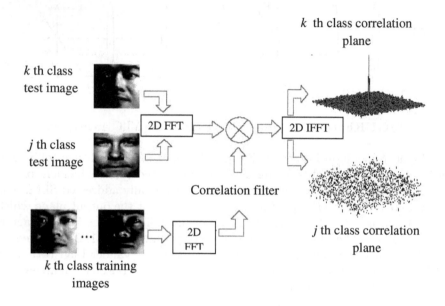

FIGURE 10.3: Basic frequency domain correlation technique for face recognition

Since the phase information is significantly more important than amplitude in an image, a pure phase correlation filter can have an optical efficiency of 100% in an optical correlation system, compared to the classical matched filter. Another advantage of the phase-only filter is the absence of sidelobes around the main correlation peak, thus reducing the chances of false alarms. The phase-only filter is thus a robust filter and has been shown to be more discriminant than the matched filter.

$$H(u,v) = \frac{R(u,v)}{|R(u,v)|} \tag{10.3}$$

Such a matched Fourier-domain filter was first synthesized by the holographic technique.

The concept of complex matched filters developed with the idea of recognising target images irrespective of rotation, scale changes, or any other distortions in the image. So, instead of storing all such distorted images to match with the unknown target, the complex matched filter synthesized to cater to all such distortions need only be stored. The complex matched filter is thus trained using a set of training images, and the role of the correlation filter is to find out the similarity of the test or target image with any of the trained images. Since the filter is synthesized with the set of reference images, the reference set of images need not be stored. The synthesis of a matched filter is carried out prior to correlation, which can cater to the real-time distortions.

10.3.1 Performance evaluation metrics of correlation

Several metrics for determining the goodness of correlation are available, for example, sharpness of correlation peak represented by [3] correlation peak intensity (CPI), peak-to-correlation energy ratio (PCE), the peak-to-sidelobe ratio (PSR), signal-to-noise ratio (SNR), etc.

10.3.1.1 Peak-to-correlation energy ratio

Peak-to-correlation energy ratio (PCE) is a ratio of the correlation peak to the energy of the correlation plane. Hence, PCE may be written as

$$\text{PCE} = \frac{\text{Correlation peak intensity (CPI)}}{\sum_{ij} I_{cp}(i,j)} \tag{10.4}$$

where $I_{cp}(i,j)$ is the intensity at (i,j)th point in the plane.

10.3.1.2 Signal-to-noise ratio (SNR)

SNR [5], [6] is defined as the ratio of the expected correlation peak intensity to the variance of the peak.

$$\text{SNR} = \frac{C_0 - \bar{c}}{\sigma_{noise}} \tag{10.5}$$

where C_0 is the correlation peak intensity, $\bar{c}$ and σ_{noise} are the average and standard deviation, respectively, of the data over the entire correlation plane excluding the peak itself, and the plane excluding the peak itself. The filters are synthesized such that the SNR value is maximised. The SNR falls off as the targets vary in scale or are subject to in-plane rotation and out-of-plane rotation.

10.3.1.3 Discrimination ratio

Discrimination ratio (DR) is used to classify a true and false class image and is defined as the ratio of the CPI of the true class image to the CPI of the false class. So

$$DR = \frac{CPI_{true}}{CPI_{false}} \tag{10.6}$$

10.3.1.4 Peak-to-sidelobe ratio

The most used metric in frequency domain correlation techniques is the peak-to-sidelobe ratio (PSR). PSR is measured from the correlation plane. A rectangular region, say (20×20) pixels) centred at the peak, is extracted and used to compute the PSR. A (5×5) rectangular region centred at the peak is masked out and the remaining annular region, shown in Figure 10.4, defined as the sidelobe region, is used to compute the mean and standard deviation of the sidelobes. In an ideal case a correlation peak with high PSR [4] value is obtained when any Fourier-transformed test image of the kth class is correlated with kth correlation filter. The PSR is then calculated as

$$\text{PSR} = \frac{\text{peak-mean}}{\text{standard deviation}} \tag{10.7}$$

PSR images for true and false recognition are shown in Figure 10.4 where (a) is the true recognition and (b) is false recognition.

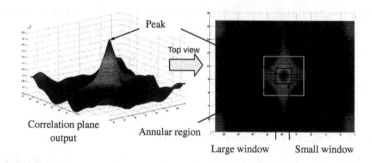

Correlation plane output Annular region Large window Small window

FIGURE 10.4: Pictorial representation of PSR metric evaluation from correlation plane output

The major correlation peak is below the threshold PSR value if the test image belongs to other class of gallery, i.e., jth class, where, $j \in C$ and $j \neq k$. However, the performance of the system in terms of recognition rate depends on the design of the correlation filter.

10.3.2 Types of frequency domain correlation filters

Development of correlation filters can be categorized broadly in two different classes: linear constrained correlation filters and linear unconstrained correlation filters. Initially, the linear constrained filters were developed. To overcome the problems of constrained filters, their unconstrained versions were developed.

Many constrained correlation filter designs require that the filter designer specify the output of the filter for each training image provided. For N training images, this results in N constraints, which is typically much less than the number of free parameters and d is the dimensionality of the filter. For this reason, many of these designs optimise some filter performance criterion while satisfying the N constraints. This class of designs is referred to as constrained linear correlation filters. The general form of a constrained linear filter $\mathbf{h}$ is

$$\mathbf{h} = \bar{\mathbf{Q}}^{-1}\mathbf{A}(\mathbf{A}^{+}\bar{\mathbf{Q}}^{-1}\mathbf{A})^{-1}u \qquad (10.8)$$

where the bold characters represent matrices, $\mathbf{A}$ is a matrix whose N columns are the N vectorized frequency-domain training images ($\mathbf{x}_i$s), $\bar{\mathbf{Q}}$ is a diagonal preprocessor matrix, and u is an ($N \times 1$) vector of the specified correlation output values for each training image. Special cases of the preprocessor $\bar{\mathbf{Q}}$ result in different linear constrained correlator filters. These cases are listed in Table 10.1.

TABLE 10.1: Different values of diagonal preprocessor matrix $\bar{\mathbf{Q}}$ result constrained correlation filters

Filter Type	Value of $\mathbf{Q}$ from Eq.10.8
ECPSDF [7]	$\mathbf{Q} = \mathbf{I}$ (Identity matrix)
MVSDF [8]	$\bar{\mathbf{Q}} = \bar{\mathbf{O}}$
MACE [9]	$\bar{\mathbf{Q}} = \bar{\mathbf{D}}, \ \bar{\mathbf{D}} = \sum_{i=1}^{N} \bar{\mathbf{D}}_i,$ where, $\bar{\mathbf{D}}_i = \bar{\mathbf{X}}_i \bar{\mathbf{X}}_i^{*}$
OTSDF [10]	$\bar{\mathbf{Q}} = \alpha\bar{\mathbf{O}} + \sqrt{1-\alpha^2}\bar{\mathbf{D}}$
MINACE [11]	$\bar{\mathbf{Q}} = max(\alpha\bar{\mathbf{O}}, \sqrt{1-\alpha^2}\bar{\mathbf{D}}_1, \cdots, \sqrt{1-\alpha^2}\bar{\mathbf{D}}_N)$

In Table 10.1, if $\bar{\mathbf{Q}}$ is replaced by $\bar{\mathbf{I}}$ (identity matrix), the design equation reduces to a synthetic discriminant function (SDF) filter. The drawback of the SDF filter is that it cannot tolerate significant input noise. To achieve robustness to noise, Kumar [8] introduced the minimum variance synthetic discriminant function (MVSDF). The design equation of MVSDF is given in Table 10.1 by replacing $\bar{\mathbf{Q}}$ by $\bar{\mathbf{O}}$ where $\bar{\mathbf{O}}$ is a diagonal matrix containing the power spectral density of the noise. MVSDF minimises the correlation output noise variance (ONV) while satisfying the correlation peak amplitude constraints.

The first constraint is that the MVSDF also controls only one point in the correlation map, just like a special case of SDF. The second is that the variance of the noise matrix must be known beforehand in order to design the filter. However, even if the latter is known exactly, MVSDF is impractical because it requires inverting a large noise covariance matrix [12]. The minimum average correlation energy (MACE) filter was an attempt to control the entire correlation plane. The attempt was made to reduce correlation function levels at all points except at the origin of the correlation plane and thereby obtained a very sharp correlation peak [9]. This is equivalent to minimising the energy of the correlation function while satisfying intensity constraints at the origin. The closed form solution of the MACE filter is obtained by replacing $\bar{\mathbf{Q}}$ with $\bar{\mathbf{D}}$, given in Table 10.1, where $\bar{\mathbf{D}}_i$ is the power spectrum of the ith training image, and $\bar{\mathbf{D}}$ contains the average training power spectrum. In Table 10.1, $\bar{\mathbf{X}}_i = diag\{\mathbf{x}_i\}$. However, MACE filters often suffer from two main drawbacks. Firstly, there is again no built-in immunity to noise. Second, the MACE filter is often excessively sensitive to intra-class variations. Nevertheless, this filter leads to a useful frequency domain design approach for object recognition. The optimal trade-off synthetic discriminant function (OTSDF) filter [10] includes a trade-off parameter α that allows the user to emphasize low output noise variance (ONV) (α closer to 1) or low average correlation energy (ACE) (α closer to 0). Setting $\alpha = 1$ yields the minimum variance SDF, which has minimum ONV but usually exhibits broad correlation peaks; in contrast, setting $\alpha = 0$ yields the MACE filter, which has minimum ACE and produces sharp peaks on the training images but is highly sensitive to noise and distortion.

The minimum noise and correlation energy (MINACE) filter [11], [13] achieves an alternative compromise between these two extremes by using an envelope equal to the greater of the noise and training image power spectra at each frequency. It may be noted that the trade-off parameter α appearing in the MINACE formulation in Table 10.1 is not part of the traditional MINACE filter design as reported in [11]; rather, the value of $\bar{\mathbf{O}}$ is varied directly, since the input noise level is typically unknown. This difference is merely semantic; in practice, the same effect is achieved by varying either $\bar{\mathbf{O}}$ or α. In both the OTSDF and the MINACE designs, because both the input noise level and the trade-off can be effected by scaling $\bar{\mathbf{O}}$ relative to $\bar{\mathbf{D}}$, a single parameter α simultaneously accomplishes both of these goals.

The hard constraints on correlation values at the origin are not only unnecessary but can be counter-productive [14], [15]. Relaxing or removing such constraints might lead to a larger filter solution space. Also, the matrix inversion in the constrained design may be ill-conditioned when highly similar training images are included. For these reasons, several unconstrained linear filter designs have been proposed. These designs maximise some measure of the average output on true-class training images while minimising other criteria such as ONV and ACE. The maximum average correlation height (MACH) filter [16] is one such design, which achieves distortion tolerance by maximising the similarity of the shapes of true-class correlation outputs over the training images. This maximisation is realised by minimising a dissimilarity metric known as the average similarity measure (ASM) of true class images. Replacing $\bar{\mathbf{S}}$ by $\bar{\mathbf{D}}$ results in unconstrained MACE (UMACE) filter [17] solution.

The unconstrained OTSDF (UOTSDF) filter [18] is a design that minimises a trade-off between true-class ACE and ONV (as in the OTSDF design). An optimal trade-off approach which relates correlation plane metrics was introduced [19] resulting in an OTMACH filter where α, β and γ are the non-negative optimal trade-off (OT) parameters.

Design equations of some important filters are given in Table 10.2.

TABLE 10.2: Unconstrained linear filter designs

Filter Type	Filter (h)
MACH [16]	$\bar{S}^{-1}m$
UMACE [17]	$\bar{D}^{-1}m$
UOTSDF[18]	$\{\alpha O + \beta D\}^{-1}m$
OTMACH [19]	$\{\alpha O + \beta D + \gamma S\}^{-1}m$
EMACH [20]	Dominant eigenvector of $\{\alpha \bar{I} + (1-\alpha^2)^1/2\bar{S}^\beta\}^{-1}\bar{C}^\beta$
EEMACH [21]	Dominant eigenvector of $\{\alpha \bar{I} + (1-\alpha^2)^1/2\bar{S}^\beta\}^{-1}\hat{\bar{C}}^\beta$

In addition to OTMACH filter, different variations of MACH filters were proposed. In [20] the extended MACH (EMACH) filter design is addressed by reducing the dependence on the average training image **m**. A tunable parameter β given in Table 10.2 is used to control this reduction. Two new metrics are used in the design: (1) the all-image correlation height (AICH), which takes into account the filter output on **m** as well as on individual training images, and (2) a modified average similarity measure (MASM), which measures the average dissimilarity to the optimal output shape. This optimal shape takes into account the reduced dependence on **m** realised by the new AICH metric. The EMACH design also includes an ONV criterion to help maintain noise tolerance. A trade-off parameter α (given in Table 10.2) is used to control the relative importance of the ONV and MASM criteria, where higher values of α correspond to greater emphasis on ONV and vice versa. If the covariance matrix $\bar{C}^\beta$ is approximated by only its dominant eigenvectors and eigenvalues to yield a new matrix $\hat{\bar{C}}^\beta$, the resulting filter solution is referred to as the eigen-extended MACH (EEMACH) filter [21].

In addition to linear correlation filters, several nonlinear correlation filters were developed. Some of the design equations are given in Table 10.3. A linear correlation output is an array of scalar output values from a linear discriminant applied to the input image at every shift. While limited in capability by their linear nature, linear CFs have the important advantage of efficient frequency-domain computation; however, special cases of nonlinear discriminant functions have been proposed for which the attractive computational properties of linear filters may be retained by specialized implementation schemes. Such systems are referred to as nonlinear correlation filters. There are mainly two classes of designs.

The first class of designs, quadratic correlation filters (QCFs), is characterized by solving for a quadratic discriminant function in d-dimensional space, where d is the number of pixels in the image. This quadratic discriminant can then be efficiently implemented as a set of linear filters via eigen-decomposition. Several methods were proposed for solving the diagonal matrix in QCF design. Some advanced correlation filters are indicated in Table 10.3.

TABLE 10.3: Some advanced correlation filters

Filter	Design Equation
GMACH [22, 20]	$\mathbf{h} = \{\delta\Omega + \alpha\mathbf{O} + \beta\mathbf{D} + \gamma\mathbf{S}\}^{-1}\mathbf{m}$ where, Ω is $d^2 \times N$ matrix with rank N.
WaveMACH [23]	$\mathbf{h} = \{\bar{\mathbf{S}}^{-1}\mathbf{m}\}\|\mathbf{H}(u,v)\|^2$ where $\mathbf{H}(u,v)$ is Mexican hat filter.
ARCF [24]	$\mathbf{h} = (\bar{\mathbf{D}} + \epsilon\bar{\mathbf{I}})^{-1}\bar{\mathbf{X}}[\bar{\mathbf{X}}^+(\bar{\mathbf{D}} + \epsilon\bar{\mathbf{I}})^{-1}\bar{\mathbf{X}}]u$ where $\epsilon = 0$ indicates MACE filter and $\epsilon = \infty$ represents SDF filter.
CMACE [25]	$\mathbf{h} = \mathbf{V}^{-1}\mathbf{A}\{\mathbf{A}^+\mathbf{V}^{-1}\mathbf{A}\}^{-1}u$ in feature space where $\mathbf{V} = \frac{1}{N}\sum_{i=1}^{N}\mathbf{V}_i$, $\mathbf{V}_i \triangleq$ correntropy matrix.
ActionMACH [26]	3D version of MACH filter.
MMCF [27]	$\mathbf{h} = \{\lambda\bar{\mathbf{I}} + (1-\lambda)(\bar{\mathbf{D}} - \mathbf{A}\mathbf{A}^+)\}^{-1/2}\tilde{\mathbf{A}}a$ where $\tilde{\mathbf{A}} = [\tilde{\mathbf{x}}_1, \tilde{\mathbf{x}}_2, \cdots, \tilde{\mathbf{x}}_N]$, and, $\tilde{\mathbf{x}} \triangleq \{\lambda\bar{\mathbf{I}} + (1-\lambda)(\bar{\mathbf{D}} - \mathbf{A}\mathbf{A}^+)\}^{-1/2}\mathbf{x}_i$ and a[27] is evaluated by sequential minimum optimization technique.

In contrast, the second class of designs, polynomial correlation filters (PCFs), are sets of linear filters applied to multichannel input images, whose outputs are subsequently summed to form a single output. Two variants of PCF were proposed: 1) constrained PCF (CPCF) where CPCF design minimises a weighted sum of ONV and ACE analogous to the OTSD design, using the trade-off parameter α in a similar manner, and 2) unconstrained PCF (UPCF) where UPCF design maximises the average correlation height (ACH) while minimising a weighted sum of ONV and ACE.

The mathematical background of some basic correlation filter designs are given below.

10.3.2.1   SDF filter design

Traditionally, in the design of SDF-type correlation filters, linear constraints are imposed on the training images to yield a known value at specific locations in the correlation plane. The classical SDF [7], [28] filter is designed for a two-class problem where the correlation value at the origin is set to 1 (it may be selected as other values for multi-class problems) for training images from one class, generally authentic or true class, and to 0 for the training images from false class. SDF can be formulated by a single matrix-vector notation,

$$\mathbf{A}^+\mathbf{h} = u \tag{10.9}$$

where $\mathbf{A} = [\mathbf{x}_1, \mathbf{x}_2, ..., \mathbf{x}_N]$ is a $(d \times N)$ matrix with N training Fourier transformed vectors as its columns, $u = [u_{(1)}, u_{(2)}, \ldots, u_{(N)}]^T$ is an $N \times 1$ vector containing the desired peak values at the correlation plane origin for the desired class, and d is the total number of pixels present in one image. Here $\mathbf{h}$ is the desired filter of size $(d \times 1)$ and the superscript $+$ indicates the complex conjugate transpose. Consider that the conventional SDF is matched to a composite image $\mathbf{h}$ where

$$\mathbf{h} = \mathbf{A}c \tag{10.10}$$

where the coefficient vector $c = [c_{(1)}, c_{(2)}, \cdots, c_{(N)}]^T$ of the linear combination chosen to satisfy the deterministic constraints indicated in Equation (10.9). Thus from Equation (10.10) and Equation (10.9) the SDF filter $\mathbf{h}_{\text{SDF}}$ can be formulated by substituting c as:

$$\mathbf{h}_{\text{SDF}} = \mathbf{A}(\mathbf{A}^+\mathbf{A})^{-1}u \tag{10.11}$$

10.3.2.2 MACE filter design

The MACE filter is designed to ensure a sharp correlation peak and allowing easy detection in the full correlation plane as well as control of the correlation peak value. To achieve good detection, it is necessary to reduce correlation function levels at all points except at the origin of the correlation plane. Specifically, the value of the correlation function must be at a user-specified value at the origin but it is free to vary elsewhere. This is equivalent to minimising the energy of the correlation function while satisfying intensity constraints at the origin. The correlation peak amplitude constraint for the MACE filter is the same as that for the SDF filter given in Equation 10.9. Now the correlation plane in response to $\mathbf{x}_i$ for the MACE filter $\mathbf{h}$ can be expressed in matrix-vector form as

$$\mathbf{g}_i = \bar{\mathbf{X}}_i^*\mathbf{h} \tag{10.12}$$

where $\bar{\mathbf{X}}_i$ represents a $d \times d$ diagonal matrix containing ith training vector $\mathbf{x}_i$ along its diagonal. Hence the energy of the ith correlation plane can be formulated as

$$|\mathbf{g}_i|^2 = |\bar{\mathbf{X}}_i^*\mathbf{h}|^2 \tag{10.13}$$

where $\bar{\mathbf{D}}_i = \bar{\mathbf{X}}_i\bar{\mathbf{X}}_i^*$ is a $(d \times d)$ diagonal matrix containing a power spectrum corresponding to $\mathbf{x}_i$. Now for all $i = 1, 2, \cdots, N$ the ACE is

$$\text{ACE} = \frac{1}{N}\sum_{i=1}^{N}|\mathbf{g}_i|^2 = \frac{1}{N}\sum_{i=1}^{N}\mathbf{h}^+\bar{\mathbf{D}}_i\mathbf{h} = \mathbf{h}^+\bar{\mathbf{D}}\mathbf{h} \tag{10.14}$$

where $\bar{\mathbf{D}}$ represents a $d \times d$ diagonal matrix containing average power spectrum along its diagonal and is given by

$$\bar{\mathbf{D}} = \frac{1}{N} \sum_{i=1}^{N} \bar{\mathbf{D}}_i = \frac{1}{N} \sum_{i=1}^{N} \bar{\mathbf{X}}_i \bar{\mathbf{X}}_i^* \qquad (10.15)$$

Therefore, to synthesize the MACE filter the attempt is made to minimise ACE given in Equation 10.14 while meeting the linear constraints in Equation 10.9. The solution to this problem can be found by using the method of Lagrange multipliers. Similar derivation of the constrained optimization problem can be found in [9], [8]. The optimum solution of Equation 10.14 can be shown to be

$$\mathbf{h}_{\text{MACE}} = \bar{\mathbf{D}}^{-1} \mathbf{A} (\mathbf{A}^+ \bar{\mathbf{D}}^{-1} \mathbf{A})^{-1} u \qquad (10.16)$$

10.3.2.3 MVSDF design

MVSDF minimises the correlation output noise variance (ONV) $\mathbf{h}^+ \bar{\mathbf{O}} \mathbf{h}$, where $\bar{\mathbf{O}}$ is the diagonal matrix whose diagonal entries are the noise power spectral density, while satisfying the correlation peak amplitude constraints. The solution of MVSDF is expressed as

$$\mathbf{h}_{\text{MVSDF}} = \bar{\mathbf{O}}^{-1} \mathbf{A} (\mathbf{A}^+ \bar{\mathbf{O}}^{-1} \mathbf{A})^{-1} u \qquad (10.17)$$

10.3.2.4 Optimal trade-off (OTF) filter design

Due to the minimisation criteria of ACE of the MACE filter, the sharp correlation peak is made possible by suppressing the side lobes for the target images. However, a MACE filter can result in poor intraclass recognition of images which are not included in the training set, as this filter emphasizes the high-frequency components. Moreover, a MACE filter is often excessively sensitive to noise as there is no in-built immunity to noise. To get the sharp correlation peak with suppressed noise, a MACE filter is combined with MVSDF, thus emphasizing low spatial frequencies to reduce noise. The technique resulted in evolving an optimal trade-off function (OTF) [29]. The optimum solution of OTF is given by

$$\mathbf{h}_{\text{OTF}} = \bar{\mathbf{T}}^{-1} \mathbf{A} (\mathbf{A}^+ \bar{\mathbf{T}}^{-1} \mathbf{A})^{-1} u \qquad (10.18)$$

where $\bar{\mathbf{T}} = \alpha \bar{\mathbf{D}} + \sqrt{1 - \alpha^2} \bar{\mathbf{O}}$, $0 \leq \alpha \leq 1$. α is used as controlling trade-off parameter, i.e., for $\alpha = 0$ leads to MVSDF and $\alpha = 1$ leads to a MACE filter.

10.3.2.5 MACH filter design

Another approach to a design correlation filter is to remove the hard constraint at the correlation plane, and hence, in general these types of filters are termed unconstrained correlation filters. The unconstrained correlation filter improves distortion tolerance since during the design phase of such filters the training images are not treated as deterministic representations of the object but as samples of a class whose characteristic parameters are used in encoding the filter [17]. In order to do that an optimal shape of correlation plane $\mathbf{f}$ (in vector form of dimension $d \times 1$) is needed so that the deviation of the ith correlation plane in Equation 10.12 from the ideal shape vector $\mathbf{f}$ will be minimised. This deviation can be quantified in terms of average squared error (ASE) as

$$\text{ASE} = \frac{1}{N} \sum_{i=1}^{N} |\mathbf{g}_i - \mathbf{f}|^2 \tag{10.19}$$

Minimising ASE by setting $\nabla_{\mathbf{f}}(ASE) = 0$ the optimum shape vector is obtained as

$$\mathbf{f}_{\text{opt}} = \frac{1}{N} \sum_{i=1}^{N} \mathbf{g}_i = \frac{1}{N} \sum_{i=1}^{N} \bar{\mathbf{X}}_i^* \mathbf{h} = \bar{\mathbf{M}}^* \mathbf{h} \tag{10.20}$$

where $\bar{\mathbf{M}} = \frac{1}{N} \sum_{i=1}^{N} \bar{\mathbf{X}}_i$.

Equation 10.20 represents the average correlation plane, and $\bar{\mathbf{M}}$ is the average training image expressed in diagonal form. The average correlation plane $\bar{\mathbf{M}}^* \mathbf{h}$ offers minimum ASE out of all possible reference shapes and hence the least distortion in squared error sense is achieved. The average similarity measure (ASM) is obtained from Equation 10.19 by substituting $\mathbf{f} = \mathbf{f}_{\text{opt}} = \bar{\mathbf{M}}^* \mathbf{h}$ and $\mathbf{g}_i = \bar{\mathbf{X}}_i^* \mathbf{h}$

$$\text{ASM} = \frac{1}{N} \sum_{i=1}^{N} |\bar{\mathbf{X}}_i^* \mathbf{h} - \bar{\mathbf{M}}^* \mathbf{h}|^2 = \mathbf{h}^+ \bar{\mathbf{S}} \mathbf{h} \tag{10.21}$$

where

$$\bar{\mathbf{S}} = \frac{1}{N} \sum_{i=1}^{N} (\bar{\mathbf{X}}_i - \bar{\mathbf{M}})(\bar{\mathbf{X}}_i - \bar{\mathbf{M}})^* \tag{10.22}$$

is a $(d \times d)$ diagonal matrix measuring the similarity of the training images to the class mean in the frequency domain. Another filter design criteria is formulated as maximising the correlation peak intensity of the average correlation plane instead of specifying values at the correlation planes for each training image. Mathematically the peak intensity of the average correlation plane can be written as

$$|g(0,0)|^2 = |\mathbf{m}^+ \mathbf{h}|^2 = \mathbf{h}^+ \mathbf{m} \mathbf{m}^+ \mathbf{h} \tag{10.23}$$

where

$$\mathbf{m} = \frac{1}{N} \sum_{i=1}^{N} \mathbf{x}_i \tag{10.24}$$

represents the mean vector corresponding to training vectors $\mathbf{x}_i$ for all $i = 1, 2, \cdots, N$. Now the behaviour of the average correlation plane is explicitly optimised by minimising ASM and maximising peak value. Hence the criteria to be optimised to improve distortion tolerance is

$$J(\mathbf{h}) = \frac{\mathbf{h}^+\mathbf{m}\mathbf{m}^+\mathbf{h}}{\mathbf{h}^+\bar{\mathbf{S}}\mathbf{h}} \qquad (10.25)$$

and is referred to as the average correlation height criterion. The filter of interest $\mathbf{h}$ maximises this criterion and thus is called a maximum average correlation height (MACH) filter. The MACH filter maximises the relative height of average correlation peak with respect to expected distortions. Since $J(\mathbf{h})$ in Equation 10.25 results in a small denominator, the filter $\mathbf{h}$ will reduce ASM given in Equation 10.21. The optimum filter is found by setting the gradient of $J(\mathbf{h})$ with respect to $\mathbf{h}$ to zero and is given by [17].

$$
\begin{aligned}
\nabla_h\{J(\mathbf{H})\} &= \frac{(\mathbf{h}^+\bar{\mathbf{S}}\mathbf{h})(2\mathbf{m}\mathbf{m}^+\mathbf{h}) - \mathbf{h}^+\mathbf{m}\mathbf{m}^+\mathbf{h}(2\bar{\mathbf{S}}\mathbf{h})}{(\mathbf{h}^+\bar{\mathbf{S}}\mathbf{h})^2} = 0 \\
&= \frac{\mathbf{m}\mathbf{m}^+\mathbf{h}}{\mathbf{h}^+\bar{\mathbf{S}}\mathbf{h}} - \frac{\mathbf{h}^+\mathbf{m}\mathbf{m}^+\mathbf{h}(\bar{\mathbf{S}}\mathbf{h})}{(\mathbf{h}^+\bar{\mathbf{S}}\mathbf{h})^2} = 0 \qquad (10.26)
\end{aligned}
$$

This can be simplified to

$$\mathbf{m}\mathbf{m}^+\mathbf{h} = \lambda\bar{\mathbf{S}}\mathbf{h} \qquad (10.27)$$

where

$$\lambda = \frac{\mathbf{h}^+\mathbf{m}\mathbf{m}^+\mathbf{h}}{\mathbf{h}^+\bar{\mathbf{S}}\mathbf{h}} \qquad (10.28)$$

is the scalar identical to $J(\mathbf{h})$. Considering $\bar{\mathbf{S}}$ is invertible[1], Equation 10.27 can be rewritten as

$$\bar{\mathbf{S}}^{-1}\mathbf{m}\mathbf{m}^+\mathbf{h} = \lambda\mathbf{h} \qquad (10.29)$$

Equation 10.27 represents a generalized eigenvalue problem and from Equation 10.29 it can be stated that $\mathbf{h}$ is the eigenvector of $\bar{\mathbf{S}}^{-1}\mathbf{m}\mathbf{m}^+$ with corresponding eigenvalue λ. Since λ in Equation 10.28 is identical to $J(\mathbf{h})$, the eigenvector corresponding to the largest eigenvalue λ is to be selected to maximise $J(\mathbf{h})$. Since $\mathbf{m}\mathbf{m}^+$ is the outer product of a vector, $\bar{\mathbf{S}}^{-1}\mathbf{m}\mathbf{m}^+$ has only one non-zero eigenvalue. The corresponding eigenvector is then the obvious choice for the optimum filter and can be found by substituting $\mathbf{m}^+\mathbf{h} = \mu$ (a scalar) in Equation 10.29 so that

$$\mu\bar{\mathbf{S}}^{-1}\mathbf{m} = \lambda\mathbf{h} \qquad (10.30)$$

or,

$$\mathbf{h}_{\text{MACH}} = \frac{\mu}{\lambda}\bar{\mathbf{S}}^{-1}\mathbf{m} \qquad (10.31)$$

where $\mathbf{h}_{\text{MACH}}$ is the desired MACH filter, the transformed class-dependent mean image.

---

[1]It is assumed that the training vectors are linearly independent of each other.

10.3.2.6   UMACE filter design

Replacing $\bar{\mathbf{S}}$ by $\bar{\mathbf{D}}$ in Equation 10.31, the closed form solution of the UMACE filter is obtained as

$$\mathbf{h}_{\text{UMACE}} = \frac{\mu}{\lambda}\bar{\mathbf{D}}^{-1}\mathbf{m} \tag{10.32}$$

where $\frac{\mu}{\lambda}$ is a constant term and does not affect the performance of the filter. Generally $\frac{\mu}{\lambda}$ is set to 1.

10.3.2.7   OTMACH filter design

It has been shown in [19], [17] that MACH filter and its other variants, most notably, the optimal trade-off MACH (OTMACH) is a very powerful correlation filter algorithm. In practice, other performance measures like ACE, ONV are also considered to balance the system performance for different applications. The optimal trade-off approach was developed by relating correlation plane metrics such as ONV, ACE, ASM, and ACH. The performance of the OTMACH filter can be improved by minimising the energy function $E(\mathbf{h})$ of the correlation filter $\mathbf{h}$, given by,

$$E(\mathbf{h}) = \alpha(ONV) + \beta(ACE) + \gamma(ASM) - \delta(ACH) \tag{10.33}$$
$$= \alpha\mathbf{h}^{+}\bar{\mathbf{O}}\mathbf{h} + \beta\mathbf{h}^{+}\bar{\mathbf{D}}\mathbf{h} + \gamma\mathbf{h}^{+}\bar{\mathbf{S}}\mathbf{h} - \delta|\mathbf{m}^{+}\mathbf{h}|^2 \tag{10.34}$$

These considerations lead to the expression for the OTMACH filter, as

$$\mathbf{h}_{\text{OTMACH}} = \frac{\mathbf{m}}{\alpha\bar{\mathbf{O}} + \beta\bar{\mathbf{D}} + \gamma\bar{\mathbf{S}}} \tag{10.35}$$

where α, β, and γ are the non-negative optimal trade-off parameters.

10.3.3   Hybrid digital-photonic correlation

Vander Lugt correlator (VLC) requires very stringent alignment of the $4f$ optical setup. To overcome the drawbacks of VLC, a hybrid digital-photonic correlation architecture is used as shown in Figure 10.5.

The Fourier-transformed image is digitally multiplied with the pre-synthesized filter and displayed onto an SLM. The SLM is illuminated with a coherent collimated laser beam. Optical Fourier transform of the product-image displayed on the SLM is performed using a Fourier transform (FT) lens. The obtained correlation peaks are analysed for detection of the test or target image. The SLM shown in the figure is of the reflective type and hence a polarizing beam-splitter (PBS) is used to view the correlation peaks on a CCD camera. The hybrid digital-optical correlator offers several advantages over the VLC geometry:

(a) The input SLM is not required, and hence a precise spectrum is available for subsequent digital mixing with the stored templates.

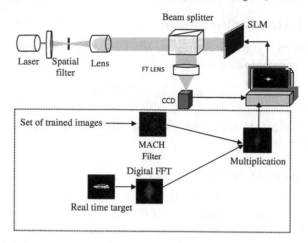

FIGURE 10.5: Optical arrangement of the hybrid digital-optical correlation architecture

(b) Unlike with 4f optical VLC architecture, the necessity to maintain a match of an optically formed spectrum to the frequency plane SLM where the filter is to be displayed is not necessary.

(c) A two-dimensional FT at video rates and video resolution is easily computed.

(d) The optical design and alignment are much simplified from those of an all-optical correlator since only a single optical Fourier transform is required.

(e) It has reduced mechanical alignment difficulties and improved robustness of the device to external mechanical disturbances.

The hybrid approach takes care of the variation of speed between performing the FT of the input signal and that required for template search.

10.3.4 Joint transform correlation

Joint transform correlator (JTC) was first demonstrated by Weaver and Goodman [1] a few years after VLC was reported. It was experimentally demonstrated again by several others after [2]. In JTC, a complete databank of reference images needs to be stored. The input test pattern and each stored reference are jointly displayed in the input plane of a Fourier transform (FT) lens to get the joint power spectrum (JPS) at the back focal plane of the lens. The captured JPS is then inverse Fourier transformed to achieve correlation. Let $f(x, y)$ be the image input to a JTC, consisting of the test or target image $t(x + a, y - b)$ and a stored image $r(x - a, y + b)$ serving as the reference image and separated by a distance $2(a^2 + b^2)^{1/2}$; then

$$f(x, y) = t(x + a, y - b) + r(x - a, y + b) \tag{10.36}$$

The Fourier transform $F(u,v)$ of $f(x,y)$ produces the JPS given by

$$
\begin{aligned}
I(u,v) = |F(u,v)|^2 &= T^2(u,v) + R^2(u,v)\\
&+ R(u,v)T^*(u,v)e^{[j2(-au+bv)]} + T(u,v)R^*(u,v)e^{[j2(au-bv)]}
\end{aligned} \tag{10.37}
$$

where (u,v) are the frequency-domain coordinates and $T(u,v)$ and $R(u,v)$ are the Fourier transforms of $t(x,y)$ and $r(x,y)$, respectively. The symbol $*$ denotes the complex conjugate. A second FT of $I(u,v)$ gives the correlation output

$$
\begin{aligned}
C(x,y) = t(x,y) \otimes t(x,y) &+ r(x,y) \otimes r(x,y)\\
&+ r(x-2a,y+2b) \otimes t(x-2a,y+2b)\\
&+ t(x+2a,y-2b) \otimes r(x+2a,y-2b)
\end{aligned} \tag{10.38}
$$

where the symbol $\otimes$ denotes the correlation operator.

The first two terms of correlation output $C(x,y)$ produce the on-axis autocorrelation terms formed by the correlation of reference and target signals with themselves and gives rise to dc or zero-order correlation, whereas the third and fourth terms are the correlation terms of the reference and target. Thus auto-correlation peaks are produced, only if there is complete matching between the target and the reference pattern. The JTC technique fails if the test pattern undergoes distortions like rotation, scaling, etc. The schematic of the JTC architecture is shown in Figure 10.6.

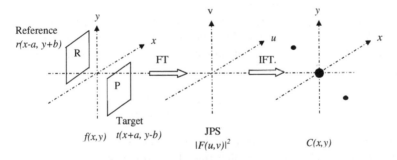

FIGURE 10.6: Schematic diagram of the JTC architecture

JTC in general suffers from low detection efficiency, particularly when embedded in noise or in an intense background or when the target is subject to rotations, scaling, etc. Nonetheless, it has several advantages like easy real-time implementation and higher space-bandwidth product. Another disadvantage of JTC is the very intense zero-order correlation peak, which suppresses the intensity of the auto-correlation peaks.

One of the problems associated with the classical JTC is the presence of an undesirable strong zero order autocorrelation peak at the output plane, which almost overshadows the desired cross correlation peaks. This situation is more bizarre in the case of a noisy environment because the zero order peaks may over-saturate the output detector and cause spurious detection. To over-come some of the detection problem, a binary JTC technique is used. In this technique the joint power spectrum is binarised by hard clipping nonlinearly at the Fourier plane before applying the inverse Fourier transform operation. Only two binary values +1 and −1 are permitted. Binary JTC is found to yield superior correlation peak intensity, better correlation width, and higher discrimination sensitivity. The binary JTC, however, suffers from the low processing speed due to the calculation of the threshold value used for binarisation of the joint power spectrum (JPS). Moreover, the binarisation produces harmonic correlation peaks and therefore some energy is distributed among these higher order harmonic terms. In addition, the higher-order harmonic terms may yield false alarms or may result in misses, thereby complicating the target detection process.

To alleviate some of the problems of false alarm, the technique of median thresholding for binarisation is sometimes used. It has been observed that for large input scene noise, the performance of the binary JTC using the noise-dependent threshold is better than using the fixed threshold. In this case, the threshold value is updated for every new input scene. The evaluation of the median value for large arrays can be computationally intensive in terms of both software and hardware implementation. For this reason, another method called subset median thresholding is evaluated for thresholding the joint power spectrum. It also takes into account the effect of the input scene noise to compute the threshold value.

In another approach, the JPS is multiplied by an amplitude modulated filter (AMF) function before the inverse Fourier transform is attempted for getting better correlation performance than the classical and binary JTC technique. The AMF is defined as

$$H_{amf}(u, v) = \frac{1}{|R(u, v)|^2} \tag{10.39}$$

where $|R(u, v)|$ is the Fourier transform of the reference signal $r(u, v)$. However, the factor $|R(u, v)|^{-2}$, which may be associated with one or more poles may force the gain of the AMF to approach infinity. The problem imposes a serious restriction on the realisation of this technique. Moreover, the advantage of having AMF is somewhat masked by high optical gain for lower values of the reference signal, which may actually degrade the noise performance of the JTC.

A JTC technique that uses a fractional Fourier plane for correlation was also reported. In this technique, a phase-only spatial light modulator (SLM) is used at the Fourier plane. The analysis/simulation results are applicable to only those joint power spectra of reference images that do not contain any zeros. To alleviate the problems and at the same time to increase the autocorrelation peak intensity, a fringe adjusted JTC technique was developed in which a real valued filter called a fringe adjusted filter (FAF) is used. The

FAF function is defined as

$$H_{faf}(u,v) = \frac{B(u,v)}{A(u,v) + |R(u,v)|^2} \qquad (10.40)$$

where $A(u,v)$ and $B(u,v)$ are either constants or functions. The value of $B(u,v)$ is chosen so as to avoid the optical gain greater than unity. A small value of $A(u,v)$ can solve the pole problem, and also at the same time a high correlation peak is obtained. The function $A(u,v)$ is so chosen that it suppresses the effect of noise in the band limit signal. In this technique, a larger and sharper auto-correlation peak may be obtained using proper amplitude matching. The FAF function contains only the intensity and has no phase term. So it is simple and suitable for optical implementation. As the computation time for calculating FAF is very small, it has no significant detrimental effect on the processing speed of the system. One of the disadvantages of this method is that it requires an additional SLM to display the filter function at the Fourier plane. A general flow chart for obtaining JTC for pattern recognition is shown in Figure 10.7.

Another class of correlation technique is known as chirp-encoded JTC. The technique produces three output correlation functions at different planes. The autocorrelation functions on the optical axis are focused on one of the output planes and the off-axis cross-correlation is produced in two separate output planes. In this system the reference signal is placed in different input planes, which encode the joint power spectrum with a function (that is, the chirp function) for each correlation term.

In general, the reference image used in a JTC is presented to the test system from the computer memory while the test scene is input from the outside world and may or may not contain the target of interest. The input scene illumination can therefore be entirely different from that of the reference. This poses a serious drawback in the JTC technique where the illumination balance of the target with that of the reference determines the correlation performance. In a real-time situation, it is almost impossible to employ such a system in a non-cooperative environment where no real control over the test scene is possible. To alleviate the detrimental effects of illumination on the correlation output of a JTC, a polarisation-encoded two channel JTC architecture is proposed. An intensity-compensated filter to sharpen the correlation peak for a JTC has also been employed by using the inverse reference power spectrum. The technique can be implemented by using the inverse of the pre-processed reference spectrum, which is independent of input parameter. A hybrid system has also been developed that performs the JTC with a multi reference by using a self-generated threshold function. The intensity problem of JTC can be solved by an intensity compensation filter.

10.3.4.1 Photonic joint transform correlator

The schematic diagram of the classical JTC technique is shown in Figure 10.8. The target scene is captured by a CCD camera (CCD_0) and is displayed

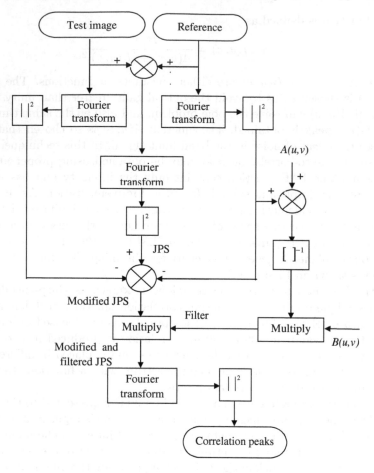

FIGURE 10.7: Flowchart for obtaining JTC

jointly at the input plane P_1 along with the reference scene on a spatial light modulator (SLM$_1$).

A coherent parallel light beam illuminates the SLM$_1$. The joint image displayed at the SLM is then Fourier transformed by lens L_1 and the joint diffraction pattern is produced at the plane P_2. The joint power spectrum is recorded by a CCD camera ((CCD$_1$) located at plane P_2. A second SLM (SLM$_2$) is located at plane P_3 to read out the joint power spectrum. The correlation function is produced at plane P_4 by taking the inverse Fourier transform of the joint power spectrum located at the plane P_3. For obtaining Fourier transform a parallel coherent laser beam is required, which is obtained from the primary laser source by using a beam splitter and a mirror.

Important parameters of the JTC are (a) cross correlation peak intensity, (b) the ratio of correlation peak intensity to the maximum correlation side lobe

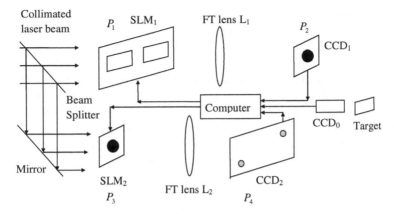

FIGURE 10.8: Schematic diagram of photonic JTC

intensity, (c) full correlation width at half maximum, (d) correlation width, and (e) ratio of the correlation peak amplitude to the rms deviation of the noise amplitude.

It has already been mentioned that the classical joint transform image correlation technique suffers from low light efficiency, large correlation side-lobes, large correlation width, and low discrimination ability. These parameters can be significantly improved in the single SLM JTC technique where the amplitudes of input signal and the reference signal are binarised to two values, $+1$ and -1. There would also be a reduction in the memory space required to store the binary reference signals compared to storing space necessary for the continuous tone reference images. A single SLM, as shown in Figure 10.9, can be used to display the thresholded input signals and the thresholded Fourier transform interference intensity.

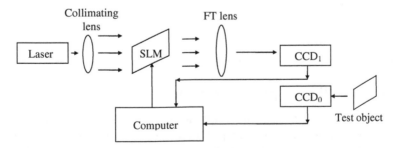

FIGURE 10.9: Schematic diagram of single SLM photonic JTC

The thresholded input and reference signals enter the SLM, which operates in the binary mode. The interference between the Fourier transforms of the input signals is obtained using a lens, and a CCD_1 is used to produce the transform interference intensity distribution. The interference intensity is then thresholded to +1 and −1 values. The binarised interference intensity is then written on the same SLM. The inverse Fourier transform of the thresholded interference intensity produces the correlation signals. CCD_0 is used to capture the image of the test object.

Many variations of the basic scheme are now available for the task of pattern recognition using JCT. Each of the architectures have their respective drawbacks. Although JTC is more robust and easily implementable through a two-step process, it produces a highly intense zero-order diffraction pattern that at times overshadows the presence of correlation peaks in the detection plane. To overcome this problem, studies have suggested a non-zero order differential JTC.

A two-stage JCT architecture as shown in Figure 10.10 gives some flexibility in accommodating various operational algorithms. A Fourier transform produces the JPS, which is captured using a CCD camera in JTC. This JPS, presented through an SLM, is illuminated by a collimated laser light and the correlation peaks captured through a CCD camera.

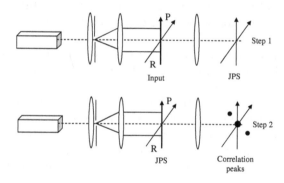

FIGURE 10.10: Optical arrangement of the JTC architecture

10.4 Intelligent photonic processing

Basically there are two main approaches in the photonic implementation of pattern recognition, namely, the photonic correlator which has been described in previous sections and through the simulation of artificial neural networks. The study of autonomous surveillance or image identification using neural control systems is a key enabling technology for future information management systems. The ability of systems to detect, filter out, and process data to arrive at some optimum decision or inference under uncertain, non-stationary, or ambiguous situations can be referred to as intelligent information processing. Although there have been many methods and technologies, the systems mostly follow the logic of working of human brain. The functions of human brain is best simulated by artificial neural networks.

10.5 Artificial neural network (ANN)

An artificial neural network (ANN) consists of an ensemble of highly parallel and interconnected processing elements (PEs) called neurons, analogous to the human brain or the biological neural network [30]. A connection or link between a pair of PEs has an associated numerical strength, called synaptic weight or adaptive coefficient. An ANN architecture is formed by organizing PEs into fields or layers and linking them with weighted interconnections. There are two types of primary connections, excitatory and inhibitory. Excitatory interconnections increase a PE's activation and are typically represented by positive (rewarding) signals. Inhibitory interconnections decrease a PE's activation and are normally represented by negative (punishing) signals. Some of the features of ANN are:

1. Adaptability: The ability of ANN to automatically learn to respond to a newly encountered input pattern by adapting the synaptic weights connecting the PEs in the network is termed adaptability.

2. Distributed storage and associated memory: Any information within an ANN is distributed and encoded in the connections and not stored in a specific location. Also, ANN has an associative memory which accesses information by content.

3. Fault tolerance: Since information is distributed across many PEs in the network, information is not lost due to damage of a few PEs or links. ANNs have the ability to tolerate hardware malfunctions and hence are quite fault-tolerant.

Figure 10.11 displays the anatomy of a generic PE. This model of connected neurons was later named a perceptron [31]. The input signal comes either from the environment or from the output of other PEs and it forms the input vector $\mathbf{X} = (x_1, ...x_i...x_n)$. The weights associated with each input form the weight vector $\mathbf{W}_j = (w_{1j}, ...w_{ij}...w_{nj})$. For the jth PE, where w_{ij} represents the connection strength from the ith input to the jth PE. There may also be a threshold θ weighted by w_{0j} associated with each PE. All weights, also referred to as interconnection strengths, are used to compute the output value y_j of the jth PE. y_j computation is typically performed by taking the dot product of $\mathbf{X}$ and $\mathbf{W}_j$, subtracting the threshold weight, and passing the result through a threshold function. The instantaneous input-output relation can be expressed

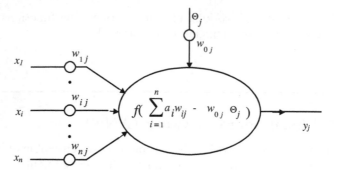

FIGURE 10.11: A simple processing element (PE)

as

$$x_j = \sum_{i=1}^{n} w_{ij} y_j - w_{0i} \theta_i \qquad (10.41)$$

The short term average of the input-output relation is expressed as

$$y_i = f(\lambda x_j) \qquad (10.42)$$

where λ is a positive number and $f(.)$ is the threshold function.

The threshold function $f(.)$, also referred to as the transfer function, maps the output value of a PE to a pre-specified range. Commonly used threshold functions are: (i) nonlinear ramp function, (ii) relay function, and (iii) sigmoid function, as shown in Figure 10.12. The most useful threshold function is, however, the sigmoid threshold function. The sigmoid (S-shaped) function is a bounded, monotonic, and non-decreasing function that provides a graded nonlinear response. A typical sigmoid function is defined as

$$S(x) = (1 + e^{-x})^{-1} \qquad (10.43)$$

Incorporating sigmoid function output y_i can be expressed in terms of inputs x_i as

$$y_i = \frac{1}{1 + e^{-\lambda x_i}} \qquad (10.44)$$

and it approaches the unipolar threshold as $\lambda \longrightarrow \infty$.

Learning is the most important feature of an ANN. All learning methods can be classified into two categories: (a) supervised learning and (b) unsupervised learning. Supervised learning is a process that incorporates an external guidance for incorporating global information. The supervisor decides when to turn off the learning and how often to present the inputs for training and supplying information according to performance error. Unsupervised learning, also referred to as self-organization, is a process that incorporates no external guidance and relies only upon local information. Techniques of supervised learning are broadly classified as:

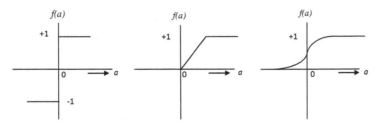

FIGURE 10.12: Transfer functions: (a) step function, (b) linear threshold function, (c) sigmoid function

(a) **Error-correction learning:** This technique adjusts the connection weights between the input and output PEs in proportion to the difference between the desired and computed values of each PE in the output layer. If the desired value of the jth output PE is c_j and the computed value of the same PE is y_j, then a general equation adapting the inter-field weight correction between the ith input layer PE x_i, and the jth output layer PE y_j is defined as

$$\Delta w_{ij} = \alpha x_i [c_j - y_j] \qquad (10.45)$$

where Δw_{ij} is the change in interconnection strength from x_i to y_j and α is the learning rate, typically lies between, $0 < \alpha < 1$.

(b) **Reinforcement learning:** This technique of learning is similar to error-correction learning except that the weights are reinforced for properly performed actions and punished for poor performance. The error-correction learning requires an error value for each output layer PE (i.e., an error vector) whereas the reinforcement learning requires only one value to describe the output layer's performance [32]. The general reinforcement learning equation is defined as

$$\Delta w_{ij} = \alpha [r - \theta_j] e_{ij} \qquad (10.46)$$

where, r is the scalar success/failure value provided by the environment, θ_j is the reinforcement threshold value for the jth output PE and e_{ij} is termed the canonical eligibility between the ith input PE, and the jth output layer PE. The e_{ij} value depends on a previously selected probability distribution and is used to determine whether the computed output value equals the desired output value.

(c) **Stochastic learning:** This technique uses a random process relating a probability distribution and an energy relationship to adjust the memory connection weights [33], [34]. Stochastic learning makes a random weight change, determines the resultant energy created after the change, and keeps the weight change according to a given criteria, such as: (i) if the ANN energy is lower after the random weight change, (i.e., the system performs better), then accept the change, (ii) if the energy is not lower after the random weight change accept the weight change according to the pre-chosen probability distribution, otherwise (iii) reject the change. The acceptance of some changes despite poorer performance allows the ANN to escape local energy minima in favour of a deeper energy minima, by a process known as simulated annealing [35]. This process slowly decreases the number of probabilistically accepted weight changes that result in a poorer performance.

(d) Unsupervised learning: This general form of learning, also referred to as self-organization, is a process that incorporates no external guidance and relies upon only local information. The unsupervised learning method self-organises presented data and discovers its emergent collective property. Examples of unsupervised learning include: (a) Hebbian learning and (b) competitive learning.

(a) Hebbian learning was formulated as the initial concept of correlation learning, and is concerned with the adjustment of a connection weight according to the correlation of the values of two interconnected PEs. Simple Hebbian correlation learning where the weight value w_{ij} is the correlation of PE x_i with PE y_j, is defined as

$$\frac{dw_{ij}}{dt} = -w_{ij} + x_i y_j \qquad (10.47)$$

Two extensions of simple Hebbian learning are (i) signal Hebbian learning and [36] (ii) differential Hebbian learning [37]. Signal Hebbian learning is the correlation of activations filtered through a sigmoid function. Differential Hebbian learning is the correlation of the changes in the signals between PEs.

(b) Competitive learning technique is utilised in a pattern classification procedure for conditioning intra-field connections. It is described as either neighbour-inhibiting (competitive) or neighbour-exciting (cooperative). Competitive learning works during PE activations in the following manner: (i) an input pattern is presented to a layer of PEs, (ii) each PE competes with the others by sending positive signals to itself (self-exciting) and negative signals to all its neighbours (neighbour-inhibiting), and (iii) eventually the PE with the greatest activation is singularly active and all others are nullified.

10.5.1   Training of the ANN

Training the ANN can be done through either by the supervisory or unsupervisory mode. Based on this classification, ANN networks, like the perceptron model, multilayer perceptron, Hopfield model, and Boltzmann machine belong to the class of supervised ANNs, whereas Kohonen's self-organizing feature map and Carpenter/Grossberg model belong to the class of unsupervised ANNs.

Supervised ANNs require supervised training wherein the ANN is supplied with a known sequence of inputs $(x_1, x_2, ...x_k....)$ and the desirable or correct outputs $(y_1, y_2, ...y_k....)$ expected after processing in the ANN. For this, the

ANN undergoes an iterative process such that the output obtained is compared with the desired output, and the difference if any, is corrected by using some learning algorithm by modifying the synaptic weights. The process is repeated until the actual output reaches an acceptable value close to the desired output.

The perceptron model [38] is a single-layer network consisting of one or more PEs. It is a supervised ANN model, in which the inputs are either binary or continuous-value. The inputs are used to train the network. It is basically a form of linear classifier, which maps its input x to the output $f(x)$. When the input pattern $(x_1, x_2, ...x_n)$ is applied at the input, it gets multiplied with the corresponding interconnection weights $w_1, w_2, ...w_n$. The addition of all the weighted inputs is thresholded to represent output y of the PE. The model is shown in Figure 10.13.

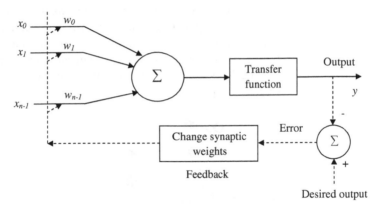

FIGURE 10.13: Perceptron model

If θ is the threshold function for the PE, then the output y at time t may be written as

$$y(t) = \sum_{i=0}^{n} w_i(t)x_i(t) - \theta \tag{10.48}$$

In a two class $\{0, 1\}$ classification problem, the output y of the ANN is $+1$, if the weighted sum is greater than the threshold value and the input belongs to Class 1, whereas the output y of the ANN is -1, if the weighted sum is less than the threshold value, and the input then belongs to Class 0. The output y of the ANN at time t as written in Equation 10.48 becomes

$$y(t) = \begin{cases} +1 & \text{if } y(t > 0); \\ -1 & otherwise \end{cases} \tag{10.49}$$

The synaptic weights at time $(t+1)$ are modified on the basis of the difference $\triangle$ between the desired $d_i(t)$ and obtained output $y_i(t)$ in such a way that

$$w_i(t+1) = w_i(t) + \eta\triangle \tag{10.50}$$

where $0.1 < \eta < 1.0$ controls the adaptation rate of the weights.

10.5.2 Linear pattern classifier and multilayer perceptron

The usefulness of the simple neuron model is exemplified in a pattern recognition technique known as linear classification. To classify an input into say, two possible classes A or B by a linear decision boundary. Two inputs to the PE are (x_1 and x_2), and the output is y. It is assumed that the output $y = 0$ identifies a particular case. It is also assumed that the threshold element is a simple relay function. Then for a one-dimensional case, the following equation can be written

$$x_2 = \frac{w_1}{w_2}x_1 + \frac{w_0}{w_2}\theta = 0 \qquad (10.51)$$

Therefore, by controlling the slope and intercept of the straight line by changing the weights, the two classes A and B can be linearly separated. However, there are many situations where the division between the classes are complex in nature and cannot be separated by a straight line. Such a situation is an Ex-OR problem. In no case will a straight line be able to separate two classes. However, combining two PEs into another PE can solve such a problem as shown in Figure 10.14. PE1 detects when the input pattern corresponds to (0,1) and PE2 detects when the pattern is (1,0).

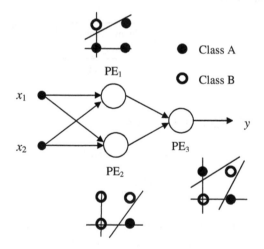

FIGURE 10.14: Ex-OR classification problem

Combining these two effects by another PE (designated as PE3) allows classification of the input correctly. However, the second layer or the PE3 does not know which of the real inputs are on or not. In fact, the hard limiting threshold function removes the information that is needed if the network is to learn successfully, that is, no weight change in the input of the PE1 and PE2 will be able to change the input of the PE3. The problem is taken care of by changing the hard-limiting function by a sigmoid function. This simple exercise establishes the necessity of a multilayer perceptron network model. The three-layer model has an input layer, an output layer, and a layer in between known as a hidden layer. The thresholding function is usually a sigmoid or linear ramp threshold. There can be many hidden layers in a multilayer perceptron model.

For a one-layer neural network of N neurons or PEs, there would be approximately N^2 interconnections. Hence the state of the ith neuron can be expressed as

$$y_j(t) = \sum_{j=0}^{n-1} W_{ij}(t)x_i(t) - \theta_j \qquad (10.52)$$

where W_{ij} is the interconnection weight matrix between the ith and jth neuron. The right side term of Equation 10.52 is basically a matrix-vector outer-product.

There are different interconnection schemes for the PEs. The two primary schemes are the intra-field and the inter-field interconnections. Intra-field connections, also referred to as lateral connections, are connections between the PEs of same layer. Inter-field connections are the connections between the PEs in different layers. Inter-field connection signals propagate in one of two ways—either feed forward or feedback. Feed-forward signals only allow information to flow amongst PEs in one direction, whereas feedback signals allow information to flow amongst PEs in either direction. The layer that receives input signals from the environment is called the input layer, and the layer that emits signals to the environment is called the output layer. Any layer that lies between the input and the output layer is called the hidden layer and has no direct contact with the environment.

A multilayer perceptron [39] is a feed-forward ANN where the first layer is the input layer, as opposed to the nth layer, which is the output layer. All layers in between the input and output layers are hidden layers, as shown in Figure 10.15. A node in a particular layer is connected to all nodes of the previous layer as well as all nodes of the subsequent layer. The number of hidden layers and the number of nodes in the hidden layers are determined by the proper internal representation of the input patterns best suited for classification of linearly inseparable problems.

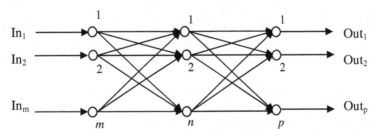

FIGURE 10.15: Multilayer perceptron model

For multilayer perceptrons, where a hidden layer exists, training algorithm such as back-propagation algorithm is used. Since this is a supervised ANN, the input and the desired output are both fed to the network. An error function δ is defined which computes the difference between the desired and obtained output. In order to train the ANN with the given input and desired output, the error function is to be made minimum by adjusting the connecting synaptic weights between the nodes. The back-propagation algorithm can be described by the following steps:

Step 1: All weights and thresholds are set to small random values.

Step 2: The input vector $(x_1 x_1 ... x_m)$ is presented to the m nodes and the desired output vector $(d_1 d_2 ... d_n)$ is fed into the n output nodes.

Step 3: The actual output vector $(y_1 y_2 ... y_n)$ at any instant t is calculated using the sigmoidal non-linear threshold θ as

$$y_j(t) = \sum_{j=0}^{n} w_{ij}(t) x_i(t) - \theta_j \tag{10.53}$$

which becomes the input for the next layer.

Step 4: The weights on the links are adjusted using a recursive algorithm with the error term δ_j used for output node j, as

$$w_{ij}(t+1) = w_{ij}(t) + \eta \delta_j y_j \tag{10.54}$$

where η is the gain term.

The error term can be written for output nodes as $\delta_j = k y_j (1 - y_j)(d_j - y_j)$ and for hidden nodes as $\delta_j = k y_j (1 - y_j) \Sigma_k (\delta_k w_{jk})$, where the sum is over the k nodes in the hidden layer above node j.

To avoid the network getting stuck at a local minima and to make the convergence faster, another momentum term α can be introduced in a form such that Equation 10.54 can be replaced as

$$w_{ij}(t+1) = w_{ij}(t) + \eta \delta_j y_j + \alpha [w_{ij}(t) - w_{ij}(t-1)] \tag{10.55}$$

10.5.2.1 Hopfield model

Hopfield model is a single-layer recurrent ANN. In this type of ANN, each node or PE is connected to every other node, thus making it a fully interconnected network. The number of input nodes is equal to the number of output nodes. The output of each node is fed back to all the other nodes. The interconnection weights are the same in both directions between nodes i and j, i.e., $w_{ij} = w_{ji}$.

Each node calculates the weighted sum of the inputs, and thresholds it to determine the output. Depending on the type of input, Hopfield ANNs can either be discrete or continuous type. In the discrete type, the inputs have either (0,1) or (-1, +1) values. The interconnection weight between nodes i and j, i.e., w_{ij}, is assigned values as

$$w_{ij} = \begin{cases} \sum_{s=1}^{m} x_i^s x_j^s & \text{if, } i \neq j; \\ 0 & \text{otherwise} \end{cases} \tag{10.56}$$

Positive weights are assumed to be excitatory and strengthen the connections, whereas negative weight are inhibitory and weaken the interconnection. The model of Hopfield network is shown in Figure 10.16.

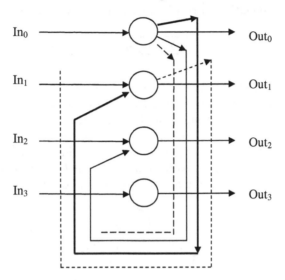

FIGURE 10.16: Hopfield model of ANN

An objective function, also referred to as an energy function, is associated with each state of the discrete Hopfield ANN, and is given by

$$E = -\frac{1}{2}\sum_{j}\sum_{i} s_i s_j w_{ij} \tag{10.57}$$

where s is the state of activation and he state of activation of the ith node is given by

$$s_i = \begin{cases} +1 & \text{if, node=1} \\ -1/0 & \text{otherwise} \end{cases} \tag{10.58}$$

The objective function E is minimised using the gradient descent rule, i.e., the difference of the global energy of the on and off states of the nodes are calculated and if the energy difference is positive, the node turns on, or else it is off. Eventually, when the energy is at a minimum, the node outputs remain unchanged, the network converges, and a stable state is reached.

In the continuous type of Hopfield ANN, the inputs are not limited to discrete on and off states only, but have continuous analogue values. Continuous Hopfield ANN uses a sigmoid non-linearity instead of a binary threshold. The nodes are governed by the equation given as

$$C_j \frac{ds_j}{dt} = \sum_{i} w_{ji} y_i - \frac{s_j}{R_j} + I_j \tag{10.59}$$

where s_j is the state of activation of jth PEC_j and R_j are positive constants, I_j is the input bias term, and y_i is the output of node i after sigmoidal threshold is applied.

The energy equation in continuous Hopfield ANN becomes

$$E = -\frac{1}{2}\sum_j \sum_i s_i s_j w_{ij} - \sum_j s_j I_j \qquad (10.60)$$

The updating of the interconnection weights can be done either in the synchronous or asynchronous way. Asynchronous updating of weight means the output of any one neuron is calculated at a time and updated immediately. In synchronous updating of weights, the total weighted sum of all inputs is calculated and thresholded to get the corresponding outputs, and the weights are updated accordingly. It has been observed that synchronous updating can lead to oscillatory states, unlike asynchronous updating of weights.

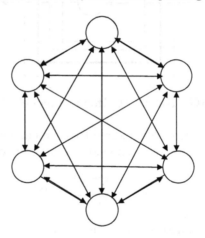

FIGURE 10.17: Hopfield auto-associative model

An associative memory can be realised by the Hopfield network. The model for implementing such an associative model is shown in Figure 10.17. A pattern is fed into the network by setting all nodes to certain specific values, i.e., the input pattern is stored in the interconnection weight matrices. The network is then made to undergo a number of iterations for weight updating, using asynchronous or synchronous technique. After quite a few iterations, when the output nodes are read, it is found that the pattern associated with the fed-in pattern is obtained. These Hopfield nets are found to serve as content-addressable memory systems, wherein even if some of the connections of the network are destroyed, the performance is degraded, but some network capabilities of recalling the associated pattern is retained. Since there is only one absolute minimum of the objective function in a Hopfield network for a particular trained pattern, and the minimum is reached after several iterations, the network must always converge to the trained pattern even if some distorted pattern is presented to it later or even if some of the nodes are destroyed.

10.5.2.2 Boltzmann machine

A stochastic version of the Hopfield model, the Boltzmann machine is also a fully interconnected ANN. Apart from input and output nodes, the remaining are hidden nodes as shown in Figure 10.18.

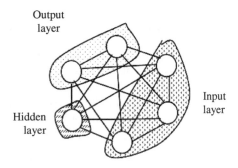

FIGURE 10.18: Boltzmann machine

The input and output nodes are selected randomly, whereas the remaining nodes are considered to be hidden nodes. Unlike the Hopfield model, which optimises the energy or objective function through a gradient descent rule to converge to a stable state, the Boltzmann machine employs simulated annealing to converge and hence does not converge to a local minima like the Hopfield ANN.

At any instant of time t, the net input to the ith node is given by

$$net_i(t) = \sum_j w_{ij}s_j(t) - \theta_i \qquad (10.61)$$

where $s_j(t)$ is the state of the jth node at time t. An asynchronous updating of the nodes occurs in this type of ANN. The probabilistic decision rule for the Boltzmann machine is such that the probability of node i having state s_j is

$$P_i = \frac{1}{1 + e^{(-\Delta E_i/T)}} \qquad (10.62)$$

where $\Delta E_i = \sum_i w_{ij}s_i$.

Such a probabilistic approach enables the Boltzmann machine to settle down to a global energy state [41]. The temperature T is gradually lowered and the probability of states calculated with each iteration. When the probability of the states with energy E_β does not change in subsequent iterations, the ANN is said to settle down to a thermal equilibrium condition.

10.5.3 Unsupervised ANN

On the other hand, the unsupervised ANNs do not require any supervised training, but are trained using graded or reinforcement training and self-organization training. In the case of graded training, the ANN is given data inputs, but no desired outputs. Instead, after each trial, the network is given a 'performance grade' or evaluation rule to tell how good the output is after upgrading. ANN automatically learns the unknown input in the process, whereas with self-organization training, the network is given only the data inputs and the ANN organizes itself into some useful configuration.

10.5.3.1 Kohonen's self-organising feature map

Kohonen's self-organising feature map is a type of ANN which has two layers, one input layer consisting of a linear array of nodes, and an output layer consisting of a rectangular array of nodes. Each node in the input layer is connected to each node of the output layer. But feedback is restricted only to immediate neighbouring nodes in the output layer. The model of Kohonen's self-organising ANN is shown in Figure 10.19.

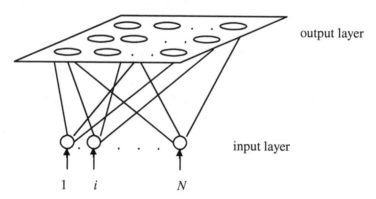

FIGURE 10.19: Kohonen's self-organising map

Like the brain, which uses spatial mapping to model complex data structures, Kohonen's self-organising ANN uses the technique of vector quantisation for performing data compression on vectors to be stored in the network. During the learning process, the nodes in the output layer are organised into local neighbourhoods forming feature maps. After the input $x_i(t)$ is presented at node i at time t to the network, the distance d_j between input node i and output node j is calculated using

$$d_j = \sum_{i=0}^{n-1} [x_i(t) - w_{ij}(t)]^2 \tag{10.63}$$

where $w_{ij}(t)$ is the connection weight from input node i to output node j.

The output node has the minimum distance d_j, called Euclidean distance. The weights to node j around the neighbourhood are updated. The updating of weights takes place in the following manner:

$$w_{ij}(t+1) = w_{ij}(t) + \eta(t)((x_i(t) - w_{ij}(t))) \qquad (10.64)$$

where $\eta(t)$ is a gain term $(0 < \eta(t) < 1)$ that decreases with time.

10.5.3.2   Carpenter-Grossberg model

Carpenter-Grossberg model is based on the adaptive resonance theory (ART). It was developed to model a massively parallel self-organizing unsupervised ANN. It displays sensitivity to information that is repeatedly shown to the network and can discriminate irrelevant information. It contains an input layer and an output layer, each consisting of a finite number of nodes. At any time, few of the output nodes are in use and the rest wait until needed. Once all the nodes are used up, the network stops adapting. The input and output nodes are extensively connected through feedforward and feedback weights. Initially, the feedforward w_{ij} and feedback t_{ij} values are taken as $w_{ij}(0) = \frac{1}{n}$ and $t_{ij} = 1$, respectively, where $1 \leq i \leq n$ and $1 \leq j \leq m$). The model is shown in Figure 10.20.

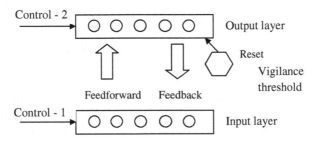

FIGURE 10.20: Carpenter-Grossberg model of adaptive resonance theory

Control-1 and Control-2 are external control signals to the input and output layer, controlling the flow of data through the network. There is a reset circuit between the input and output layers, which is used to reset the output layer and is also responsible for comparing the inputs to a vigilance threshold ρ, having values between zero and one. This vigilance threshold determines how closely an input matches a stored exemplar pattern and whether a new class of pattern should be created for an input pattern using a previously uncommitted output node. The output is calculated using

$$y_j = \sum_{i=1}^{n} w_{ij}(t)x_i \qquad (10.65)$$

The best matching exemplar governed by the equation $y_j^* = max_j[y_j]$ is then selected. The feedforward and feedback weights are updated accordingly. The test and training patterns are said to resonate when they are sufficiently similar. If there are no uncommitted nodes left in the output layer, then a new input gives no response.

10.5.4 Photonic implementation of artificial neural network

The attractiveness of the photonic implementations of a neural network has been recognised for some time. Wide varieties of architectures have been proposed and constructed in the past few years and therefore it is a daunting task to define a generic photonic neural network processor. The photonic architectures primarily exploit the parallelism and connectivity of light. Light beams do not interact in linear media such as in free-space or in glass and have the great advantage of implementing photonic interconnections in two or three dimensions. Therefore an architecture consisting of planes of neurons separated by optical systems can implement the connections between the other neuron planes. The neural planes may consist of nonlinear photonic processing elements, whereas the interconnection system may consist of hologram and/or spatial light modulators. An individual pixel or a group of pixels in an image may be considered as processing elements or neurons.

Practical systems are, however, dominated by system level and implementation technology considerations (such as connectivity constraints) as well as the computational complexity of the problem being solved. At the outset, therefore, direct comparison between optics and electronics is of little use. All electronic implementations of the neural associative structures are nowadays increasingly limited by the number and bandwidth of the interconnections and by the data storage and retrieval rates rather than by the processing power. Most of the photonic neural network systems, however, contain both optical and electronic subsystems mainly because most practical photonic processors have electronic input. Such systems must communicate with a host computer or output devices and therefore there are always possibilities of utilising the potential application area of both optics and electronics. Furthermore, implementation of any basic neural network architecture requires that a nonlinearity must be added to the system either by including nonlinear optical material or electronic circuitry at the processing nodes. Selection of a nonlinear optical material involves operational trade-off between the speed of response and magnitude of the nonlinear effect.

A good match between photonics and neural networks has been known for many years [43]. Wavefronts representing paired 2D patterns are interfered on a holographic recording medium. If one of the wavefronts is incident on the hologram, the other wavefront is reconstructed. Multiple pairs of patterns thus can be stored in this way. Some early experiments based on this principle in holographic systems were performed [44], [45]. The results of such research in the photonic implementation of neural networks have inspired workers to add gain, nonlinear feedback, and competition in creating a class of holographic memory systems. The scheme of implementing the perceptron network is done by the holographic method as shown in Figure 10.21. The input pattern is stored as $(N \times N)$ array onto a holographic plate. If the hologram is illuminated with a parallel beam of light and if the transmittance of the hologram is W_{ij} and that of the SLM is x_i, then a summed product of the transmittances $y_j(t) = \sum_j w_{ij} s_i(t)$ is obtained at the output plane.

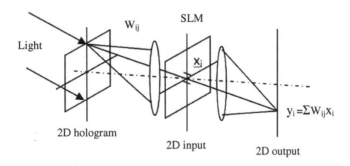

FIGURE 10.21: Holographic architecture for implementing perceptron network

Figure 10.22 shows a hybrid photonic architecture for implementing a perceptron model [46]. As seen, light from a laser source is expanded by means of a beam expander (BE), collimated, and then incident on SLM1, on which the input pattern is displayed. Through a second SLM2, the interconnection weight matrix (IWM) is displayed. The input and IWM get multiplied using the lenslet array LA. The product is summed up and captured by a CCD camera and then thresholded. The thresholded output can be viewed on the PC monitor or sent back as input for the next iteration.

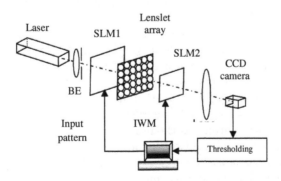

FIGURE 10.22: Hybrid photonic architecture for implementing perceptron network

The input and desired output patterns can be converted to their n-bit binary form and displayed on the SLM as inputs to the nodes of an ANN. The IWM at each node can take on values either (0,1) indicating inhibitory and excitatory or ($\bar{1}$, 0, 1) indicating inhibitory, null, and excitatory. Thus the IWM can be coded accordingly as per the bi-state or tri-state logic.

10.6 Associative memory model for 2D pattern recognition

One of the useful neural network models is the associative memory model. Insight into neural network associative memory springs from the introduction of the feedback associative structure during the development of the bidirectional associative memory (BAM), where the input/output state vectors have binary quantised components [42]. The BAM is hetero-associative but also extends to the unidirectional auto-associations. Forward and backward information flow is possible to produce a two-way associative search for a pattern pair. The construction of the associative memory model is based on an iterative thresholded matrix-vector outer product and hence can be implemented either electronically or by photonic technique.

Many attempts have been made to implement neural network models using photonic techniques, particularly in implementing the optical/photonic associative memory and Hopfield model [47], [48]. It has been shown that a straightforward adaptation of an optical matrix-vector multiplier can be used to implement an artificial neural network. Moreover, several photonic techniques of implementing associative networks for pattern recognition have been suggested, which include the integrated image technique, the joint transform correlator technique, and the holographic methods [49], [50], [51], [52]. Techniques of implementing the winner-take-all model have been proposed [53], based on a matrix-vector multiplier.

In its simplest form, the associative memory is designed to store input output pairs of patterns. When a certain input is recognised as one of the stored input then the output associated with it is retrieved, even if the input is imperfect and/or noisy. For a particular case, a pattern can associate with itself by making the same input and output responses. In that case presentation of an incomplete pattern as the input results in the recall of the complete pattern. Recall of this nature is called auto associative. Ideal storage and recall occurs when the stored vectors are randomly chosen, that is the vectors are uncorrelated. Specific storage mechanism based on the Hebbian model of learning is generally adopted to calculate the weights of interconnections which are fixed for a specific set of pattern concerned. The memory is therefore explicitly taught what it should know in a distributed manner.

10.6.1 Auto-associative memory model

One of the simplest auto-associative memory models is the Hopfield model, which has been used in a wide range of applications for pattern recognition as well as in combinatorial optimisation problems. The architecture has well defined time domain behaviour. This model assumes that the Hopfield memory consists of N neurons whose states take on the bipolar values +1 or -1. The state of the whole system is represented by a bipolar vector $A = \{a_1 a_2 .... a_N\}$. The elements a_i, $[i = 1, ...N]$ represents the states of the neurons. The associative search is performed in two phases. The first phase follows a learning procedure to calculate the connection matrix W. The second phase recalls a stored pattern vector from its imperfect input.

Let $\{A^1, A^2, ..., A^M\}$ be a set of M training patterns to be memorised, where $A^k = \{a_1^k, a_2^k, ..., a_N^k\}$. According to the Hebbian learning rule, the connection matrix W is defined as

$$W = \sum_{k=1}^{M} \left[(A^k)^\tau A^k - I_N \right] \qquad (10.66)$$

where τ denotes the transpose of the vector and I_N is an $N \times N$ identity matrix. It is noted that W is symmetric with zero diagonal elements.

The recall phase uses the asynchronous updating mode and therefore the next state $a_j(t+1)$ is determined by

$$a_j(t+1) = \Phi \left[\sum_{j=1}^{N} w_{ij} a_j \right] \qquad (10.67)$$

where $\Phi[.]$ is a threshold function and $\Phi(x) = +1$ for $x > 0$ and $\Phi(x) = -1$ for $x < 0$.

For any symmetric connection matrix W starting anywhere, the system can converge to a stable state by using the recall rule given. The learning rule, however, cannot ensure that all training patterns are stable states in the memory which is mandatory for any associative memory to operate properly. According to the recall rule a pattern A is a stable state if and only if for every neuron $a_i(t) = a_i(t+1)$. This may involve two cases: (1) $a_i = +1$, then $\sum_{j=1} w_{ij} a_j > 0$, and (2) $a_i = -1$, then $\sum_{j=1} w_{ij} a_j < 0$. Therefore, it may be noted that the pattern vector A is a stable state in the memory, if

$$\sum_{j=1}^{N} w_{ij} a_j a_i > 0 \qquad (10.68)$$

For M training patterns $A^1, A^2, ..., A^M$ are all stable states in the memory if

$$\sum_{j=1}^{N} w_{ij} a_j^k a_i^k > 0, \quad i = 1, 2, ..., N, \quad k = 1, 2, ..., M \qquad (10.69)$$

Conventionally, in a binary connection memory matrix the information to be stored is coded in the form of N-dimensional binary vectors. The storage of such vectors in the memory matrix is done by using the sum of the outer product rule as in the case of the Hopfield model. The difference lies in the facts that the interconnection weights are quantised into binary memory points (bits) and an input-dependent thresholding operation is used. The information to be stored is coded in the form of N dimensional binary vectors A^k. The storage of M such vectors into the memory matrix is to be mathematically written as

$$w_{ij} = 1, \quad \text{if} \quad \sum_{k=1}^{M} a_i^k a_j^k > 0$$

and

$$w_{ij} = 0, \quad \text{if otherwise} \tag{10.70}$$

If the memory as described in the above equation is prompted with any stored pattern vector A^t, the output of the ith neuron, y_i is found as

$$y_i = \sum_{j=1}^{N} w_{ij} a_{j^t} = S a_i^t + f_i(A^t) \tag{10.71}$$

where S is the vector strength (the number of 1) of the input vector A^t and $f_1(A^t)$ is the difference between the quantities $\sum w_{ij} a_{j^t}$ and $S a_i^t$, which is considered as the memory output signal and the second term as the output noise. If y_i is now thresholded at S, the correct auto association is obtained when $f_i(A^t)$ is less than S for all $i = 1, 2, ..., N$. One may now obtain

$$a_i^t = \Phi_S(y_i) \tag{10.72}$$

where $\Phi_S(x) = 1$, if $x \geq S$ and $\Phi_S(x) = 0$, if $x \leq S$.

It may be noted that unless the set of binary vectors are judiciously chosen (i.e., an orthogonalised set), the ability of such a binary memory matrix to yield correct auto-association would be seriously questioned. It has also been established that using the dilute vector coding, i.e., limiting the vector strength of the binary stored vectors to a small fraction of N, the retrieval error rate of the binary matrix memories can be reduced significantly with reasonable storage capacity, the optimum vector strength being of the order of $\log_2 N$. In connection with the examining of performance of a neural network memory with binary interconnection, it has been shown that incorporating a local inhibition scheme for dilute coding, nearly $0.25N$ patterns can be stored with error free recall, if N is very large and the stored patterns are completely uncorrelated. This concept of dilute vector coding is quite a severe restriction on the selection of the input pattern vectors and evidently is of little use in any real system.

To overcome this problem an intermediate class patterns vector C^k may be introduced which has known and restricted number of bits set to binary

1. Instead of storing the auto associative patterns in the memory, we store the pattern associated to its defined class pattern. Under this situation the proposed binary interconnection memory matrix T is thus defined element-wise as

$$t_{ij} = +1, \quad \text{if } \sum_{k=1}^{M} a_i^k c_j^k > 0 \tag{10.73}$$

and

$$t_{ij} = -1, \quad \text{if } \sum_{k=1}^{M} a_i^k c_j^k < 0 \tag{10.74}$$

where c_j^k is an element of the vector C^k.

If the memory is initiated by an imperfect pattern vector X^t, the output is the class pattern truly associated to any input pattern vector which is nearest to X^t. The recall of the class vector, that is the matrix response y_i where $[i = 1, 2, ..., N]$ is determined by taking the product of the imperfect pattern vector with the memory matrix T. The states of the neurons are computed by thresholding the matrix response. We used here a dynamic thresholding operation which selects all largest y_i and sets them to binary 1 and returns all other remaining values to zero. Therefore, the recalled class pattern vector may be found as

$$c_i = \Phi[y_i] \quad \text{for all } i \tag{10.75}$$

where the threshold function $\Phi[.] = 1$ for $y_i = (y_i)_{max}$ and $\Phi[.] = 0$ otherwise.

The binary class pattern vector C^k associated to the pattern which is closest to the test input vector X^t can be found from the last equation. The error-free class vector is now fed back to the memory and another multiplication with the transpose of the memory matrix t, followed by thresholding, recalls the true pattern vector, say A^k, which is closest to X^t, which is expressed mathematically as

$$z_j^k = \sum_{i=1}^{N} \tag{10.76}$$

and

$$a_j^k = \Phi[z_j^k] = 1 \text{ or } 0 \tag{10.77}$$

z_j^k represents the response of the jth output neuron and defines the jth element of the retrieved vector A^k.

Figure 10.23(a) shows the input training patterns consisting of four letters: **A, S, P,** and **D**, where **A** and **S** are associated. Input test pattern is distorted **A** as shown in the left-hand side of Figure 10.23(b). After a few iterations the distorted test pattern **A** is generated as a true pattern **A**. Since **A** is associated with **S**, the final retrieved pattern is S as shown in the right-hand side of Figure 10.23(b). This shows the capability of Hopfield model in association of patterns and removal of distortions in the test pattern.

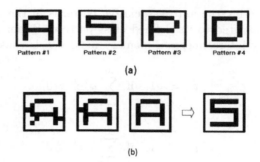

(a)

(b)

FIGURE 10.23: (a) Set of training input patterns where the associations are (A,S) and (P,D), and (b) output of associated pattern S even if the given input pattern A is distorted

10.6.1.1   Holographic storage of interconnection weight matrix

An efficient method of storage and retrieval of data related to the learning patterns is of fundamental importance in almost all neural associative formulations. Holographic memory, due to its potentially very high capacity and fast random access, offers the possibility of storing the interconnection weight matrix conveniently either in 2D or 3D format. The holographic memory is accessed in a photonic architecture for memory association.

The associative network model can be implemented using a 2D holographic storage scheme. A zonal or spatial multiplexing scheme may be utilised, where the configuration records one hologram per (x, y) location of the recording medium. As opposed to direct image or bit-by-bit storage at each location, the hologram records the interference pattern between the Fourier transform of the input bit pattern and a plane wave reference beam localised at a particular (x, y) location. One of the interesting properties of a Fourier hologram is that the location of the reconstructed image remains invariant under lateral translation of the hologram.

The input information to be stored in the memory (weight matrix) is arranged in a page-organized format. The reconstructed data are recorded in the form of 2D array of bright and dark spots representing 1 and 0 of the digital input. It has been noted that for very high information densities, constraints of the data display and detector devices favour the use of binary code.

It is necessary to store the binary input pattern vectors as digital data pages into the holographic memory and also to evolve a process of readout of data pages for the formation of interconnection weight matrix. The data input device is a page composer which consists of a 2D electrically addressed spatial light modulator (SLM). The simple matrix addressing scheme switches the SLM pixels from transparent to dark and vice-versa. The digital electrical signals are thus directly converted into a 2D array of data bits. Initially the

SLM is visualised as a matrix of unlinked wires where one vertical wire is assigned for each bit in the input pattern vector A^k and one horizontal wire represents each bit in the class vector, as shown in Figure 10.24. In the learning phase each pattern vector is presented along with its class vector. The pattern vector appears on the vertical wire whilst the class vector appears on the horizontal wire in the word parallel format. A link, i.e., a weight of $+1$ is set in the memory matrix whenever an active vertical wire crosses an active horizontal wire.

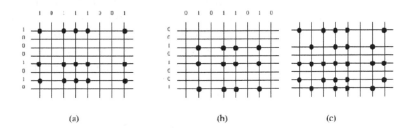

FIGURE 10.24: (a) Frame T^i showing the outer-product matrix for the vector $(A^1), C^i$, (b) frame T^j showing the outer-product matrix for the vector $(A^j), C^j$, and (c) sum of the outer product matrix $(T^i + T^j)$

The resulting binary matrix is therefore the outer-product matrix between the pattern vector A^k and its associated class vector C^k. Figure 10.24(a) shows the state of the matrix when it is taught with any arbitrary pattern vector A^i and its associated class vector C^i. This outer product matrix is considered as a frame and is labelled as T^i. Figure 10.24(b) similarly shows the outer-product matrix T^i formed with the vectors A^j and its associate C^i. The sum of the two outer-product matrices, i.e., $T = T^i + T^j$ is shown in Figure 10.24(c). According to our proposition, once formed, a link remains in place. If a new pattern requires a link in an empty position, a new link is formed. But if the position already has a link then nothing is altered. This process thus justifiably translates the learning rule as described. It is to be noted that until now the interconnection matrix T is constructed by the unipolar binary values only and thus in its true sense reflects the unipolar form. However, a bipolar binary interconnection matrix is always preferable to obtain better signal to noise ratio in associative recall formulations. One conventional approach to handle bipolar binary data is to split the T matrix into two parts as $T = T^+ - T^-$ where both T^+ and T^- are unipolar binary matrix. T^- is obtained by complementing each element of the matrix T^+ at any processing stage of the recall phase. Therefore storing the binary bipolar interconnection weights into the memory is not mandatory.

The procedure of forming the interconnection weight matrix described in connection to Figure 10.24 is implemented by the Fourier holographic technique. The basic scheme of writing different frames $(T^i, T^j$ etc.), that is the outer-product matrices, and subsequently their readout processes to form the optical sum of the stored outer-product matrices is depicted in holographic set up. A representative scheme for recording a frame T^i at any (x, y) location of the storage medium is illustrated in Figure 10.25(a). The SLM displays the frame T^i representing the object. The amplitude of the object beam at the recording plane is the Fourier transform of the data page or the frame T^i. This amplitude pattern is interfered with the reference beam at the recording plane H. The optics of the system causes the object and the reference beams to intersect at the storage medium placed at H for a particular spatial address selected corresponding to that particular plane. Figure 10.25 illustrates the repetition of the same process for frame T^j at O, in another location at the recording plane H. This process may be repeated in a similar fashion to storing several other pattern vectors appearing as different frames at the SLM plane.

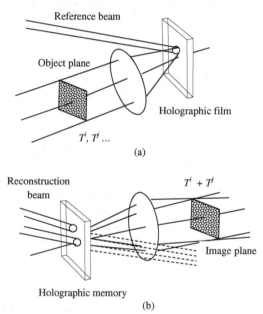

FIGURE 10.25: Scheme of (a) recording and (b) readout of outer product matrix

The readout process for reconstructing the data pages is depicted in Figure 10.25(b). Now only the reference beam is present and the reconstructed image at any plane contains all the data pages or the frames which are superposed at the image plane. It may be noted that the object plane and the image plane are the two conjugate planes. Therefore if the SLM at object plane is kept fixed during recording of the frames, the shift of the coordinates (different frames are recorded at different (x, y) location) of the respective spectrum of individual frame does not alter its reconstructed image position at the image plane. Therefore the superposed image at image plane is actually the optical sum of

the outer-product matrices which are previously stored in the hologram. At
the image plane a bright spot designates a link, i.e., a binary 1, and the dark
spot or the absence of light denotes a 0. The two-dimensional array of light
spots is read by the photodetector array. Each detector in the array would
function as a threshold detector indicating the presence (=1) or absence (=0)
of light. It may be noted that more than one pattern may require a link at
a same location in the weight matrix. This simply suggests that the bright
spots in the 2D array at image plane will have different light levels because
different locations contain the superposition of a different number of bright
spots. The threshold detection level for the detectors is set in such a way so as
to set its output to 1 for a light level associated to a single link. Therefore, all
the higher levels automatically switch the detector output to 1. However, the
photodetector ideally must exhibit a high detectivity in order to differentiate
between 1 and 0 in presence of noise. The proposed threshold detection scheme
relaxes the requirement of high dynamic range of the photodetector array.

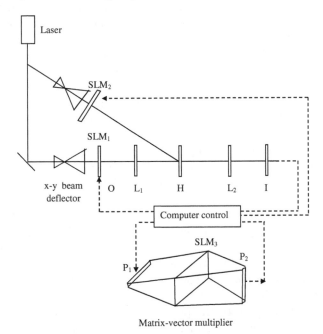

FIGURE 10.26: Scheme of (a) recording and (b) readout of outer product
matrix

Figure 10.26 shows the holographic recording geometry as well as the optical matrix-vector processor used in the recall phase of the associative memory.
In the learning phase each input stored pattern vector A^k and its associated
class vector C^k is fed to the SLM_1 placed at the object plane O during one
clock cycle. The resulting outer-product matrix between the vectors A^k and
C^k is illuminated by a plane parallel coherent laser beam which is then Fourier

transformed by the lens L. The Fourier spectrum is recorded into the holographic plate H with a parallel reference beam. The reference beam is also modulated spatially by the SLM$_2$ for illuminating different spatial regions in two dimensions on the hologram plane. In the next cycle, SLM$_1$ is updated with another vector pair and the state of SLM$_2$ is also changed to illuminate another spatial location on H. A mechanical x-y beam deflector is used to shift the location of the Fourier spectrum and to track the new spatial location selected by the reference beam. For an on-line situation, the addresses which drive the beam deflector and control the switching of SLM$_2$ are programmed in such a way, that the two operations are synchronised for tracking the object beam and the reference beam simultaneously. The above operation may be repeated M times to store M patterns.

During reconstruction phase, the hologram is illuminated by the same reference beam with all cells of the SLM$_2$ at on position. The superimposed image of all stored outer product matrices is thus reconstructed on the image plane I which is the sum of the outer product matrices of the stored patterns and their class vectors. This eventually gives the interconnection weight matrix. The resulting memory matrix T at plane I is transferred to the processing SLM, the SLM$_3$, with the help of a 2D detector array placed at plane I. The processing architecture is based on a photonic matrix-vector multiplier, where P_1 and P_2 comprise pairs of sources and detector arrays. The number of elements in each array is equal to the length of the vector to be processed. The array of sources in both the layers represent the neurons of the network. Their states, either on or off, are used to represent the unipolar binary vectors.

For experimentation, an imperfect input is fed in parallel to the sources P_1. The detectors of P_2 receive the response signals y_i^ks which result from the multiplication of the input with matrix T. The multiplication of the input vector with the weight matrix is achieved by vertical smearing of the input vector which is displayed by the sources P_1 on the plane of SLM$_3$ by means of an anamorphic lens system (not shown in figure, for simplicity). A second anamorphic lens system (also not shown) collect the light emerging from each row of SLM$_3$ on individual photodiodes of the detector array at P_2. The detector signals are then A/D converted and are transferred to the electronic processor for thresholding operation. The thresholded output represents any class vector C^p assigned to the pattern A^p which is the nearest to the input and is fed back to the sources of P_2. In this stage the detectors at P_1 receive the matrix response z_j^ks obtained as the product of the vector C^p with the matrix T^τ which transposes the memory matrix T. This after thresholding gives the true pattern A^p and completes the recall process.

10.6.2 Hetero-associative memory

Various schemes of hetero-associative formulation hitherto proposed are broadly termed as linear associative memory (LAM) model. LAM is basically a two-layer feed forward hetero-associative pattern matcher which stores arbitrary analogue pattern pairs (A^k and B^k) using correlation matrix formalism. In LAM description, the further the pattern vectors are from being orthonormal, the worse is the recall because of the presence of cross talk. The assumption of orthonormality of the input vectors to be stored has found to be unrealistic in many real environments. This situation was however improved with an encoding procedure, where pseudo-inverse (also known as the Moore-Penrose generalised inverse) of the matrix A is used in place of the transpose operation required to form the memory matrix using correlated Hebbian learning. The columns of the matrix define the input vectors A^k. The pseudo inverse solution is guaranteed only when the vectors A^k are linearly independent.

It has been already mentioned that a feedback associative structure is used to develop the bi-directional associative memory (BAM) model, where the input/output state vectors have binary quantised components. The BAM model has several distinctive merits. For example, incorporating the concept of bi-directionality it enables forward and backward information flow to produce a two-way associative search for a pattern pair. Secondly, it has been shown that for any arbitrary interconnection matrix, there always exists an energy function and thus BAM is unconditionally of the Lyapunov type. In contrast the Hopfield associative memory, though constructed in a similar manner, requires a symmetric connection matrix with zero diagonal elements for achieving global stability.

A general two-layer hetero-associative, nearest-neighbour pattern matcher encodes arbitrary pattern pairs $A^k, B^k, k = [1, 2...M]$ using Hebbian learning. The kth pattern pair is represented by the bipolar vectors and is given by,

$$A^k = \{a_i^k = (+1, -1)\} \quad [i = 1, 2, ...N]$$

$$B^k = \{b_j^k = (+1, -1)\} \quad [j = 1, 2, ...N] \tag{10.78}$$

where N is the element in the vector.

N first layer processing elements (PE) or the neurons are assigned to A^k components and N second layer PEs correspond to B^k components. The M training vector pairs are stored in the form of a weight matrix W, denoted by the sum of the outer products of the pattern pairs, and is expressed in element-wise notation as

$$w_{ij} = \sum_k a_i^k b_j^k \tag{10.79}$$

where w_{ij} is the connection strength interconnecting the ith processing element of the input layer with the jth processing element of the output layer.

The activation values of the output layer PEs are calculated as

$$b_j(t+1) = b_j(t) \quad \text{if } y_i = 0 \tag{10.80}$$

and

$$b_j = -1 \quad \text{if } y_j < 0 \tag{10.81}$$

$$b_j = +1 \quad \text{if } y_j > 0 \tag{10.82}$$

Evidently, $b_j(t+1)$ is the activation of the jth output PE at time $(t+1)$ and y_j is the preactivation value of the jth output PE as denoted by the equation

$$y_j = \sum_{i=1}^{N} a_i(t) w_{ij} \tag{10.83}$$

where $a(t)$ is the activation value of the ith input layer PE at time t.

Therefore, the recall equation for association can be written as,

$$b_j(t+1) = \phi \left[\sum_{i=1}^{N} a_i(t) w_{ij} \right] \tag{10.84}$$

where $\phi[.]$ is the hard limiter threshold function.

The case of a 2D image of size $L \times L$ can be considered. The vectors A^ks and B^ks are obtained by stacking the columns of each image to produce a vector of length $N = L^2$. The matrix W has, therefore, dimensions of $N \times N = L^2$, which is extremely large for any typical image. An alternative to this formulation is to use 2D vectors A^k and a 4D memory function. However, this approach is cumbersome, since it requires implementation of 4D operations with complex multiplexing. The alternative approach of subspace techniques, such as principal component analysis, also can be done.

Instead of taking the sum of the outer products of the library vectors it is possible to formulate a general hetero-associative memory algorithm in the form of inner products. The inner product correlation scheme is named as distributed associative memory. The input/output state vectors as well as the correlation coefficients are all binary quantized components. Moreover, an additional feedback loop is also incorporated to retrieve the associated patterns pairs as in the case of BAM. The preactivation value of the jth output PE appears as

$$y_j = \left[\sum_{i=1}^{N} a_i(t) \left\{ \sum_k a_i^k b_j^k \right\} \right] \tag{10.85}$$

where a_i^k and b_j^k are the ith and the jth elements of the kth library vector pairs A^k and B^k, respectively.

Let A be the input data vector and $a_i(t)$ the ith element of the input data vector at tth instant. The problem at hand, is to identify the associated pattern of the input data vector A. Rearranging the order of summation the above equation can be written as

$$y_j = \left[\sum_k g^k b_j^k \right] \tag{10.86}$$

where

$$g^k = \sum_{i=1}^{N} a_i^k a_i(t) = A^{k\tau} A \tag{10.87}$$

and τ is the transpose.

This equation gives the inner product of the kth library vector of the first set and the input data vector. In its truest sense if M pattern pairs $(A^k, B^k,$ with $[k = 1, .., M]$ are associated then M different g values are obtained. The matrix representation of the same is expressed as

$$XA = G \tag{10.88}$$

where X is a matrix having M rows and $G : [g^1, g^2, ...g^k]$ is an M element vector consisting of M correlation coefficients.

Each row of X represents the library vectors of the first set (A^k), N columns of which is equal to the number of elements in each vector. Now for perfect recall of any particular say pth association (A^p, B^p), the following property must be satisfied

$$a^{p\tau} \cdot A = a^{k\tau} \cdot A|_{max} = g^k|_{max} = g^p \tag{10.89}$$

It is evident that the increased noise or bit drop-outs in the input vector A reduces the auto correlation g^p to cross-correlation coefficients ratios, that is $g^k s$, $k \neq p$. These undesired cross correlation coefficients creep into the recall process which may induce the output to converge at a false state. To overcome this problem a dynamic multilevel thresholding operation is needed. This operation selects the maximum correlation coefficient among all $g^k s$, $[k = 1, ..., M]$ and sets it to state 1. All other states are set to 0. The process in turn gives preference to a particular vector associated to the given input, over the other vectors. For a particular pth association A^p, B^p the thresholding operation is thus performed as $\Psi[g^k] = 1$, if $k = p$ and 0 if $k \neq p$. Then the state of the jth output PE is represented as

$$y_j(t+1) = \sum_{k=1}^{M} \Psi[g^k]b_j^k = b_j^p \tag{10.90}$$

where b_j^p denotes the jth component of the true vector B^p. The matrix representation of the above recall equation is expressed as

$$Y^T \tilde{G} = B \tag{10.91}$$

where $\tilde{G}$ is an M element binary vector obtained by thresholding the vector G, that is $\tilde{G} = \Psi[G]$.

The matrix Y has M rows and each row of Y represents the library vectors of second set b^k. Vector B is an N element recalled vector, say B^p, truly associated to the input data vector A which may even be noisy. The true input vector A^p paired with B^p can now be recalled by feeding the noise free vector B^p back to the network as input. Therefore, the associated pattern pairs can be recalled by this algorithm involving simple inner products of binary matrices and vectors. The intermediate thresholding of the correlation coefficients restricts the propagation of noise terms to the output.

The measure of closeness between a bipolar stored library vector A^k and the input data vector A (which may be a partial or noisy one) can be obtained as

$$C[A \cdot A^k] = A^{k[T]} \cdot A = g^k = Q - d \qquad (10.92)$$

where Q is the number of bits in which A^k and A agree and d is the number in which they do not.

Therefore, the effective Hamming distance between the two vectors is

$$H[A, A^k] = N - C(A, A^k) = N - g^k \qquad (10.93)$$

Thus, minimisation of the Hamming distance between two binary bipolar vectors is identical to maximisation of the value of g^k.

It is evident that the algorithm is actually a two-stage association process. In the first stage the level of the nearest neighbour of the input data vector A among all A^k is found out by determining the highest component in G vector through a winner-take-all (WTA) operation. The elements of G, that is the g^k values, are found through the inner product. The system then enters into a second stage. Taking the thresholded g^k values, the inner product recalls the library vector B^k associated to any vector A^k which is nearest to or having the minimum Hamming distance with the input data vector A. The problem outlined always yields the truly associated pattern.

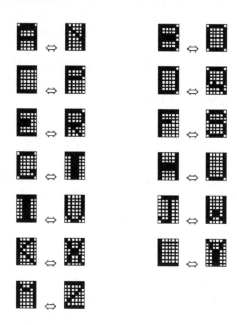

FIGURE 10.27: Stored pattern pairs

The model is utilised to carry out hetero-associative recall of binary patterns. Thirteen spatial alphabet associations [**A,N**], [**B,O**], [**C,P**] ... to [**M,Z**] are stored (Figure 10.27). The pattern vectors A^k and B^k $(k = 1, 2, ..., M)$ are obtained by stacking the rows of each pattern to produce a vector of length $N = (8 \times 6)$. The black cells are coded as binary 1 and the white cells are coded as -1. The interconnection matrix X is constructed by stacking the first set of vectors representing the characters [**A,B,..M**] and similarly the Y matrix with the characters [**N,O...Z**], i.e., with the second set, where both X and Y have elements either +1 or -1.

The architecture is prompted by the input vectors representing the noisy input patterns as shown in Figure 10.28(a). The final state of the memory, i.e., the recalled output pattern, associated to the pattern having minimum Hamming distance with the noisy input is displayed in Figure 10.28(b). In the next iteration, the recalled pattern of input is extracted, shown in Figure 10.28(c).

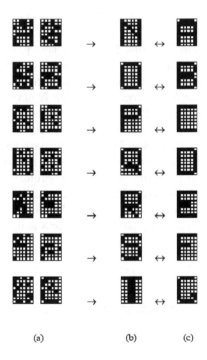

(a) (b) (c)

FIGURE 10.28: Hetero-associative recall of patterns

The performance of the two-layer BAM is faithful only if strict constraint is satisfied regarding the selection of patterns to be stored into the memory. The system demands that the pairwise Hamming distances among inputs (first set), as well as that of the output (second set) patterns must be of the same order. But the constraint is difficult to satisfy for all occasions. The recall efficiency of the outer product BAM model degrades further for strongly correlated stored patterns, i.e., when the patterns to be stored share more common bits between themselves.

10.6.3 Photonic hetero-associative memory

Photonic implementation of the hetero-associative memory model can be carried out by executing multichannel inner product rule via a digital photonic matrix-vector multiplication process. The WTA operation has been realised through an electronic feedback circuit. Since optical signals can only be at ON (=1) or OFF (=0) states, the binary bipolar vectors are converted to unipolar one. Therefore, bipolar data in each matrix and vector are split into two unipolar matrices and vectors. The equation for G can be expressed as

$$
\begin{aligned}
G &= (X^+ - X^-) \cdot (A^+ - A^-) \\
&= (X^+A^+ + X^-A^-) - (X^+A^- + (X^-A^-) \\
&= K_1 - K_2
\end{aligned}
\tag{10.94}
$$

where X^+, X^- and A^+, A^- are unipolar matrix/vectors which consist of only $\{+1, 0\}$.

The implementation is realised through two parallel channels. The similarity measurement is carried out by unipolar matrix vector multiplication process in both the channels. Channel 1 computes the expression for K_1 and channel 2 computes the K_2 vectors in parallel. It may be noted that the maximum value of any element in G, (say g^p), requires that the difference between K_1^p and K_2^p must be maximum. For the case under consideration, assume that the bipolar matrix X comprises M arbitrary pattern vectors $A^1, A^2, \ldots A^M$, and the vector A is chosen as noisy input with a minimum Hamming distance with one of the stored patterns (say A^1). Therefore, the element g^1 in G is maximum and needs to be won. Figure 10.29 outlines the implementation of a photonic architecture interconnected by WTA networks. However, the architecture required to compute partial products is not shown in Figure 10.29.

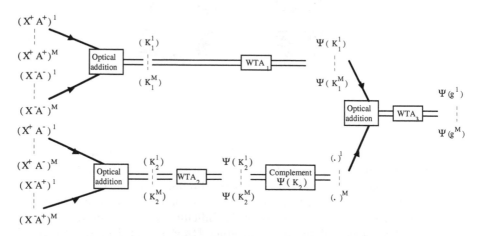

FIGURE 10.29: Operational flowchart of the hetero-associative architecture

To detect any element g^1 as a maximum it is necessary to detect the maximum-value element of K_1 (that is k_1^1) and the minimum-value element of K_2 (that is k_2^1) so that the difference $(k_1^1 - k_2^1)$ is always maximum for this case. The WTA1 circuit detects k_1^1 as maximum and sets this to 1 and the rest of k_1^k, $(k = 2, ..., M)$ are all set to 0. To set the minimum of K_2, the output of the WTA2 circuit is complemented to give exclusively one minimum ($=0$), because only one k_2^k is won by WTA2 and the other elements are set to 1. The resulting M-element binary vectors are the output of WTA1. The complemented output of WTA2 are added optically and are then fed to the WTA3 network. Evidently, WTA3 selects the node which corresponds to g^1 among all g^k in vector G.

10.6.4   Winner-take-all model

A three-layer associative memory model [54] is shown in Figure 10.30. As seen from the figure the three-layer network comprises of input and output layers as R_A and R_B, each having N neurons. R_H denotes the hidden layer with M neurons, where R_{HI} and R_{HO} are the input and output of this hidden layer. For an association (A_k, B_k), the weight $w_{ik} = a_i^k$. If G and G' represent the input and output of the hidden layer, W_1 and W_2 represent the IWM between the layers A, R_{HI}, and A, R_{HO}, respectively, then $G = W_1 A$ and $B = W_2 G'$.

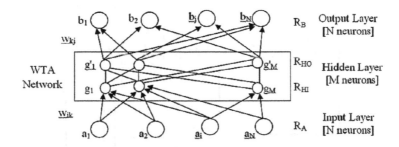

FIGURE 10.30: Three-layer associative memory model

The Hamming distance is calculated to get the associated pattern and is given by

$$H(A, A^k) = N - (A^k)^\tau A = N - g^k \qquad (10.95)$$

Therefore, finding any stored vector with minimum Hamming distance with the test input vector A is equivalent to the detection of maximum element in the activation vector G. The activation vector $G\{g_k\} : (k = 1, M)$ is the input data vector to the WTA network at the hidden layer. After the WTA procedure is completed, the state of the M neurons at the output layer R_{HO} is represented by the M element binary vector G.

The states of the output neurons RB may now be obtained by taking the linear sum of the outputs of the hidden layer R_{HO}, weighted by the interconnection strengths between the layer RHO and RB. Mathematically, the state of the jth output neuron at RB is then expressed as

$$b_j(t+1) = \sum_{k=1}^{M} g'_k w_{kj} \qquad (10.96)$$

The output state vector is thus expressed as

$$\mathbf{B} = W_2 \mathbf{G} \qquad (10.97)$$

where W_2 is the interconnection weight matrix between the layer R_{HO} and the output layer RB.

The photonic architecture for WTA is shown in Figure 10.31. The binary bipolar weight matrix is split into two unipolar matrices as $W(t) = W^+(t) - W^-(t)$. The unipolar IWM, W^+ and W^- are loaded on the spatial light modulator (SLM). The input data vector is fed into the source array S, where each element represents a neuron of the input layer R_{HI} of the WTA network. The cylindrical lens L_1 distributes light from each source array along each column of the weight matrix on the SLM. The combination of lenses L_2 and L_3 collects light from each row of the SLM onto a detector array D. The resulting product, i.e., S_1 and S_2, on the detector are fed back to the computer, read and multiplied by pre-assigned weight factors Z_1 and Z_2.

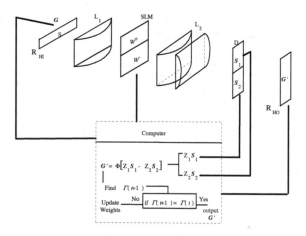

FIGURE 10.31: Photonic architecture for WTA model

The difference of the new weighted vectors produces the binary state vector of the WTA as given as

$$G' = \Phi[Z_1 S_1 - Z_2 S_2] \qquad (10.98)$$

The thresholding of the output neuron at time $(t + 1)$ gives

$$\Gamma(t+1) = \sum_{j=1}^{M} g'_{j(t+1)} \tag{10.99}$$

The weights are updated until $\Gamma(t+1) = \Gamma(t)$, the winning neurons are found. It is evident that for a particular pth association (A_p, B_p), the vector B^p in the output is always recalled when the network is initialized by any data vector A which is having maximum correlation with the stored pattern vector A_p. The recall is always perfect (i) if the inner product between the test input vector A and the target stored vector A_p is positive and (ii) if the inner product between the test input vector with other stored vectors is less than $A_p^T A$. Under these conditions, the state of the jth output neuron at RB is b_{pj} and the vector B_p associated to the vector A_p is always recalled. The vector A_p can now be obtained by feeding the true recalled vector B_p back to the network at RA with interchanging W_1 and W_2. Auto associative search of patterns is possible in a similar way when $W_1 = W_2$.

A three-layer associative memory model can be implemented efficiently in the photonic architecture using the WTA as shown in Figure 10.32. Overall processing sequences are controlled by a digital computer. The task involved during recall of any particular association (A_p, B_p) is to be summarised step wise.

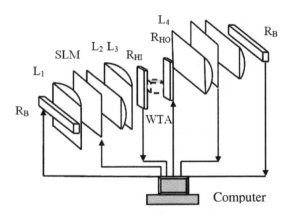

FIGURE 10.32: Photonic architecture for three-layer hetero-associative memory with WTA

The steps are:

(i) Feeding of input neurons at R_A by any initialising binary vector A; the vector may be noisy. The source array S with N elements represents the state of the input vector.

(ii) Computation of activation vector G with elements g^k, $[k = 1, ..., M]$

through the interconnection matrix W_1 in parallel using the photonic matrix-vector architecture. The lenses L are symbolic of the anamorphic lens system used in the matrix-vector architecture. The elements g^k, $[k = 1, ..., M]$ designate the activation levels of M different hidden layer neurons at R_{HI} where the detector array D_1 receives the signal for the neurons. The WTA network in the next stage selects the particular hidden neuron having highest activation level and sets the state of the corresponding neuron at layer R_{HO} to 1 and successively turns the remaining outputs to 0. The source array S_1 represents the states of the R_{HO} neuron.

(iii) The interconnection matrix W_2 between R_{HO} and the output layer R_B forward the non-zero output of the layer R_{HO} to the output layer for completing an associative recognition by recalling B^p. The detector array D_2 receives the binary recalled vector.

(iv) The true output B^p is fed back to the input layer $R_A A$. The matrix W_1 is updated to W_2 and vice versa.

(v) The steps from (i) to (iv) are repeated to obtain the true input pattern A^p which is associated with the recalled pattern B^p.

It is to be mentioned at this point that only the algorithmic or architectural issues are considered regarding the photonic WTA processors. The solution may need more refinements if the problems of actual hardware implementation are attempted.

Books for further reading

1. *Real time Optical Information Processing*: B. Javidi and J.L. Horner, Academic Press, California, 1994.

2. *Perceptrons*: M. L. Minsky and S. Papert, MIT Press, Cambridge, MA, 1962.

3. *The Organization of Behaviour*: D. O. Hebb, John Wiley & Sons, New York, 1949.

4. *The Adaptive Brain: Vol I and II*: S. Grossberg, Elsevier Science Publishers, 1986.

5. *Neural Networks*: R. Rojas, Springer-Verlag, Berlin, 1996.

6. *Self-Organization and Associative Memory* : T. Kohonen, Verlag, Berlin, 1990.

7. *Introduction to Information Optics*: F. T. S. Yu, S. Jutamulia, and S. Yin, Academic Press, California, 2001.

Bibliography

[1] C. S. Weaver and J. W. Goodman. A technique for optically convolving two functions. *App. Opt.*, 5:1248–1249, 1966.

[2] F. T. S. Yu and X. J. Lu. A technique for optically convolving two functions. *Opt. Comm.*, 2:10–16, 1984.

[3] J. L. Horner. Metrics for assessing pattern recognition performance. *App. Opt.*, 31:165–166, 1992.

[4] B. V. K. Vijaya Kumar, M. Savvides, C. Xie, K. Venkataramani, J. Thornton, and A. Mahalanobis. Biometric verification with correlation filters. *Appl. Opt.*, 43(2):391–402, 2004.

[5] B. V. K. Vijaya Kumar and E. Pochapsky. Signal-to-noise ratio considerations in modified matched spatial filters. *J. Opt. Soc. Am. A*, 3(6):777–786, 1986.

[6] B. Javidi and J. L. Horner. *Real Time Optical Information Processing*. Academic Press, California, 1994.

[7] C. F. Hester and D. P. Casasent. Multivariate technique for multiclass pattern recognition. *App. Opt.*, 9(11), 1980.

[8] B. V. K. Vijayakumar. Minimum variance synthetic discriminant functions. *J. Opt. Soc. Am. (A)*, 3(10), 1986.

[9] A. Mahalanobis, B. V. K. Vijaykumar, and D. Casassent. Minimum average correlation energy filter. *App. Opt.*, 26(17):3633–3640, 1987.

[10] P. Refregier. Filter design for optical pattern recognition: Multi-criteria optimization approach. *Opt. Let.*, 15:854–856, 1990.

[11] G. Ravichandran and D. Casasent. Minimum noise and correlation energy optical corrrelation filter. *App. Opt.*, 31:1823–1833, 1992.

[12] B. V. K. Vijayakumar. Tutorial survey of composite filter designs for optical correlators. *App. Opt.*, 31(23):4773–4801, 1992.

[13] R. Patnaik and D. Casasent. Illumination invariant face recognition and impostor rejection using different MINACE filter algorithms. In *Proc. SPIE 5816, Optical Pattern Recognition XVI*, 2005.

[14] B. V. K. Vijayakumar and A. Mahalanobis. Recent advances in composite correlation filter designs. *Asian J. Physics*, 8(3), 1999.

[15] M. Savvides, B. V. K. Vijayakumar, and P. K. Khosla. Two-class minimax distance transform correlation filter. *App. Opt.*, 41:6829–6840, 2002.

[16] S. Bhuiyan, M. A. Alam, S. Mohammad, S. Sims, and F. Richard. Target detection, classification, and tracking using a maximum average correlation height and polynomial distance classification correlation filter combination. *Opt. Eng.*, 45(11):116401–116413, 2006.

[17] A. Mahalanobis, B. V. K. Vijaykumar, S. Song, S. Sims, and J. Epperson. Unconstrained correlation filter. *App. Opt.*, 33:3751–3759, 1994.

[18] B. V. K. Vijayakumar, D. W. Carlson, and A. Mahalanobis. Optimal trade-off synthetic discriminant function filters for arbitrary devices. *Opt. Let.*, 19(19):1556–1558, 1994.

[19] P. Refregier. Optimal trade-off filters for noise robustness, sharpness of the correlation peak, and Horner efficiency. *Opt. Let.*, 16(11):829–831, 1991.

[20] M. Alkanhal, B. V. K. Vijayakumar, and A. Mahalanobis. Improving the false alarm capabilities of the maximum average correlation height correlation filter. *Opt. Eng.*, 39(5):1133–1141, 2000.

[21] B. Vijayakumar and M. Alkanhal. Eigen-extended maximum average correlation height (EEMACH) filters for automatic target recognition. *Proc. SPIE 4379, Automatic Target Recognition XI*, 4379:424–431, 2001.

[22] A. VanNevel and A. Mahalanobis. Comparative study of maximum average correlation height filter variants using LADAR imagery. *Opt. Eng.*, 42(2):541–550, 2003.

[23] S. Goyal, N. K. Nischal, V. K. Beri, and A. K. Gupta. Wavelet-modified maximum average correlation height filter for rotation invariance that uses chirp encoding in a hybrid digital-optical correlator. *App. Opt.*, 45(20):4850–4857, 2006.

[24] H. Lai, V. Ramanathan, and H. Wechsler. Reliable face recognition using adaptive and robust correlation filters. *Computer Vision and Image Understanding*, 111:329–350, 2008.

[25] K. H. Jeong, W. Liu, S. Han, E. Hasanbelliu, and J. C. Principe. The correntropy MACE filter. *Patt. Rec.*, 42(9):871–885, 2009.

[26] M. D. Rodriguez, J. Ahmed, and M. Shah. Action MACH: a spatio-temporal maximum average correlation height filter for action recognition. In *Proc. IEEE Int. Con. Computer Vision and Pattern Recognition*, 2008.

[27] A. Rodriguez, V. N. Boddeti, B. V. K. Vijayakumar, and A. Mahalanobis. Maximum margin correlation filter: A new approach for localization and classification. *IEEE Trans. Image Processing*, 22(2):631–643, 2012.

[28] D. Casasent. Unified synthetic discriminant function computational formulation. *App. Opt.*, 23(10):1620–1627, 1984.

[29] P. Refregier. Filter design for optical pattern recognition: multi-criteria optimization approach. *Opt. Let.*, 15, 1990.

[30] S. W. McClloch and W. Pitts. A logical calculus of the ideal imminent in nervous activity. *Bull. Math. Biophysics*, 5:115–133, 1843.

[31] F. Rosenblatt. *Principles of Neurodynamics*. Spartan Books, New York, 1959.

[32] B. Widrow, N. Gupta, and S. Maitra. Punish/reward: learning with a critic in adaptive threshold systems. *IEEE Trans. Syst. Man Cybern*, SMC-5:455–465, 1973.

[33] D. Ackley, G. Hinton, and T. Sejnowski. A learning algorithm for Boltzman machines. *Cognitive Sci.*, 9:147–169, 1985.

[34] S. Geman and G. Geman. Stocastic relaxation, Gibbs distributions and the Bayesian restoration of the images. *IEEE Trans. Patt. Anal. Mach. Intell*, PAMI-6:721–741, 1984.

[35] S. Kirkpatrik, C. Gelatt, and M. Vecchi. Optimization by simulated annealing. *Science*, 220:671–680, 1985.

[36] M. Koskinen, J. Kostamovaara, and R. Myllyla. Comparison of the continuous wave and pulsed time-of-fight laser range finding techniques. *Proc. SPIE: Optics, Illumination, and Image Sensing for Machine Vision VI*, 1614:296–305, 1992.

[37] B. Kosko. Competitive adaptive bidirectional associative memories. *Proc. IEEE First Int. Conf. Neural Networks*, II:759–464, 1987.

[38] F. Rosenblatt. *Principles of Neurodynamics*. Spartan Books, New York, 1957.

[39] D.E. Rumelhart, G.E. Hinton, and R.J. Williams. *Learning internal representations by error propagation*. MIT Press, Cambridge, MA, 1986.

[40] J. J. Hopfield. Neural network and physical systems with emergent collective computational abilities. In *Proc. of Nat. Aca. of Sc.*, volume 79, pages 2554–2558, 1982.

[41] T. J. Sejnowski and D. H. Ackley. A learning algorithm for Boltzmann machines. *Technical Report, CMU-CS*, 84:119, 1984.

[42] B. Kosko. Bidirectional associative memories. *IEEE. Trans. Syst. Man and Cybern.*, 18:49–60, 1988.

[43] D. Gabor. Associative holographic memories. *IBM J. Res. Dev.*, 13:156, 1969.

[44] K. Nakamura, T. Hara, M. Yoshida, T. Miyahara, and H. Ito. Optical frequency domain ranging by a frequency-shifted feedback laser. *IEEE J. Quant. Elect.*, 36:305–316, 2000.

[45] H. Akafori and K. Sakurai. Information search using holography. *App. Opt.*, 11:362–369, 1972.

[46] F. T. S. Yu, S. Jutamulia, and S. Yin. *Introduction to Information Optics.* Academic Press, California, 2001.

[47] N. H. Farhat, D. Psaltis, A. Prata, and E. Paek. Optical implementation of the Hopfield model. *App. Opt*, 24:1469–1475, 1985.

[48] D. Psaltis and N. Farhat. Optical information processing based on an associative memory model of neural nets with thresholding and feedback. *Opt. Let.*, 10:98–100, 1985.

[49] D. Fisher and C. L. Giles. Optical adaptive associative computer architecture. *Proc. IEEE COMPCON*, CH 2135-2/85:342–348, 1985.

[50] C. Guest and R. TeKolste. Designs and devices for optical bidirectional associative memories. *App. Opt.*, 26:5055–5060, 1987.

[51] K. Wagner and D. Psaltis. Multilayer optical learning networks. *App. Opt.*, 26:5061–5076, 1987.

[52] J. Ticknor and H. H. Barrett. Optical implementations in Boltzman machines. *Opt. Eng.*, 26:16–22, 1987.

[53] S. Bandyopadhyay and A. K. Datta. A novel neural hetero-associative memory model for pattern recognition. *Pattern Recognition*, 29:789–795, 1996.

[54] S. Bandyopadhyay and A. K. Datta. Optoelectronic implementation of a winner-take-all model of hetero-associative memory. *Int. J. Optoelectronics*, 9:391–397, 1994.

Chapter 11

Quantum Information Processing

11.1 Introduction

From the early days of quantum theory, it is felt that the classical ideas about information may need reorientation under a quantum theoretical framework. In quantum theory, where Heisenberg uncertainty principle is the guiding philosophy, a non-commuting observable cannot simultaneously have precisely defined values. In fact, if two observations do not commute, performing a measurement of one observable will necessarily influence the outcome of a subsequent measurement of the other observable. Hence, the act of acquiring information about a physical system inevitably disturbs the state of the system. This is because of the fact that the outcome of a measurement has some random elements due to uncertainty of the initial state of the system. Moreover, acquiring information causes a disturbance, and due to this disturbance the information cannot be copied with perfect fidelity. The information under this framework is termed as quantum information [1].

The study of quantum information as a coherent discipline began to emerge in the 1980s, although during the 1960s, the way in which quantum information differed from classical information emerged from the work of John Bell [2]. He showed that the predictions of quantum mechanics cannot be reproduced by any local hidden variable theory. Bell also showed that quantum information can be encoded in non-local correlations between the different parts of a physical system [3].

Quantum information processing of the contemporary era is closely tied to the development of several other interdisciplinary fields. Quantum computation, for example, generally deals with the study of quantum algorithms which are shown to be more efficient than traditional Turing machine algorithms [4]. As a building block of a quantum computer and its heuristic execution, quantum circuits need to be designed so as to encompass smaller units of quantum gates. Quantum gates would manipulate the quantum states as lossless reversible operators. The hardware implementations also become an open issue from the device point of view [5]. The development of quantum information processing also generates new ways of secure communications and the area is collectively known as quantum communication and quantum cryptography [6].

It is, however, necessary to elaborate on the issue of why quantum information processing is of interest and why the techniques of such a process are thought to be more powerful than conventional electronics-based processing. The reason lies in the extension of Moore's law, which indicates the future requirements of information processing capabilities. To achieve the high demands of speed and bandwidth, the logic gate size is expected to reach almost the size of an atom, at least in theory. Consequently, the exploration and utilisation of quantum mechanics becomes very important to the development of logic design of future devices of this size. However, there have been several practical limitations in the theory and practices of quantum engineering, mainly including the phenomena of quantum decoherence, the scalability, and the ability to control and couple basic building blocks of quantum information processing units.

11.2   Notations and mathematical preliminaries

Quantum theory is a mathematical model of the physical world. It is generally represented by states, observables, measurements, and dynamics. A *state* is a complete description of a physical system. In quantum mechanics, a state is a ray in a Hilbert space H^s. A ray is an equivalence class of vectors that differ by multiplication by a non-zero complex scalar. Hilbert space is a vector space over the complex numbers. In other words, H^s space is an inner product vector space, where the unit vectors are the possible quantum states of the system. It is important to introduce the orthonormal basis on H^s, in particular the quantum system that is described by two orthonormal basis states. An orthonormal set of vectors M in H^s is such that every element of M is a unit vector (vector of length 1) and any two distinct elements are orthogonal.

The notation, which is mostly used in quantum information processing, is the *bra-ket* notation introduced by Dirac. It is used to represent the operators and vectors, where each expression has two parts, a bra and a ket expressed as $\langle|$ and $|\rangle$. Each vector in the H^s is a ket $|.\rangle$ and its conjugate transpose is bra $|.\rangle$. In this notation, the inner product is represented by $\langle\psi_m|\psi_n\rangle = 1$, for $n = m$. By inverting the order, the ket-bra outer product is obtained as, $|\psi_m\rangle\langle\psi_n|$. The inner product has the properties of positivity, linearity, and skew symmetry.

An important notion in quantum information processing is density matrix [7]. Density matrices, sometimes also known as density operators, have the useful property of being able to describe the states of quantum systems which are only partially known. In more precise terms, density matrices are capable of expressing mixed states, which are a sum of multiple possible states, weighted by their respective probabilities. Mixed states are not actual physical states; they are a representation of knowledge about a given system. At any moment in time, the quantum system is in a well-defined state; but mixed states are a useful tool when only partial knowledge about the state is available.

Given that, the state of a quantum system system is at one of several possible quantum states $|\phi_i\rangle$ indexed by i, with associated probabilities P_i, the density matrix describing this state can be obtained by computing

$$\rho = \sum_i P_i.|\phi\rangle\langle\phi_i| \tag{11.1}$$

$|\phi\rangle\langle\phi_i|$ denotes an outer product, which returns a matrix The trace operator represents the possible observable states of a quantum system. Any quantum state $|\psi\rangle$ when observed collapses according to the applied measurement, with P being the probability of observing the state $|\psi\rangle$ from the set of all possible output states. Thus representing the overall state of a quantum system can be represented as the *trace* $\sum P_i|i\rangle\langle i|$ with P_i being the probability of observing the state $|i\rangle$.

Unitary transformations extend linearly to mixed states, and can therefore be defined on density matrices. Given a unitary transformation U (e.g., the matrix representation of a quantum gate), the density matrix representation of the state obtained by the transformation is given by

$$\rho' = U\rho U^{-1} \tag{11.2}$$

An *observable* is a property of a physical system that in principle can be measured. In quantum mechanics, an observable is a self-adjoint operator, where the operator is a linear map taking vectors to vectors given by

$$A : |\psi\rangle \rightarrow A|\psi\rangle \tag{11.3}$$

where A is a disjoint operator.

In quantum mechanics, the numerical outcome of a *measurement* of the observable A is an eigenvalue of A. After the measurement, the quantum state is an eigenstate of A with the measured eigenvalue. If the quantum state just prior to the measurement is $|\psi\rangle$, and the outcome a_n is obtained with probability $P(a_n) = \langle\psi|P_n|\psi\rangle$, then the (normalised) quantum state becomes $\frac{P_n|\psi\rangle}{(\langle\psi|P_n|\psi\rangle)^{1/2}}$.

While considering *dynamics* of the quantum system, time evolution of a quantum state is unitary. It is generated by a self-adjoint operator, called the Hamiltonian H_m of the system. The vector describing the system moves in time as governed by the Schroedinger equation. The classical equation in the infinitesimal quantity dt is rearranged as

$$|\psi(t+dt)\rangle = (1 - iH_m dt)|\psi(t)\rangle \tag{11.4}$$

Since H_m is self-adjoint, satisfying $U^\dagger U = 1$, the operator $U(dt) \equiv (1 -$

$iH_m)dt$. A product of unitary operators is finite and therefore, time evolution over a finite interval is also unitary, as given by

$$|\psi(t)\rangle = U(t)|\psi(0)\rangle \qquad (11.5)$$

There are two quite distinct ways for a quantum state to change. On the one hand this is unitary evolution, which is deterministic. That is, when $|\psi(0)\rangle$ is specified, the theory predicts the state $|\psi(t)\rangle$ at a later time. However, on the other hand there is measurement, which is probabilistic. The theory does not make definite predictions about the measurement outcomes; it only assigns probabilities to the various alternatives. This is one of the manifestations of dualism of quantum theory.

11.3 Von Neumann entropy

In classical information theory, a source generates messages of n letters ($n >> 1$), where each letter is drawn independently from an ensemble $X = \{x, P(x)\}$. Shannon information $H(X)$, as discussed in Chapter 1, is the number of incompressible bits of information carried per letter. The correlations between two ensembles of letters X and Y are characterised by conditional probabilities $P(y|x)$ and the mutual information is given by

$$I(X|Y) = H(X)H(X|Y) = H(Y)H(Y|X) \qquad (11.6)$$

Mutual information is the number of bits of information per letter about X that can be acquired by reading Y (or vice versa). If the $P(y|x)$ is characterised as a noisy channel, then $I(X|Y)$ is the amount of information per letter that can be transmitted through the channel, when *a priori* distribution for the X is known.

In the case of quantum information transmission, where a source prepares messages of n letters, each letter is chosen from an ensemble of quantum states. The signal alphabet consists of a set of quantum states q^x, each occurring with a specified *a priori* probability P_x. Therefore, the probability of any outcome of any measurement of a letter chosen from this ensemble can be completely characterised by the density matrix, if the observer has no knowledge about which letter is prepared. Then the density matrix is given by

$$\rho = \sum_x P_x \, \rho_x \qquad (11.7)$$

For this density matrix, the Von Neumann entropy is defined as

$$S(\rho) = -tr(\log \rho) \qquad (11.8)$$

where tr is the trace of the matrix.

Alternatively, selecting an orthonormal basis $\{|a\rangle\}$ that diagonalizes ρ, the density matrix is given by

$$\rho = \sum_a \lambda_a |a\rangle\langle a| \qquad (11.9)$$

then,

$$S(\rho) = H(A) \qquad (11.10)$$

where $H(A)$ is the Shannon entropy of the ensemble $A = \{a, \lambda_a\}$.

Incidentally, if the signal alphabet consists of mutually orthogonal pure states, the quantum source reduces to a classical one, where all of the signal states can be perfectly distinguished, and $S(\rho) = H(X)$. The quantum source is more interesting when the signal states ρ are not mutually commuting.

One important property of Von Neumann entropy is the entropy of measurement in a quantum environment. In a quantum state ρ, the observable entity A is given by

$$A = \sum_y |a_y\rangle a_y \langle a_y| \qquad (11.11)$$

where a_y is the outcome of the measurement with probability of occurrence $P(a_y)$ given by,

$$P(a_y) = \langle a_y|\rho|a_y\rangle \qquad (11.12)$$

The Shannon entropy of the ensemble of measurement outcomes $Y = \{a_y, P(a_y)\}$ satisfies, $H(Y) \geq S_{(\rho)}$ with equality when A and ρ commute. Therefore, the randomness of the measurement outcome is minimised, if an observable is measured that commutes with the density matrix.

In conclusion, the Von Neumann entropy quantifies the incompressible information content of the quantum source as much as the Shannon entropy quantifies the information content of a classical source. Von Neumann entropy plays a dual role. It quantifies not only the quantum information content per letter of the ensemble, that is, the minimum number of quantum states per letter needed to reliably encode the information, but also its classical information content. Thus quantum information processing theory is largely concerned with the interpretation and uses of Von Neumann entropy whereas the classical information theory is largely concerned with the interpretation and uses of Shannon entropy.

11.4 Quantum bits or qubits

The bit or binary digit is the fundamental building block of classical information system. Being a binary digit, it can take on the values 0 or 1, which are the two possible states of a bit. For physical realisation, the bits can then be represented by a high and low voltage level in an electrical circuit, or by two levels of intensity in a pulse of light or similar bi-valued physical phenomenon. The theory of classical information does not, however, depend on the choice of physical representation of the bits, but gives a general mathematical description of the techniques of processing and transmitting information.

In contrast, a quantum information processor stores and processes information as qubits (quantum bits). In physical terms qubits are quantum mechanical states of photons, nuclei, or atoms, or some sub-atomic phenomenon. The main difference between binary digits and qubits is that qubits can represent not only 0 and 1, but also take on superposition states, in which the system can be both in 0 and 1 state at the same time. This is a direct corollary of the superposition principle of quantum mechanics.

A qubit has two computational basis vectors, $|0\rangle$ and $|1\rangle$, of the Hilbert space H^s corresponding to the classical bit values 0 and 1. An arbitrary state $|\psi\rangle$ of a qubit is then a linearly weighted combination of the computational basis vectors. A qubit is defined as a quantum state satisfying the equation

$$|\psi\rangle = \alpha|0\rangle + \beta|1\rangle \tag{11.13}$$

where α and β are probability amplitudes and $|\psi\rangle$ denotes a quantum state called a wave function of the state, and $(\alpha^2 + \beta^2) = 1$.

In vector notation, a qubit is denoted as

$$|\psi\rangle = \alpha \begin{bmatrix} 1 \\ 0 \end{bmatrix} + \beta \begin{bmatrix} 0 \\ 1 \end{bmatrix} \tag{11.14}$$

A graphical representation of classical bits and qubits is shown in Figure 11.1. It is shown that qubits can be in an unlimited number of states, called states of superposition. In contrast to classical bits, qubits need not remain independent from other qubits: In such a case, the qubits in question are said to be entangled. The phenomenon as such is referred to as entanglement.

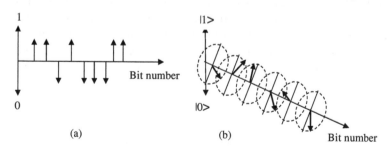

FIGURE 11.1: Visualisation of (a) classical bits and (b) qubits

A system of quantum bits can be in any (properly normalised) superposition of the classical states of the system. The state of a quantum system is

$|\psi\rangle = \sum_i \alpha_i |i\rangle$, where $|i\rangle$'s are the classical states and $\sum_i \alpha_i^2 = 1$. The qubit is thus a vector, normalised to length 1, in a two-dimensional complex vector space with inner product. In deference to the classical bit, the elements of an orthonormal basis in this space are $|0\rangle$ and $|1\rangle$.

The properties of qubits are governed by their quantum mechanical wave function. States $|0\rangle$ and $|1\rangle$ of a qubit can be seen and are often referred to as representing classical states (bits). While a classical bit can be set up only to one of the two states, namely 1 or 0, a qubit can take any quantum linear superposition of $|0\rangle$ and $|1\rangle$ and in principle can be in any of uncountably many states. This means that a large, even infinite amount of information could potentially be encoded in amplitudes of a single qubit by appropriately choosing α and β.

Equation 11.13 can be written as

$$|\psi\rangle = \cos\frac{\theta}{2}|0\rangle + e^{i\psi} \sin\frac{\theta}{2}|1\rangle \qquad (11.15)$$

After removing the overall phase factor, this expression of qubit states can be visualised as points on a spherical surface with radius 1. The numbers θ and ψ can then be identified as the angles defined in the qubit sphere shown in Figure 11.2. The basis states $|0\rangle$ and $|1\rangle$ are positioned at the north and south pole, respectively, and the two remaining mutually unbiased bases, $\frac{1}{\sqrt{2}}(|0\rangle + |1\rangle)$, $\frac{1}{\sqrt{2}}(|0\rangle - |1\rangle)$ and $\frac{1}{\sqrt{2}}(|0\rangle + i|1\rangle)$, $\frac{1}{\sqrt{2}}(|0\rangle - i|1\rangle)$ are positioned at the equator.

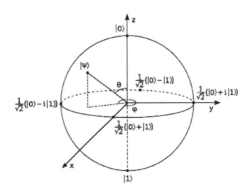

FIGURE 11.2: Representation of qubit states in a spherical surface

A measurement on a quantum system in a superposition $|\psi\rangle = \sum_i \alpha_i |i\rangle$ results in a value consistent with state $|i\rangle$ with probability α_i^2. The measurement can be performed that projects $|\psi\rangle$ onto the basis $|0\rangle, |1\rangle$. The outcome of the measurement is not deterministic; the probability of obtaining the result $|0\rangle$ is $|\alpha^2|$ and that of $|1\rangle$ is $|\beta^2|$. Moreover, a measurement on a state $|\psi\rangle$ removes all elements of the state that are inconsistent with the result of the measurement. In other words, a measurement on a state $|\psi\rangle = \sum_i \alpha_i |i\rangle$ results in a state $|\psi'\rangle = \sum_j \frac{\alpha_i}{N} |j\rangle$ where $|j\rangle$'s are the states consistent with the measurement, and $N = \sqrt{\sum \alpha_j^2}$ preserves normalisation.

The above discussion leaves the actual physical medium of a qubit completely undefined as long as objects are treated according to the quantum principles.

11.4.1   Multiple qubits

It is possible to naturally extend single qubit description to several or multiple qubit operations. The simplest case is to have two qubits. The Hilbert space for qubits is spanned by the four states $|00\rangle$, $|01\rangle$, $|10\rangle$, and $|11\rangle$, where $|00\rangle \equiv |0\rangle \otimes |1\rangle$. Hence, any two-qubit state can be expressed as a superposition of these basis states, given by

$$|\psi\rangle = \alpha_{00}|00\rangle + \alpha_{01}|01\rangle + \alpha_{10}|10\rangle + \alpha_{11}|11\rangle \qquad (11.16)$$

where α_{ij} are complex numbers. The state is normalised so that $|\alpha_{00}|^2 + |\alpha_{01}|^2 + |\alpha_{10}|^2 + |\alpha_{11}|^2 = 1$.

The important two-qubit states are the so-called Bell states, given by

$$|\Phi^\pm\rangle = \frac{1}{\sqrt{2}}(|00\rangle \pm |11\rangle) \qquad (11.17)$$

and

$$|\Psi^\pm\rangle = \frac{1}{\sqrt{2}}(|01\rangle \pm |10\rangle), \qquad (11.18)$$

The two qubits in these states exhibit correlations, since two qubits in $|\Phi^\pm\rangle$ are always in the same state, either both in $|0\rangle$ or both in $|1\rangle$, and in $|\Psi^\pm\rangle$ they are always in opposite states, one in $|0\rangle$ and one in $|1\rangle$.

The quantum state of N qubits is expressed as a vector in a space of dimension 2^N. An orthonormal basis for this space can have a state in which each qubit has a definite value, either $|0\rangle$, or $|1\rangle$. These can be labeled by binary strings such as, $|001001100....10\rangle$. The state can be represented as

$$|\psi\rangle = |\psi_1\rangle|\psi_2\rangle.....|\psi_N\rangle \qquad (11.19)$$

involving any one of all of the configurations $|00...0\rangle, |00...1\rangle, ..., |11...1\rangle$.

Instead of enlarging a system by increasing the number of qubits, it is possible to stick with a single qubit, which can have more than two orthogonal basis states. With n basis vectors $|x_n\rangle$ a general qubit system, known as *qu-nit* can be expressed as

$$|\psi\rangle = \sum_n c_n |x_n\rangle \qquad (11.20)$$

with $\sum_n |c_n|^2 = 1$.

A qu-nit cannot be easily encoded using the physical properties and cannot be fully described by a two-dimensional Hilbert space. However, degrees of freedom such as, e.g., time or spatial modes can still be used.

11.4.2 Conditional probabilities of qubits

For simplicity, a two qubits system is considered and the measurement of the energy of the first qubit gives a value of E_0. The measurement of the energy of the second qubit will give a certain value with certain probability. Let A be the event for the measurement of E_0 for the first qubit, B be the event for the measurement of E_0 for the second qubit, and C be the event with the measurement of E_1 for the second qubit. The states $|00\rangle$ and $|01\rangle$ are consistent with measurement of E_0 for the first qubit. Therefore, the probability of event A is $P(A) = \alpha^2 + \beta^2$.

State $|00\rangle$ is consistent with $A \cap B$, so that $P(A \cap B) = \alpha^2$. Similarly, $P(A \cap C) = \beta^2$. Thus, $P(B|A) = \dfrac{\alpha^2}{\alpha^2 + \beta^2}$, and $P(C|A) = \dfrac{\beta^2}{\alpha^2 + \beta^2}$, and the sum of probabilities is 1.

However, instead of measuring the energy of the second qubit, the state of the system can be calculated after measuring the energy of the first qubit. Since $P(A|A) = 1$, the system is composed of only basis state with first qubit as 0 while $P(B|A)$ and $P(C|A)$ are known. The new state satisfying the states of second qubit 0 and 1 is given by

$$|\psi'\rangle = \frac{\alpha}{\sqrt{\alpha^2 + \beta^2}}|00\rangle + \frac{\beta}{\sqrt{\alpha^2 + \beta^2}}|01\rangle \qquad (11.21)$$

11.4.3 Operations on qubits

Quantum states can be manipulated in two ways: by unitary transformations and by the measurements.

11.4.3.1 Unitary transformation

Unitary transformations can be described as matrix operations on vectors (the vector being a quantum state). The operation is given by

$$|\psi\rangle \mapsto U.|\psi\rangle \qquad (11.22)$$

where $|\psi\rangle$ is a quantum state of n qubits (a vector of dimension 2^n), and U is a unitary matrix which satisfies the equation $U^T U = I$, I being an identity matrix. Unitary transformations are reversible, and therefore $|\psi\rangle = U^{-1}U|\psi\rangle$. This means that applying them does not destroy any information.

A particular single-qubit unitary is the Hadamard transform H_t given by

$$H_t = \frac{1}{\sqrt{2}} \begin{bmatrix} 1 & 1 \\ 1 & -1 \end{bmatrix} \qquad (11.23)$$

with the property $H_t = 1$.

11.4.3.2 Measurement

The second kind of operation which can manipulate quantum states is the measurement. In contrast to unitary transformations, measurements are destructive operations and therefore not reversible. Measurement collapses a quantum state into one of the possible basis states; the probability of occurrence of a particular basis state is given by the square of the modulus of its associated coefficient. So measurement effectively extracts classical information back from quantum information, at the expense of destroying the quantum state. However, measurement is the only way of finding out anything about an unobserved quantum state. As long as it is not measured, it remains unobserved and can exist in superpositions of basis states. The moment a measurement is taken, it is forced to become a classical state.

Further, the measurements are probabilistic: the probability of obtaining each given basis state is determined by the square of the modulus of its coefficient. Measurements are usually performed along a fixed observable, i.e., an axis along which it is measured. Different observables generally give different results and assume different sets of basis states.

11.4.4 Realisation of qubits

There are many ways of realisation of qubits. One of the explored two-level qubits utilises the spin of particles with two basis states, spin-up (notation ↑) or $|0\rangle$ and spin down (notation ↓) or $|1\rangle$. However, not all two level systems are suitable for implementing the qubits, though any physical phenomena can be used provided the level spectrum is sufficiently anharmonic to provide a good two-level system. Table 11.1 indicates a summary of different technologies available for the realisation of qubits; however, it may be pointed out that most of the technologies are in various experimental stages.

TABLE 11.1: Technologies for qubit realisation

Technology	Single/Ensemble	Measurement
Atom/Ion	Single	optically induced flourescence
NMR	ensemble	frequency analysis
JJ charge	single	concurrent, charge probe
Quantum dot	single	on-chip structure
LOQC	single	polarisation via single photon

Most of the techniques in the implementation of qubits can be broadly distinguished between microscopic, mesoscopic, and macroscopic techniques. Microscopic two-level systems are localised systems evolved by a material growth process, typically by atomic or molecular impurities utilising electronic charge, or nuclear spins. Mesoscopic qubit systems typically involve geometrically defined confining potentials like quantum dots, which are usually made in semiconductor materials with quantised electronic levels. Macroscopic super-conducting qubits are mostly based on electrical circuits containing Josephson junctions (JJ) [8].

During physical implementation of qubits, the major requirement is the coherence, measured by the number of quantum gate operations that can be performed with small error. This requirement is extremely demanding and therefore qubits need to be protected from decoherence.

11.4.4.1 Atom and ion qubits

The main advantage of using microscopic quantum systems like atoms and ions is in their quantumness. One of the advanced qubit implementations is based on ions in linear traps coupled to their longitudinal motion and addressed optically [9]. The idea is to trap several neutral atoms inside a small high-precision optical cavity where quantum bits can be stored in the internal states of the atoms, as they are all coupled to the normal modes of the electromagnetic field in the cavity. By driving transitions with a pulsed laser, a transition can be achieved in one atom that is conditioned on the internal state of another atom [10].

For an ion trap, the quantum state of each ion is a linear combination of the ground state $|g\rangle$ equivalent to $|0\rangle$ and a particular long-lived metastable excited state $|e\rangle$ equivalent to $|1\rangle$. A coherent linear combination of the two levels, $\alpha|g\rangle + \beta e^{i\omega t}|e\rangle$, can survive for a time comparable to the lifetime of the excited state [11]. Read-out of the ions can be done by performing a measurement that projects onto the $\{|g\rangle, |e\rangle\}$ basis. A laser is tuned to a transition from the state $|g\rangle$ to a short-lived excited state $|e^t\rangle$.

Because of their mutual Coulomb repulsion, the ions are sufficiently well separated and they can be individually addressed by pulsed lasers. When the laser illuminates the ions, each qubit with the value $|0\rangle$ repeatedly absorbs and re-emits the laser light, thus giving visible fluoresces, and the qubits with the value $|1\rangle$ remain dark. If a laser is tuned to the frequency ω of the transition and is focused on the n-th ion, then oscillations are induced between $|0\rangle$ and $|1\rangle$. By timing the laser pulse properly and choosing the phase of the laser appropriately, any one-qubit unitary transformation can be achieved. Any desired linear combination of $|0\rangle$ and $|1\rangle$ can also be generated by acting on $|0\rangle$, by the laser pulse.

Ion traps offer the ability to exploit physical symmetries, which reduce sensitivity to environmental noise via small decoherence-free subspaces [12]. A major drawback of the ion trap qubit, however, is its slow transition speed, which is ultimately limited by the energy-time uncertainty relation [13]. The ion-traps proposed and experimented have shown the following key features: (1) States of qubits are defined by the hyperfine interactions between the nuclear and electronic states of each ion, which are laid out in a linear radio frequency trap. (2) Universal quantum gates are implemented by using laser beams to excite the collective quantised motion of the ions. (3) Quantum information exchange is done via Coulomb interactions between the ions. (4) Multi-qubit systems are implemented by appending ions on the trap.

11.4.4.2 Electronic qubits

Two main strategies based on quantum states of either single particles or of the whole circuit have been followed for making electronic qubits. In the first strategy, the quantum states are nuclear spin states, single electron spin states, or single electron orbital states. The second strategy has been developed in superconducting circuits based on Josephson junctions, which form a kind of artificial atoms. Although the quantumness can be good at low temperature, when spin states are used, the main drawback is that qubit operations are difficult to perform since single particles are not easily controlled and can be made ready for readout. The quantumness of these artificial atoms does not compare to that of natural atoms or of spins.

Using electron spins for the qubits is attractive because the spin is weakly coupled and the spin state can be transferred to a charge state for the purpose of readout [14], [15]. A quantum dot, on the other hand, as used in preparing qubits, is a lithographically-defined structure that confines electrons at the boundary layer between two materials. By varying the surrounding electrical potential, individual electrons can be positioned in a small area. A qubit can be defined based on the number of electrons in a quantum dot or the spin or energy levels of a single electron held in a quantum dot [16].

Several quantum dot devices are under development. One experimentally advanced approach uses a pair of quantum dots as a dual-rail encoded logical qubit, with a single electron in the left dot representing a logical 0, and the electron in the right dot representing a logical 1 [17], [18]. Another approach uses a linear array of single-electron quantum dots, and encodes the qubit in the spin of the excess electron. In a third approach, exchange between two neighbouring qubits is accomplished by lowering the electrical potential and allowing the electrons to tunnel. However, the system suffers from low coherence time.

11.4.4.3 Josephson junction

Josephson junction-based quantum information processing devices are superconducting systems and are used for implementing qubits [19], [20]. These junctions can be considered artificial macroscopic atoms, whose properties can be controlled by applying electric or magnetic fields, and also the bias current. Josephson junction(JJ) for implementing qubits can be done in three ways: those that represent qubits using charge, those that use flux, and those that use phase [21], [22]. Fabrication is done using conventional electron-beam lithography. In the JJ charge qubit, a sub-micron size superconducting box (essentially, a small capacitor) is coupled to a larger superconducting reservoir [23].

The simplest superconducting qubit circuit consists of a tunnel junction JJ with superconducting electrodes, connected to a current source. It is necessary to introduce the superconducting phase difference across the junction, which has the form of a damped nonlinear oscillator. The dissipation determines the qubit lifetime, and therefore circuits suitable for qubit applications must have extremely small dissipation. The electrostatic energy of the junction capacitors plays the role of kinetic energy, while the energy of the Josephson current plays the role of potential energy. It is convenient to introduce the charging energy of the junction capacitor. The potential energy corresponds to the energy of the Josephson current, and the magnetic energy corresponds to the bias current. In the absence of bias current the potential leads to oscillations known as plasma oscillation. The simplest qubit realisation is a current biased JJ with large Josephson energy compared to the charging energy. In the classical regime, the particle representing the phase either rests at the bottom of one of the wells of the potential or oscillates within the well.

The rf-superconducting quantum inference device (rf-SQUID) is another important superconducting circuit, where a tunnel Josephson junction is inserted in a superconducting loop. This circuit realises magnetic flux bias for the Josephson junction. For bias flux equal to integer number of flux quanta, the potential energy of the SQUID has one absolute minimum. For half integer flux quanta, the potential energy has two degenerate minima, which correspond to the two persistent current states circulating in the SQUID loop in the opposite directions, and thus a persistent-current flux qubit can be constructed. When two Josephson junctions are coupled in parallel to a current source, the new physical feature is the dependence of the effective Josephson energy of the double junction on the magnetic flux threading the SQUID loop.

Another Josephson junction found suitable is a circuit consisting of a small superconducting island connected via a Josephson tunnel junction to a large superconducting reservoir and known as single electron box. The island is capacitively coupled to another electrode, which may act as an electrostatic gate, and a voltage source controls the gate potential. The system is in a Coulomb blockade regime, where the electrons can only be transferred to the island one by one, the number of electrons on the island being controlled by the gate voltage. In the superconducting state, the same circuit is called a single Cooper pair box. The qubit is simply represented by the two rotation directions of the persistent supercurrent of Cooper pairs in a superconducting ring containing Josephson tunnel junctions [24], [25]. In a superconductor, electrons move in pairs and the qubit representation is the number of Cooper pairs in the box, controlled to be either 0 or 1, or a superposition of both. Cooper pairs can be introduced into a superconducting ring, where they circulate and induce a quantised magnetic flux [26], [27]. Because the flux qubit has slower gate times but has relatively longer coherence time, experimental efforts appear to be shifting toward the flux qubit approach [28]. Recently, some rather complicated JJ networks have been discussed, which have the unusual property of degenerate ground state, which might be employed for efficient qubit protection against decoherence.

11.4.4.4 Nuclear magnetic resonance (NMR) devices

Probably the most complete demonstrations of quantum devices to date are achieved by NMR experiments [29]. In an NMR system, the qubit is represented by the spin of the nucleus of an atom. When placed in a magnetic field, the spin can be manipulated via microwave radiation. In solution, a carefully designed molecule is used. Some of the atoms in the molecule have nuclear spins, and the frequency of radiation to which they are susceptible varies depending on their position in the molecule. Therefore, different qubits are addressed by frequency. No special cooling apparatus is required for this ensemble system. However, its scalability is believed to be quite limited due to falling signal/noise ratio as the number of qubits increases [30], [31].

Instead of solution NMR, in an all-silicon NMR-based qubit processor, qubits are stored in the nuclear spin. Readout is done via magnetic resonance force microscopy (MRFM). Initialisation is done via electrons whose spins are set with polarised light. Operations are done via microwave radiation directed at the device. A micromagnet provides a high field gradient, allowing individual atoms to be addressed by frequency.

11.4.4.5 Photonic qubits

As previously mentioned, any two-level quantum-mechanical system can be used as a qubit representation. Therefore, many possibilities are open for qubit implementation by using various physical phenomena associated with photons [32]. Controlled gate operations are especially promising because they allow an implementation of photon-photon interactions using only linear optical elements.

(a) Using polarisation: Because of the relative ease with which polarisation can be manipulated in an experimental setup using half- and quarter-wave plates, and polarisers, polarisation of the photons is a common choice for the implementation of qubits. The logic basis states $|0\rangle$ and $|1\rangle$ are represented by the orthogonal horizontal and vertical polarisation $|H\rangle$ and $|V\rangle$. The diagonal/anti-diagonal polarisation, $|D\rangle/|A\rangle$, and the left-and right-circular polarisation, $|L\rangle/|R\rangle$, correspond to the two remaining mutually unbiased bases $\frac{1}{\sqrt{2}}(|0\rangle \pm |1\rangle)$ and $\frac{1}{\sqrt{2}}(|0\rangle i \pm |1\rangle)$.

(b) Using phase: Phase qubits is also a common choice. The logic basis states $|0\rangle$ and $|1\rangle$ are represented by the relative phase difference between two arms in an interferometer, where $|0\rangle$ corresponds to zero phase difference and $|1\rangle$ corresponds to a phase difference of π. The basis states can then be written as $\frac{1}{\sqrt{2}}(|1\rangle \otimes |0\rangle \pm |0\rangle \otimes |1\rangle)$, where $|1\rangle \otimes |0\rangle$ denotes the presence of one photon in the first arm of the interferometer and none in the other. In practice, a pair of interferometer is used, one for encoding and the other for decoding. The phase difference is set by phase modulators in the arms of the interferometers.

For encoding to work, the coherence length of the photons must be longer than the path mismatch of the arms of the interferometers.

(c) Using time: Time coding is usually used for long-distance communication over optical fibres because it is fairly robust against decoherence effects in the fibres. Timing information can be used as qubits where two time slots correspond to the logic basis states $|0\rangle$ and $|1\rangle$. Hence, the basis states can be written as $|1\rangle \otimes |0\rangle$ and $|0\rangle \otimes |1\rangle$, where $|1\rangle \otimes |0\rangle$ denotes having one photon in the first time slot and none in the second. The qubits are created by sending photons through an unbalanced interferometer, creating a superposition between two different time slots corresponding to whether the photon traversed the short or the long arm in the interferometer.

(d) Using spatial mode: Another possibility for generation of qubits is to use two spatial modes where having a photon in one mode would correspond to the state $|0\rangle$ and having it in the other mode would correspond to the state $|1\rangle$. The basis states can be written $|1\rangle \otimes |0\rangle$ and $|0\rangle \otimes |1\rangle$, where $|1\rangle \otimes |0\rangle$ denotes having one photon in the first spatial mode and none in the other. Superpositions between the two are achieved by a variable coupler, and decoding is done by a reversed setup.

(e) Using frequency: Using the frequency of a photon is also a possibility, where the basis states are represented by two different frequency states, $|\omega_1\rangle$ and $|\omega_2\rangle$. In general, the problem of dispersion during propagation limits the interest of this particular qubit implementation.

11.5   Entangled states and photon entanglement

By definition, a composite system is called entangled if it cannot be written as the composition of the states of its component systems. Defined mathematically, two quantum states are entangled, if it is not possible to find the four parameters a_1, a_2, b_1, b_2 that satisfy the equation given by

$$|\psi\rangle = (a_1|0\rangle + b_1|1\rangle) \otimes (a_2|0\rangle + b_2|1\rangle) \tag{11.24}$$

Two entangled particles have no determined state of their own prior to a measurement, but it is only the relation between them that is known, e.g., that they are always in the same state. Upon measurement of one of the particles, its state will collapse to one of the possible measurement outcomes, and then the state of the other particle will be completely determined as well. This holds true even if the two particles are separated in space, consequently giving a nonlocal influence on one particle by measuring the other [33].

Entanglement can occur in a vast diversity of ways. The simplest situation is when only two quantum systems are entangled. The general two-system entangled states are called Bell states [34]. These states are maximally entangled, meaning that the collapse of any of the two systems to one of the two states $|0\rangle$ or $|1\rangle$ completely determines the state of the other system as well. It is also important to note that a measurement in another basis, e.g., $\frac{1}{\sqrt{2}}(|0\rangle \pm |1\rangle)$, does not affect the correlations between the systems; a measurement of one of the systems will again completely determine the state of the other system. This is a key property of entanglement: that the correlations remain under local operations. Moreover, if a partial trace is performed over the first system, which corresponds to only having access to information about the second system, the resulting state of the second system is a mixed state, i.e., a state about which complete information is not available, despite the fact that full information is available for the two-system state (called pure state).

Entangled states can be realised in many physical systems. The photonic technique for creating entanglement is by using spontaneous parametric down-conversion (SPDC) in nonlinear crystals. When using SPDC to generate photon pairs, the entanglement can either be created directly or by post-selection using a phase matching process. The two down-converted photons are then emitted on two cones centred around the pump beam. If the polarisation of one beam is rotated by 90° and the two beams are subsequently combined on a beam splitter, a polarisation-entangled state is obtained for those events when the two photons exit in separate output ports of the beam splitter. Alternatively, a source can be used to create product states where the two photons are then sent to two input ports of a beam splitter. Selecting the events when the photons exit different output ports gives a polarisation-entangled state of the form

$$|\psi\rangle = \frac{1}{\sqrt{2}}(|HV\rangle + |VH\rangle) \tag{11.25}$$

11.6   Quantum computing

Quantum computing is a rapidly evolving field. Researchers are motivated by the enormous computational potential compared to electronic or photonic computing machines. Current research is centred around algorithms and complexity theory on the one hand, and on the other hand, quantum gate generation and qubit storage technologies. The architectures in designing practical machines that can run quantum algorithms efficiently is also being explored. At the outset it may be mentioned that a quantum computer may operate according to different physical principles than those for a classical computer. However, it cannot do anything that a classical computer cannot perform. What is computable will be the same, whether a classical computer or a quantum computer is used.

A quantum computer works by storing information in physical carriers, where the information undergoes a series of unitary (quantum) evolutions and measurements. The information carrier is usually taken to be qubits. In general, a quantum computer consists of at least two addressable quantum states, and furthermore, the qubit can be put in arbitrary superposition states, too. The unitary evolutions on the qubits that make up the computation can be decomposed in single-qubit operations and two-qubit operations. Both types of operations or gates are necessary if the quantum computer is to outperform any classical computer.

Since a quantum computer works in terms of qubits, an assembly of N qubits can be used with an assumed initial zero state such as $|0\rangle|0\rangle....|0\rangle$, or $|x = 0\rangle$. Then, a unitary transformation U is applied to N qubits. Assuming that the transformation U is constructed as a product of standard quantum gates, the unitary transformations act on just a few qubits at a time. After U is applied, all states of the qubits are measured by projecting onto the $\{|0\rangle, |1\rangle\}$ basis.

The algorithm to be used is a probabilistic algorithm. That is, by running exactly the same program twice, different results are probable, because of the randomness of the quantum measurement process. Thus, quantum computer with an algorithm actually generates a probability distribution of possible outputs. The measurement outcome is, however, the output of the computation in terms of classical information that can be printed out.

The critical issue in quantum computation is the problem of decoherence, or loss of the quantum state. Decohenerce is a function of time, gate errors, and qubit transport. Therefore the quantum technologies need to achieve error rates well below the critical threshold to avoid undue overhead from error-correction processes. A set of general rules that need to be fulfilled in order to perform quantum computing has been formulated by DiVincenzo [35]. The hardware has to meet some stringent specifications, such as

1. Register of 2-level systems (qubits), $n = 2^N$ states $|101..01\rangle$ (N qubits)

2. Initialisation of the qubit register: e.g., setting it to $|000..00\rangle$

3. Tools for manipulation: qubit gates need to flip the state.

4. Readout of single qubis

5. Long decoherence times and error correction to maintain coherence.

6. Transport of qubits, and to transfer entanglement between different coherent systems (quantum-quantum interfaces).

7. Creation of classical-quantum interfaces for control, readout, and information storage.

11.6.1 Quantum logic gates and circuits

A classical (deterministic) computer evaluates a function $f : \{0,1\}^n \to \{0,1\}^m$. This means that given n-bits of input, the computer must produce

m-bits of output that are uniquely determined by the input, for a particular specified n-bit argument. A function with an m-bit value is equivalent to m functions, each with a one-bit value; then the basic task performed by a computer is the evaluation of $f : \{0,1\}^n \to \{0,1\}$. There are 2^n possible inputs, and for each input there are two possible outputs, then the number of possible functions is 2^{2^n}. The evaluation of any such function can be reduced to a sequence of elementary logical operations.

11.6.1.1 Classical logic circuits

It is possible to divide the possible values of the input $x = x_1 x_2 ..... x_n$ into two sets; one set may consist of values for which $f(x) = 1$, and a complementary set is for values for which $f(x) = 0$. A function $f(x)$ can be constructed from three elementary logical connectives: NOT, AND, OR expressed by symbols ¬, ∧ and ∨, respectively.

Logical OR (∨) operation for the function f of all the $f^{(a)}$s, where $x^{(a)}$ is such that $f(x^{(a)}) = 1$, can now be stated as

$$f^{(a)}(x) = \begin{cases} 1, & x = x^{(a)} \\ 0, & \text{otherwise} \end{cases} \tag{11.26}$$

then,

$$f(x) = f^{(1)}(x) \vee f^{(2)}(x) \vee f^{(3)}(x) \vee .... \tag{11.27}$$

Similarly, the logical AND (∧) operation of n bits is given by

$$f(x) = f^{(1)}(x) \wedge f^{(2)}(x) \wedge f^{(3)}(x) \wedge .... \tag{11.28}$$

The expression obtained is called the disjunctive normal form of $f(x)$.

In computation, a useful operation is COPY, which takes one bit to two bits as, $COPY : x \to xx$. The set of other elementary logical operations is NAND, which may be constructed by using NOT and AND operations. Similarly, by using COPY and NAND, a NOT operation can be performed. By using COPY, NAND performs AND and OR as well. It may be concluded that the single logical connective NAND, together with COPY, is sufficient to evaluate any function f. Another alternative possible choice of the universal connective is NOR. The computation, in general, acts on a particular fixed input, but families of circuits that act on inputs of variable size are also possible.

11.6.1.2 Quantum gates

The circuit model of classical computation can be generalised to the quantum circuit model of quantum computation. A classical computer processes bits. It is equipped with a finite set of gates that can be applied to sets of bits. A quantum computer processes qubits. It may be assumed that during computation, a quantum computer is also equipped with a discrete set of fundamental components, called quantum gates. Each quantum gate is a unitary transformation that acts on a fixed number of qubits. Just as any classical computation be broken down into a sequence of classical logic gates that act on only a few classical bits at a time, so too can any quantum computation can be broken down into a sequence of quantum logic gates that act on only a few qubits at a time. The main difference is that whereas classical logic gates manipulate the classical bit values, 0 or 1, quantum gates can manipulate arbitrary multi-partite quantum states, including arbitrary superpositions of the computational basis states, which are frequently also entangled. Thus the logic gates of quantum computation are considerably more varied than the logic gates of classical computation.

In a quantum computation, a finite number n of qubits are initially set to the value $|00...0\rangle$. A circuit is executed that is constructed from a finite number of quantum gates acting on these qubits. Finally, a Von Neumann measurement of all the qubits (or a subset of the qubits) is performed, projecting each onto the basis $\{|0\rangle, |1\rangle\}$. The outcome of this measurement is the result of the computation.

A quantum gate, being a unitary transformation, is reversible. In fact, a classical reversible computer is a special case of a quantum computer. A classical reversible gate implements a permutation of n-bit strings $x^{(n)} \rightarrow y^{(n)} = f(x^{(n)})$ and can be used as a unitary transformation that acts on the computational basis $\{|x_i\rangle\}$, according to

$$U : |x_i\rangle \rightarrow |y_i\rangle \tag{11.29}$$

This action is unitary because the second strings $|y_i\rangle$ are all mutually orthogonal. A quantum computation constructed from such classical gates takes $|0...0\rangle$ to one of the computational basis states, so that the final measurement is deterministic. The most general unitary transformation that can be performed on n qubits is an element of $U(2^n)$. Using universal gates, circuits can be constructed that compute a unitary transformation that comes close to any element in $U(2^n)$.

Physically, in quantum computing, as long as a generic interaction between two qubits can be devised, and that interaction can be accurately implemented, it is possible to compute anything, no matter how complex it is. Nontrivial computation is ubiquitous in quantum theory. Aside from this general result, it is also of some interest to exhibit particular universal gate sets that might be particularly easy to implement physically.

A quantum computer can easily simulate a probabilistic classical computer: it can prepare $\frac{1}{\sqrt{2}}(|0\rangle + |1\rangle)$ and then project to $\{|\rangle, |1\rangle\}$, generating a random bit. The fundamental difficulty is that the Hilbert space of n qubits is huge, of dimension 2^n, and hence the mathematical description of a typical vector in the space is exceedingly complex.

11.6.1.3 Quantum gate circuits

The first class of important quantum operators are the one-qubit operators realised in the quantum circuit as the one-qubit (quantum) gates. For two-qubit transformations, the most important operation is the controlled-NOT (CNOT) operation [36]. Quantum computational networks that involve such n-bit operations can be constructed with these gates. The main reason for its importance is that any multiple qubit operation can be constructed from CNOT operations together with single qubit operations. CNOT uses one qubit as the control qubit and the other as the target qubit and operates by (depending on the value of the control qubit) flipping the target qubit or keeping it in its present state. The operations can be represented by the state transformations as

$$|00\rangle \rightarrow |00\rangle, |01\rangle \rightarrow |01, |10\rangle \rightarrow |11\rangle, |11\rangle \rightarrow |10\rangle$$

In matrix notation, using the basis $\{|00\rangle, |01\rangle, |10\rangle, |11\rangle\}$, where the matrix performs the same mapping as a CNOT gate, the operation can be considered as

$$CNOT = \begin{bmatrix} 1 & 0 & 0 & 0 \\ 0 & 1 & 0 & 0 \\ 0 & 0 & 0 & 1 \\ 0 & 0 & 1 & 0 \end{bmatrix} \tag{11.30}$$

The CNOT gate is unitary, and can be easily recognised by identifying the states where rows/columns of the matrix forms an orthogonal set of unit vectors. Also useful is the two-qubit transformation known as the XOR or CNOT transformation; it acts as

$$CNOT : |a, b\rangle \rightarrow |a, a \oplus b\rangle \tag{11.31}$$

where a and b are the classical values called the control (or source) bit and the target bit of CNOT, respectively.

The Bell state mentioned earlier can be expressed in terms of CNOT operation by a compact equation, as

$$|\Phi\rangle = \frac{|0, b\rangle + (-1)^a |1, NOT(b)\rangle}{\sqrt{2}} \tag{11.32}$$

where $a, b \in \{0, 1\}$.

It is worth pointing out that the CNOT gate can be regarded as a one-bit copy machine. Provided its data input is initialised permanently with $|0\rangle$, then the CNOT gate emits a copy of the control input on each output. By composing these primitive transformations, or quantum gates, other unitary transformations can be executed.

Any state in the qubit Hilbert space H^s can be reached by applying the proper combination of identity operators. In practice, rotations around all three coordinate axes x, y, and z of the qubit sphere can be performed, since any unitary qubit operation U can be expressed as $U = e^{i\alpha} R_z(\beta) R_y(\gamma) R_x(\delta)$, where $R_{x,y,z}$ are the rotation matrices around respective axis and $\alpha, \beta, \gamma, \delta$ are angles. Depending on the specific physical realisation of the qubits, these rotations are obtained by using different physical components.

Qubits are generally denoted by horizontal lines and the single-qubit unitary transformation U is denoted as shown in Figure 11.3 (a). The circuit shown in Figure 11.3 (b) (to be read from left to right) represents the product of H_t applied to the first qubit, followed by CNOT with the first bit as the source and the second bit as the target. The corresponding quantum circuit which can be used to produce the Bell states can be seen in Figure 11.3 (c). On the basis states $a, b = 0, 1$, where $a \oplus b$ denotes addition modulo 2, and is denoted as shown in Figure 11.3 (d). Thus this gate flips the second bit if the first is 1, and acts trivially if the first bit is 0, such that $(CNOT)^2 = 1$. This gate also performs a NOT on the second bit if the first bit is set to 1, and it performs the copy operation if a is initially set to zero. A parallel connection of gates corresponds to the Kronecker product of unitary matrices of respective gates. The serial connection of gates corresponds to the matrix multiplication (in reverse order) of the matrices of these gates.

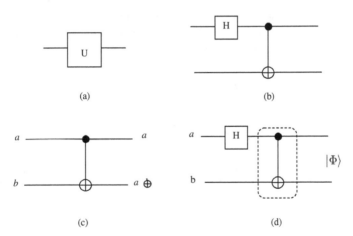

(a)

(b)

(c)

(d)

FIGURE 11.3: Notation used for representing qubit gates

The SWAP gate can be obtained from a sequence of three CNOT gates. Just as there can be universal gates for classical computing, such as the NAND gate (which has two inputs and one output), so too can there be universal gates. However, the smallest such gates require three inputs and three outputs. Two well-known examples are the FREDKIN (controlled-SWAP) gate and the TOFFOLI (controlled-CNOT) gate. The TOFFOLI gate is also called the controlled-controlled-NOT gate since it can be understood as flipping the third input bit if, and only if, the first two input bits are both 1. In other words, the values of the first two input bits control whether the third input bit is flipped. The FREDKIN gate can also be seen as a controlled-SWAP gate since it swaps the values of the second and third bits, if and only if the first bit is set to 1 [37]. The SWAP gate is shown in Figure 11.4 (a), along with a FREDKIN gate (Figure 11.4 (b)) and TOFFOLI gate Figure 11.4 (c)).

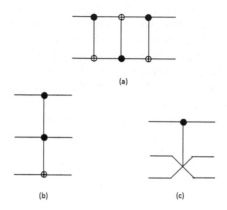

FIGURE 11.4: Notation used for representing some quantum gate circuits

The distinction between classical and quantum operation can be presented when register operation is visualised as shown in Figure 11.5, where output of a register with classical bit and qubit input is shown.

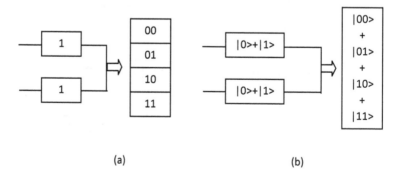

FIGURE 11.5: Register operation using (a) classical bits and (b) qubits

11.6.2 Quantum computing architecture

There are two main branches of fundamentally different computer architectures, namely logically irreversible and logically reversible. Ordinary computers are irreversible because they dissipate both energy and information. The use of quantum-effect devices does not change the fact that the computers where each gate is logically irreversible and where discarded information is constantly erased and turned into heat. A computer with quantum device components therefore does not make a quantum computer. A quantum information processor has to be built on fundamentally reversible schemes with reversible gates where no information is discarded, and where all internal processes in the components are elastic. This issue is connected with the problem of the minimum energy needed for performing a calculation (connected with the entropy change created by erasing the final result, i.e., reading the result and then clearing the register). The characteristic of a quantum computer is that it is reversible and quantum coherent, meaning that one can build entangled non-classical multi-qubit states.

Quantum computing architectures can be divided into two categories: those in which the qubits are represented by constantly moving phenomena (photons) and those in which qubits are represented by static phenomena (nuclear or electron spins). For phenomena that move, gates are physical devices which affect qubits as they flow through the gate. Photonic implementations of quantum computing architecture generally fall into this category. For stationary phenomena, qubits occupy a physical place and gate operations from an application are applied to them. The stationary notion, however, applies only during gate operation. Some stationary technologies, such as scalable ion trap, permit the physical qubit carrier to be moved prior to application of a gate. The key reason to make the distinction between stationary and moving implementations is dynamic control. In a moving qubit device, the order and type of gates must typically be fixed in advance, often at device construction time; different program execution is achieved by classical control of switches that route qubits through different portions of the circuit. A stationary qubit device has more flexibility to reconfigure gates.

Another significant distinction in quantum computing architecture is the choice of ensemble computing or singleton computing. In ensemble computing, generally implemented on static qubit systems. There may be many identical quantum processors all receiving the same operators and executing the same program on the same data (except for noise). Singleton systems have the ability to directly control a single physical entity that is used to represent the qubit. On the other hand, ensemble systems are easier to experiment with, as techniques for manipulating and measuring large numbers of atoms or molecules are well understood. Hence, the largest quantum computing system demonstrations have all been on bulk-spin NMR, which uses an ensemble of molecules to compute.

In order to compute reliably, and to be able to observe the result of computation, computing technologies must support a readout process. This readout, called measurement, observes the state of the quantum bit and produces a classical result. Reliably computing on a quantum system will mean that many, if not most, of the total quantum operations will be the measurements. From an architectural perspective, if measurements must be performed serially or are inordinately slow, then measurement will be the bottleneck in computation. Furthermore, if additional clean qubits are required for measurement to take place, then frequent initialisation processes will be another bottleneck.

Similarly, if technologies restrict where measurement can occur, then those restrictions will need to be designed into the architecture.

The architecture should have the facility for quantum error correction (QEC), which traditionally depends on interleaving measurement. However, it is possible to perform QEC without measurement, at a cost of using additional qubits, which grows with the number of applications of error correction. Additional techniques known as decoherence-free subspaces are especially useful when nearby qubits are subject to collective error processes. In photonic systems, the principal source of error is loss of photons. In this case, erasure codes (in contrast to error correcting codes) work well, because it is easy to determine which qubit has been lost. For example, using a quantum computer, factoring a 576-bit number in one month using Shors algorithm requires, first of all, that a coherent state be maintained for a month across several thousand logical qubits, through the use of quantum error correction. Secondly, a logical clock speed of 0.3 Hz to 27 Hz is required.

The field of quantum computer architecture can be said to be in its infancy. Researchers who have focused on the fundamental technologies are now beginning to examine how to build complete quantum computing systems. At the moment, all scalable quantum computing technologies are proposals and significant advances in manufacturing will be required to bring them to reality. Nevertheless, some proposals have less onerous technological hurdles in front of them than others. Furthermore, certain proposed technologies integrate better with existing classical silicon-based computing.

11.7   Quantum communication

In quantum communication systems, special requirements appear on the different constituents of the conventional communication system. At the sender end, a source of single quantum system is required, with fine control so that information can be encoded onto one of the related degrees of freedom [38]. In a photonic quantum communication system, these quantum systems are photons. Hence, bright, efficient, and reliable high-quality sources of photons are necessary for such a system. At the receiver the information encoded onto the quantum systems needs to be decoded, again requiring precise control of degrees of freedom of quantum systems. Finally, the quantum systems need to be detected. In a photonic system these requirements turn out to be the manipulation of specific properties of photons and the need for detectors capable of detecting light at the photon level. Although such detectors exist, their efficiency differs drastically depending on the wavelength of the photons. The natural development of the general bidirectional (duplex) communication system is to connect several systems into a network. This way, any user in the network can communicate with any other user. Communication networks are very well developed for classical communication, but quantum networks are still in their infancy and are the topic of intense research. In the present day, it is only within quantum cryptography that small-scale quantum networks have been implemented.

11.7.1 Quantum cryptography

Secure communication is one of the areas of key importance for modern society to which quantum cryptography seems to bring significant contributions. Quantum cryptography is the first area in which quantum laws were directly exploited to bring an essential advantage in secure information communication [39], [40]. An important new feature of quantum cryptography is that security of quantum key generation and quantum cryptographic protocols are based on a more reliable fact of the laws of quantum mechanics than in the case of classical cryptography, whose security is based on unproven assumptions concerning the computational difficulty of some algorithmic problems.

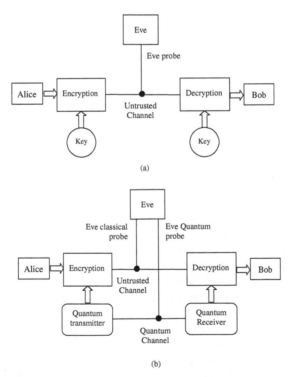

FIGURE 11.6: (a) Classical and (b) quantum cryptography system

Quantum cryptography may work like classical cryptography under certain conditions. Figure 11.6(a) shows a classical channel and Figure 11.6(b)

shows the modifications necessary to make a quantum communication system keeping the classical channel intact.

In the basic system of classical cryptography Alice tries to send a keyed message through an unsecured channel to Bob. An adversarial eavesdropper, Eve tries to learn or change the message as much as possible without being detected. The cryptographic scheme can be modified by introducing quantum transmitter and receiver which can make an eavesdropper's attempts especially difficult.

Alice prepares a quantum system in a specific way unknown to Eve and sends it to Bob. The quantum states cannot be copied and cannot be measured without causing, in general, a disturbance and therefore the system is most secure [41]. The questions are how much information can Eve extract from that quantum system, and what is the cost of that information in terms of the disturbance to the quantum system. Two concepts are therefore crucial here: information and disturbance. What is normally called quantum cryptography can help to solve the problem just mentioned, of the distribution of a random key in a secure manner [40].

Quantum cryptography cannot be used to communicate information in a secure way but it can indeed be used by two users to share a cryptographic key between each other. That is why quantum cryptography is called quantum key distribution [42], [43]. By definition, a quantum key distribution system is an information communication system that can create a perfectly secure symmetric key at the transmitter and at the receiver. It needs two communication channels as shown in Figure 11.6(b). One is a quantum channel, for the transmission of the quantum bits, and the other is a classical public channel, for the communication of classical messages between the transmitter and the receiver.

Two extreme cases can be considered within the scheme where Alice sends a state $|\psi\rangle$ to Bob. The first is that Eve has no information about how $|\psi\rangle$ was prepared. The only thing Eve can then do is to choose some basis $\{e_i\}$ for orthonormal states and use the corresponding projection measurement on $|\psi\rangle$. In such a case, $|\psi\rangle$ collapses into one of the states $|e_i\rangle$; therefore, the only information Eve has learned is that $|\psi\rangle$ is not orthogonal to $|e_i\rangle$. At the same time, $|\psi\rangle$ can get destroyed to a large extent. The second extreme case is that Eve knows that $|\psi\rangle$ is one of the states of the basis $\{e_i\}$. In such a case, by measuring $|\psi\rangle$ with respect to the basis $\{e_i\}$, Eve gets full information about $|\psi\rangle$ because $|\psi\rangle$ collapses into itself. No disturbance to $|\psi\rangle$ occurs.

The most difficult case for quantum cryptography is the third case, where Eve knows that $|\psi\rangle$ is one of the states $|\psi_1\rangle.....|\psi_n\rangle$ that are mutually nonorthogonal and p_i is the probability that $|\psi\rangle$ is the state Alice sends. In this case the question is how much information Eve can get by a measurement and how much disturbance she can cause. The information gain of Eve can be indicated by Shannon entropy, $\sum_{i=1}^{n} p_i \lg p_i$ which is a measurement of her ignorance about the system before the transmission takes place. She can try to decrease this entropy by some measurement and she can get some mutual information. In such a general case, the problem of Eve's information gain versus disturbance is one of the central ones for security of quantum cryptography protocols. In the case of eavesdropping, Bob does not get a pure state but a mixed state specified by a density matrix ρ_i for the case when Alice sends $|\psi\rangle$. The disturbance detectable by Bob is given by $D = 1 - \langle\psi_i|\rho_i|\psi_i\rangle$.

One of the most famous quantum computing algorithms is Shor's algorithm. The algorithm is significant because it implies that public key cryptography might be easily broken, given a sufficiently large quantum computer. An algorithm evolved by Rivest, Shamir, and Adleman known as RSA algorithm, for example, uses a public key N which is the product of two large prime numbers. One way to crack RSA encryption is by factoring N, but with classical algorithms, factoring becomes increasingly time consuming as N grows large. By contrast, Shor's algorithm can crack RSA in polynomial time.

11.7.1.1 Quantum key distribution(QKD)

How to send quantum states in general and qubits in particular is another general problem. Transmission of polarised photons is so far one of the basic tools in quantum cryptographic protocols for sending qubit states. Alice chooses a random bit string and a random sequence of polarisation bases (rectilinear or diagonal). She then sends the other user (Bob) a train of photons, each representing one bit of the string in the basis chosen for that bit position, a horizontal or 45-degree photon standing for a binary zero $|\nearrow\rangle$ and a vertical or 135 degree photon standing for a binary one $|\nwarrow\rangle$. As Bob receives the photons, he decides, randomly for each photon and independently of Alice, whether to measure the photon's rectilinear polarisation or its diagonal polarisation, and interprets the result of the measurement as a binary zero or one. A random answer is produced and all information lost when one attempts to measure the rectilinear polarisation of a diagonal photon, or vice versa. Thus Bob obtains meaningful data from only half the photons he receives, that is, those for which he guessed the correct polarisation basis. Bob's information is further degraded by the fact that, realistically, some of the photons would be lost in transit or would fail to be counted by Bob's imperfectly efficient detectors.

Subsequent steps of the protocol take place over an ordinary public communications channel, assumed to be susceptible to eavesdropping but not to the injection or alteration of messages. Bob and Alice first determine by public exchange of a message, about which photons are successfully received and detected with the correct basis. If the quantum transmission has been undisturbed, Alice and Bob should agree on the bits encoded by these photons, even if this data has never been discussed over the public channel. Each of these photons presumably carries one bit of random information (e.g., whether a rectilinear photon was vertical or horizontal), known to Alice and Bob but to no one else.

Because of the random mix of rectilinear and diagonal photons in the quantum transmission, any eavesdropping carries the risk of altering the transmission in such a way as to produce disagreement between Bob and Alice on some of the bits on which they think they should agree. If all the comparisons agree, Alice and Bob can conclude that the quantum transmission has been free of significant eavesdropping, and those of the remaining bits that were sent and received with the same basis also agree. These can safely be used for subsequent secure communications over the public channel.

There are mainly two types of QKD schemes: One is the prepare-and-measure scheme and the other is entanglement-based QKD. BB84 and B92 are QKD of the first type. In BB84, Alice sends each qubit in one of the four states of two complementary bases, and in B92 Alice sends each qubit in one of two non-orthogonal states, totalling six states. Alice sends each qubit in one of the six states of three complementary bases. Ekert91 and BBM92 are the QKD of the second type, where entangled pairs of qubits are distributed to Alice and Bob, who then extract key bits by measuring their qubits. In BBM92 each party measures half of the pair in one of two complementary bases. In Ekert91, Alice and Bob estimate Eve's information based on the Bell's inequality test; whereas in BBM92, similar to BB84, Alice and Bob try to eliminate Eve's information about the final key. In such protocols, there is an entanglement source which generates entangled particles (generally photons). This source is usually implemented in photonics using a nonlinear down-conversion, which results from shining the laser beam into some particular crystal.

As already mentioned, QKD needs a quantum channel and a classical channel. The quantum channel can be insecure whereas the classical channel is assumed to be authenticated. Fortunately, in classical cryptography, unconditionally secure authentication schemes exist. Moreover, those unconditionally secure authentication schemes are efficient. To authenticate an N bit message, only an order $\log N$ bits of the shared key are needed. Since a small amount of pre-shared secure bits is needed between Alice and Bob, the goal of QKD is key growing, rather than key distribution. In the conventional information theory, however, key growing is an impossible task. Therefore, QKD provides a fundamental solution to a classically impossible problem.

11.7.1.2 Generic features of a protocol

Some of the generic features of a protocol can be stated with reference to BB84 protocol, which is one of the much used type of quantum cryptography protocols. Each party chooses either of the two possible bases randomly, which are rectilinear or diagonal. Each bit is defined by one of the mutually orthogonal quantum states, which means that the sequence of bases of each of the parties is in principle independent of one another. Key features are, transmission, measurement, shifting, and error corrections.

During transmission, Alice selects a block of N classical bits of the message (key) and encodes it into quantum states depending on the selected basis for each bit. By polarisation encoding, the qubits are translated to photon polarisation states and then the photons are transmitted. During measurement, the qubits pass the quantum channel and are measured one after another by Bob and Eve according to their own random basis for each single qubit measurement. The result is a sequence of classical bits. The measurement is nothing but the projection of qubit vector on the basis. In a noiseless channel where there is no modification of the transmitted qubit, if the transmitter and receiver choose an identical basis, the measurement outcome is the same classical bit. But in the case of a different basis, the outcome of measurement is expressed in terms of probability.

During the sifting operation, Alice and Bob declare their choice of basis for state preparation and measurement, respectively, on a public channel. In the case of a lossless and noiseless quantum channel, if Bob wants to be sure that he measures the exact bit which Alice has sent, he should only consider the results for the bits which are measured with the same basis as Alice's. Hence, the rest will be discarded. In this step, even though Eve is able to know Alice and Bob's selection of basis, she does not know the measurement result for Bob. Hence, for those bits which she uses a wrong basis while Bob's basis was correct, there is no way for her to redo the measurement in another (correct) basis. It should be noted that the communication requires an authenticated classical channel in parallel with the quantum channel. By authenticated, we mean a link which is not necessarily secure but where all the messages can be authenticated. In practice, since the channel is never noise-free, it is always necessary to use some error correction. During this step, more bits are discarded from the remaining bit sequence so as to mitigate the possible information leakage at the cost of lower rate.

After BB84 was created, people started modifying it to propose new protocols featuring new enhancements or advantages. In BB92, for example, it was shown that non-orthogonal states can also be used for polarisation encoding. In contrast with these four-state protocols, the six-state protocol introduced lower tolerable channel noise. After the cryptography steps, which include discarding a considerable portion of the initial (raw) exchanged key, the length of the remaining sequence is the actual key length. This gives the secure key exchange rate when divided by time. This rate depends on the quantum channel bit error rate (QBER), which is an important performance measure for quantum communication systems.

11.7.2   Quantum teleportation

Quantum teleportation allows transmission of quantum information to a distant place in spite of the impossibility of measuring or broadcasting information to be transmitted. It is an interesting demonstration of quantum entanglement as a communication resource. Assume that Alice and Bob share a pair of entangled particles A and B and let Alice be given a particle C in an unknown state $|\psi\rangle$. Now Alice wishes that $|\psi\rangle$ may go to Bob instead of it going to her. The constraints is that Alice cannot measure the particle to learn the information which she may relay or transmit to Bob over a classical channel, because such a measurement would almost certainly destroy the information irreversibly.

However, Alice can make the particle C in the unknown state $|\psi\rangle$ to interact, in the proper way, with her part A of the entangled pair, and then she can measure a shared property of both C and A. As the result, the state of Bob's particle B of the entangled pair instantaneously becomes a replica of $|\psi\rangle$ with a rotation (in case of polarisation states). At the same time, Alice's particles C and A lose their information but she obtains, as the result of her measurement, two purely random bits of classical information that tell her which operation (rotation, in the case of polarisation) Bob should perform on his particle B to make the particle get into the state $|\psi\rangle$. Alice then has to communicate these two bits to Bob over a classical channel, because Bob's particle B remains indistinguishable from a purely random qubit until he performs the required rotation to get the particle into the state $|\psi\rangle$. Alice can therefore divide information encoded in $|\psi\rangle$ into two parts: classical and non-classical, and send them to Bob through two different channels: a classical one and a quantum channel. The quantum channel may be a EPR (Einstein, Podolski, and Rosen) channel which consists of atleast two Hilbert spaces. One can also say that by performing a proper measurement on C and A a part of quantum information in C is immediately transmitted to B and Alice gets the remaining part of information on $|\psi\rangle$ in the classical form. This can then be used by Bob to make his particle B get to the unknown state $|\psi\rangle$.

At the end, neither Alice nor Bob knows the state $|\psi\rangle$, but both of them know that at the end of the teleportation not Alice but Bob has a particle in the state $|\psi\rangle$. This way, the unknown quantum state $|\psi\rangle$ can be disassembled into and later reconstructed from two classical states and an entangled, purely quantum state. It may be observed that in this way, an intact and unknown quantum state can be teleported from one place to another by a sender who does not need to know for teleportation itself either the state to be teleported or the location of the intended receiver. Finally, it may be observed that the above process cannot be used to transmit information faster than light, but it can be argued that a part of the information that was present in the particle C is transmitted instantaneously, except for two random bits that needed to be transported at the speed of light, at most. In addition, if Bob already possesses the state $|\psi\rangle$, then teleportation can be used by Bob to determine $|\psi\rangle$ more completely, by making measurements on both copies of $|\psi\rangle$. Thus teleportation is possible without Alice knowing the exact position of Bob. It is sufficient to broadcast classical bits to all the possible locations where Bob could be.

Teleportation can be modelled by a 3-qubit system without measurement and classical transmission as shown in Figure 11.7. The unknown qubit 1 to be teleported is $|\psi\rangle = a|0\rangle + b|1\rangle$. The state of qubits 1 and 2 are now used to control CNOT gates to restore the original single-qubit state on qubit 3. As a

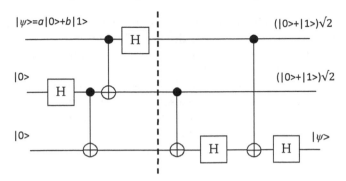

FIGURE 11.7: Teleportation model by a 3-bit system

result of the teleportation, a specific but unknown state has been transferred from qubit 1 to qubit 3, leaving qubits 1 and 2 in superposition states.

11.8 Concluding remarks

It is still much too early to give definite answers to the most of the basic questions of quantum information processing from the practical point of view. Regarding the evolution of all-powerful quantum computers, at the moment, the designs of quantum gates and processors are in a very rudimentary stage. However, various ingenious ideas have evolved to build at least restricted working prototypes and simple quantum algorithms using qubits. Yet there are still too many hard device problems to solve. It has also been fully realised that the success of such experiments may not be sufficient for scaling up such computers to large enough so as to make them more useful than classical ones. On the experimental side, the main current challenge is actually not to build a complete quantum computer right away, but rather to move from the experiments that exhibit quantum interference and entanglement to the experiments that can control these phenomena.

However, some experimental demonstration has been achieved in case of quantum communication using encrypted messages. In fact, successful quantum key distribution has been carried out in an experimental free-space atmospheric link between the Canary Islands of La Palma and Tenerife. In the system, multi-photon pulses emitted by the transmitter contribute to the key. The optics of the QKD transmitter (Alice) consisted of four laser diodes, whose orientation was rotated by 45 degrees relative to the neighbouring ones. The output beams of all diodes were overlapped by a concave-convex pair of conical mirrors and coupled into a single-mode optical fibre running to the transmitter. Decoy pulses were randomly interspersed in the signal sequence by firing two randomly chosen diodes simultaneously. For the empty decoy pulses, the electrical pulse driving the laser diode was suppressed. The mean photon number for all decoy states was monitored with a calibrated photon detector at one of the output ports of a 50:50 fibre beam splitter before coupling to the optical system (telescope). Single-photon polarisation analysis was performed inside the transmitter telescope to correct changes along the fibre. The current outdoor experiment definitely shows the feasibility of secure key exchange which might open up many possibilities of secure communications.

Books for further reading

1. *Physics and Applications of the Josephson*: A. Barone and G. Paterno, John Wiley, New York, 1982.

2. *The Physics of Quantum Information-Quantum Cryptography, Quantum Teleportation, Quantum Computation*: A. Ekert, D. Bouwmeester and A. Zeilinger, Springer, New York, 2000.

3. *Quantum Computing*: J. Gruska, McGraw-Hill, New York, 1999.

4. *Introduction to Quantum Computation and Information*: S. Popescu (eds.) H. K. Lo, T. Spiller, World Scientific, Singapore, 1998.

5. *Quantum Information*: H. Weinfurter and A. Zeilinger (eds.) G. Alber, Springer, Berlin, 2001.

6. *Probabilistic and Statistical Aspects of Quantum Theory*: A. S. Holevo, North-Holland, Amsterdam, 1982.

7. *Quantum Computation and Quantum Information*: I. L. Chuang and M. A. Nielsen, Cambridge University Press, Cambridge, 2000.

8. *Quantum Entropy and Its Use*: D. Petz and M. Ohya, Springer, Berlin, 1993.

Bibliography

[1] A. Galindo and M. A. Martin-Delgado. Information and computation: Classical and quantum aspects. *Rev. Mod. Phys.*, 74(2):347–357, 2002.

[2] J. S. Bell. On the Hinstein-Podolsky-Rosen paradox. *Physics*, 1(3):195, 1964.

[3] A. Zeilinger, D. M. Greenberger, M. A. Horne. *Going beyond Bell's theorem, in Bell's Theorem, Quantum Theory, and Conceptions of the Universe*. Kluwer Academic Publishers, 1989.

[4] D. Deutsch. Quantum theory, the Church-Turing principle and the universal quantum computer. *Proc. Royal Soc. London*, 400:97–117, 1985.

[5] C. C. Cheng and A. Scherer. Fabrication of photonic bandgap crystals. *J. Vac. Sci. Tech. (B)*, 13(6):2696–2700, 1995.

[6] D. Bouwmeester, A. Ekert, and A. Zeilinger. *The Physics of Quantum Information: Quantum Cryptography, Quantum Teleportation, Quantum Computation*. Springer, New York, 2000.

[7] K. Blum. *Density matrix: theory and applications*. Plenum, New York, 1996.

[8] Y. Makhlin, G. Schwon, and A. Shnirman. Quantum-state engineering with Josephson-junction devices. *Rev. Mod. Phys*, 73 (2):357–400, 2001.

[9] H. C. Nager, W. Bechter, J. Eschner, F. Schmidt-Kaler, and R. Blatt. Ion strings for quantum gates. *App. Phys. B*, 66(5):603–608, 1998.

[10] J. I. Cirac and P. Zoller. New frontiers in quantum information with atoms and ions. *Phys. Today*, 57:38, 2004.

[11] J. P. Home, D. Hanneke, J. D. Jost, J. M. Amini, D. Leibfried, and D. J. Wineland. Complete methods set for scalable ion trap quantum information processing. *Science*, 325:1227–1241, 2009.

[12] R. Blatt and D. Wineland. Entangled states of trapped atomic ions. *Nature*, 453:1008–1012, 2008.

[13] K. Molmer and A. Sorensen. Multiparticle entanglement of hot trapped ions. *Phys. Rev. Let.*, 82:1835, 1999.

[14] S. Bose. Quantum communication through an unmodulated spin chain. *Phys. Rev. Let.*, 91:207901, 2003.

[15] T. J. Osborne and N. Linden. Propagation of quantum information through a spin system. *Phys. Rev. A*, 69, 2004.

[16] H. Kamada and H. Gotoh. Quantum computation with quantum dot excitons. *Semicond. Sci. Tech.*, 19:392, 2004.

[17] A. Loss and D. P. DiVincenzo. Quantum computation with quantum dots. *Phys. Rev. A*, 57:120, 1998.

[18] A. Zrenner, E. Beham, S. Stufler, F. Findeis, M. Bichler, and G. Abstreiter. Coherent properties of a two-level system based on a quantum-dot photodiode. *Nature*, 418:612, 2002.

[19] A. Shnirman, G. Scheon, and Z. Hermon. Quantum manipulation of small Josephson junctions. *Phys. Rev. Let.*, 79:2371, 1997.

[20] B. D. Josephson. Possible new effects in superconductive tunneling. *Phys. Let.*, 1:251, 1962.

[21] R. R. Clark. A laser distance measurement sensor for industry and robotics. *Sensors*, 11:43–50, 1994.

[22] J. E. Mooij, T. P. Orlando, L. Levitov, L. Tian, C. H. Vander Wal, and S. Lloyd. Josephson persistent current qubit. *Science*, 285:1036, 1999.

[23] D. P. Goorden and F. K. Wilhelm. Theoretical analysis of continuously driven Josephson qubits. *Phys. Rev. B*, 68:012508, 2003.

[24] K. Nakamura, T. Hara, M. Yoshida, T. Miyahara, and H. Ito. Optical frequency domain ranging by a frequency-shifted feedback laser. *IEEE J. Quant. Elect.*, 36:305–316, 2000.

[25] V. Bouchiat, P. Joyez, H. Pothier, C. Urbina, D. Esteve, and M. H. Devoret. Quantum coherence with a single Cooper pair. *Phys. Scripta*, T76:165, 1898.

[26] T. Yamamoto, Y. Pashkin, O. Astafiev, Y. Nakamura, and J. S. Tsai. Demonstration of conditional gate operation using superconducting charge qubits. *Nature*, 425:941, 2003.

[27] E. N. Bratus, J. Lantz, V.S. Shumeiko, and G. Wendin. Flux qubit with a quantum point contact. *Physica C*, 368:315, 2002.

[28] R. W. Simmonds, K. M. Lang, D. A. Hite, D. P. Pappas, and J. M. Martinis. Decoherence in Josephson qubits from junction resonances. *Phys. Rev. Let.*, 93, 2004.

[29] M. N. Leuenberger, D. Loss, M. Poggio, and D. D. Awschalom. Quantum information processing with large nuclear spins in GaAs semiconductors. *Phy. Rev. Let.*, 89:207601, 2002.

[30] A. B. Vander-Lugt. Signal detection by complex spatial filtering. *IEEE Trans. on Inf. Th.*, IT-10:139–145, 1964.

[31] L. M. K. Vandersypen, M. Steffen, G. Breyta, C. S. Yannoni, M. H. Sherwood, and I. L. Chuang. Experimental realization of Shors quantum factoring algorithm using nuclear magnetic resonance. *Nature*, 414:883, 2001.

[32] P. Kok, W. J. Munro, K. Nemoto, and T. C. Ralph. Linear optical quantum computing with photonic qubits. *Rev. Mod. Phys*, 79:135, 2007.

[33] V. Vedral, M.B. Plenio, M.A. Rippin, and P. L. Knight. Quantifying entanglement. *Phys. Rev. Let.*, 78 (12):22752279, 1997.

[34] R. F. Werner and M. M. Wolf. Bell inequalities and entanglement. *Quant. Inf. Comp.*, 1 (3):125, 2001.

[35] D. P. DiVincenzo. The physical implementation of quantum computation. *Fortschritte der Physik*, 48:771, 2000.

[36] V. Bubzek, M. Hillery, and R. F. Werner. Optimal manipulations with qubits: universal-not gate. *Phy. Rev*, A 60 (4):2626–2629, 1999.

[37] E. Fredkin and T. Toffoli. Conservative logic. *Int. J. Theor. Phys*, 21:219, 1982.

[38] A. Zeilinger in G. Alber et al. (ed.) H. Weinfurter. *Quantum Information*. Springer, Berlin, 2001.

[39] C. H. Bennett and S. J. Wiesner. Communication via one- and two-particle operators on Einstein–Podolsky–Rosen states. *Phy. Rev. Let.*, 20:2881–2884, 1992.

[40] N. Gisin, G. Ribordy, W. Tittel, and H. Zbinden. Quantum cryptography. *Rev. Mod. Phys.*, 74:145–160, 2002.

[41] D. Mayers. Unconditional security in quantum cryptography. *ACM J.*, 48:351–406, 2001.

[42] S. Chiangga, P. Zarda, T. Jennewein, and H. Weinfurter. Towards practical quantum cryptography. *App. Phy. B: Lasers and Optics*, 69:389–393, 1999.

[43] V. Scarani, H. Bechmann-Pasquinucci, N. J. Cerf, M. Dusek, N. Lutkenhaus, and M. Peev. The security of practical quantum key distribution. *Rev. Mod. Phys.*, 81:1301–1309, 2009.

Chapter 12

Nanophotonic Information System

12.1 Introduction

Nanophotonics is a subset of nanotechnology that uses the ability to manipulate both the electromagnetic and electronic properties of nanoscale photonic devices. The area basically deals with the interactions between light and matter at a scale shorter than the wavelength of light itself. One way to induce interactions between light and matter in nanometre size is to confine light in such dimensions. These interactions, taking place within the light wavelength and sub-wavelength scales, are determined by the physical, chemical, and structural nature of artificially or naturally nanostructured materials. Different approaches to confining the electromagnetic or quantum mechanical waves are the key issues in nanophotonics. The technology that is driven by the ever-growing need to shrink the size of electronic chips has now evolved to a precision such that the control of flow of photons within such information systems is possible. By combining the need for integration of photonics with the promise of nanofabrication technology, the field of nanophotonic information processing has emerged.

A plethora of functional devices in the form of light sources, signal transmitters, and detectors, along with signal interconnects are effectively realised in different laboratories by nanophotonic techniques. Moreover, by capitalising on the transfer of optical excitation from one nanometric system to another via optical near field, a nanophotonic information system has also been conceived. Strong coupling between light and matter, efficient control of spontaneous emission, and enhanced nonlinear phenomena are some of the interesting phenomena that can be observed when light and matter interact at the scale of nanometre.

12.2 Nanophotonic devices

Broadly speaking, nanophotonics is the study of the behaviour of light of wavelengths around 300 to 1200 nanometres that includes ultraviolet, visible, and near-infrared light, which interacts with structures of nanometre-scale (around one hundred nanometres or below). The fundamental role of all nanophotonic devices is to increase the efficiency of transformation of energy from other forms to that of light. Devices can control the flow of light and sometimes localise or confine light within the nanovolume. Thus, localisation of light in photonic crystals, nanoscale optical wave guides, micro-resonators, plasmonics, or localisation of quantum mechanical waves in nanostructures such as quantum wells, wires, and dots are some examples of nanophotonic devices. Such devices may employ materials in the form of colloids, or synthesised through lithography, ion-beam milling, sputtering, chemical vapour deposition, or laser ablation.

In the context of nanophotonic devices, two issues are of major importance. These are the generation and detection of light. Many experiments have been carried out to evolve nano-lasers, yet it seems the main attention is focussed on the development of nano-LEDs. These ultra-compact laser and LED sources at the wavelength and sub-wavelength scale, are the result of a demand for use as an indispensable and key building block in an optical integrated chip.

Detection of photon serves as an indispensable node in the whole chain of photonic information technology. The basic detector size has now been dramatically reduced, yet miniaturised devices can detect modulated photonic digital information in the form languages, pictures, or motions. The key issue of using nanophotonics lies in offering better solutions for enhanced sensitivity, bandwidth, and speed.

Two other issues are of certain importance in evolving and using nonophotonic devices are related to the transportation, modulation, and transformation of light in the nanoscale regime. Development of optical fibre and its application in long-distance transportation of photonic signal is a milestone in photonics. The emphasis of research has now shifted to another important issues: efficient information transport in ultra-compact integrated photonic chips, where functionalities such as information transport, modulation, exchange, processing, computing, and detection have to be in-built. These nanophotonic devices can offer very useful solutions to the problem of light-matter interaction and transformation taking place at sub-wavelength size scales.

Theoretically, operations with a nanophotonic device sometimes interpreted as a system that consists of several functional areas: a localised photon field (optical near field) for driving carriers in the systems, and a free photon (radiation) field for extracting some information. The important point of such a nanophotonic system is that the spatial distribution of photons is localised in a nanometric space rather than the matter itself. From this point, valuable device operations of these devices are expected. Relating to a theoretical viewpoint, some restrictions under the long-wavelength approximation are allowed in a nanophotonic device system.

In this chapter, only two much studied and useful nanophotonic devices which are now well established and available commercially shall be discussed. These are: photonic crystals and plasmonic devices.

12.2.1   Photonic crystal

The photonic crystal (PhC) is one of the platforms that facilitates miniaturization of photonic devices and their large-scale functional integration [1]. These man-made periodic dielectric structures can be designed to form frequency bands (photonic bandgaps) within which the propagation of electromagnetic waves is forbidden irrespective of the propagation direction. Even though there are many investigations of these structures in the 100 nm regime, the practical realisation and utilisation for nanophotonics applications is yet to be exploited because of fabricational constraints of nanostructures. Mostly, the device structure, as of now, is about half of the wavelength of light and hence cannot be strictly called a nanophotonic device as per the defined scope of nanotechnology.

Ideal photonic crystals are made of an infinite number of arbitrarily shaped scatters, forming either one, two, or three dimensional periodic lattices designed to influence the behaviour of photons in much the same way that the crystal structure of a semiconductor affects the properties of electrons [2], [3]. Similar to a microscopic atomic lattice that creates a semiconductor bandgap in semiconductors, photonic crystals can be viewed as a photonic analogue of semiconductors that modify the propagation of light by establishing photonic bandgap. Therefore, by replacing relatively slow electrons with photons as the carriers of information, the speed and bandwidth of systems based on photonic crystals can be drastically increased. The basic phenomenon is based on diffraction, and therefore the periodicity of the photonic crystal structure has to be in the same or atleast half of the length-scale of the wavelength of the electromagnetic waves (i.e., around 300 nm for photonic crystals operating in the visible spectrum). The periodicity, whose length scale is proportional to the wavelength of light in the bandgap, is analogous to the crystalline atomic lattice.

Building photonic bandgap in the optical regime requires micro and nanofabrication techniques, which is quite difficult depending upon the desired wavelength of the bandgap and also on the dimensions. Lower frequency structures requiring larger dimensions are easier to fabricate [4]. However, one- or two-dimensional photonic crystals require periodic variation of the dielectric constant in one or two directions and hence are relatively easy to build compared to three-dimensional ones. A 1D structure is considered periodic in one direction (say, z direction), and homogeneous in the other plane (say, x, y plane). A 2D structure is considered periodic in the x, y plane and unchanged in z direction and the propagation is examined in the plane of periodicity. The electromagnetic analogue of an ordinary crystal is a 3D photonic crystal, a dielectric that is periodic along three different axes [5], [6]. Examples of 3D PhCs include opal and inverse-opal structures, woodpiles [7], and layer-by-layer construction [8]. There is also another periodic structure known as a quasi-2D photonic crystal, that is geometrically more complex than 2D photonic crystals, but not as complex as full 3D photonic crystals. Typically planar photonic crystal (PPC) and photonic crystal slabs are considered as quasi-2D structures [9]. Figure 12.1 shows 1D, 2D, and 3D structures of photonic crystal, where the grey areas with different shades represent materials of different dielectric constants [10].

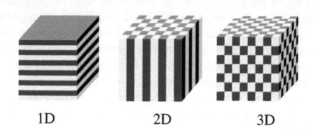

1D   2D   3D

FIGURE 12.1: Representative 1D, 2D, and 3D photonic crystal structure (redrawn from [11])

If the photonic crystal exhibits the propagation of electromagnetic waves in any direction with any polarisation, then such devices have complete photonic bandgap. A layered dielectric medium cannot have a complete bandgap, but by making the spatial periodicity in three dimensions, PhCs can have complete bandgap, and therefore can control propagation of light in all directions. These structures can block certain wavelengths of light at any angle in the plane of the device, and can even prevent light entering from certain angles in the third dimension (i.e., perpendicular to the surface). A 3D lattice structure needs to be fabricated to gain complete control of the light in all three dimensions.

Photonic crystals can be designed as a mirror that reflects a selected wavelength of light from any angle with high efficiency. They can also be integrated within the photoemissive layer to create a light-emitting diode that emits light at a specific wavelength and in a specific direction.

12.2.1.1 Photonic bandgap

Due to the regular arrangement of atoms in a crystal lattice, the periodicity of electronic potential in semiconductor materials results in the existence of the forbidden energy bands for electrons known as electronic bandgap. Similarly, the periodicity of a dielectric lattice of photonic crystals results in the photonics bandgap, forbidden energy bands for photons. The photonic bandgap prevents light from propagating in certain directions with specific frequencies or certain range of wavelengths [8].

A periodic structure is mainly characterised by three parameters: a spatial period defining the lattice constant, the fractional volume and the dielectric contrast between the constituent materials. Photonic bandgap crystals have a further property of gap dimensionality, which is directly related to the number of dimensions of the structure that are periodic. Within these periodic structures the electromagnetic mode distributions and their dispersion relations differ significantly from those of free space. A structure that is periodic in only one dimension will have a 1D photonic bandgap, while another correctly designed structure that is periodic in all three dimensions can display a fully 3D photonic bandgap [2].

Maxwell's equations include built-in scalability, which explains the formation of photonic bandgaps in wavelengths ranging from the visible to microwave. This formulation is the basis for establishing the so-called *master equation* for solving modes of the crystal [12]. The following assumptions are made while dealing with master equation: (1) the mixed dielectric medium in question is void of free charges and currents, so $\rho = 0$ and $\mathbf{J} = 0$; (2) field strengths are small enough that higher order terms in the electric susceptibility χ can be ignored; (3) the material is macroscopic and isotropic; (4) the frequency dependence of the dielectric constant (material dispersion) can be neglected; and (5) the material is assumed to have purely real and positive $\varepsilon(\mathbf{r})$ for all positions in space $\mathbf{r}$. With these assumptions in place, the time and spatially varying Maxwell's equations are:

$$\nabla \cdot [\varepsilon(\mathbf{r})\mathbf{E}(\mathbf{r}, t)] = 0$$

$$\nabla \times \mathbf{E}(\mathbf{r}, t) + \mu_0 \frac{\partial \mathbf{H}(\mathbf{r}, t)}{\partial t} = 0$$

$$\nabla \cdot \mathbf{H}(\mathbf{r}, t) = 0$$

$$\nabla \times \mathbf{H}(\mathbf{r}, t) - \varepsilon_0 \varepsilon(\mathbf{r}) \frac{\partial \mathbf{E}(\mathbf{r}, t)}{\partial t} = 0 \tag{12.1}$$

where $\varepsilon(\mathbf{r})$ represents the dielectric function in space.

The solutions are assumed to vary sinusoidally in time (harmonically), resulting in complex exponential solutions of the form

$$\mathbf{H}(\mathbf{r}, t) = \mathbf{H}(\mathbf{r})e^{-j\omega t}$$

$$\mathbf{E}(\mathbf{r}, t) = \mathbf{E}(\mathbf{r})e^{-j\omega t} \tag{12.2}$$

where $\mathbf{H}(\mathbf{r})$ and $\mathbf{E}(\mathbf{r})$ are the spatial distributions of field contained in a mode of the crystal.

These solutions result in

$$\nabla \cdot [\varepsilon(\mathbf{r})\mathbf{E}(\mathbf{r})] = 0$$

$$\nabla \times \mathbf{E}(\mathbf{r}) - j\omega\mu_0 \mathbf{H}(\mathbf{r}) = 0$$

$$\nabla \cdot \mathbf{H}(\mathbf{r}) = 0$$

$$\nabla \times \mathbf{H}(\mathbf{r}) + j\omega\varepsilon_0 \varepsilon(\mathbf{r})\mathbf{E}(\mathbf{r}) = 0 \tag{12.3}$$

These divergence equations show that the solutions are modes that are transverse electromagnetic waves. That is, for plane-wave solutions with wave vector $\mathbf{k}$ and field component $\mathbf{a}$, $\mathbf{a} \cdot \mathbf{k} = 0$. This means that if propagation occurs in the x-direction, the EM components are aligned with $\mathbf{a} = y$ or z. The two transverse polarisations are the transverse-electric (TE, E orthogonal to $\mathbf{k}$) and transverse-magnetic (TM, H orthogonal to $\mathbf{k}$). By decoupling the curl equations and substituting for $\mathbf{E}(\mathbf{r})$ and with $1/c = \sqrt{\varepsilon_0 \mu_0}$, the master equation is arrived at as

$$\nabla \times \left(\frac{1}{\varepsilon(\mathbf{r})} \nabla \times \mathbf{H}(\mathbf{r}) \right) = \left(\frac{\omega}{c} \right)^2 \mathbf{H}(\mathbf{r}) \qquad (12.4)$$

The equation is a function of the angular frequency of light ω and the dielectric function in space $\varepsilon(\mathbf{r})$. This equation also solves for the states $\mathbf{H}(\mathbf{r})$ of a PhC and their eigenvalues in a plane-wave basis, and helps in plotting the dispersion diagrams for finding bandgaps and resonant modes.

It is important to consider the scalability of Maxwell's equations, as the band diagrams are plotted normalised to a scale factor usually referred to as a. In PhCs, a is usually set as the lattice constant of the crystal, and all other dimensions are defined relative to the lattice constant. Because of this normalization scheme, all eigenvalue solutions to the master equation are relative to the periodicity of the structure. The free-space wavelength of any eigenvalue solution can then be found by taking the normalised frequency and dividing the real lattice spacing dimension by this value [13]. In contrast to a homogeneous isotropic material, an 1D photonic crystal structure consists of a periodic succession of layers of permittivity ε_1 and ε_2 and period a is shown in Figure 12.2.

An electromagnetic wave of wavelength $\lambda \ll a$ is allowed to interact with the structure and the resulting phenomenon can be explained by the Snell-Descartes law of geometrical optics. Within a one-dimensional periodic structure, the electromagnetic dispersion relation has frequency regions in which propagating electromagnetic modes are forbidden. In such forbidden frequency gaps or Bragg frequencies, electromagnetic waves attempting to propagate will experience evanescent exponential attenuation due to Bragg reflections.

If $\lambda \gg a$, then the electromagnetic wave perceives an homogeneous permittivity given by the average between ε_1 and ε_2 weighted by the respective thickness. It is between these two regimes, when $\lambda \sim a$, that the periodic structure behaves like a Bragg mirror, where all the successive reflections are in phase and the wave attenuates rapidly as it penetrates the medium.

The propagation properties in such 1D periodic media are well described by the dispersion relation $\omega(k)$, where the frequencies are given as a function of the wave vector $\mathbf{k}$. The dispersion relation is obtained starting from the Maxwell's equation considering the periodicity of the permittivity $\varepsilon(z) = \varepsilon(z + a)$. This periodic permittivity can be decomposed using a Fourier expansion, as

$$\varepsilon(z) = \sum_m \varepsilon_m e^{-jKmz} \qquad (12.5)$$

where $K = \frac{2\pi}{\Lambda}$

The solution of the Maxwell's equations for this periodic system are the Bloch waves given by

$$E(z) = e_m e^{-j(k_z + mK)z}$$
$$H(z) = h_m e^{-j(k_z + mK)z} \qquad (12.6)$$

where e_m and h_m are periodic in nature.

The wave vector, $k_z + mK$, of the the Bloch mode is invariant under $\pm K$ transformation. Restricting the Brillouin zone [14] in $k-$space, defined between the interval $-\frac{K}{2} < k \leqslant \frac{K}{2}$, the effective refractive index is given by $n_{eff}(\omega) = \frac{k_z(\omega)}{k_0}$ where $k_0 = \frac{2\pi}{\lambda}$. The dispersion relation, or the band diagram given in Figure 12.2(b), can be drawn for different values of ε_1 and ε_2. When $\varepsilon_1 \neq \varepsilon_2$, there is no propagative solution for $k = \frac{K}{2}$, and therefore a photonic bandgap exists. No bandgap exists when $\varepsilon_1 = \varepsilon_2$. At the photonic bandgap, a 1D photonic crystal works as a Bragg mirror and discrete translational symmetry of a 1D photonic crystal makes it possible to classify the electromagnetic modes with respect to their wave vectors **k**. The modes can be explained in terms of a Bloch form of semiconductor materials, which usually consist of a plane wave modulated by a periodic function related to the periodicity of the crystal. Bloch modes have a fundamentally important feature, that is, the different values of k do not necessarily lead to different modes. The smallest region within the Brillouin zone for which the k directions are not related by symmetry is called the irreducible Brillouin zone.

Similarly, a structure can be conceived as transiting from a unidimensional periodic structure to a system with periodicity in two directions and invariant in the third one. Such a structure is called a 2D photonic crystal. This geometry can confine light in the two directions of the plane because of the presence of bandgaps. The confinement in the third direction (perpendicular to the plane of symmetry) is achieved by index contrast using total internal reflection [15]. The simplest possible 2D photonic crystal is based on a silicon block perforated with air holes of periodicity a, as shown in Figure 12.3(a).

In such a structure, the light line equation can be written in the normalised form as,

$$\frac{\omega a}{2\pi c_a} = \frac{1}{2n_{si}} \cdot \frac{ka}{\pi} \qquad (12.7)$$

The Brillouin zone centred at the origin Γ, of such a lattice is shown in Figure 12.3(b). The irreducible zone is the grey triangular wedge. The special points at the centre, corner, and the faces are conventionally denoted as Γ, M, and X. The dispersion diagram for different size of hole is shown in the bottom panel in Figure 12.4.

In the figure, normalised frequency in y axis is in $\frac{a}{\lambda} = \frac{\omega a}{2\pi c}$, and in x axis is normalised propagation constant $\frac{ka}{\pi}$. The light line in the reduced scheme is artificially folded back into the first Brillouin zone. By adding a periodic lattice of holes into this silicon block, dispersion is modified and the bandgap

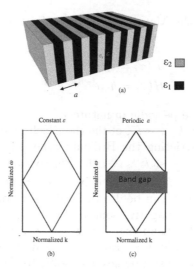

FIGURE 12.2: (a) 1D structure with unidirectional periodicity of permittivity, (b) photonic band structure for propagation perpendicular to the structure for the same permittivity, and (c) formation of photonic bandgap when permittivity is periodic, as shown in (a)

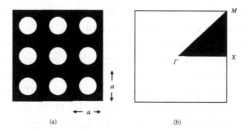

FIGURE 12.3: (a) Square lattice of photonic crystal and (b) Brillouin zone of square lattice (irreducible zone is shaded)

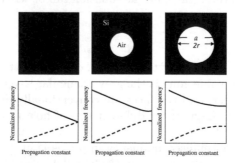

FIGURE 12.4: Dispersion diagram for light propagation along x-axis for various hole sizes

opens at the edge of the Brillouin zone. It is seen that as holes become bigger, the bandgap becomes wider and shifts towards higher frequencies. The latter can be attributed to the increased overlap between light and low-dielectric constant material (air) as holes becomes bigger.

For a complete analysis of the optical properties of uniform, unpatterned material (e.g., silicon block), it is sufficient to study light propagation along one spatial direction (say, x direction) only, since all directions are equivalent. This is not the case when studying multi-dimensional periodic dielectric lattices. The introduction of a periodic lattice reduces the symmetry of the system, and it becomes necessary to study light propagation along various directions in order to describe the optical properties of patterned media. As in the case of electronic bandgap, it is sufficient to study light propagation along high symmetry directions of the periodic photonic crystal lattice. In the case of a square lattice, these directions are labelled $\Gamma X, XM$, and $M\Gamma$ in Figure 12.5, where the dispersion diagram for the modes propagating in a 2D square lattice photonic crystal is shown. It is seen that a complete bandgap exists between M point in the dielectric band and X point in the air band, and that it is narrower than the stop-band calculated along the ΓX direction.

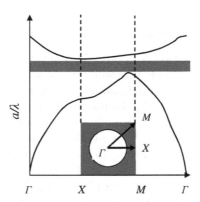

FIGURE 12.5: Band diagram calculated along the high-symmetry direction of the 2D photonic crystal

A key point for 2D photonic crystals is the discrimination between two polarisations of the electromagnetic fields: transverse-electric (TE) modes where the magnetic field is normal to the plane and the electric field lies on the plane, and transverse-magnetic (TM) modes where the magnetic field is in the plane and the electric field is normal to the latter. The band structure for the TE and TM modes can be completely different. In particular, it may occur that a photonic bandgap exists for one polarisation while no bandgap exists for the other one. Frequency bands for which propagation is forbidden both for TE and TM modes are called total bandgaps. Designing photonic crystals with complete bandgaps for both TE and TM polarisations is more challenging, however, it is possible. The dissimilarity of band diagram along high symmetry direction of 2D photonic crystal is shown in Figure 12.6.

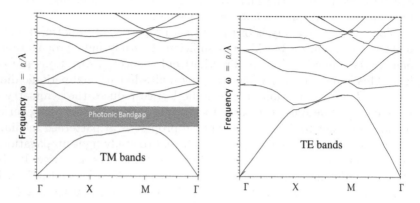

FIGURE 12.6: Band diagram calculated along the high-symmetry direction of 2D photonic crystal

12.2.1.2 Planar photonic crystal

Since the fabrication of 2D photonic crystal structure is not easy, it is necessary to consider a planar photonic crystal structure which is essentially a 2D photonic crystal with a finite third dimension. In this type of a structure, an optically thin semiconductor slab (thickness is roughly $\lambda/2$) is surrounded with a low-refractive index material (usually air), and perforated with a 2D lattice of holes. The localisation of light in all three dimensions is made possible by the combination of two mechanisms. In the vertical direction of such structure, light is confined to the slab by means of total internal reflection (TIR) due to the high index contrast between the high-index slab and the low-index environment. In the lateral direction light is controlled by means of distributed Bragg reflection due to the presence of a 2D lattice of holes. In the planar photonic crystal (PPC) the third dimension is not periodic or infinite, and therefore photons incident to the interface between the semiconductor slab and air with small angles (angles smaller than the critical angle for TIR) can escape from the slab and couple into the continuum of radiation modes [16]. These photons leak energy from the slab and therefore represent the loss mechanism of the planar photonic crystal.

Since radiative modes exist at all frequencies, including the bandgap region, they close the bandgap, and the complete bandgap does not exist in planar photonic crystals. The forbidden frequency range still exists, however, for the guided modes of the slab; that is, for the photons confined in the patterned slab. Because of this, any defects introduced into the photonic crystal lattice can couple the guided modes into the radiative modes and can scatter the light guided in the slab. These defects enhance coupling to the leaky modes and increase the losses of the PPC device.

In contrast to the infinite 2D case, finite-thickness slabs can support modes with higher-order vertical oscillations. If the slab is made too thick, the presence of these modes can result in the closing of the bandgap. 2D analysis of a vertically extended structure (an infinitely thick slab) would result in band diagrams that are shifted toward the lower frequencies, as the guided modes are not completely confined in the slab, but also extend into the air. A schematic diagram of PCC is shown in Figure 12.7.

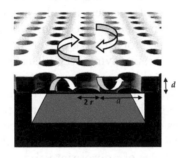

FIGURE 12.7: Schematic diagram of planar photonic crystal

The properties of a planar photonic crystal, such as position and width of the bandgap, depends strongly on several important parameters, such as the type of lattice (e.g., triangular, square, honeycomb), the thickness and effective refractive index of the slab, and the environment surrounding the slab, the periodicity of the lattice and the size of the holes. The bandgap becomes wider as the holes become bigger. Also, due to increased overlap with low-dielectric material (air), band edges are shifted towards higher frequencies when the hole size is increased. This results in widening of the photonics bandgap. When holes are made too big, the bandgap for TE-like modes can be closed and the bandgap for TM-like modes can be open. Slab thickness, however, does not have strong influence on the width of the bandgap, but affects the position of the bandgap only. Band-edges are, however, shifted towards higher frequencies when the slab is made thinner. It is also of interest to explore the PPC properties when the air is replaced by material with a refractive index higher than 1. As expected, the structure with bigger holes is more sensitive to the changes in the refractive index of the environment.

For some applications it is necessary to make PPCs in dielectric materials, that have a small refractive index. One promising candidate for PPCs that operate in the visible wavelength range is silicon-nitride (Si_3N_4), with a refractive index nearly 2.02. Due to small refractive index contrast between core and environment, the bandgap is narrower, particularly in the case of smaller holes. This allows the device to operate in a visible region.

12.2.1.3   Defects in photonic crystal structure

The photonic bandgap can be used to confine light in reduced volumes. Interesting consequences arise when defects are introduced into the crystal lattice [17]. Following an analogy to semiconductor crystals that can possess donor and acceptor defect states in the electronic bandgap, the inclusion of defects in photonic crystals create states inside the photonic bandgaps, which can be utilised to selectively guide light or allow it to resonate in small volumes. These defects can be point defects of various sizes, line defects, and other defects may arise unintentionally. Point defects can only occur in 1D structures; however, in 2D structures, line defects and point defects can occur. In 3D structures, planar defects can occur over and above point and line defects. Basically, when the properties of one or more lattice points are modified, an isolated region of broken symmetry (defect) is created, surrounded by the periodic photonic crystal lattice. There are two simple ways of perturbing a lattice for defect creation: add extra dielectric material where it does not belong, or remove some of the dielectric material that should be there. The first case is referred to as dielectric defect and the second case is the air defect.

In a situation where a defect is introduced into a 2D photonic crystal structure by removing some holes of appropriate size in the lattice, to support a mode surrounded by the photonic bandgap, then the light gets trapped into it. As a result, an optical cavity is achieved. The radius of one lattice hole can be modified, making it smaller or larger relative to the surrounding lattice holes or by filling in the hole altogether. The broken symmetry creates a resonant state within the bandgap, the frequency of which is the characteristic resonance frequency of light associated with the cavity geometry. At resonance, photons are trapped inside the cavity until they are eventually lost by leakage, absorption, or scattering. The time constant τ associated with the photon energy decays in the cavity is directly related to a measurable quantity called quality factor, commonly referred to as Q, which relates the decay time constant to the measurable quantities of resonance frequency ω_0, and the full-width half maximum (FWHM) of the resonance peak in the transmission. Thus Q represents the rate of energy loss relative to the stored energy in the cavity and is given by $Q = \omega_0 \tau = \frac{\omega_0}{\Delta \omega_{1/2}}$, where ω_0 is the cavity resonance frequency, and $\Delta \omega_0$ is the resonance width. The longer light resonates inside the cavity, the higher the quality factor. A small FWHM gives a narrower resonance and a higher Q.

Moreover, a linear defect can be created by removing a row of holes. Light propagates through this defect, confined by total internal reflection in the vertical direction, and Bragg-like reflection, due to the photonic crystal bandgap, in the lateral direction. This kind of photonic crystal waveguide where just one line of holes is missing is usually called a W_1 waveguide. W_1 photonic crystal waveguides can be either mono or multimode. A W_2 has two rows missing. Adding more rows to a waveguide adds more guiding bands within the gap, whereas shrinking the waveguide width creates wider bandwidth guiding bands. In addition to small-area bends, photonic crystal waveguides can be used to couple light to point defects. Variations can be made to build different kinds of waveguides following the same principle.

12.2.1.4 Photonic crystal fibre

Compared to the index-guiding fibres, photonic crystal fibres (PCFs) are rather new and fairly difficult to fabricate [18]. They are named because of their microstuctured cladding, which acts as an artificial photonic crystal. The appearance of PCFs may vary; however, fibres with solid core and hollow core are very common [19]. It is useful to categorise the PCFs by their guiding mechanism, instead of their appearance, such as the photonic bandgap guiding and the quasi guiding. This categorisation does not depend on the properties of the core, but on the photonic properties of the cladding. Both mechanisms confine light mode in the core and allow for low-loss guiding over long distances [18]. Some of the typical structures of photonic crystals and holey fibres are shown in Figure 12.8.

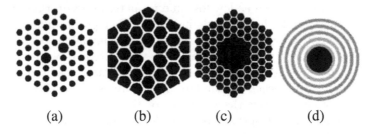

(a)   (b)   (c)   (d)

FIGURE 12.8: Drawings of various structures: (a) birefringent PCF; (b) ultra-small core PCF; (c) hollow core PCF; and (d) hollow core Bragg fibre. The white regions represent silica, the black regions are hollow, and the gray regions are other materials (glasses or polymers)

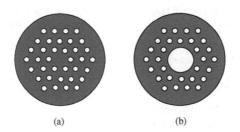

(a)   (b)

FIGURE 12.9: Microstructured optical fibres with air holes

So far, PCFs have been studied in two different forms [19], one which is similar to standard total internal reflection guidance, and in the other form, light guidance is based on the presence of photonic bandgaps. An example of each of these is shown in Figure 12.9(a). The structure on the left in the figure is an example of a glass with air holes in a periodic fashion surrounding a glass core where there are no air holes. This configuration has produced phenomena often referred to as endlessly single-mode, where wavelengths over a very broad range work as single-mode. These structures have also shown the ability to operate in a single-mode regardless of the scale of the structure. However, this is far different from standard single-mode fibres, which typically must have a small core radius to operate in this fashion [20].

In the structure on the right in Figure 12.9, the light is guided in a hollow air core. In terms of traditional fibre optic theory, this device, where the index of the core is less than that in the cladding should not even guide the light. However, due to the periodic structure in the cladding, light with an energy or wavelength which falls inside the bandgap of the cladding is confined to the core. The core therefore acts as a defect in the periodic photonic crystal structure. The cladding in this case can be thought of as 3D in which, at certain regions of the spectrum, light is diffracted from all radial directions back to the core. These hollow-core fibres offer some very desirable advantages over standard fibres such as low-loss due to material absorption and scattering, since most of the light travels in air. Another promising feature of this structure is that the group velocity dispersion is anomalous over most of the band, allowing for very short pulses to be supported. However, this hollow core structure is yet to match the low loss of optimized traditional fibres.

When light is launched in standard optical fibres, light with a given propagation constant β along the direction of the fibre axis remains constant throughout the entire length. In order to form a mode which is guided in the core of the structure, β value must satisfy the relation $\beta \leq n_{cl}k_0$ where n_{cl} is the refractive index of the cladding and $k = \frac{2\pi}{\lambda}$. Effective refractive index for the mode can be defined as $n_{eff} = \frac{\beta}{k_0}$, which allows the mode to propagate. In a normal fibre, where the core index is greater than the cladding, guided modes can propagate in a range of effective indices between the index of the core and the cladding. This is also the case for the structure shown in Figure 12.9(a), where the index of the cladding can be thought of as being reduced by the presence of the air holes. For the hollow core fibre, depending on the design of the structure, bands of allowed and disallowed effective indices or propagation constants can define the light propagation.

It is interesting to study the similarity and the dissimilarity of an ordinary fibre and a PCF. In the glass core/air hole structure, the primary difference between PCF and the ordinary fibre is the distinction between a guided and an unguided mode. In traditional fibres guided modes have real propagation constants and are lossless, while non-guided modes have a complex β, where the imaginary part is related to the loss. The normalised propagation constant β_{norm} is given by

$$\beta_{norm} = \frac{(\frac{\beta}{k})^2 - n_{cl}^2}{n_{co}^2 - n_{cl}^2} \tag{12.8}$$

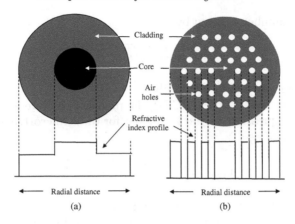

FIGURE 12.10: Behaviour of glass core/air hole fibre

Figure 12.10 shows the refractive index profile of a single mode fibre and a PCF. In the micro-structured fibres, all modes experience some tunnelling effects due to the periodicity of the holes and thus have complex β and effective indices. In Figure 12.11 there are four primary sections of interest with regards to the behaviour of light in the above structure. The region below the A-line represents where the fibre becomes multimode, in the region above the B-line (region CF_1), the fundamental guided mode fills the entire section of the fibre which follows traditional fibre optic theory. In the region below the C-line, leaving out the multimode area (area CF_2), the mode is strongly confined in the core. The area in between the B-line and the C-line, leaving out the area CF_2, is of major interest and is sensitive to other parameters such as the hole width, the pitch, and number of rings in the structure.

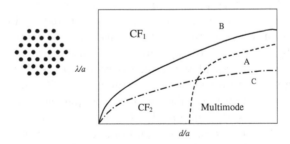

FIGURE 12.11: Behaviour of glass core/air hole fibre (d is the hole width and λ is the pitch) (redrawn from [21])

Without using analysis of mode propagation that make use of electromagnetic theory, a simplified analysis for mode propagation can be made with the help of V number. The parameter V-number is often used to characterize a fibre. The V-number is given by

$$V(\lambda) = \frac{2\pi r}{\lambda}\sqrt{n_{core}^2 - n_{clad}^2} \tag{12.9}$$

where r is the core radius of the fibre.

In photonic crystal fibres (PCFs), the radius of the core is not well defined. However, an expression for the V-number for these microstructured fibres can be approximated as

$$V_{pcf}(\lambda) = \frac{2\pi r \Lambda}{\lambda}\sqrt{n_c^2(\lambda) - n_{cl}^2(\lambda)} \tag{12.10}$$

where $n_c(\lambda)$ is the effective core index for the fundamental mode, and $n_{cl}(\lambda)$ is, similarly, the effective index of the fundamental space-filling mode and a function of λ.

For a photonic crystal fibre that has a single missing hole as its core, the value for V_{pcf} below which the fibre is single is π. It can be calculated that all fibres that have approximately $\frac{d}{\lambda} \leq 0.45$, where d is the hole diameter, are single-mode for all wavelengths. For cores formed by more than a single missing hole, the different single-mode criteria can be obtained. The endlessly single-mode behaviour of the photonic crystal and microstructure fibres also allows for large mode area fibre that is single-mode. Step index fibre becomes multimode when the core radius is increased since the refractive index difference between the core and the cladding cannot be decreased without limits. However, the single-mode behaviour of a photonic crystal fibre is independent of the core size and instead depends only on the cladding geometry.

Since total internally reflected guidance of traditional waveguides cannot explain a mode being guided in air surrounded by a higher index cladding in a PCF, a vectorial master equation can be considered for the mode guidance in a PCF. The solution of the equation in a structured medium involves using Bloch's theorem, which states that the solution can be written as a plane wave modulated by a function corresponding to the structure of the medium.

In photonic bandgap fibre the microstructured cladding in a bandgap fibre represents a two-dimensional photonic crystal. Because of the boundary conditions, only a limited number of optical frequencies are able to propagate in the photonic crystal structure. In the case of a one-dimensional photonic crystal, a band structure of such a photonic crystal can be derived using the same mathematical methods as in solid-state physics. The two-dimensional case is usually too complex to be solved analytically. A numerical method needs to be employed to calculate the band structure of such a 2D photonic crystal.

The photonic analogue to the electronic band structure of a crystal can be studied in terms of density-of-photonic states (DOPS), equivalent to density of state of electronic structures. The density of photonic states is, by definition, the number of allowed electromagnetic modes per unit frequency interval and propagation-constant interval-unit. In a photonic crystal fibre field, the DOPS is typically plotted against both the normalised frequency $\frac{k}{\lambda}$ and normalised wave-vector component along the propagation direction $\frac{\beta}{\lambda}$ or the effective refractive index, n_{eff} of a given perfect photonic structure (i.e., with no core defect). The DOPS is also a mapping of the solutions of the propagation equation and shows all modes supported in the periodic structure. It also indicates clearly the regions for which there are no solutions to the equation (i.e., no light can be guided in any constituent of the structure at the associated pair of $\frac{k}{\lambda}$ and $\frac{\beta}{\lambda}$).

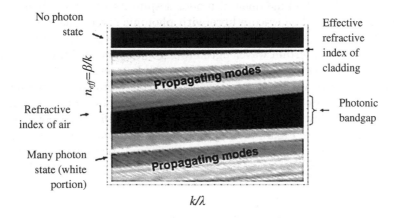

FIGURE 12.12: Sketch of a typical density-of-photonic state (DOPS) diagram

The DOPS is shown in Figure 12.12. The plot indicates several white areas where no optical states are present; these areas are photonic bandgap areas, and the white coloured regions correspond to zero DOPS. There is no solution to the propagation equation with respect to pair $\frac{k}{\lambda}$ - $\frac{\beta}{\lambda}$ and therefore, no photonic bandgap can exist. In Figure 12.12 the DOPS is shown for a specific frequency. Any optical mode with an effective refractive index within this bandgap will be confined in the core because the light simply cannot leak through the cladding. Depending on the manufacturing quality of the micro-structure, this guiding mechanism allows for relatively small losses. For guidance to be achieved in a hollow core, the core modes need to lie within the white region of the PBG that is under the vacuum line (refractive index of the air) as plotted with a dash-dotted line in Figure 12.12. However, when the core mode arrives at the edge of the PBG, strong coupling is achieved between the core and the cladding mode, due to the very large overlap integral between them. This leads to energy transfer from the core to the cladding mode and ultimately to an increase of the core mode propagation loss at the edge of the bandgap. The bandwidth for bandgap PCF is thus limited.

One interesting application of PCF is supercontinuum generation. Supercontinuum (SC) sources are a new type of light source that provides a combination of high output power, a broad spectrum, and a high degree of spatial coherence that allows tight focusing. A supercontinuum source typically consists of a pulsed laser and a PCF working in the nonlinear regime, in which a combination of nonlinear effects broadens the narrow-band laser radiation into a continuous spectrum without destroying the spatial coherence of the laser light. Solid-core PCFs are designed to be single-mode at all wavelengths by choosing an appropriate combination of hole size and hole spacing, something that is not possible with conventional fibre technology. This feature is exploited in the supercontinuum fibre to ensure that the supercontinuum radiation remains guided in the fundamental mode, despite its very large bandwidth. A properly designed hollow-core fibre with photonic crystal cladding surrounding the core can also be used, where the core acts as a perfect, loss-free mirror, confining light of certain wavelengths to the core. The choice of index for the core material is not important and it is possible to create optical fibres with gas-filled or even evacuated cores, with very little interaction between the light and the fibre material. This can result in unusually low optical non-linearity, making these fibres suitable, for example, for the delivery of powerful, ultrashort optical pulses. Figure 12.13, however, shows a comparison of single-mode continuous power output of PCF supercontinuum source with single-mode fibre coupled white light source. When these possibilities are combined in an

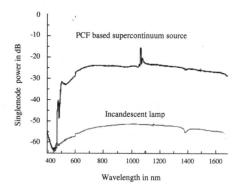

FIGURE 12.13: Comparison of a PCF based supercontinuum source with a fibre coupled white light source

optimal manner, one of the most promising applications of photonic crystal fibres appears in the rapidly developing area of compact high power doped PCF fibre lasers and amplifiers. In future optical networks, one of the enabling technologies will be tunable component elements or subsystem modules, including reconfigurable routers, switches, etc. Thus, the development of a technology platform that allows construction of tuning components is critical.

12.2.2 Plasmonic devices

The emerging field of plasmonics promises the generation, processing, transmission, sensing, and detection of signals at optical frequencies along metallic surfaces which are much smaller than the wavelengths they carry. Plasmonic technology has applications in a wide range of fields, including biophotonics, sensing, chemistry, and medicine. But perhaps the area where it will have the most profound impact is in photonic communications, since plasmonic waves oscillate at optical frequencies and thus can carry information at optical bandwidths.

In plasmonics, the important signal processing step is performed by a phenomenon known as surface plasmon polaritons (SPPs) instead of photons. SPPs are electromagnetic waves coupled to charge density oscillations at the interface between a material of negative permittivity and one of positive permittivity (e.g., a metal and an insulator, respectively). Surface plasmon is a trapped surface mode which has electromagnetic fields decaying into both media but which, tied to the oscillatory surface charge density, propagates along the interface. If photons oscillating at an optical frequency (say around 190 THz) irradiate onto a metal-insulator interface, it may pass its energy in part to the electrons at the interface, and the electrons may start to oscillate themselves at that frequency. After signal processing has been performed, the SPPs are converted back to the optical domain. Exciting an SPP with photons is only possible if both energy and momentum conservation are met. This is done by properly choosing the incidence angle and material composition at the interfaces.

Signals in a plasmonic system can have transverse dimensions smaller than 100 nm, and thus plasmonic devices can be under the family of nanophotonic devices. These may also speed up photonic signal processing by virtue of their compactness. Further, an electrically operated plasmonic device is not speed-limited. Intensities in plasmonic devices are very strong due to the small cross-section, and thus plasmonic nonlinear devices are possible within shorter devices and with lower optical powers. Therefore, given their large bandwidth and compact size, plasmonic devices are of particular interest in applications where speed, footprint, and CMOS compatibility matter, such as in photonic integrated circuits.

Since the surface plasmon wavelengths can reach to nanoscale at optical frequencies, plasmonics can go beyond the sub-diffraction limit. Therefore in recent years, techniques using plasmons have been exploited in many applications by manipulating and guiding light at resonant frequencies. Plasmon resonance of the surface plasma waves depends, however, on the metal particle size and geometry.

The periodic movement of ions or electrons close to the surface of a solid (typically an ionic crystal or a metal) can create an interface polarisation. As a consequence, ions or free charge carriers in the solid are subject to restoring Coulomb forces, which lead to an acceleration of charge and hence to an oscillatory motion. The periodic movement of polarisation constitutes the source for an electromagnetic field, which is bound to the solid's surface and propagating along it.

More generally, an electromagnetic surface excitation has its physical origin in a mechanical displacement of charge carriers, atoms, or molecules. This displacement leads to the formation of a time-dependent polarisation or magnetisation and the creation of an associated time-dependent electromagnetic field close to the interface. Hence, mechanical and electromagnetic excitation are not independent from but coupled to each other. This coupled state is often referred to as an interface polariton where the expression *polariton* is supposed to emphasise the presence of the electromagnetic field outside the solid; however, there must also be a non-vanishing field inside the solid [22]. Many different kinds of interface polaritons may exist depending on the nature of the excitation inside the solid, e.g., magnon-polaritons, phonon-polaritons, or surface plasmon polaritons.

12.2.2.1 Surface plasmon polariton

A surface plasmon polariton (SPP) is a collective charge excitation at an interface formed by a conductor and a dielectric. This excitation can be interpreted as an electromagnetic wave which is trapped to the interface because of the presence of free charge carriers provided by the conductor. Therefore, inside the conductor this excitation has a plasma-like character while inside the dielectric it more resembles a free electromagnetic wave. The term *surface plasmon polariton* (SPP) is intended to reflect this double-sided trait.

Plasmons are plasma oscillations in metals. The treatments for plasma oscillations in metal can be considered using the jellium model of plasma. A metal may be considered to be composed of positive ions forming a regular lattice and conduction electrons which move freely through this ionic lattice. In the jellium model the ionic lattice is replaced by a uniform positive background whose density is equal to the mean electronic density but of opposite sign. The electrons then behave as a gas whose density may fluctuate due to external excitations, thermal vibrations, etc. If the negative charge density is reduced locally, the positive background is no longer screened by the electronic negative charges and produces an attractive force on the neighbouring electrons. These electrons move to the positive region and accumulate with a density greater than necessary to obtain charge neutrality. The Coulomb repulsion between the electrons now produces motion in the opposite direction, etc. This process continues causing longitudinal oscillations of the electronic gas to setup plasma oscillations. A plasmon is a quantum of the plasma oscillation.

The interface between a metal and a dielectric may also support charge density oscillations, called surface plasmons. These occur at a different frequency of the bulk plasmon oscillations and are confined to the interface. The periodic surface charge density sets up a macroscopic electric field in the two media with components along the y and z directions. Since the surface density alternates in sign, this sums in the z direction to give an exponential decay in the magnitude of the electric field. The phenomenon is schematically depicted in Figure 12.14.

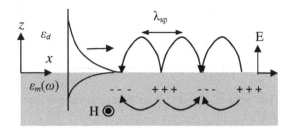

FIGURE 12.14: (a) Schematic of the SPP electromagnetic field at an interface defined by $z = 0$, and (b) the SPP field component $|E_z|$ decays exponentially into both conductor ($z < 0$) and dielectric ($z > 0$) while propagating along the y-direction

A theoretical description of SPPs at an interface formed by a conductor and a dielectric is based on classical field theory [23]. The permittivity of the involved materials is a parameter that determines the SPP field structure. The classical Drude model is used in most of the cases involving permittivity.

The electromagnetic field of the surface plasmon polariton at a single interface system can be derived from Maxwell's equations. Formally, a single interface system (a so-called stratified medium) can be considered as an inhomogeneous medium with a permittivity $\varepsilon = \varepsilon(z, \omega)$. For such a system, all possible solutions of Maxwell's equations can generally be classified into s polarised and p polarised electromagnetic modes. Physically, an s polarised wave whose electric field vector is parallel to the interface merely creates a motion of charge parallel to the interface. For an ideal gas of charge carriers, however, no restoring force can build up and consequently no propagating wave can form. On the other hand, a p polarised wave (whose magnetic field vector is parallel to the interface) results in a charge accumulation at the interface itself as the electric field vector has a non-vanishing component perpendicular to the interface. Since the charge carriers are in fact trapped inside the solid, they cannot escape, and consequently a restoring force will build up. Therefore, the surface plasmon polariton can be regarded as a longitudinal wave in which the charge carriers move in a direction parallel to the propagation direction [24].

The geometry used for exciting SPPs consists of a semi-infinite dielectric (medium 1) which is later assumed to be air and a semi-infinite conductor (medium 2) interfaced at $z = 0$. A p polarised electromagnetic wave incident from the (y, z) plane can be coupled into a p-polarised surface evanescent wave (the SPP) travelling in the y direction along the interface and decaying exponentially into both media.

Because the SPP is a p polarised electromagnetic wave, the **H** field vector

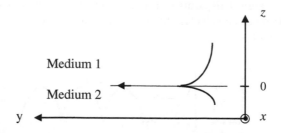

FIGURE 12.15: Schematic of the geometry used for deriving the SPP electromagnetic field

has to be perpendicular to the (y, z)-plane as shown in the geometry of Figure 12.15, where a flat interface is defined by two media, 1 (dielectric) and 2 (conductor) at $z = 0$. The SPP with wave vector k propagates along this interface (parallel to the y-axis) while being damped exponentially. Perpendicular to the z-direction, the SPP field vectors H and E decay exponentially [25].

The SPP is a p-polarised electromagnetic wave, therefore the **H** field vector has to be perpendicular to the (y, z) plane in the geometry. Assuming an exponential decay perpendicular to the interface **H** field vector can be expressed in dielectric material as

$$\mathbf{H}(y, z, t) = e^{j(ky-\omega t)}.A_1 e^{-\beta_2 z} e^x, z < 0 \qquad (12.11)$$

and in conductor material as

$$\mathbf{H}(y, z, t) = e^{j(ky-\omega t)}.A_2 e^{\beta_1 z} e^x, z > 0 \qquad (12.12)$$

where $A_1 e^{-\beta_2 z}$ and $A_2 e^{\beta_1 z}$ are envelope functions in the dielectric and conductor materials, respectively.

The electric field $\mathbf{E}(y, z, t)$ of the SPP can be derived from $\mathbf{H}(y, z, t)$. The SPP decay constants perpendicular to the interface is $\beta_i = \sqrt{k^2 - \varepsilon_i.\omega^2/c^2}$ into both media where $i = 1$ (dielectric) and $i = 2$ (conductor). The vacuum speed of light is denoted by c. The decay length L_i into medium i perpendicular to the interface is given by $L_i = 1/Re(\beta_i)$. The decay length along the propagation direction is further given by $L_y = 1/Im(k)$ and it describes the SPP damping along its propagation direction. This damping is due to electromagnetic field energy dissipation inside the conductor into Joule heat.

Taking into account the stated boundary conditions of continuity of the tangential electric and the perpendicular magnetic field components at the interface, which translates as the continuity of envelope functions, the following equation can be obtained as

$$\beta_2(\omega)\varepsilon_1(\omega) + \beta_1\varepsilon_2(\omega) = 0 \qquad (12.13)$$

and

$$A_1 = A_2 \qquad (12.14)$$

where the dispersion in both media is considered.

The first of these equations is an implicit form for the SPP dispersion relation $\omega = \omega(k)$, as can be seen when inserting the value of β_i. If a solution indeed exists, medium 2 is called a surface-active material, and the SPP dispersion relation is written as

$$k(\omega) = \frac{\omega}{c}\sqrt{\frac{\varepsilon_1(\omega)\varepsilon_2(\omega)}{\varepsilon_1(\omega) + \varepsilon_2(\omega)}} \qquad (12.15)$$

The above equation contains the complex quantities $\varepsilon_1(\omega)$ and $\varepsilon_2(\omega)$. Since the speed of light c and the frequency ω are real quantities, the wave vector k has to be complex in accordance with the identification of $\frac{1}{Im(k)}$ as the SPP decay length along propagation direction. The SPP dispersion relation is schematically depicted in Figure 12.16. From this figure, it can be inferred that for low frequencies, the SPP dispersion relation gets the same slope as the free electromagnetic wave dispersion $\omega = \frac{ck}{\sqrt{\varepsilon_1}}$ in medium 1. For high frequencies, the SPP dispersion relation saturates at $\frac{\omega_p}{\sqrt{1+\varepsilon_1}}$, where ω_p is the so-called plasma frequency of the conductor (medium 2).

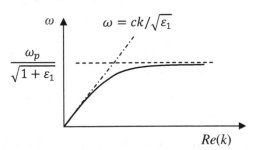

FIGURE 12.16: Dispersion relation of SPP

In particular, the dispersion relations of the evanescent (non-radiating) SPP and a freely propagating electromagnetic wave do not cross. This means that the latter cannot excite SPPs by simply irradiating a conductor's surface because this would violate the conservation of momentum and energy. Hence, additional methods are needed for generating SPPs with electromagnetic radiation. Some important conclusions can now be drawn.

At $z = 0$, the field yields that the components H_x and H_y are continuous. For E_z, one obtains $E_z(1)/E_z(2) = \varepsilon_2/\varepsilon_1$ at the interface. Generally, the permittivities ε_1 and ε_2 are complex, therefore both amplitude and phase of E_z change at the interface.

Similarly, the following relations can be derived as

$$\frac{E_z}{E_y} = -\sqrt{\frac{\varepsilon_2}{\varepsilon_1}} \qquad (12.16)$$

for $z > 0$, and for $z < 0$ as

$$\frac{E_z}{E_y} = \sqrt{\frac{\varepsilon_1}{\varepsilon_2}} \qquad (12.17)$$

Thus, depending on the values of the dielectric functions ε_1 and ε_2, the two electric field components can have very different amplitudes.

Inserting the SPP dispersion relation given by $k(\omega)$ into expression of β_i for the perpendicular decay constants, the latter take the simple form

$$\beta_1 = \frac{\omega}{c}\sqrt{\frac{-\varepsilon_1^2}{\varepsilon_1 + \varepsilon_2}} \qquad (12.18)$$

and

$$\beta_2 = \frac{\omega}{c}\sqrt{\frac{-\varepsilon_2^2}{\varepsilon_1 + \varepsilon_2}} \qquad (12.19)$$

In conclusion, the structure of the SPP electromagnetic field at an interface is unambiguously determined by the values of the permittivities ε_1 and ε_2 of the two media defining the interface. By using specific values of the permittivity of metals and semiconductors, the SPP decay lengths $L_1 = \frac{1}{Re(\beta_1)}$, $L_2 = \frac{1}{Re(\beta_2)}$, and $L_y = \frac{1}{Im(k)}$ can be obtained.

Frequencies in the THz range correspond to small values of $Re(k)$ where the SPP dispersion curve approaches the light line $\omega = ck$. For $Re(k) \to \infty$, the dispersion curve approaches asymptotically the limit $\frac{\omega_p}{\sqrt{1 + \varepsilon_1}}$. The conductor plasma frequency is denoted by ω_p and ε_1 is the permittivity of the dielectric adjoining the conductor.

12.2.2.2 Surface plasmon polaritons at two-interface system

For a two-interface system as shown in Figure 12.17, similar analysis can be carried out. A two-interface system has a thin dielectric slab of thickness d sandwiched between a semi-infinite dielectric (medium 1), and a semi-infinite conductor (medium 2). Interfaces occur at $z = -d$ (medium 2/dielectric film) and at $z = 0$ (dielectric film/medium 1).

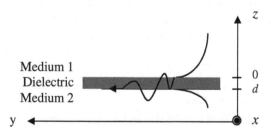

FIGURE 12.17: Schematic of the geometry used for deriving the SPP in a two-interface system

The permittivities can be written as $\varepsilon(z,\omega) = \varepsilon_1(\omega)$ for $z > 0$, $\varepsilon_d(\omega)$ for $-d \leq z \leq 0$ and $\varepsilon_2(\omega)$ for $z < -d$. Considering this configuration and the case $d << \lambda$, the only mode that can still exist and propagate along the slab is the TM_0 mode. In fact, in the limiting cases $d = 0$ and $d \to \infty$, the dispersion relation of the TM_0 mode turns out to be

$$k = \frac{\omega}{c}\sqrt{\frac{\varepsilon_1\varepsilon_2}{(\varepsilon_1 + \varepsilon_2)}} \qquad (12.20)$$

and

$$k = \frac{\omega}{c}\sqrt{\frac{\varepsilon_d\varepsilon_2}{(\varepsilon_d + \varepsilon_2)}} \qquad (12.21)$$

These relations are nothing but the SPP dispersion relation derived for a single-interface system. However, even for very thin films ($d << \lambda$) on top of the conductor, the field geometry is drastically altered compared to the single interface system. This is ascribed to the additional boundary conditions that have arisen compared to those of the single-interface system.

Outside the dielectric film (for z lying between $-d$ and 0, the fields decay exponentially. Compared to the single-interface system, the decay length L_2 into the conductor is almost not affected by the presence of the dielectric film on top. The decay length L_1 into medium 1 (a dielectric), however, is drastically decreased compared to the single interface system, that is, the SPP mode becomes strongly confined to that of the conductor surface. The SPP damping along propagation direction increases. This is attributed to additional dielectric losses inside the film.

12.2.2.3 Coupling between free electromagnetic waves and surface plasmon polaritons

The dispersion relations of freely propagating electromagnetic waves and SPPs do not intersect at any frequency. This signifies that one cannot generate SPPs by simply irradiating a metal surface with light. Instead, coupling between propagating waves and surface evanescent waves requires additional techniques [26]. All these techniques have in common is that they can be used for coupling free electromagnetic radiation with surface evanescent waves (in-coupling) or vice versa (out-coupling). This symmetry is in fact a direct consequence of the invariance of Maxwell's equations under time reversal.

(a) **Prism coupling:** One of the easy way of in-coupling is a prism coupling method which is based on first transforming free electromagnetic waves into an evanescent electromagnetic field that may then couple to SPPs. On the other hand, the way the evanescent field is generated and the interpretation of the coupling process in k space is entirely different. A dielectric slab of refractive index n_g is used as a spacer layer between a prism of refractive index n_p and a metal of complex permittivity $\varepsilon_m = n_m^2$. If $n_p > n_g$ and if the angle α of the incident radiation is greater than the critical angle α_c arcsin$(\frac{n_g}{n_p})$, total internal refection (TIR) will take place at the prism base. Usually, in experiments, $n_g = 1$ (air gap) and in the case of total internal reflection (TIR) the reflected field penetrates a distance $1/(\frac{\omega}{c}\sqrt{n_p^2\sin^2\alpha - n_g^2})$ into the medium with the lower refractive index in the form of an evanescent wave. The penetration depth is on the order of the employed wavelength. Figure 12.18 shows the configuration of prism coupling.

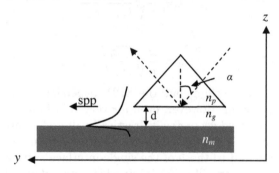

FIGURE 12.18: General schematic of prism coupling of free electromagnetic waves to SPPs

For the SPP propagating along the metal/dielectric interface, the dispersion relation is given by

$$k(\omega) = k_{ssp}(\omega) = \frac{\omega}{c}\sqrt{\frac{\varepsilon_m\varepsilon_g}{\varepsilon_m + \varepsilon_g}} \tag{12.22}$$

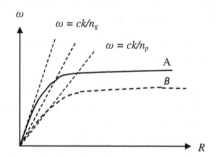

FIGURE 12.19: Dispersion relation of SPP for prism coupling

The dispersion relationship is shown in Figure 12.19. The evanescent wave arising from TIR inside the prism one obtains for the projection of the wave vector on the prism base is given by

$$k(\omega) = k_{tir}(\omega) = \frac{\omega}{c}n_p\sin\alpha(\omega) \tag{12.23}$$

This evanescent wave can couple with the SPP if $k_{ssp}(\omega) = k_{tir}(\omega)$. For a specific frequency, this relation can only be fulfilled for a specific value $\alpha(\omega)$, the so-called resonance angle, given by

$$\alpha_{res}(\omega) = \arcsin\left(\frac{1}{n_p}\sqrt{\frac{\varepsilon_m(\omega)\varepsilon_g}{\varepsilon_m(\omega) + \varepsilon_g}}\right) \tag{12.24}$$

where $\varepsilon_g = n_g^2$. One can always find a resonance angle as long as $90° \geq \alpha(\omega) \geq \alpha_c$, that is, $1 \geq \sin\alpha(\omega) \geq n_g/n_p$ remains valid.

(b) Grating coupling: The other in-coupling method is grating coupling [27]. If light hits a grating with a grating constant a at an angle θ, its wave vector along the grating surface is given by

$$k = \frac{\omega}{c}\sin\theta \pm \frac{2\pi}{a}n \tag{12.25}$$

where $n = 0.1.2\ldots$ Therefore, a grating with period a can transfer momentum to free, incident electromagnetic waves. In reciprocal space, this transfer results in a shift of the free wave dispersion relation $\omega = ck/\sin\theta$ (solid line) by amounts of $\frac{2\pi}{a}$ where $n = 0, \pm1, \pm2, ..$ The shifted dispersion relations (dashed lines) are depicted for three values of n. Points of intersection between the shifted free wave dispersion relation and the SPP dispersion relation (solid curve) imply coupling.

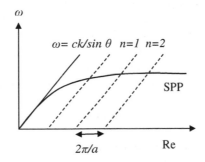

FIGURE 12.20: The shifted dispersion relations (dashed lines) are depicted for three values of n

Because the grating can provide the impinging free waves with additional momentum arising from the grating's periodic structure, the linear free wave dispersion relation changes into a set of straight lines, as shown in Figure 12.20, which can then match the SPP dispersion relation. Although with such a grating coupling high efficiencies for field amplitude can be achieved, the method has two major disadvantages. First, it requires a structuring of the surface which is often difficult and unwanted [28]. Second, only one frequency (determined by the grating constant) can be efficiently coupled into SPPs and therefore grating coupling is not well suited for broadband measurements [29].

The latest interest in SPPs goes hand in hand with major advances in nanotechnology and it is sometimes referred to as the plasmonic revolution, one of its primary goals being the fabrication of nanoscale photonic circuits operating at optical frequencies [30], [31]. In such circuits, light could be beamed and channelled on sub-wavelength scales and lead to smaller and faster devices [32], [33].

12.3 Photonic integration and nanophotonic system

Generally speaking, photonic integration means a large-scale integration of nanostructured photonic devices in an integrated circuit in the same way as microelectronics. Such large-scale photonic integration requires miniaturisation of optical and photonic components and devices, their compact interconnections, dense packaging, and cooperative actions in a monolithic chip. Nanophotonics is naturally expected to be a reasonable solution to address these problems, and can offer a promising technological roadmap towards smaller and faster devices for information communication technology.

A nanophotonic integrated chip acts as an interface between an electrical to microelectronic optical converter and optical to electrical converter. The interconnection will preferably use photonics technology, leading to high-speed and dense-bit-rate photonic modulation, broad-band communication, and low-loss transport of optical signals. Nanophotonic channel-drop filters for dense wavelength division multiplexer/demultiplexer, high-efficiency and high-sensitivity detector are prerequisites in such a scheme. However, a true nanophotonic integrated scheme is still at a premature stage as the overall architecture is still not very clear and not well established.

There are many options in evolving nanoscale photonic intercconnection devices. Those can either be based on photonic crystals, where photonic bandgaps and defects (point defects for cavities and line defects for waveguides), or total internal reflection in a silicon nanowire, or ring resonators are promising candidates. A basic information processing block for building channel-drop filters, using extremely low-loss waveguide and cavities with very high quality factor (Q-factor) are now made. Many other functional devices such as splitter, filter, memories, and switches have been explored. However, several key devices that are indispensable for information processing with photonic integrated circuit are still under development and the routes to build them are not clear. These include photonic diodes, isolators, logic gates, and the photonic equivalent of transistors, which might play a central role in a photonic CPU or photonic integrated circuit as much as a conventional transistor does in current microelectronic ICs. These developments are related to the questions of fast response speed, low energy consumption, and large bandwidth. The interconnect problem will not be solved by copper, carbon nanotubes, or code-division multiple-access radio frequency (RF), because of bandwidth to power and signal integrity limitations. Finding a suitable method for integrating optical components and CMOS transistors that minimises fabrication costs, utilises standard processing techniques, and does not impact the functionality of both technologies is still an active area of research.

The extreme temperature sensitivity of nanophotonic building blocks, which cease to operate correctly with temperature fluctuations makes integration even more difficult. Instead, it is possible that future systems will be based on monolithically integrated nanoscale electronic-photonic circuits, with the information processing primarily relying on electrons and the majority of the information transfer above a particular architecture-dependent length scale accomplished using photons. An interesting aspect of architectural research in the field of nanophotonics is that there is not a natural progression of scaling parameters that will necessarily dictate future designs as is the case in CMOS. Because nanophotonics is an emerging technology, the potential is limitless for creating new devices that solve previous challenges.

To obtain a very high degree of communication bandwidth in information communication system WDM and TDM can be combined. Exceptional bandwidth density comes from the simultaneous use of TDM and WDM and the nanometer sized width of a single link information systems. Latency characteristics fall into two categories: switching speeds of the optical devices and the velocity of the optical signal in a link. Depending on the material used to construct the optical network, the latter has been shown to be about 10.45 ps/mm in single crystalline silicon links. This equates to double speed-up over a highly-optimised copper connection. Using optical links made of silicon nitride, the latency can be reduced to approximately 6 ps/mm at a cost of increased link width and spacing.

The energy consumption of a nanophotonic link is another issue. This can be divided into two main components: electrical energy spent in transmitting and receiving the optical data, and the energy required to power a laser for supplying the wavelengths of light to the modulators. When an optical signal travels between a source and destination node, it encounters multiple points of power loss. The primary reasons for signal attenuation are due to roughness in the optical link (due to fabrication imperfections), absorption of light by the link's material, and insertion loss as the light passes other devices while travelling to its destination node.

At the end of a nanophotonic information communication link the receiver's photodetector requires enough optical power to mitigate the potential for bit errors. This power level dictates the characteristics of the laser at the front end, which must supply enough power to potentially multiple wavelengths in a waveguide to account for all of the insertion loss it will experience prior to reaching the detector. Depending on the network architecture and number of communication nodes, relatively high laser power in future chip multiprocessors is required. The second component of energy consumption, however is from the electrical power dissipation in the transmitter and receive components.

The area requirements of a basic nanophotonic link (i.e., a link consisting of modulators at the front-end, links for the data to travel though, and receivers at the back-end) are dictated by the size of the modulators, width and spacing requirements of the link and size of the receiver components. A modulator's size is dependent on the amount of optical insertion loss it adds to the system (due to bending losses) and the material from which it is fabricated. The electrical driver circuit, depending on the required drive strength and technology node, should fit within the small dimensions of the modulator. The dimensions of a link depend on its material composition. Lastly, the receiver is composed of two components, the photodetector and a series of amplifying stages for converting the optical signal into a digital level voltage. The size of a germanium based photodetector is limited by the width of a single photonic link in one dimension and a required length for absorbing the light in the second dimension. Experts predict lithographic resolution as fine as a few nm will be available in near future, which is about 100 times smaller than the telecommunication wavelength of 1550 nm. These lithographic techniques can be used to create sub-wavelength features with optical properties controlled by the density and geometry of the pattern and its constituent materials.

Moreover, nanophotonic communication network among tens and eventually hundreds of processors on a die requires a very complex array of optical waveguides. Depending on the topology of the network architecture, waveguide crossings or multiple waveguide layers may be required. The latter is similar to the different metal layers in a CMOS metal stack. Single crystalline waveguides cannot be deposited and thus multiple layers are infeasible. Silicon nitride waveguides have the benefit of being fabricated with back-end-of-line techniques with multiple deposited layers that eliminate crossings.

Nanophotonic information systems [34] that can operate beyond the diffraction limit possess unique information processing features, such as:

(i) Dissipation of low energy: As a rule, a nanophotonic system dissipates exciton energy from an electric dipole in the upper state to an electric dipole allowed energy level in the lower state. If $h(\omega)$ denotes the exciton-phonon coupling energy and $D(\omega)$ denotes the density of states, then the rate of relaxation Γ in a nanophotonic system is given by $\Gamma = 2\pi\hbar^2|h(\omega)^2D(\omega)$.

(ii) Operation with a single photon: When there is more than one exciton to be generated from a nanophotonic device such as quantum dot, the electric dipole's allowed energy level in the lower state tries to lower its energy still more so as to correspond to the binding energy of the exciton molecule. Due to this, the energy level gets detuned from the input signal and hence the exciton generation no longer takes place. Moreover, the exciton transfer process is a resonant process. It is observed from experiments that if more than one exciton is transferred from one nanodevice to another, the transfer process is rendered off-resonant and hence the transfer does not take place at all. Moreover, with the energy transfer, a single exciton can remain stable in a given state only if

the binding energy of the exciton molecule is large and in such a case, only a single photon gets emitted by a fast relaxation process.

(iii) Resistance against non-invasive attacks: The only plausible solution to this problem is to resort to optical near-field technology, which enables one to operate novel nanophotonic systems beyond the diffraction limit. Moreover, the nanophotonic systems, even macroscopic free space in a vacuum may enact the role of interconnects. Thus, it may be possible that any signal be transferred by the photon exchange between nanometric particles without using any interconnects. As the signal intensity is decided by the energy dissipation inside the nanometric particles, the possibility of non-invasive attack is completely ruled out. For novel nanophotonic systems beyond the diffraction limit, in general, even macroscopic free space in a vacuum may enact the role of interconnects.

The most important aspect of a nanophotonic system is that in spite of not having a central controller in the system, efficient transfer of optical excitations can be realised, which addresses the autonomous behaviour of optical excitations. This can lay the foundation for the realisation of self-organization and distributes complex information and communications technology based systems on an Internet scale. Using such a distributed and autonomous network system, unbalanced traffic load and energy consumption can be avoided. One can also ensure overall sustainability and reliability as the above mentioned network topology no longer depends on single points of failure. Moreover, as the output signal can be increased due to degraded optical near-field interactions, robustness against errors can be achieved. Thus, on employing nanophotonics technologies, one can easily handle the huge demand of energy efficient data transfer with more robustness, overall sustainability, and reliability.

Books for further reading

1. *Introduction to Nanophotonics*: S. V. Gaponenko, , Cambridge University Press, Cambridge, 2010.

2. *Principles of Nano-Optics*: L. Novotny and B. Hecht, Cambridge University Press, Cambridge, 2012.

3. *Photonic Crystals: Molding the Flow of Light*: J. D. Joannopoulos, S. G. Johnson, J. N. Winn, and R. D. Meade, Princeton University Press, 2nd edition, 2008.

4. *Plasmonics—Fundamentals and Applications*: S. A. Maier, Springer, Heidelberg, 2007.

5. *Introduction to Metal-Nanoparticle Plasmonics*: M. Pelton and G. W. Bryant, John Wiley, New York, 2013.

6. *Monolithic Nanoscale Photonics-Electronics Integration in Silicon and Other Group IV Elements*: H. Radamson and L. Thylen, Academic Press, London, 2014.

7. *On-Chip Photonic Interconnects*: C. J. Nitta et al., Morgan and Claypool Publishers, California, 2013.

8. *Photonic Network-on-Chip Design*: K. Bergman et al., Springer, Heidelberg, 2013.

9. *Nanophotonic Information Physics*: M. Naruse (ed.), Springer-Verlag, Berlin Heidelberg, 2014.

10. *Photonic Crystal Fibres*: F. Poli, A. Cucinotta, and S. Selleri, Springer, Dordrecht, 2007.

11. *Photonic Crystals*: I. A. Sukhoivanov and I. V. Guryev, Springer-Verlag, Berlin Heidelberg, 2009.

12. *Photonic Crystals Towards Nanoscale Photonic Devices*: J. M. Lourtioz et al., Springer-Verlag, Berlin, 2008.

13. *Progress in Nanophotonics*: M. Ohtsu (ed.), Springer-Verlag, Berlin, Heidelberg, 2011.

14. *Surface Plasmon Nanophotonics*: M. L. Brongersma and P. G. Kik (ed.), Springer, Dordrecht, 2007.

15. *Fundamentals of Photonic Crystal Guiding*: J. Yang and M. Skorobogatiy, Cambridge University Press, 2008.

16. *Photonic Crystal Fibres*: A. Bjarklev, J. Broeng, and A. S. Bjarklev, Kluwer Academic, Boston, 2003.

17. *Plasmonics: Fundamentals and Applications*: S. A. Maier, Springer, Germany, 2007.

18. *Surface Plasmons* in *Volume 111 Tracts in Modern Physics*, H. Raether (ed.), Springer-Verlag, New York, 1988.

19. *Surface Plasmon Nanophotonics*: M. L. Brongersma and P. G. Kik, Springer, UK, 2007.

20. *Nanophotonics with Surface Plasmons* in *Advances in Nano-optics and Nano-photonics*: V. M. Shalaev and S. Kawata, Elsevier Science, UK, 2007.

21. S. V. Gaponenko, *Introduction to Nanophotonics*, Cambridge University Press, 2005.

22. *Handbook of Nanophysics. Nanoelectronics and Nanophotonics*: K. D. Sattler (ed.), CRC Press, London, 2011.

23. *Principles of Nanophotonics*: M. Ohtsu, K. Kobayashi, T. Kawazoe, T. T. Yatsui, and M. Naruse, CRC Press, New York, 2008.

24. *Fundamentals of Photonic Crystal Guiding*: M. Skorobogatiy and J. Yang, Cambridge University Press, 2009.

25. *Plasmonics-From Basics to Advanced Topics*: S. Enoch and N. (ed.): Springer-Verlag, Berlin, Heidelberg, 2012.

26. *Nanophotonic Information Physics*: M. Naruse (ed.), Springer, Dordrecht, 2014.

27. *Optical Properties of Photonic Crystal*: K. Sakoda, Springer-Verlag, 2001.

28. *Foundations of Photonic Crystal Fibres*: F. Zolla, G. Renversez, A. Nicolet, B. Kuhlmey, S. Guenneau, and D. Felbacq, Imperial College Press, London, 2005.

Bibliography

[1] S. Fan, M. F. Yanik, Z. Wang, S. Sandhu, and M. L. Povinelli. Advances in theory of photonic crystals. *J. Lightwave Tech.*, 24:4493–4501, 2006.

[2] E. Yablonovitch and T. J. Gmitter. Photonic band structure: The face-centered-cubic case. *Phy. Rev. Let.*, 63:1950–1953, 1989.

[3] S. G. Johnson, P. R. Villeneuve, S. Fan, and J. D. Joannopoulos. Linear waveguides in photonic crystal slabs. *Phys. Rev. B*, 62:8212–8218, 200.

[4] C. C. Cheng and A. Scherer. Fabrication of photonic bandgap crystals. *J. Vac. Sci. Tech. (B)*, 13(6):2696–2700, 1995.

[5] S. Y. Lin, J. G. Fleming, D. L. Hetherington, B. K. Smith, R. Biswas, K. M. Ho, M. M. Sigalas, W. Zubrzycki, S. R. Kurtz, and J. Bur. A three-dimensional photonic crystal operating at infrared wavelengths. *Nature*, 394:251–253, 1998.

[6] S. Noda, K. Tomoda, N. Yamamoto, and A. Chutinan. Full three-dimensional photonic bandgap crystals at near-infrared wavelengths. *Science*, 289(5479):604–606, 2000.

[7] H. S. Sozuer and J. P. Dowling. Photonic band calculations for woodpile structure. *J. Mod. Opt.*, 41:231–236, 1994.

[8] K. M. Ho, C. T. Chan, C. M. Soukoulis, R. Biswas, and M. Sigalas. Photonic band gaps in three dimensions: new layer-by-layer periodic structures. *Solid State Comm.*, 89(5):413–416, 1994.

[9] H. Y. D. Yang. Theory of photonic band-gap materials. *Electromagnetics*, 19(3):Special Issue, 1999.

[10] G. Kurizki and J. Haus, Photonic band structures. *Special Issue, J. Mod. Opt.*, 41(2), 1994.

[11] J. D. Joannaopoulos, S. D. Johnson, J. N. Winn, and Meade R. D. *Photonic Crystals Molding the Flow of Light*. Princeton University Press, Princeton, NJ, 2008.

[12] S. Guo and S. Albin. Simple plane wave implementation for photonic crystal calculations. *Opt. Exp.*, 11:167–175, 2003.

[13] Z. Zhang and S. Satpathy. Electromagnetic wave propagation in periodic structures: Bloch wave solution of Maxwell's equation. *Phy. Rev. Let.*, 60:2650–2653, 1990.

[14] L. Brillouin. Wave propogation in periodic structures. Dover Publications, New York, 1953.

[15] P. P. Villeneuve and M. Piche. Photonic bandgaps in two-dimensional square and hexagonal lattices. *Phy. Rev. (B)*, 46:4969–4972, 1990.

[16] S. Shi, C. Chen, and D. W. Prather. Plane-wave expansion method for calculating band structure of photonic crystal slabs with perfectly matched layers. *J. Opt. Soc. Am. (A)*, 21:1769–1775, 2004.

[17] D. R. Smith, R. Dalichaouch, N. Kroll, S. Scholtz, S. L. McCall, and P. M. Platzman. Photonic band structure and defects in one and two dimensions. *J. Opt. Soc. Am. (B)*, 10(2):314–321, 1993.

[18] J. C. Knight et al. All-silica single-mode optical fiber with photonic crystal cladding. *Opt. Let.*, 21,19:1547–1549, 1996.

[19] P. J. Russell. Photonic crystal fibers. *J. Lightwave Tech.*, 24(12):4729–4749, 2006.

[20] T. A. Birks, J. C. Knight, and P. J. Russell. Endlessly single-mode photonic crystal fiber. *Opt. Let.*, 22,13:961–963, 1997.

[21] B. T. Kuhlmey, R. C. McPhedran, M. de Sterke, C. P. A. Robinson, G. Renversez, and D. Maystre. Microstructured optical fibers: where's the edge? *Opt. Exp.*, 10(22):1285–1290, 2002.

[22] W. A. Murray and W. L. Barnes. Plasmonic materials. *Adv. Mater*, 19:3771–3776, 2007.

[23] J. M. Pitarke, V. M. Silkin, E. V. Chulkov, and P. M. Echenique. Theory of surface plasmons and surface-plasmon polaritons. *Rep. on Progress in Phy.*, 70:1, 2007.

[24] P. Berini. Plasmon-polariton modes guided by a metal film of finite width. *Opt. Let.*, 24,15:1011–1013, 1999.

[25] P. Berini. Plasmon-polariton waves guided by thin lossy metal films of finite width: Bound modes of asymmetric structures. *Phys. Rev. B*, 63,12:125417, 2001.

[26] A. Giannattasio and W. L. Barnes. Direct observation of surface plasmon polariton dispersion. *Opt. Exp.*, 13(2):428–434, 2005.

[27] F. Pincemin and J. J. Greffet. Propagation and localization of a surface plasmon polariton on a finite grating. *J. Opt. Soc. Am. (B)*, 13:1499–1509, 1996.

[28] R. H. Ritchie, E. T. Arakawa, J. J. Cowan, and R. N. Hamm. Surface-plasmon resonance effect in grating diffraction. *Phy. Rev. Let.*, 21(22):1530–1533, 1968.

[29] J. Gomez-Rivas, M. Kuttge, P. Bolivar, P. Haring, H. Kurz, and J. A. Sanchez-Gill. Propagation of surface plasmon polaritons on semiconductor gratings. *Phys. Rev. Let.*, 93, 2004.

[30] A. Polman. Plasmonics applied. *Science*, 322:868–869, 2008.

[31] W. L. Barnes, A. Dereux, and T. W. Ebbesen. Surface plasmon subwavelength optics. *Nature*, 424,6950:824–830, 2003.

[32] S. A. Kalele, N. R. Tiwari, S. W. Gosavi, and S. K. Kulkarni. Plasmon-assisted photonics at the nanoscale. *J. Nanophotonics*, 1:012501–012520, 2007.

[33] M. Kobayashi, T. Kawazoe, S. Sangu, and T. Yatsui. Nanophotonics: design, fabrication, and operation of nanometric devices using optical near fields. *J. Selected Topics in Quan. Elect.*, 8:839–862, 2002.

[34] Y. Fainman, K. Ikeda, M. Abashin, and D. Tan. Nanophotonics for information systems. *J. of Physics Conf.*, 206, 2010.

Index

2D wavelet transform, 237
3D Fourier transform, 222

absorption, 75
access network, 305
acousto-optic modulator, 190
add-drop multiplexer, 303
add/drop filter, 303
Ampere's law, 46
array logic, 330
artificial neural network, 393
associative memory model, 408
atom and ion qubits, 441
attenuation in optical fibre, 287
auto-associative memory model, 409
avalanche photodiode, 164

bacteriorhodopsin storage, 200
band-limited channel, 20
Bayes estimator, 27
Bayes theorem, 6
Bayes's criteria, 24
Bayesian interpretation of visual perception, 102
Boltzmann machine, 403
Brillouin scattering, 77

camera model, 110
Carpenter-Grossberg model, 405
channel capacity, 19
charge-coupled device, 166
chemical laser, 156
CIE V-λ function, 96
CNOT gate, 450
coherence, 42

coherence length, 44
communication channels, 16
computer vision, 103
conditional entropy, 9
conditional probabilities of qubits, 439
conditional probability, 5
constitutive relations, 47, 48
continuous wavelet transform, 230
convolution based multiplier architecture, 355
convolution theorem, 219
correlation filter designs, 380
correlation filters, 371
cumulative distribution function, 6
current density in semiconductor, 121
current flow in homojunction, 127

Daubechies wavelet, 235
deformable mirror devices, 191
detection of signal, 22
dielectric constant, 47
diffraction, 66
digital holography, 202
digital optics, 330
digital-photonic correlator, 385
discrete channel, 19
discrete Fourier transform, 220
dispersion in optical fibre, 286
distributed feedback laser, 152
double heterostructure, 151

electric displacement, 46

electrically addressed spatial light modulators, 186
electro-optic effect, 179
electro-optic modulators, 180
electroluminescence, 132
electromagnetic spectrum, 54
electromagnetic theory, 46
electromagnetic wave equation, 49
electron and hole mobilities, 121
electronic qubits, 442
electrophoresis and e-book display, 213
energy velocity, 42
entropy, 7
entropy of continuous channel, 18
erbium-doped fibre amplifier, 306
estimation techniques, 26
etalon switching devices, 195

Faraday's law, 46
fast Fourier transform, 221
Fechner's law, 98
Fermat's principle, 61
Fermi-Dirac distribution, 117
fibre distributed data interface, 296
fibre optic communication, 280
flat panel displays, 203
flexible flat panel display, 212
forerunner velocity, 42
Fourier optics, 223
Fourier transform, 217
Fourier transform profilometry, 270
Fourier transform properties of lens, 223
fractional Fourier transform, 225
Fraunhofer diffraction, 68
Fredkin gate, 451
frequency domain correlation, 373
Fresnel equation, 75
fringe projection technique, 266

gas lasers, 155
Gauss laws, 46

Gaussian channel, 20
geometric transformations, 108
geometrical primitives, 107
gestalt theory, 100
group velocity, 42
grouping laws of visual perceptions, 99

Harr wavelets, 234
Hartley transform, 227
Helmholtz equation, 65
Helmholtz principle of visual perception, 101
hetero-associative memory, 416
heterojunction, 128
holographic storage, 200
holographic weight matrix, 412
homojunction, 124
Hopfield model, 400
Hough transform, 241
human visual system, 85
Huygens-Fresnel principle, 65
hybrid photonic multiprocessor, 360

image acquisition, 106
image format, 104
image registration, 261
imitations of electronic computation, 323
incoherent source, 45
information, 1
information transmission, 9
infrared spectrum, 56
injection laser diode, 149
interference, 69
intrinsic wave impedance, 54

joint transform correlator, 386
Jones vector, 72
Josephson junction, 442
junction photodiodes, 162

Kerr effect, 179
knowledge-based system, 2

Kohonen's self-organising feature
  map, 404

laser amplifier, 133
laser based range acquisition, 262
laser cavity, 140
laser Doppler anemometry, 270
laser oscillators, 139
laser triangulation technique, 265
LED and OLED display panels,
  209
light-emitting diode, 143
liquid crystal flat panel displays,
  206
liquid crystal light modulators,
  182
liquid crystals, 183
liquid dye laser, 157
log-polar Fourier transform, 222
long-haul photonic system, 292
low-level processing, 251

magneto-optic spatial light
  modulator, 188
mass of a star, 40
matched filter, 374
material dispersion, 49
matrix-vector multiplication
  architecture, 357
maximum average correlation
  height filter, 382
maximum likelihood estimates, 28
Maxwell's equations, 46
metamaterials, 49
metrics for correlation filter, 375
Mie scattering, 77
minimax criteria, 25
minimax estimates, 27
minimum average correlation
  energy, 377
minimum average correlation
  energy filter, 381
minimum noise and correlation
  energy filter, 378
minimum variance synthetic

discriminant function,
  377, 382
modes in an optical fibre, 282
momentum of photon, 39
morphological operations, 256
morphological operations, 255
multilayer perceptron, 398
multiresolution analysis, 236
mutual information, 18

nanophotonic devices , 467
nanophotonic information
  systems, 496
nature of light, 35
negabinary carry-less arithmetic
  processor, 349
negabinary number system, 341
negabinary-based logic processor,
  351
negative refractive index material,
  49
Neyman-Pearson criteria, 25
noiseless coding, 11
nonlinear correlation filters, 379
nonlinear materials, 48
nuclear magnetic resonance
  devices, 444
number systems for photonic
  computing, 335

optic nerves, 91
optical clock and bus, 304
optical fibre, 280
optical imaging, 59
optical programmable array logic,
  331
optical shadow casting
  architecture, 345
optically addressed spatial light
  modulator, 187
organic light-emitting diode, 154

p-i-n photodiode, 163
partial coherence, 45
peak-to-sidelobe ratio, 373
perception laws, 98

perceptron, 397
permeability, 47
permittivity, 47
phase and polarisation encryption, 309
phase velocity of wave, 41
phase- measuring profilometry, 268
photo-refractive storage, 198
photochromic storage, 199
photoconductors, 160
photodetectors, 158
photoelectric effect, 36
photometric imaging, 77
photon, 37
photon energy, 38
photon entanglement, 445
photon mass, 39, 40
photonic bistable devices, 192
photonic computing, 326
photonic computing architecture, 344
photonic cross-connect, 301
photonic crystal, 469
photonic crystal defects, 478
photonic crystal fibre, 479
photonic flip-flop processor, 353
photonic free-space communication, 292
photonic hetero-associative memory, 422
photonic Hough transform, 247
photonic image encryption, 310
photonic integration, 494
photonic morphological processing, 258
photonic multistage interconnection, 333
photonic network, 295
photonic neural network, 406
photonic profilometric techniques, 264
photonic qubits, 444
photonic secure communication, 309

photonic sources, 131
photonic storage in 3D, 198
photonic wavelet transform, 246
photonics, 34
photonics bandgap, 470
photopic, scotopic, and mesopic vision, 94
planar photonic crystal, 476
plane wave solution, 52
plasma display panel, 211
plasmon resonance, 485
plasmonics, 485
Pockels effect, 179
point-to-point fibre link, 288
polarisation, 48, 71
polariton, 486
polynomial correlation filters, 380
population inversion, 135
probability density function, 6
probability theory, 3
programmable logic devices, 331
ptimal trade-off filter, 382
Purkinje shift, 96

quadratic correlation filters, 379
quantum bits or qubits, 435
quantum communication, 454
quantum computing, 446
quantum computing architecture, 452
quantum cryptography, 455
quantum gates, 448
quantum information processing, 431
quantum key distribution, 457
quantum logic gates, 447
quantum protocol, 458
quantum teleportation, 459
quantum well laser, 153
quantum well structure, 129
qubit properties, 437
qubits, multiple, 438

Radon transform, 239
Raman scattering, 76

random variable, 4
range image acquisition, 261
ray optics, 59
ray transfer matrix, 62
realisation of qubits, 440
reflectance, 78
refraction and reflection, 62
residue number system, 337
retina, 88
rods and cones, 90
routing topologies, 299

scattering, 76
Schwarzschild radius, 40
SEED based all-photonic
        multiprocessor, 362
self-electro-optic effect, 196
semiconductor junction, 123
semiconductor materials, 122
Shannon's entropy, 8
Shannon's theorems, 2, 12
Shannon-Hartley law, 22
short-time Fourier transform, 229
signal distortion in optical fibre,
        286
signed-digit number, 338
silicon photonics, 169
solid-state lasers, 157
SONET/SDH ring, 298
spatial coherence, 44
spatial light modulator, 182
speed of light, 40
Steven's power law, 99
Stiles-Crawford effect, 95
structure of eye, 85
supercontinuum source, 483
surface plasmon, 485
surface plasmon polariton, 485,
        486
SWAP gate, 451
synchronous optical network, 296

synthetic discriminant function,
        377
synthetic discriminant function
        filter, 380

temporal coherence, 44
terahertz radiation, 57
Toffoli gate, 451
transformation of images into
        graphics, 260
transport network, 305
two photon absorption, 199
types of optical fibre, 284

ultraviolet spectrum, 57
unconstrained filters, 378
unified arithmetic and logic
        processing architecture,
        347
unitary transformation of qubits,
        439

Vander Lugt correlator, 372
vertical-cavity surface-emitting
        lasers, 153
visible spectrum , 57
visual cortex, 93
visual perception, 85, 97
visual perception in 3D, 101
visual signal processing, 91
Von Neumann bottleneck, 323
Von Neumann entropy, 434

wave optics, 63
wave propagation, 54
wavelength division multiplexing,
        298
wavelet analysis, 227
wavelet transform, 229
Wavelet transform in frequency
        domain, 231
Weber's law, 98
winner-take-all model, 423

9 780367 574185